# Drilling Engineering

# Drilling Engineering

Edited by **Alexis Federer**

SYRAWOOD
PUBLISHING HOUSE
New York

Published by Syrawood Publishing House,
750 Third Avenue, 9th Floor,
New York, NY 10017, USA
www.syrawoodpublishinghouse.com

**Drilling Engineering**
Edited by Alexis Federer



International Standard Book Number: 978-1-68286-349-7 (Hardback)



Printed in China.

# Contents

# Preface

Drilling engineering is an important field of study that focuses on understanding the significant concepts and emerging technologies in drilling and excavations. The book aims to provide a comprehensive overview of the various techniques and practice involved in the process of drilling. Most of the topics included in this book such as geology, geochemistry, seismology, micropaleontology, etc. are bound to provide a multidisciplinary perspective of this field. The case studies in this book have been contributed by an international panel of experts. It attempts to assist those who are seeking detailed information in the field of drilling engineering.

This book unites the global concepts and researches in an organized manner for a comprehensive understanding of the subject. It is a ripe text for all researchers, students, scientists or anyone else who is interested in acquiring a better knowledge of this dynamic field.

I extend my sincere thanks to the contributors for such eloquent research chapters. Finally, I thank my family for being a source of support and help.

**Editor**

# 1

# The Pacific Equatorial Age Transect, IODP Expeditions 320 and 321: Building a 50-Million-Year-Long Environmental Record of the Equatorial Pacific Ocean

**by Mitchell Lyle, Heiko Pälike, Hiroshi Nishi, Isabella Raffi, Kusali Gamage, Adam Klaus, and the IODP Expeditions 320/321 Scientific Party**

## Introduction

In March 2009, the R/V *JOIDES Resolution* returned to operations after its extended refit and began with a drilling program ideally suited to its drilling strengths, the Pacific Equatorial Age Transect (PEAT, IODP Exp 320/321; Fig. 1A). The PEAT drilling program was developed to understand how a major oceanic region evolved over the Cenozoic Era (65–0 Ma) and how it interacted with global climate. It specifically targeted the interval between 52 Ma and 0 Ma and drilled a series of sites that originated on the paleoequator. These sites have since been moved to the northwest by plate tectonics.

The equatorial Pacific is an important target for paleoceanographic study because it is a significant 'cog' in the Earth's climate machine, representing roughly half of the total tropical oceans that in turn represent roughly half of the total global ocean area. Prior drilling in both the Deep Sea Drilling Project (DSDP) and the Ocean Drilling Program (ODP) outlined the changes that have occurred through the Cenozoic (e.g., van Andel et al., 1975; Pisias et al., 1995). Not only did the earlier work fail to cover sufficient time intervals but also many of the sites were cored with 'first-generation' scientific drilling technology with incomplete and disturbed sediment recovery and thus cannot be used for detailed studies.

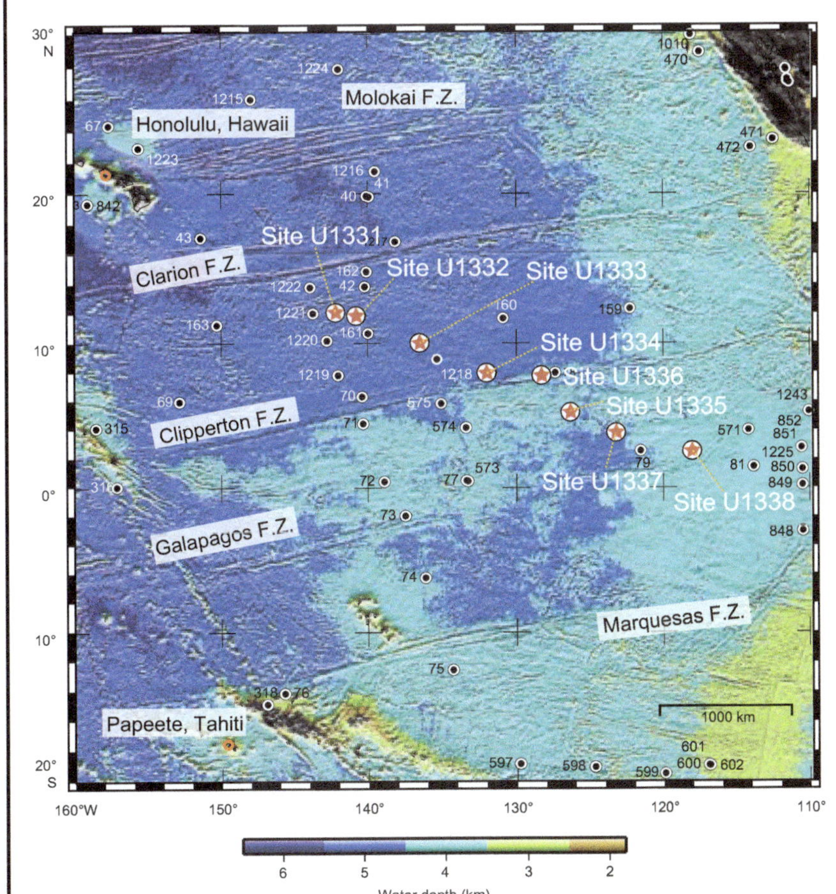


Figure 1A. Present locations of Pacific Equatorial Age Transect (PEAT) drill sites (red stars) on a bathymetric map of the central Pacific. Also shown are locations of previous DSDP and ODP drilling (solid black circles), as well as Honolulu, Hawaii and Papeete, Tahiti (open red circles). F.Z. = fraction zone.

ODP Legs 138 and 199 provide the best sample material from previous drilling, but each leg recovered sections spanning less than 10 million years suitable for cyclostratigraphy (the use of earths orbital cycles, recorded in sediments, as a measure of time). Up until the PEAT program it was difficult to achieve more than a reconnaissance of the environmental changes that have occurred in the equatorial Pacific. The PEAT program was designed to augment previous drilling and collect undisturbed sediments that could be spliced into a continuous, high-resolution environmental record of the eastern equatorial Pacific for the entire period from 56–0 Ma to present.

## Why Study the Eastern Equatorial Pacific?

As the world's largest ocean, the Pacific Ocean is intricately linked to major global changes that took place during the Cenozoic. The equatorial Pacific is a major area for trapping of incoming solar radiation (Bryden and Brady, 1985), a major zone of high primary productivity (Chavez and Barber, 1987; Westberry et al., 2008), and an important region for $CO_2$ exchange from the deep ocean to the atmosphere (Takahashi et al., 1997).

It is also the source of one of the strongest modern inter-annual climate oscillations, the El Niño-Southern Oscillation (ENSO, Philander, 1983; Cane and Zebiak, 1985). Furthermore, previous work has shown that the equatorial Pacific west of the East Pacific Rise (~100°W) coherently responds over distances >1000 km on timescales as short as ENSO (Philander, 1983) and as long as millions of years (Mayer et al., 1986; Shackleton et al., 1995; Pälike et al., 2005). It also has been established, largely via scientific drilling, that there have been large-scale, global changs in climate over the Cenozoic that affected the equatorial Pacific (van Andel et al., 1975; Mayer et al., 1986; Pisias et al., 1995; Zachos et al., 2001a; ODP Leg 199 Shipboard Scientific Party, 2002; Zachos et al., 2008).

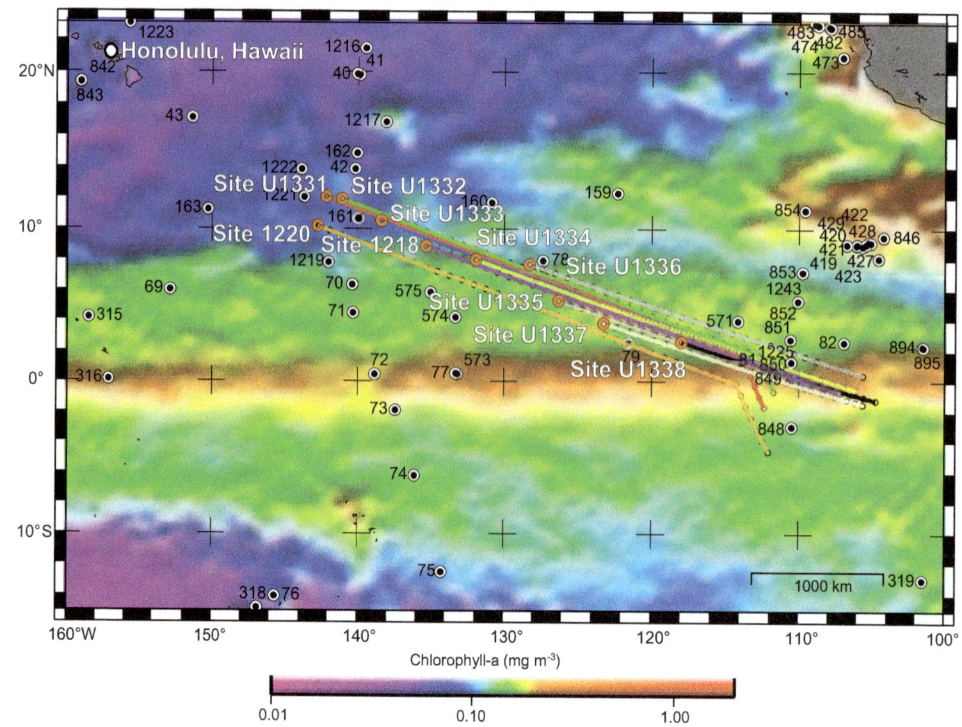


Figure 1B. Backtracked positions for PEAT sites (open red circles) , using stage poles from Koppers et al. (2001) superimposed on a satellite-derived (GlobColour) map of present ocean chlorophyll-a concentrations (December 2005 monthly mean) where red = high productivity and blue = low productivity. Solid black circles = previous DSDP and ODP sites.

The circulation of the equatorial surface ocean is inescapably linked to the trade wind system. The equatorial Pacific is the classic "world ocean" example of this linkage; it is dominated by wind-driven circulation and is largely unfettered by ocean boundaries. Here, the equator itself is characterized by a narrow zone of divergence that results from the change in the sign of the Coriolis Effect and that gives rise in the modern world to a band of high biologic productivity within a 2° latitudinal band of the equator (Fig. 1B). The strength of the equatorial circulation and of this divergence is linked to the strength of the trade winds, which are in turn strongly tied to the global climate system. Variations in global climate, inter-hemispheric differences in temperature gradients, and marked changes in the ocean boundaries are all imprinted on the biogenic-rich sediments accumulating in the equatorial zone.

Finally, the equatorial Pacific may have responded to the closing of Tethys gateways, potentially a significant Cenozoic climate driver. Closure of the Panama gateway and the constriction of the Indonesian Passage should both have affected the Pacific, and indeed, evidence for oceanographic change associated with these gateway restrictions are recorded in

**PEAT Science Objectives:**

1. To detail the nature and changes of the carbonate compensation depth (CCD) over the Cenozoic in the paleoequatorial Pacific
2. To determine the evolution of paleoproductivity of the equatorial Pacific over the Cenozoic
3. To validate and extend the astronomical calibration of the geological timescale for the Cenozoic, using orbitally-forced variations in sediment composition known to occur in the equatorial Pacific, and to provide a fully integrated and astronomically calibrated bio-, chemo-, and magnetostratigraphy at the equator
4. To determine temperature (sea-surface and bottom water), nutrient profiles, and upper water column gradients
5. To better constrain Pacific plate tectonic motion and better locate the Cenozoic equatorial region in plate reconstructions, primarily via paleomagnetic methods
6. To make use of the high level of correlation between tropical sedimentary sections and existing seismic stratigraphy to develop a more complete model of equatorial circulation and sedimentation
7. To provide information about rapid biological evolution and turnover rates during times of climatic stress
8. To improve our knowledge of the reorganization of water masses as a function of depth and time, as the PEAT drilling strategy also implies a paleo-depth transect
9. To develop a limited N-S transect across the paleoequator, caused by the northward offset of the proposed sites by Pacific plate motion, providing additional information about N-S hydrographic and biogeochemical gradients
10. To obtain a transect of mid-ocean-ridge basalt (MORB) samples from a fixed location in the absolute mantle reference frame, and to use a transect of basalt samples along the flowline that have been erupted in similar formation-water environments to study low-temperature alteration processes by seawater circulation

Neogene equatorial Pacific and Caribbean sediments (Keigwin, 1982; Romine and Lombari, 1985; Lyle et al., 1995; Haug and Tiedemann, 1998; Roth et al., 2000; Cane and Molnar, 2001; Lyle et al., 2008).

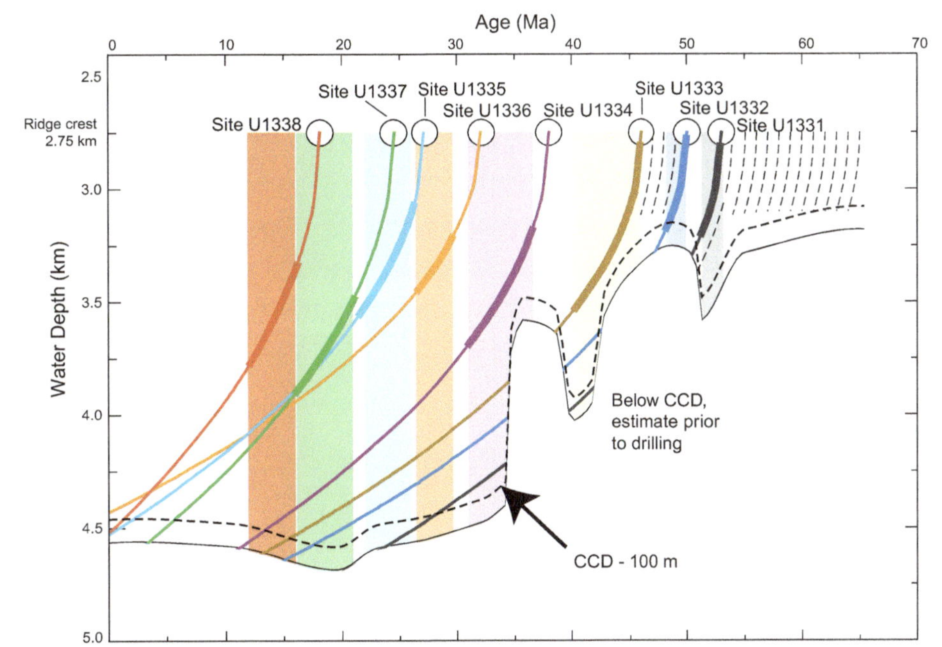


Figure 2. Targeting drill sites prior to coring based on calcium carbonate compensation depth (CCD) history (van Andel, 1975), with additional data from Leg 199. Colored boxes = critical time interval targeted for each site. Colored subsidence lines are critical time intervals where we also expected carbonate to be deposited (i.e., when site is above CCD). Subsidence curves use a subsidence parameter calculated from estimated basement age of PEAT sites and their present-day depth (k = ~0.35). Additional subsidence due to sediment loading was not modeled.

## Design of the PEAT Drilling Program

The primary design criterion of PEAT drilling was to recover sediments deposited in the equatorial zone during different time slices of the Cenozoic and assemble them into an equatorial Pacific 'megasplice' covering the interval from 56 Ma to present. The sedimentary records from "off-splice" latitudes are not ignored, but they give important insight into the strength of winds, currents, upwelling, productivity, and changes in carbonate compensation depth (CCD) once the chronostratigraphy is properly calibrated (Hovan, 1995; Lyle, 2003; Moore et al., 2004). The off-equatorial sediments are also important for calibration of paleomagnetic stratigraphy with well-developed equatorial Pacific biostratigraphy (Schneider, 1995; Lanci et al., 2005). Text box on the previous page provides the scientific objectives for the PEAT program.

Tectonic motions of the Pacific plate help to make the equatorial Pacific an attractive target for recovery of environmental records. The Pacific plate has moved with a northward latitudinal component of around 0.25° m.y.$^{-1}$ for the last 43 million years, and it moved slightly faster to the north prior to that time (Koppers et al., 2001). The north-west movement of the Pacific plate transports the equatorial sediments gradually out from under the zone of highest primary productivity at the equator, resulting in a broad mound of biogenic sediments (Fig. 1B). The transport of crust away from this equatorial zone of rapid sedimentation into regions with lower sedimentation rates keeps older equatorial sediment sections from being buried deeply beneath younger sediments. However, this tectonic movement requires that a complete environmental record of the equatorial region must be spliced together from different drill sites. Assembling a complete equatorial record requires periodic shifts to new drill site locations that contain sediments of the appropriate age deposited within the equatorial zone (Fig. 1).

While the tectonic transport of each drill site complicates reconstruction, the diminished overburden resulting from transport out of the relatively fast sedimentation regime near the equator also minimizes potential burial diagenesis and allows for good preservation of biogenic sediments. In addition, because of the shallow overburden, most of the sediment column can be cored by the advanced piston coring technique to recover sediments with minimal drilling disturbance. The northward rate of tectonic displacement, however, is not so large that a traverse of the equatorial zone (within two degrees latitude of the equator) was too rapid to record a reasonable period of equatorial ocean history. Typically drill sites remain within the equatorial zone for 10–20 m.y. before passing beyond the northern edge of high biogenic sedimentation.

## The Flow Line Strategy and Equatorial Carbonate Compensation Depth

The PEAT drilling program pursued a "flow line" rather than the "timeline" strategy pursued by previous ODP drilling legs for two reasons. A latitudinal transect (timeline) best resolves the structure of the equatorial current system, but for only a limited time window. Ocean crust cools and sinks as it ages, and the sea-floor on which the sediments are deposited approaches the lysocline and CCD within a few million years, especially during the Paleogene when the CCD was shallow. Thus, the best preserved part of sections recovered in such timeline transects is restricted by the depth at which carbonate dissolution significantly increases, as well as by the northward movement of sediment sections out of the region of high equatorial productivity. This limitation was exemplified by the results from ODP Leg 199, which recovered only limited amounts of carbonate prior to the Eocene/Oligocene boundary (e.g., at ODP Site 1218 on 42-Ma-aged crust; Coxall et al., 2005).

Most paleoceanographic indices are measured on carbonates, so only a few million years at a time can be studied in detail via the timeline approach. It would take too long to drill the number of timeline transects needed to complete a Cenozoic history of the equatorial Pacific. Fortunately, the coherent response of the equatorial Pacific to climate events covers vast areas, so that one site drilled near the equator can be used to understand changes over much of the region. When this flow line strategy is linked to previous drilling, a synoptic view of the Pacific can be developed. The most recent ODP Legs 199 and 138 drilled along a line of equal oceanic crustal age, thus obtaining an approximate north-south transect across the major east-west currents during time intervals of particular interest.

For PEAT, we planned a flow line strategy to collect carbonate-bearing equatorial sediment sections through the Cenozoic (Fig. 2), making use of the Pacific plate motion to add an oblique latitudinal transect across all time slices, and also exploiting crustal subsidence to collect limited paleodepth transects for certain time slices. We drilled a series of sites in the paleoequatorial region spanning key intervals of Cenozoic climate evolution. These intervals include the extremely warm early Eocene, the cooling of the late Eocene through Oligocene, the relatively warm climates (or low ice volume) of the early Miocene, and sections deposited during development of major Southern and Northern Hemisphere ice sheets (Fig. 3). Each site was chosen close to the geographic paleoequator at critical age intervals on ocean crust slightly older than the intervals of particular interest.

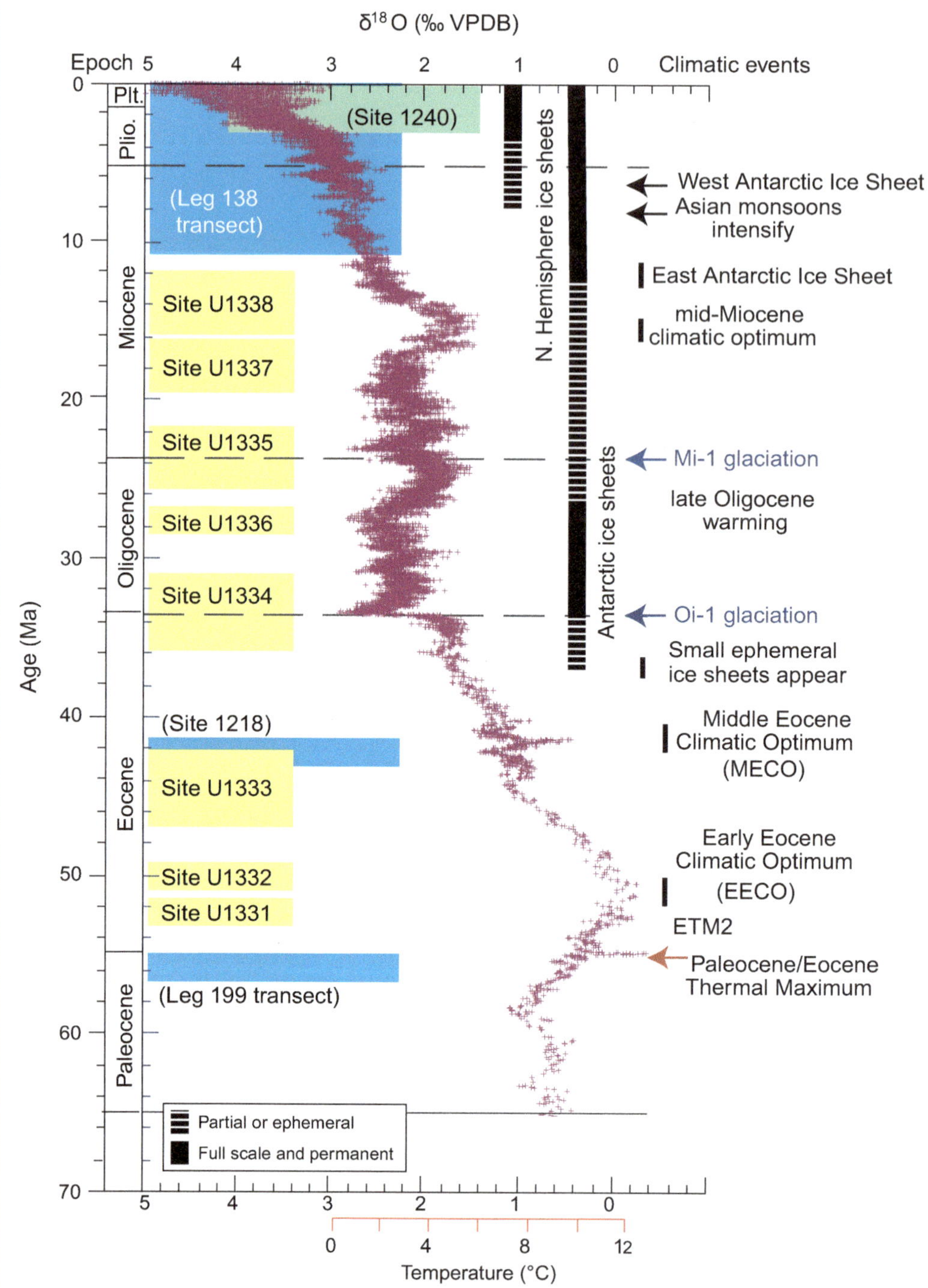


Figure 3. Evolution of oxygen stable isotopes ($\delta^{18}O$) through the Cenozoic and related major phases of climate change (modified from Zachos et al., 2001a, 2008). Yellow boxes = time slices of interest for the PEAT program, green and blue boxes = ODP legs and sites previously drilled in the equatorial Pacific region. These additional sites will be used with the PEAT sites to obtain a nearly continuous Cenozoic record of the equatorial Pacific region. VPDB = Vienna Peedee belemnite. Oi-1 = Oligocene isotopic Event 1, Mi-1 = Miocene isotopic Event 1 (Miller et al., 1991). ETM2 = Eocene thermal maximum 2.

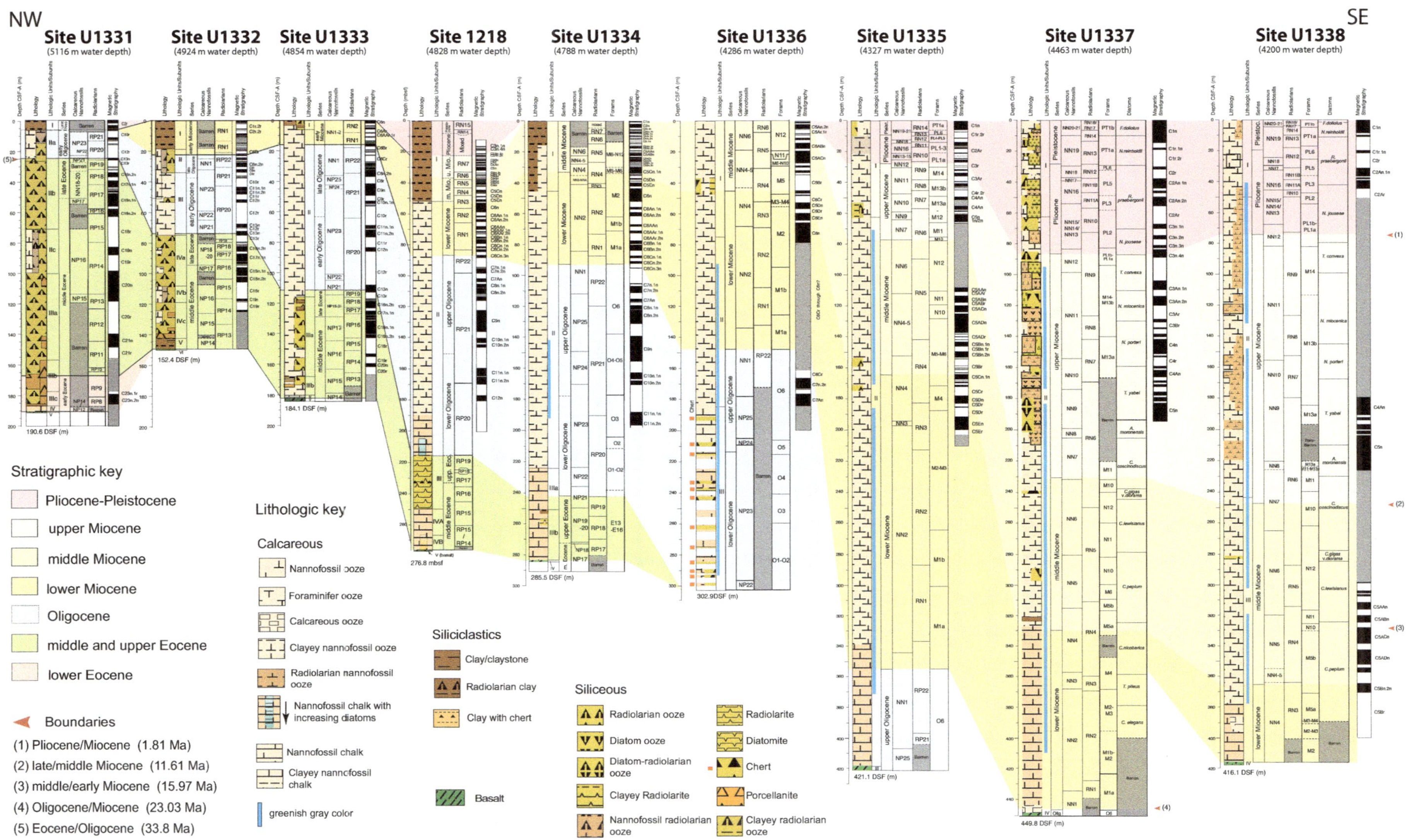


Figure 4. Stratigraphic summary plot for PEAT drill sites and ODP Site 1218. For each site the stratigraphic column and lithology are plotted against drilled depth, together with the magnetostratigraphy and biostratigraphic zonation schemes. The Eocene intervals are shaded in green, Oligocene in light blue, Miocene in dark and light shades of yellow, and Pliocene-Pleistocene in pink.

In this way we were able to track the paleoceanographic conditions at the paleoequator in the best preserved sediments obtainable. We can also make use of the high level of correlation between tropical sediment sections and seismic stratigraphy to develop a more complete model of equatorial productivity and sedimentation.

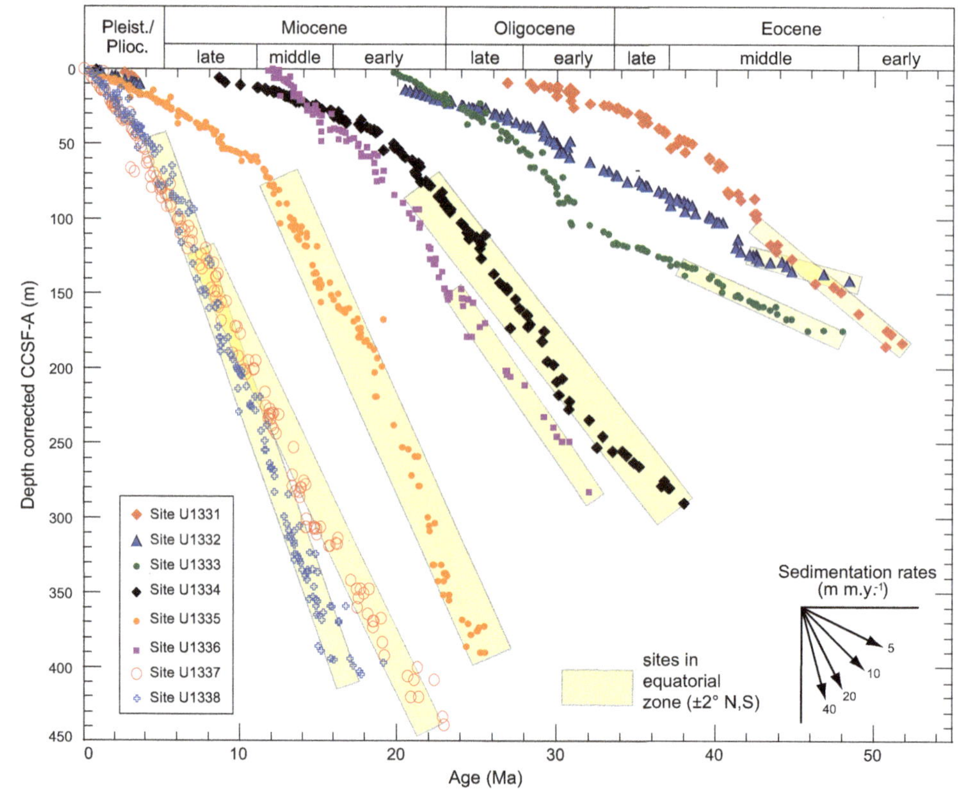


Figure 5. Summary of chronostratigraphic datum levels identified at PEAT sites plotted versus corrected composite depth (spliced depth compacted back to depth measured by the drill string). A steeper slope corresponds to a higher sedimentation rate. Yellow boxes mark approximate times when each drill site was within the equatorial zone (±2° N or S of the equator) based on estimated absolute motion of the Pacific tectonic plate.

## Drilling Results

Detailed descriptions of PEAT drilling can be found in the Exp 320 and Exp 321 Preliminary Reports (see Related Web Links), and in the Exp 320/321 Initial Reports (in press). Eight sites (U1331 to U1338) were drilled; their basement ages span from 52 Ma to 18 Ma. PEAT shipboard science has determined that the sediments recovered fill gaps from previous drill-ing and can be used to create a high-resolution megasplice of equatorial Pacific sedimentation. Cross-calibration of magneto-, bio-, and ultimately orbital stratigraphy will significantly improve chronological estimates of sedimentation and ages of significant events. The study of fluxes of different sediment components will then add a new dimension of information about biogeochemical cycling.

The PEAT program recovered sediments similar in lithology to previous DSDP and ODP expeditions to the central equatorial Pacific region (Lyle et al., 2002). Figure 4 summarizes the lithostratigraphy of the northwest—southeast transect of sites drilled during Expedition 320/321 together with the sedimentary sequence from ODP Site 1218, which is also included in the PEAT flow line strategy. As expected due to the decreasing age of crust toward the southeast, the Eocene sequence (Fig. 4, green shading) thins from northwest to southeast, pinching out east of Site U1334, the last site drilled on Eocene crust. In contrast, the Miocene sequence (Fig. 4, yellow shading) thickens substantially from northwest to southeast. The Miocene section is thickest at Site U1337, which targeted crust of latest Oligocene age, and thus is the drill site that spent the most time within the Miocene equatorial zone. The Oligocene sequence (Fig. 4, blue shading) is thickest in the middle of the PEAT transect (U1334 and U1336) and thins in both directions, marking the Oligocene equatorial zone.

The study of paleoceanographic processes—and the variations and evolution over time of mass accumulation rates across the PEAT transect—depend on a detailed knowledge of sedimentation rates. The integrated bio- and magnetostratigraphies obtained for all expedition sites are the starting point to allow us to fully exploit and understand the complex interplay of productivity, dissolution, and spatial biogenic sedimentation patterns. The sedimentation rates vary from site to site over time depending on crustal subsidence, crustal age, and the length of time spent in the equatorial region (Fig. 5).

Our results reveal the change of linear sedimentation rate in both the latitudinal and age transect components of the PEAT program. The comparison between sites reveals that the highest sedimentation rates occur within the Oligocene and Miocene equatorial zones (Sites U1334 to U1338), with sedimentation patterns similar to the modern equatorial region (highest deposition at the equator). However, sedimentation rates within the Eocene equatorial zone were not significantly higher than those outside of the equatorial zone. This result will be confirmed with revised estimates post-cruise of Pacific plate motion vectors.

Time-dependent changes in sediment production and preservation strongly affected the Eocene sedimentary record. The linear sedimentation rates of the middle Eocene were high for the pelagic realm, frequently over 10 m m.y.$^{-1}$, with a maximum of 18 m m.y.$^{-1}$ at Site U1331. Rates for the middle Eocene at Sites U1332 and 1333 were similar (6–8 m m.y.$^{-1}$). The sedimentation rates during the late Eocene decreased to 3.5–6 m m.y.$^{-1}$ at Sites U1331 through U1333. Sediment-ation rates were highest (>20 m m.y.$^{-1}$) during the early to late Oligocene at Sites U1333 and 1334, and in the early and middle Miocene at Sites U1337 and U1338.

All sites have either a hiatus or reduced sedimentation rates for the youngest sediments because they have moved out of the Neogene equatorial zone and into regions with low modern deposition rates. The data from the PEAT sites,—when combined with available data from ODP Leg 138 for 0–10 Ma and ODP Leg 199 for intervals between 32 Ma and 42 Ma (Site 1218) and >52 Ma (Sites 1219 to 1221),—will produce a continuous history of sedimentation rates in the equatorial Pacific region for the past 56 m.y.

The combined results of ODP Leg 199 and the PEAT program provide the ability to study important intervals of climate change during the Cenozoic within the equatorial Pacific, and significant post-cruise research is aimed at these intervals. Important climate intervals include the early Eocene climatic optimum (EECO, Zachos et al., 2001a; Lyle et al., 2002; Sites U1331 and U1332), the middle Eocene climatic optimum (MECO; Bohaty and Zachos, 2003; Bohaty et al., 2009; Site U1333), the middle through late Eocene carbonate accumulation events (Lyle et al., 2005; Sites U1333 and U1334), the Eocene-Oligocene (EO) transition (Coxall et al., 2005; Site U1334), the late Oligocene warming (Pälike et al., 2006a; Site U1336), the Oligocene-Miocene (OM) transition (Zachos et al., 2001b; Pälike et al., 2006b; Sites U1335, U1336, and U1337), and the middle Miocene glaciation intensification event (Holbourn et al., 2005; Sites U1337 and U1338).

## Initial Results and Future Directions

The highest shipboard priorities for a paleoceanographic drilling program are the development of a detailed sediment stratigraphy and the identification of a continuous sediment section that can be spliced together from multiple holes drilled at each site. In contrast, most of the scientific insight comes after drilling ceases and the scientific party has a chance to analyze samples collected from the cores along the spliced sedimentary sections. The sediment sampling was completed at the end of October 2009, and the analyses are just beginning. Nevertheless, broad-scale patterns can be discerned, and initial data have provided tantalizing indications of future results.

Because of the drilling design, the PEAT program was successful in collecting carbonate sediments of the late Eocene age and crossing the Eocene-Oligocene boundary. Carbonate sediments were also recovered for significant parts of the Eocene where it had been impossible from previous equatorial Pacific drilling to study the proxy climate information stored in carbonates. In addition, the Neogene PEAT sites are the first essentially complete Miocene sediment sections from the equatorial Pacific. These Miocene sediment sections will provide the first high-resolution studies of this poorly understood Cenozoic interval. Stable isotope studies on all the new sedimentary sequences will provide the backbone of information to understand the interrelationships between development of polar ice and equatorial circulation.

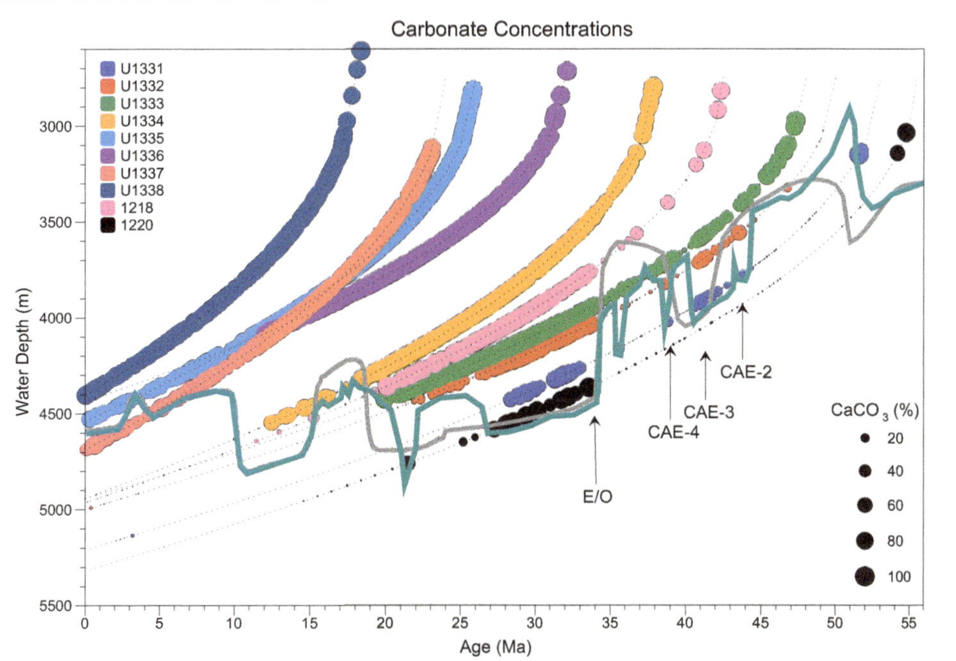


Figure 6. $CaCO_3$% vs. depth and age from initial analysis of PEAT results. The carbonate compensation depth (CCD) through the Cenozoic is shown to be very dynamic during the Eocene, with many excursions of ~500 m. The Neogene has generally a deeper CCD and milder excursions.

We also expect to recover important data about sea-surface temperature. Reconnaissance studies of alkenones (C. Beltran, unpublished) show a 4°C cooling of the equatorial Pacific since the middle Miocene. Most of the PEAT sediment splices are now being scanned using X-ray

Fluorescence (XRF) to collect high-resolution profiles of chemical data for much of the time interval. Furthermore, quantitative studies of microfossil assemblages will give new insights into the changes in the equatorial Pacific ecosystem, including the development of a diatom-based ecology in the late middle Miocene, with significant monospecific diatom intervals during the transition (Kemp and Baldauf, 1993). Finally, downhole logging will enable refinement of the equatorial seismic stratigraphy developed by Mayer et al. (1985) from Deep Sea Drilling Project Leg 85.

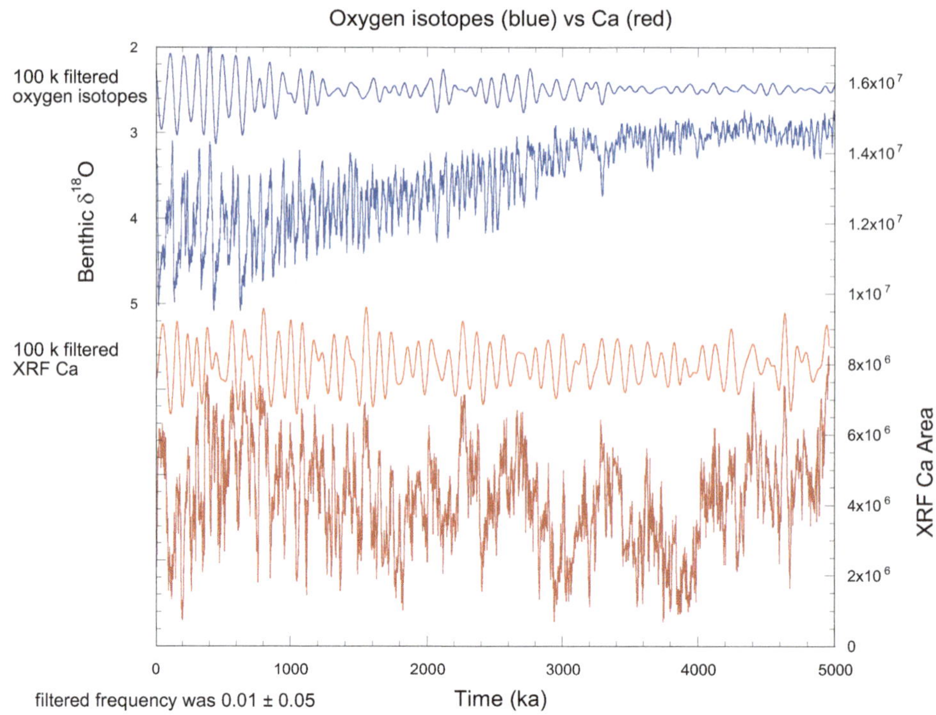


Figure 7. Oxygen isotope stack of Lisiecki and Raymo (2005) plotted against Ca peak area from XRF scanning (a measure of $CaCO_3$ content) at Site U1338. Also plotted are 100-kyr bandpass filter records (Gaussian filter with frequency of 0.01 $kyr^{-1}$ ±0.005). Initial analysis suggests that the 100-kyr period in the XRF Ca area record is a dissolution cycle because it is not present in the shallower sediments recovered by ODP Leg 138. There is a strong 'eccentricity' cyclicity in the deep Pacific even when polar ice, represented by oxygen isotopes, was waxing and waning with a 41-kyr period prior to 1 Ma.

One of the key achievements of the shipboard scientific program was better constraint of Cenozoic stratigraphy, showing the potential to achieve detailed bio-, magneto-, and chemostratigraphies for the Cenozoic from the early Eocene to the present, within an astronomically tuned age model. Shipboard results indicate that we can achieve this objective based on the observation that even decimeter-scale features in the sedimentary record from the drilled sites can be correlated over large distances across the Pacific sea-floor (Pälike et al., 2005). The PEAT program will leave a lasting legacy through detailed correlation of all major fossil groups, a detailed magnetostratigraphy with over 800 dated reversals, and sedimentary cycles that can be correlated across large distances in the Pacific Ocean.

One of the primary objectives of the PEAT program is to detail the nature and changes of the CCD throughout the Cenozoic in the paleoequatorial Pacific (see text box on page 5), with potential links to organic matter deposition (Olivarez Lyle and Lyle, 2006). The choice of drilling locations, specifically targeting positions on the palaeoequator—to track carbonate preservation during crustal subsidence through time (Fig. 2)—followed the initial work on DSDP sites by van Andel et al. (1975). The first PEAT reconstruction of the Cenozoic CCD (Fig. 6) was augmented by additional results from ODP Leg 199 (Lyle et al., 2005; Rea and Lyle, 2005). One of the very significant contributions of Leg 199 drilling was the latitudinal mapping of CCD variations with time. During the Eocene, a generally shallow CCD appeared to be deeper outside a zone ±4° from the equator, opposite the pattern established during the Neogene (Lyle, 2003). The PEAT cores allow us to refine our knowledge of temporal and spatial variation in sediment accumulation rates resulting from plate movement, varying biologic productivity at the equatorial divergence, and carbonate preservation (Fig. 6). The shipboard determinations of $CaCO_3$ concentrations reveal the carbonate accumulation events of Lyle et al. (2005) as sharp carbonate concentration fluctuations at ~44 Ma, 41 Ma, 39 Ma, and 36 Ma across Sites U1331 through U1334 and ODP Site 1218, followed by a sharp transition into much higher carbonate accumulation rates from the Eocene into the Oligocene. PEAT shipboard results reveal a complex Eocene latitudinal pattern, where Sites U1331, U1332, and U1334 track the equatorial CCD that well matches the signal observed from ODP Site 1218. On the other hand, Site U1333, which is slightly to the north of the equatorial zone during the E-O transition, shows significantly more carbonate accumulation.

The early Eocene equatorial CCD was much shallower than previously thought. Site U1332, drilled on 50-Ma crust, recovered very little carbonate in the basal sediment section, in contrast to Site U1331 that is just ~two million years older. The estimated equatorial Pacific CCD at ~49 Ma is <3000 m paleodepth. Surprisingly, the late Oligocene (23–27 Ma) CCD was also found to be 300 m shallower than previously estimated. This shallower CCD, at a paleodepth of approximately 4.5 km, along with associated reduced carbonate fluxes to the sea-floor, may be linked to a late Oligocene warming before the O/M boundary. The O/M boundary interval was first fully recovered in the equatorial Pacific at ODP Site 1218 (Fig. 3; see also suppl. Fig. 3 in Pälike et al., 2006a). Neogene carbonate minima are well documented in the Neogene PEAT sites, including a CCD minimum between 17 Ma and 18 Ma, a 'carbonate crash' interval around 10 Ma, and a newly delineated CCD minimum at about 4 Ma that occurs concurrently with enhanced deposition of diatomaceous sediments. The design of our drilling locations in combination with existing data will allow us to generate a three-dimensional view of Cenozoic CCD evolution during post-cruise research and to explore the linkage between Cenozoic changes in atmospheric $CO_2$ and global warmth.

Post-cruise research will undoubtedly enhance our understanding of the strength and timing of the CCD events and how they relate to other globally important Earth systems. These studies are intended in part to develop the tie between these events and orbital insolation changes. Reaching a sample resolution high enough to detect orbital insolation variations is an important PEAT objective, necessary to improve the Cenozoic age model and to confirm that events across the equatorial Pacific are synchronous.

Initial XRF scanning results from the Neogene (Lyle et al., unpublished) using the new Texas A&M XRF scanner at the IODP Gulf Coast Repository demonstrate how important information will result from detailed studies of the PEAT sediment (Fig. 7). Shown is a comparison between the 0–5 Ma XRF Ca peak area in Site U1338 and the Lisiecki and Raymo (2005) LR04 benthic oxygen isotope stack. The Ca peak area is correlated to the CaCO3 content in the U1338 sediments. The age model used in this example for U1338 is the linear shipboard age model, which has not been further tuned.

The benthic isotope record clearly shows a progression from low amplitude 41-kyr obliquity cycles to higher amplitude 41-kyr cycles at 2.7 Ma, and finally to the dominance of 100-kyr eccentricity cycles by 1 Ma. The development of the 100-kyr power within the oxygen isotope record is most easily observed in the 100-kyr bandpass filtered isotope record. For the oxygen isotope record older than 1 Ma, the spectral power in the 100-kyr band is only about 0.2 times that of the 41-kyr band. The evolution of the benthic isotope record may be caused by the development of Northern Hemisphere ice sheets or at least increased sensitivity to

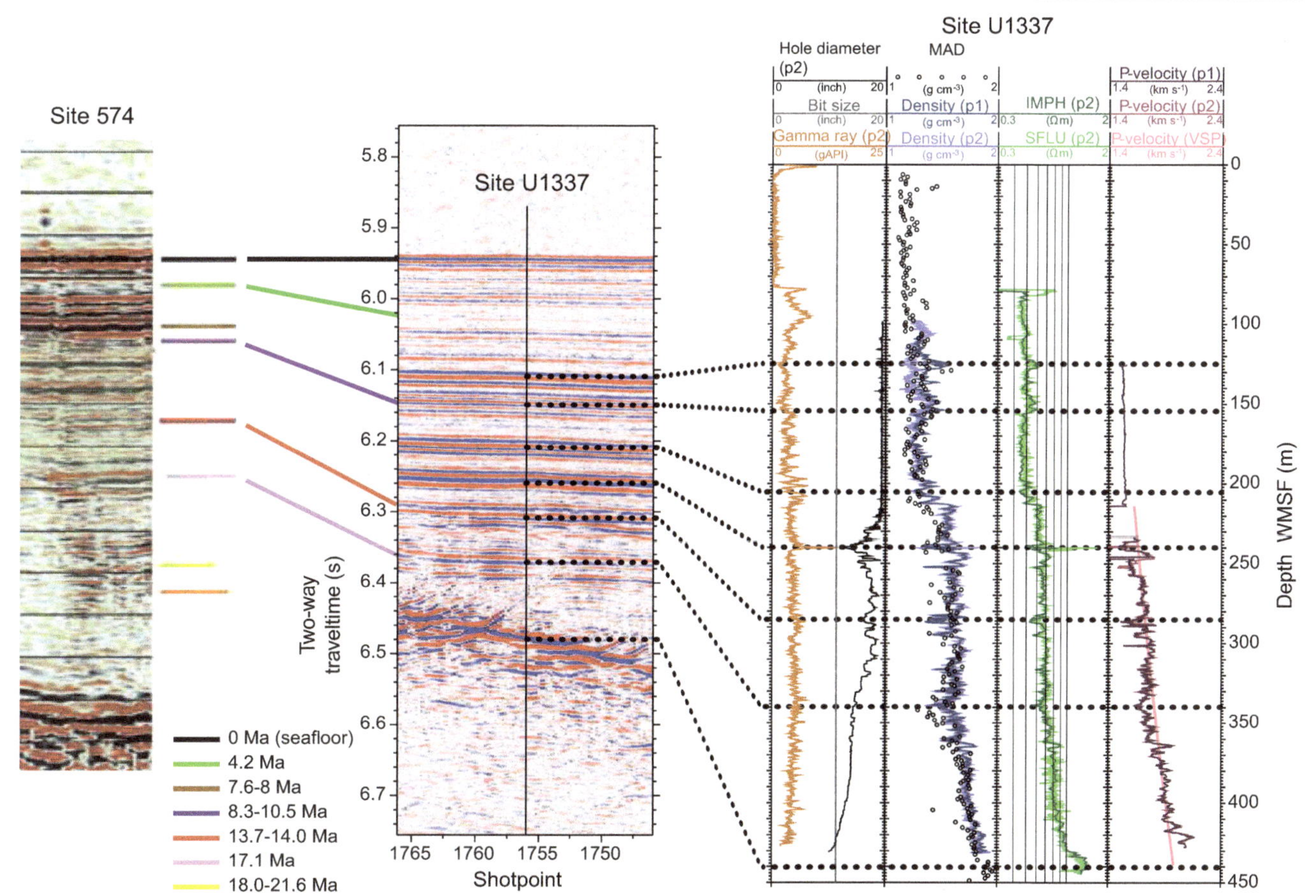


Figure 8. Correlation between the seismic reflection records from DSDP Leg 85 Site 574, IODP Expedition 321 Site U1337 (line 4 of the AMAT-03 site survey in the PEAT-7 area), and logging data from Hole U1337A. Correlations between the seismic profile at Site U1337 and the logs are based on velocities from VSP experiments and those measured by the velocity log, as well as the correlations to the age of sediments associated with the sediment physical property anomalies. The seismic horizons are associated with major fluctuations of carbonate; they appear to be chronostratigraphic, as originally suggested by Mayer et al. (1986).

high latitude insolation prior to the late Pleistocene (Lisiecki and Raymo, 2005).

In contrast, the U1338 Ca record retains spectral power in the 100-kyr band throughout the five-million-year record, suggesting that there is a linkage between carbonate burial and eccentricity (Pälike et al., 2006a). For the interval older than 1 Ma, the 100-kyr power in the Ca record is roughly six times greater than the 41-kyr power. It is interesting to note that records for 0–6 Ma from ODP Leg 138 eastern Pacific sites did not record high 100-kyr power (Hagelberg et al., 1995), but they do find high variability associated with obliquity (41 kyr) and precession (23 kyr and 19 kyr). The significant level of 100-kyr power in the older, deeper PEAT site suggests that dissolution (changes in $CO_2$ storage) may play a significant role in the development of the ~100-kyr $CaCO_3$ cycle in the central Pacific. Furthermore, it leads to the speculation that the abyssal carbon cycle played a role in 'looking in' the glacial cycles to a ~100-kyr rhythm.

Another major objective of PEAT drilling was to ground-truth the equatorial Pacific seismic stratigraphy so that seismic reflection records can be used to connect the sediment column described at each drill site to form a regional model. The PEAT expeditions have collected important new physical property data so that we can confirm the Mayer et al. (1985) seismic stratigraphy and also tie the eastern Pacific seismic stratigraphy with that of the central Pacific.

The equatorial Pacific is a classic 'binary' sediment system, with variable amounts of biogenic calcium carbonate and biosiliceous sediment components but very little clay. It is also well known that carbonate contents of equatorial Pacific sediments can be estimated from the bulk density, because carbonates have lower porosity and higher grain density than biosiliceous sediments (Mayer, 1991). Consequently, physical properties records contain meter-scale cyclicity that will ultimately be useful for orbital-tuning time scales, which is one of the PEAT objectives. Mayer et al. (1985) developed a seismic stratigraphy for the central Pacific at Site 574 on DSDP Leg 85. They noted that major seismic horizons were caused by density variations associated with low carbonate intervals. They proposed that the seismic horizons were isochrons because they were caused by paleoceanographic changes in deposition and/or dissolution of calcium carbonate.

Mayer et al. (1985) did not have logs to measure *in situ* velocities in support of their interpretation. One of the important PEAT experiments therefore was to use a combination of downhole measurements (vertical seismic profile (VSP) and standard logs) with physical properties measurements on core. We were able to run the VSP log at Site U1337 (Fig. 8) and Site U1338. Figure 8 is an initial comparison between the Site 574 seismic stratigraphy of Mayer et al. (1985) and the shipboard results for Site U1337. The events correlate in age, as would be predicted by Mayer et al. (1985). Site 574 is at essentially the same latitude as Site U1337 but is located more than 1000 km to the west. The extent of the correlatable seismic horizons across the Pacific helps to define the magnitude of the paleoceanographic events that caused them. Post-cruise studies will focus upon better defining the seismic strati-graphy at both Sites U1337 and U1338, allowing new tie points for seismic stratigraphic study of the equatorial Pacific sediment bulge (Mitchell et al., 2003).

## Outlook for the Future

The initial results from PEAT drilling illustrate the fundamental thrusts of the post-cruise science and provide a taste of new scientific insights to be reported in the next few years. We expect these insights to include a fundamental improvement of the Cenozoic time scale, an exploration of the unstable Eocene CCD and its relation to atmospheric $CO_2$, a much better understanding of the interactions between the carbon cycle and climate, and a better understanding of the history of major pelagic nutrient cycles and productivity. All of these studies will give important insights on how different Earth systems have interacted in the past and may respond in the near future.

## Acknowledgements

We thank the ship's crew and drilling crew on the *JOIDES Resolution*, and the technical staff of the U.S. Implementing Organization who made it possible to translate our plans into recovered core. M. Lyle also acknowledges support from the U.S. Science Advisory Committee for his participation and for NSF grant OCE-07253011.

## References

Bohaty, S.M., and Zachos, J.C., 2003. Significant Southern Ocean warming event in the late middle Eocene. *Geology*, 31(11):1017–1020, doi:10.1130/G19800.1.

Bohaty, S.M., Zachos, J.C., Florindo, F., and Delaney, M.L., 2009. Coupled greenhouse warming and deep-sea acidification in the middle Eocene. *Paleoceanogr.*, 24: PA2207, doi:10.1029/2008PA001676, 16 pp.

Bryden, H.L., and Brady, E.C., 1985. Diagnostic model of the three-dimensional circulation in the upper equatorial Pacific Ocean. *J. Phys. Oceanogr.*, 15:1255–1273, doi:10.1175/1520-0485(1985)015<1255:DMOTTD>2.0.CO;2.

Cane, M.A., and Molnar, P., 2001. Closing of the Indonesian Seaway as a precursor to east African aridification around 3-4 million years ago. *Nature*, 411:157–162, doi:10.1038/35075500.

Cane, M.A., and Zebiak, S.E., 1985. A theory for El Niño and the Southern Oscillation. *Science*, 228:1085–1087, doi:10.1126/science.228.4703.1085.

Chavez, F.P., and Barber, R.T., 1987. An estimate of new production in the equatorial Pacific. *Deep Sea Res.*, 34:1229–1243, doi:10.1016/0198-0149(87)90073-2.

Coxall, H.K., Wilson, P.A., Pälike, H., Lear, C., and Backman, J., 2005. Rapid stepwise onset of Antarctic glaciation and deeper calcite compensation in the Pacific Ocean. *Nature*, 433:53–57, doi:10.1038/nature03135.

Hagelberg, T.K., Pisias, N.G., Mayer, L.A., Shackleton, N.J., and Mix, A.C., 1995. Spatial and temporal variability of late Neogene equatorial Pacific carbonate, Leg 138. *In* Pisias, N.G., Mayer, L.A., Janecek, T.R., Palmer-Julson, A., and van Andel, T.H. (Eds.), *Proc. ODP, Sci. Results, 138*, College Station, Texas (Ocean Drilling Program), 321–336.

Haug, G.H., and Tiedemann, R., 1998. Effect of the formation of the Isthmus of Panama on Atlantic Ocean thermohaline circulation. *Nature*, 393:673–676, doi:10.1038/31447.

Holbourn, A., Kuhnt, W., Schulz, M., and Erlenkeuser, H., 2005. Impacts of orbital forcing and atmospheric carbon dioxide on Miocene ice-sheet expansion. *Nature*, 438(7067):483–487, doi:10.1038/nature04123.

Hovan, S.A., 1995. Late Cenozoic atmosphere circulation intensity and climatic history recorded by eolian deposition in the eastern equatorial Pacific Ocean, Leg 138. *In* Pisias, N.G., Mayer, L.A., Janecek, T.R., Palmer-Julson, A., and van Andel, T.H. (Eds.), *Proc. ODP, Sci. Results, 138*, College Station, Texas (Ocean Drilling Program), 615–626.

Keigwin, L.D., 1982. Isotopic paleoceanography of the Caribbean and East Pacific: role of Panama uplift in Late Neogene time. *Science*, 217(4557):350–353, doi:10.1126/science.217.4557.350.

Kemp, A.E., and Baldauf, J.G., 1993. Vast Neogene laminated diatom mat deposits from the eastern equatorial Pacific Ocean. *Nature*, 362:141–143, doi:10.1038/362141a0.

Koppers, A.A.P., Morgan, J.P., Morgan, J.W., and Staudigel, H., 2001. Testing the fixed hotspot hypothesis using Ar-40/Ar-39 age progressions along seamount trails. *Earth Planet. Sci. Lett.*, 185(3-4):237–252, doi:10.1016/S0012-821X(00)00387-3.

Lanci, L., Pares, J.M., Channell, J.E.T., and Kent, D.V., 2005. Oligocene magneto-stratigraphy from Equatorial Pacific sediments (ODP Sites 1218 and 1219, Leg 199). *Earth Planet. Sci. Lett.*, 237:617–634, doi:10.1016/j.epsl.2005.07.004.

Lisiecki, L.E., and Raymo, M.E., 2005. A Pliocene-Pleistocene stack of 57 globally distributed benthic $\delta^{18}O$ records. *Paleoceanogr.*, 21:PA1003, doi:1010.1029/2004PA001071, 17 pp.

Lyle, M., 2003. Neogene carbonate burial in the Pacific Ocean. *Paleoceanogr.*, 18(3):1059, doi:10.1029/2002PA000777.

Lyle, M., Barron, J., Bralower, T.J., Huber, M., Olivarez Lyle, A., Ravelo, A.C., Rea, D.K., and Wilson, P.A., 2008. The Pacific Ocean and Cenozoic evolution of climate. *Rev. Geophys.*, 46:RG2002, doi:10.1029/2005RG000190.

Lyle, M., Dadey, K., and Farrell, J., 1995. The Late Miocene (11-8 Ma) eastern Pacific carbonate crash: evidence for reorganization of deep water circulation by the closure of the Panama Gateway. *In* Pisias, N.G., Mayer, L.A., Janecek, T.R., Palmer-Julson, A., and van Andel, T.H. (Eds.), *Proc. ODP, Sci. Results, 138*, College Station, Texas (Ocean Drilling Program), 821–837.

Lyle, M.W., Olivarez Lyle, A., Backman, J., and Tripati, A., 2005. Biogenic sedimentation in the Eocene equatorial Pacific: the stuttering greenhouse and Eocene carbonate compensation depth. *In* Lyle, M., Wilson, P., Janecek, T.R., and Firth, J. (Eds.), *Proc. ODP, Sci. Results, 199*, College Station, Texas (Ocean Drilling Program), 35 pp.

Lyle, M., Wilson, P.A., Janecek, T.R., and the ODP Leg 199 Shipboard Scientific Party, 2002. *Proc. ODP Init. Repts., 199*, College Station, Texas (Ocean Drilling Program).

Mayer, L.A., 1991. Extraction of high-resolution carbonate data for palaeoclimate reconstruction. *Nature*, 352:148–50, doi:10.1038/352148a0.

Mayer, L.A., Shipley, T.H., Theyer, F., Wilkens, R.H., Winterer, E.L., Seismic modeling and paleoceanography at Deep Sea Drilling Project Site 574. In: Mayer, L.A., Theyer, F., Thomas, E., et al., *Init. Repts. DSDP* 85: Washington, U.S. Govt. Printing Office, 947–970, 1985.

Mayer, L.A., Shipley, T.H., and Winterer, E.L., 1986. Equatorial Pacific seismic reflectors as indicators of global oceanographic events. *Science*, 233:761–764, doi:10.1126/science.233.4765.761.

Miller, K.G., Wright, J.D., and Fairbanks, R.G., 1991. Unlocking the Ice House, Oligocene-Miocene oxygen isotopes, eustasy, and margin erosion. *J. Geophys. Res.*, 96(B4):6829–6848, doi:10.1029/90JB02015.

Mitchell, N.C., Lyle, M.W., Knappenberger, M.B., and Liberty, L.M., 2003. The lower Miocene to present stratigraphy of the equatorial Pacific sediment bulge and carbonate dissolution anomalies. *Paleoceanogr.*, 18:1038, doi:10.1029/2002PA000828.

Moore, T.C., Backman, J., Raffi, I., Nigrini, C., Sanfilippo, A., Pälike, H., and Lyle, M., 2004. The Paleogene tropical Pacific: clues to circulation, productivity, and plate motion. *Paleoceanogr.*, 19:PA3013, doi:10.1029/2003PA000998.

ODP Leg 199 Shipboard Scientific Party, 2002. Leg 199 Summary. *In* Lyle, M.W., Wilson, P.A., Janecek, T.R. (Eds.), *Proc. ODP, Init Repts., 199*, College Station, Texas (Ocean Drilling Program), 1–87.

Olivarez Lyle, A., and Lyle, M.W., 2006. Missing organic carbon in Eocene marine sediments: is metabolism the biological feedback that maintains end-member climates? *Paleoceanogr.*, 21:PA2007, doi:10.1029/2005PA001230.

Pälike, H., Moore, T., Backman, J., Raffi, I., Lanci, L., Parés, J.M., and Janecek, T., 2005. Integrated stratigraphic correlation and improved composite depth scales for ODP Sites 1218 and 1219. *In* Wilson, P.A., Lyle, M., and Firth, J.V. (Eds.), *Proc. ODP, Sci. Results, 199*, College Station, Texas (Ocean Drilling Program), 42 pp.

Pälike, H., Norris, R.D., Herrle, J.O., Wilson, P.A., Coxall, H.K., Lear, C.H., Shackleton, N.J., Tripati, A.K., and Wade, B.S., 2006a. The heartbeat of the Oligocene climate system. *Science*, 314(5807):1894–1898, doi:10.1126/science.1133822.

Pälike, H., Frazier, J., and Zachos, J.C., 2006b. Extended orbitally forced palaeoclimatic records from the equatorial Atlantic Ceara Rise. *Quat. Sci. Rev.*, 25:3138–3149. doi:10.1016/j.quascirev.2006.02.011.

Philander, S.G.H., 1983. El Niño Southern Oscillation phenomena. *Nature*, 302:295–301, doi:10.1038/302295a0.

Pisias, N.G., Mayer, L.A., and Mix, A.C., 1995. Paleoceanography of the eastern equatorial Pacific during the Neogene: synthesis of Leg 138 drilling results. *In* Pisias, N.G., Mayer, L.A., Janecek, T.R., Palmer-Julson, A., and van Andel, T.H. (Eds.), *Proc. ODP, Sci. Results, 138*, College Station, Texas (Ocean Drilling Program), 5–21. doi:10.2973/odp.proc.sr.138.101.1995.

Rea, D.K., and Lyle, M., 2005. Paleogene calcite compensation depth in the eastern subtropical Pacific: answers and questions. *Paleoceanogr.*, 20:PA1012, doi:10.1029/2004PA001064, 9 pp.

Romine, K., and Lombari, G., 1985. Evolution of Pacific circulation in the Miocene: radiolarian evidence from DSDP Site 289. *In* Kennett, J.P. (Ed.), *The Miocene Ocean: Paleoceanography and Biogeography, GSA Memoir 163,* Boulder, Colo. (Geological Society of America), 273–291.

Roth, J.M., Droxler, A.W., and Kameo, K., 2000. The Caribbean carbonate crash at the middle to late Miocene transition: linkage to the establishment of the modern global ocean conveyor. *In* Leckie, R.M., Sigurdsson, H., Acton, G.D., and Draper, G. (Eds.), *Proc. ODP, Sci. Results, 165,* College Station, Texas (Ocean Drilling Program), 249–273, doi:10.2973/odp.proc.sr.165.013.2000.

Schneider, D.A., 1995. Paleomagnetism of some Leg 138 sediments: detailing Miocene magnetostratigraphy. *In* Pisias, N.G., Mayer, L.A., Janecek, T.R., Palmer-Julson, A., and van Andel, T.H. (Eds.), *Proc. ODP, Sci. Results, 138,* College Station, Texas (Ocean Drilling Program), 59–72.

Shackleton, N.J., Crowhurst, S., Hagelberg, T., Pisias, N.G., and Schneider, D.A., 1995. A new late Neogene time scale: application to Leg 138 Sites. *In* Pisias, N.G., Mayer, L.A., Janecek, T.R., Palmer-Julson, A., and van Andel, T.H. (Eds.), *Proc. ODP, Sci. Results, 138,* College Station, Texas (Ocean Drilling Program), 73–101.

Takahashi, T., Feely, R.A., Weiss, R.F., Wanninkhof, R.H., Chipman, D.W., Sutherland, S.C., and Takahashi, T.T., 1997. Global air-sea flux of $CO_2$: an estimate based on measurements of sea-air $pCO_2$ difference. *Proc. Natl. Acad. Sci.*, 94:8292–8299, doi:10.1073/pnas.94.16.8292.

van Andel, T.H., 1975. Mesozoic/Cenozoic calcite compensation depth and global distribution of calcareous sediments. *Earth Planet. Sci. Lett.*, 26:187–194, doi:10.1016/0012-821X(75)90086-2.

van Andel, T.H., Heath, G.R., and Moore, T.C., 1975. *Cenozoic History and Paleoceanography of the Central Equatorial Pacific Ocean. A Regional Synthesis of Deep Sea Drilling Project Data.* The Geological Society of America, Memoir 143, 134 pp.

Westberry, T.M., Behrenfeld, J., Siegel, D.A., and Boss, E., 2008. Carbon-based primary productivity modeling with vertically resolved photo-acclimation. *Glob. Biogeochem. Cycles*, 22:GB2024, doi:10.1029/2007GB003078.

Zachos, J.C., Dickens, G.R., and Zeebe, R.E., 2008. An early Cenozoic perspective on greenhouse warming and carbon-cycle dynamics. *Nature*, 451:279–283, doi:10.1038/nature06588.

Zachos, J.C., Pagani, M., Sloan, L., Thomas, E., and Billups, K., 2001a. Trends, rhythms, and aberrations in global climate 65 Ma to present. *Science*, 292:686–693, doi:10.1126/science.1059412.

Zachos, J.C., Shackleton, N.J., Revenaugh, J.S., Pälike, H., and Flower, B.P., 2001b. Climate response to orbital forcing across the Oligocene-Miocene boundary. *Science*, 292(5515), 274–278, doi:10.1126/science.1058288.

## Authors

**Mitchell Lyle**, Department of Oceanography, Texas A&M University, TAMU 3146, College Station, TX 77840-3146, U.S.A., e-mail: mlyle@ocean.tamu.edu.

**Heiko Pälike**, School of Ocean and Earth Science, National Oceanography Centre, European Way, Southampton, SO14 3ZH, U.K., e-mail: heiko@noc.soton.ac.uk.

**Hiroshi Nishi**, Department of Natural History Sciences, Hokkaido University, Kita 10, Nishi 8, Kita-ku, Sapporo, 060-0810, Japan, e-mail: hnishi@mail.sci.hokudai.ac.jp.

**Isabella Raffi**, Dipartimento di Geotecnologie per l'Ambiente e il Territorio–DiGAT, Universitá "G. D'Annunzio", Campus Universitario, via dei Vestini 31, 66013 Chieti Scalo, Italy, e-mail: raffi@unich.it.

**Kusali Gamage**, Integrated Ocean Drilling Program, Texas A&M University, 1000 Discovery Drive, College Station, TX 77845-9547, U.S.A., e-mail: gamage@iodp.tamu.edu.

**Adam Klaus**, Integrated Ocean Drilling Program, Texas A&M University, 1000 Discovery Drive, College Station, TX 77845-9547, U.S.A., e-mail: aklaus@iodp.tamu.edu.

## IODP PEAT (Exp. 320/321) Scientific Party

**Expedition 320**: Heiko Pälike (co-chief scientist), Hiroshi Nishi (co-chief scientist), Adam Klaus (staff scientist), Gary Acton, Paul Bown, Margaret Delaney, Tom Dunkley Jones, Kirsty Edgar, Helen Evans, Peter Fitch, Nikolaus Gussone, Jens Herrle, Kiseong Hyeong, Shin-ichi Kamikuri, Junichiro Kuroda, Lizette Leon-Rodriguez, Theodore Moore Jr., Brandon Murphy, Hideto Nakamura, Christian Ohneiser, Carl Richter, Rebecca Robinson, Ken Sawada, Howie Scher, Hiroyuki Takata, Thomas Westerhold, Trevor Williams, Paul Wilson, and Yuhji Yamamoto.

**Expedition 321**: Mitchell Lyle (co-chief scientist), Isabella Raffi (co-chief scientist), Kusali Gamage (staff scientist), Louise Anderson, Jan Backman, Catherine Beltran, William Busch, James Channell, Pawan Dewangan, Hitoshi Hasegawa, Ed Hathorne, Hiroki Hayashi, Ann Holbourn, Steven Hovan, Koichi Iijima, Takashi Ito, Katsumori Kimoto, Alberto Malinverno, Daniel Murphy, Kaoru Ogane, Oscar Romero, Leah Schneider, Appy Sluijs, Jun Tian, Akira Tsujimoto, Bridget Wade, Roy Wilkens, Shinya Yamamoto, and Toshitsugu Yamazaki.

## Related Web Links

http://iodp.tamu.edu/scienceops/expeditions/equatorial_pacific.html
http://www.iodp.org/Pacific-Equatorial-Age-Transect-PEAT-1-2-Media-Kit/

# IODP Expedition 321T: Cementing Operations at Holes U1301A and U1301B, Eastern Flank of the Juan de Fuca Ridge

by Andrew T. Fisher, and IODP Expedition 321T Scientific Party

## Introduction

IODP Expedition 301 (Fisher et al., 2005a) was part of a series of expeditions and experiments to quantify hydrogeologic, lithologic, biogeochemical, and microbiological properties, processes, and linkages on the eastern flank of the Juan de Fuca Ridge, North Pacific (Fig. 1). Operations during Expedition 301 included replacement of one existing subseafloor borehole observatory ("CORK", Hole 1026B), drilling two basement holes and installing two new long-term observatories (Holes U1301A and U1301B), coring the upper ~300 m of basement and shallow sediments above basement, and collection of *in situ* hydrogeologic and geophysical data from basement. Subsequent expeditions using a remotely operated vehicle (ROV) and submersible have serviced borehole observatories, collected pressure and temperature data and fluid and microbiological samples, and replaced components as needed to maintain these systems for future use. Another drilling expedition is planned for 2010 (see backcover for schedule) to emplace three more borehole observatories and initiate cross-hole tests, and additional ROV and submersible ex-peditions will conduct long-term experiments and recover subseafloor data and samples.

Borehole observatories installed during IODP Expedition 301 were designed (i) to seal open holes so that thermal, pressure, and chemical conditions could equilibrate following the dissipation of the drilling disturbance; (ii) to facilitate collection of fluid and microbiological samples and temperature and pressure data using autonomous samplers and data logging systems; and (iii) to serve as long-term monitoring points for large-scale crustal testing (Fisher et al., 2005b). Unfortunately, the CORKs installed in Holes U1301A and U1301B were not sealed as intended (Fisher et al., 2005a). Initially, both holes were drawing cold bottom seawater into basement, but as described below, conditions in these holes are dynamic, and flow conditions have changed over time. The primary objective of Expedition 321T was to seal these observatories by pumping cement into the reentry cones surrounding the CORK wellheads, allowing the remaining components of the full experimental program to be completed during subsequent drilling and submersible expeditions.

## Background

Ocean Drilling Program (ODP) Leg 168 completed a drilling transect of eight sites across 0.9–3.6 Ma seafloor east of the Juan de Fuca Ridge. It resulted in collection of sediment, rock, and fluid samples; determination of thermal, geochemical, and hydrogeologic conditions in basement; and installation of a series of CORK observatories in the upper crust (Davis et al., 1997). Two of the Leg 168 observatories were placed in 3.5–3.6 Ma seafloor near the eastern end of the drilling transect, in Holes 1026B and 1027C (Fig. 1). IODP Expedition 301 returned to this area and drilled deeper into basement, sampled additional sediment, basalt, and microbiological materials, replaced the borehole observatory in Hole 1026B, and established two multilevel observatories at Site U1301 for use in long-term, three-dimensional hydrogeologic experiments.

Hole 1026B was drilled to 295 meters below seafloor (mbsf), cased across the sediment/basement interface, and extended to 48 meters sub-basement (msb) during ODP Leg 168 (Davis et al., 1997). The original CORK installed in Hole 1026B included a data logger, pressure sensors, thermistors at multiple depths, and a fluid sampler, all of which (except the fluid sampler) were recovered in 1999. The Hole 1026B CORK was incompletely sealed after being installed in 1996, and because basement fluids are overpressured in this location with respect to ambient hydrostatic conditions (Davis, et al., 1997), this hole discharged fluid for years until it was replaced during IODP Expedition 301. As of the start of Expedition 301, warm (~64°C) altered basement fluid vented freely through the top of the wellhead. The original ODP Leg 168 CORK installed in Hole 1026B was replaced successfully during IODP Expedition 301.

IODP Site U1301 was positioned 1 km south-southwest of ODP Site 1026, where sediment thickness is 260–265 m above a buried basement high (Fig. 1). Hole U1301A was drilled without coring to 370 mbsf (107 msb). The casing was extended into the upper 15 m of basement, but poor hole conditions prevented installation of longer casing, coring, or deeper drilling. A depth check prior to CORK deployment in Hole U1301A revealed that much of the lower part of the hole had filled in with rocks from the rubbly formation around the hole.

Hole U1301B was posi-tion-ed 36 m away. It penetrated to a total depth of 583 mbsf (318 msb). Uppermost basement was drilled without coring, and casing was installed to 85 msb. Basement was cored from 86 msb to 318 msb. The upper 100 meters of the cored interval in Hole U1301B were irregular in diameter, often much larger than the maximum inflation diameter of packers to be used for hydrogeologic testing and CORK observatories. However, the lower 100 meters of the hole were stable and to gauge, allowing collection of high-quality wireline logs and providing several horizons suitable for setting drill string and CORK casing packers.

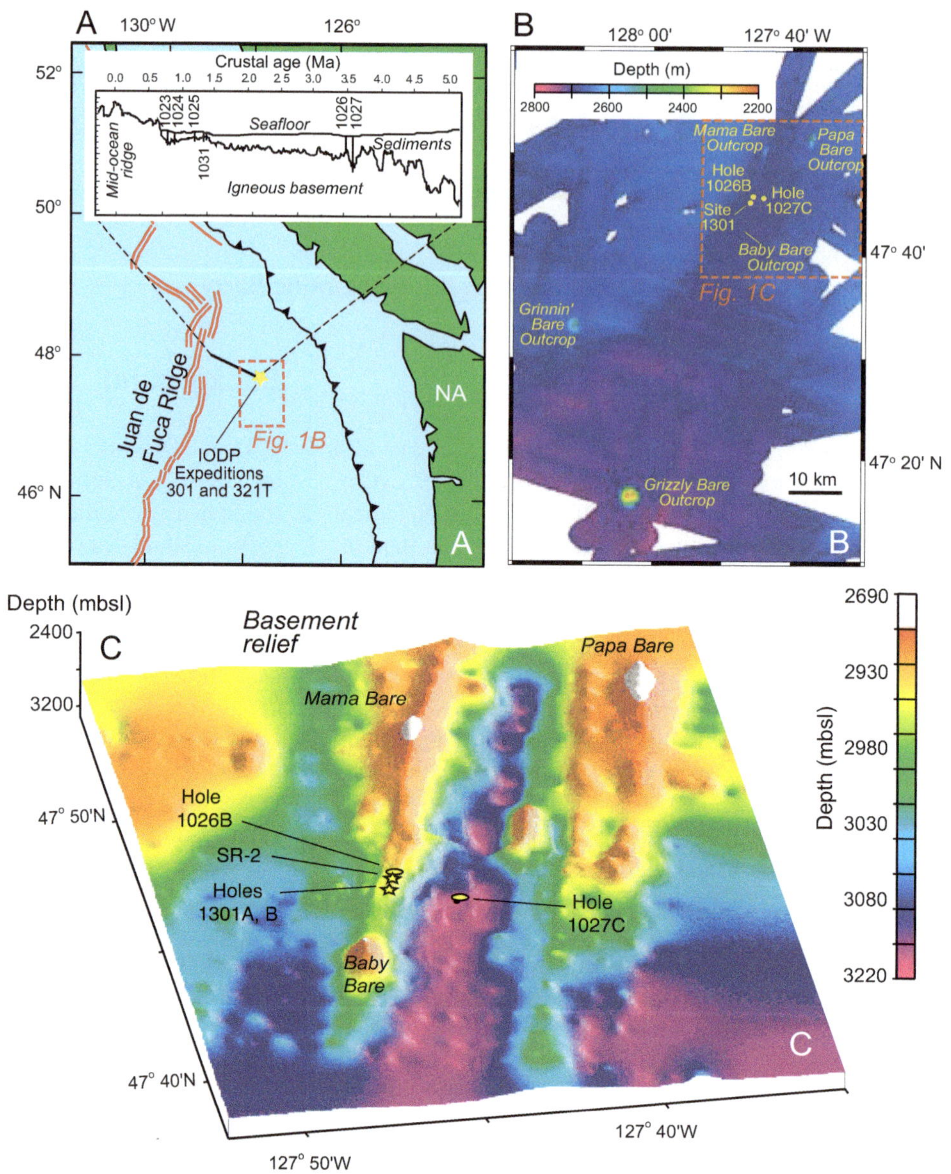


Figure 1. Maps showing the IODP Expedition 301 and 321T field area. [A] Index map showing locations of ODP Leg 168 drilling transect (thick black line), IODP Expedition 301 and 321T work area (gold star), the Juan de Fuca Ridge, and nearby continental areas. Inset profile shows locations of ODP Leg 168 sites. [B] Regional bathymetric map showing locations of ODP Holes 1026B and 1027C and IODP Site U1301. [C] Map of basement relief created from bathymetric and seismic data, showing locations of ODP and IODP drill sites. Areas marked in white are present-day basement outcrops. Figure modified from Zühlsdorff et al., 2005.

Both of the Site U1301 boreholes contained four nested casing strings: a 0.50-m casing (20-inch diameter) in the uppermost sediments, a 0.41-m casing (16-inch diameter) extending just across the sediment/basement interface, a 0.27-m casing (10.75-inch diameter) extending into basement, and a 0.11-m inner CORK casing (4.5-inch diameter) that houses instrument strings and plugs (Fig. 2). The two largest casing strings were sealed by collapse of unconsolidated sediments, and the 0.41-m string was also cemented across the sediment/basement interface. The annulus between 0.41-m and 0.27-m casing strings at Site U1301 was supposed to contain a rubber, mechanical casing seal near the seafloor, but this component was not available for use during Expedition 301 as planned. An attempt was made to seal the 0.27-m casing strings at depth with cement, but rubbly basement prevented the cement from sealing between casing and the borehole wall. Operations were additionally complicated in Hole U1301B by the separation of the unwelded 0.27-m casing string into two sections, which left a gap just above the sediment/basement interface (Fig. 2b). The CORK installed in Hole U1301A included a casing packer (as part of the 0.11-m inner casing) that was set inside the 0.27-m casing. In contrast, the CORK installed in Hole U1301B included two casing packers set in an open hole intended to hydraulically isolate sections of the upper crust (Fig. 2b).

Expedition 301 CORKs and the preexisting CORK in Hole 1027C were visited with an ROV soon after drilling in September 2004, and again by submersible in September 2005. Data recovered during these dives showed that the Hole 1026B observatory was sealed and operating as intended, although the pressure in Hole 1026B was recovering slowly from the thermal perturbation associated with eight years of upflow of warm formation fluid. The CORKs in Holes U1301A and U1301B were incompletely sealed, allowing cold ocean-bottom water to flow into the formation following CORK installation. The flow of this water into the crust at Site U1301 caused a measurable pressure perturbation at Site 1027, 2.4 km away, comprising an unintended (but scientifically useful) cross-hole test (Fisher et al., 2008).

Attempts to seal Hole U1301B using a cement delivery system with the submersible *Alvin* in summers 2006 and 2007 were unsuccessful, as the submersible could not deliver a sufficient quantity of cement to the cone. Shimmering fluid was observed discharging from Hole U1301A during and after summer 2007 dive operations. No such evidence for upflow from the borehole was observed during earlier visits, suggesting that Hole U1301A had "turned around" sometime between 2006 and 2007 servicing operations. In fact, downhole temperature loggers recovered from Hole U1301A in summer 2008 provide a detailed record of this flow reversal. Remarkably, Hole U1301B continued to draw fluid rapidly into basement as of summer 2008, even though it is located just 36 m from Hole U1301A, which was vigorously discharging warm formation fluid to the ocean. Additional research will be needed to understand the pressure and thermal interactions between Holes U1301A and U1301B as well as the implications for local and regional crustal hydrogeology.

The CORK landing platform installed at Hole U1301A included a solid surface that was perforated by eight 25-cm-diameter holes. Screens were welded below these holes prior to deployment so that instrumentation being deployed or manipulated by a submersible or ROV would not fall through the holes. In contrast, the landing platform at Hole U1301B has a series of radial support arms covered by the same screen material as was welded to the bottom of the platform at Hole U1301A. By the time the landing platform was deployed at Hole U1301B during Expedition 301, shipboard personnel realized that later cementing would be necessary, and a slot was cut through the platform screen to facilitate this operation.

## Operations and Results

A special formulation of cement, including "Cello-Flake" lost circulation material (LCM) was loaded onto the *JOIDES Resolution* during a brief port call prior to Expedition 321T. The use of cement with LCM is common in industry drilling and borehole installations, particularly in fractured rock, but this technology had not been used previously in scientific ocean drilling. In addition, a special bottom hole assembly

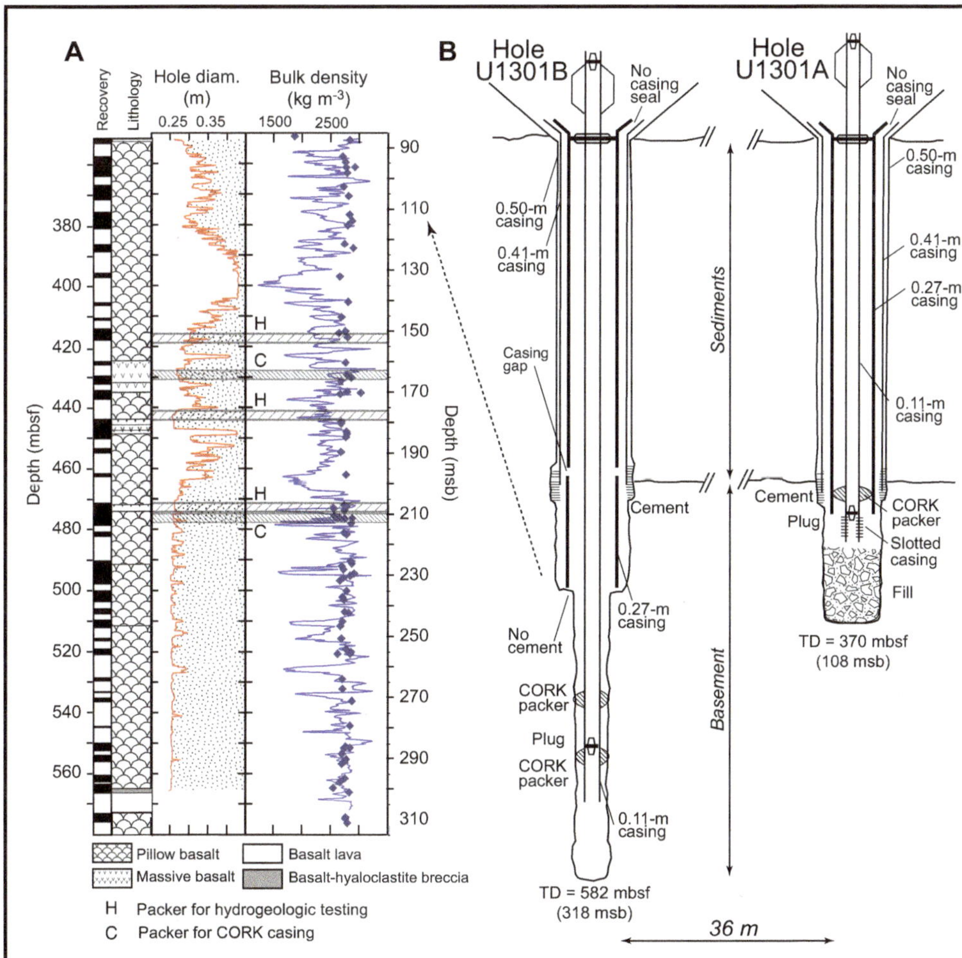


Figure 2. Summary of basement and borehole observatory characteristics in Holes U1301B and U1301A (modified from Fisher et al., 2008). [A] Basement recovery, primary lithology, borehole size, and bulk density from Hole U1301B (Fisher et al., 2005a). Basement cores were collected from ~355 mbsf to 575 mbsf (~100–320 msb), with recovery indicated by black intervals next to depth column. Bulk density log (blue line) and analyses of pieces of rock (diamonds) show evidence for considerable porosity in uppermost basement and a layered basement structure, especially below 470 mbsf (220 msb), with alternating more and less dense intervals. Horizontal bands indicate depths where CORK casing packers and the drill string packer used for hydrogeologic testing were set against the borehole wall. [B] Casing and CORK configurations in Holes U1301B and U1301A. Observatories in both holes lack a critical seal between 0.27-m and 0.41-m casing, which are needed to prevent the flow of cold bottom water into shallow basement or the flow of warm formation fluid out into the ocean. TD = total depth.

(BHA) was designed for Expedition 321T cementing operations, including a cementing "stinger" with a beveled edge that could be used to push through one of eight 25-cm-diameter holes in the landing platform in Hole U1301A.

Cementing operations occurred first at Hole U1301B, where reentry was easier because of the preexisting hole in the landing platform in the cone. Once pipe was run, reentry required 2.25 hrs, after which 60 barrels (bbls) of cement blended with LCM was pumped into the reentry cone. Seafloor operations were observed with the shipboard television camera system, and the cone was visibly filled with cement. The ship was offset to Hole U1301A, where the cement stinger was maneuvered to push through a screened hole in the CORK landing platform. This second reentry was accomplished in 2.75 hrs, after which 114 bbls of cement blended with LCM was pumped into the reentry cone. The ship was offset back to Hole U1301B, the hole was reentered again, and another 70 bbls of cement and LCM was pumped into the cone. Both reentry cones were inspected and found to be filled with cement surrounding the CORK observatory wellheads. Additional cement was observed to have poured over the edges of the cones and into the depressions in the seafloor around the cones. Cementing operations during 321T were completed more quickly, and with greater ease, than anticipated because of a combination of careful planning and preparation, the skill of the shipboard personnel in managing delicate operations, and fortuitous weather and sea conditions.

A scientific group returned to Holes U1301A and U1301B by submersible three weeks after Expedition 321T, downloaded pressure data, and evaluated the success of cementing efforts. As of summer 2009, both holes U1301A and U1301B had "turned around" and returned to overpressured conditions. Hole 1301A continued to leak fluid, as some of the cement pumped into the cone during Expedition 321T drained down into the annular gap between 0.41-m and 0.27-m casings. Additional cementing will be attempted during summer 2010 drilling operations, with a more aggressive use of LCM, to seal hole U1301A more completely. In contrast, Hole U1301B appears to be sealed between 0.41-m and 0.27-m casing, but the fluid overpressure in basement is greater than anticipated, and the top CORK plug was lifted by this pressure. Dive weights were placed on top of the plug during summer 2009 submersible operations, and additional pressure data will be collected from this system during summer 2010 servicing, which will allow a more complete evaluation of the state of this borehole observatory system.

## References

Davis, E.E., Fisher, A.T., Firth, J.V., Andersson, E.M., Aoike, K.,Becker, K., Brown, K.A., Buatier, M.D., Constantin, M., Elderfield, H., Gonçalves, C.A., Grigel, J.S., Hunter, A.G., Inoue, A., Lawrence, R.M., Macdonald, R.D., Marescotti, P., Martin, J.T., Monnin, Ch., Mottl, M.J., Pribnow, D.F.C., Stein, J.S., Su, X., Sun, Y., Underwood, M.B., Vanko, D.A., Wheat, G., 1997. *Proc. ODP, Init. Repts.*, 168: College Station, TX (Ocean Drilling Program), doi:10.2973/odp.proc.ir.168.1997.

Fisher, A.T., Davis, E.E., and Becker, K., 2008. Borehole-to-borehole hydrologic response across 2.4 km in the upper oceanic crust: implications for crustal-scale properties. *J. Geophys. Res.*, 113(B7):B07106, doi:10.1029/2007JB005447.

Fisher, A.T., Urabe, T., Klaus, A., and the Expedition 301 Scientists, 2005a. *Proc. IODP*, 301. College Station, Texas (Integrated Ocean Drilling Program Management International, Inc.), doi:10.2204/iodp.proc.301.2005.

Fisher, A.T., Wheat, C.G., Becker, K., Davis, E.E., Jannasch, H., Schroeder, D., Dixon, R., Pettigrew, T.L., Meldrum, R., Macdonald, R., Nielsen, M., Fisk, M., Cowen, J., Bach, W., and Edwards, K., 2005b. Scientific and technical design and deployment of long-term, subseafloor observatories for hydrogeologic and related experiments, IODP Expedition 301, eastern flank of Juan de Fuca Ridge. In Fisher, A.T., Urabe, T., and Klaus, A. (Eds.) *Proc. IODP*, 301. College Station, Texas (Integrated Ocean Drilling Program), doi:10.2204/iodp.proc.2301.2103.2005.

Zühlsdorff, L., Hutnak, M., Fisher, A.T., Spiess, V., Davis, E.E., Nedimovic, M., Carbotte, S., Villinger, H., and Becker, K., 2005. Site surveys related to IODP Expedition 301: ImageFlux (S149) and RetroFlux (TN116) expeditions and earlier studies. In Fisher, A.T., Urabe, T., Klaus, A., and the Expedition 301 Scientists, *Proc. IODP*, 301. College Station, Texas (Integrated Ocean Drilling Program Management International, Inc.), doi:10.2204/iodp.proc.301.102.2005.

## Author

**Andrew T. Fisher**, University of California Santa Cruz, Earth and Planetary Sciences Department and Instsitute for Geophysics and Planetary Physics, 1156 High Street, E&MS A232, Santa Cruz, CA 95064, U.S.A., e-mail: afisher@ucsc.edu.

**and IODP Expedition 321T Scientific Party**

# 3

# Drilling of Early Cretaceous Oceanic Anoxic Event 1a in Southern France

by Sascha Flögel and Wolfgang Kuhnt

## Introduction

The massive concentration of black, laminated, organic carbon-rich shales at certain time periods during the Cretaceous period (~140–65 Ma) led to the concept of Oceanic Anoxic Events (OAEs). These events are characterized by unusually enhanced preservation of organic matter across environments ranging from the deep oceans to shelf seas. Enhanced productivity of siliceous and organic-walled primary producers and/or strongly dysaerobic or anoxic conditions in all major oceans were both suggested as likely causes (Meyer and Kump, 2008). Fundamental chemical and biological changes in the world ocean must have been associated with these events.

While the geographic extent of most black shale events in the Cretaceous is still under debate, the two main OAEs in the early Aptian (Selli Event: OAE1a, ~119 Ma) and at the Cenomanian-Turonian boundary (Cenomanian/Turonian Boundary Event: OAE2, ~94 Ma) have proven to be of global distribution (Grötsch et al., 1998). Estimates of organic matter accumulation rates and recent findings of biomarkers for cyanobacteria, a proxy for photic zone anoxia in organic-carbon rich sediments of both OAEs, indicate high levels of productivity as observed in extreme upwelling environments today (Meyer and Kump, 2008). Carbon-isotope studies demonstrate that both the Aptian and the Cenomanian-Turonian black shales are associated with a positive carbon-isotope excursion in marine pelagic and shallow-water carbonate, marine organic matter, and terrestrial higher-plant material. These positive carbon-isotope excursions thus offer a means of corrclation between sediments deposited in the oceans and on the continents. The Aptian event is also associated with an initial negative carbon-isotope excursion, recently interpreted as due to dissociation of methane hydrates. It has been suggested that both phenomena, OAEs and methane-release events, were results of increased global temperatures caused by high $CO_2$-levels in the atmosphere (Wagner et al., 2007). Massive sequestration of organic carbon during an OAE, however, would draw down $CO_2$ and produce global cooling with respect to pre-OAE levels (Forster et al., 2007).

After decades of research, timing and causes of onset and spread of Cretaceous OAEs still remain a topic of vigorous debate. Deciphering the driving mechanisms and time scales for the onset and spread of OAE1a is the main objective of this study. Due to the lack of complete, unweathered sections of the early Aptian, a well-studied type locality of early Aptian deposition in southern France was selected for coring.

## La Bédoule section and coring

The drilling campaign took place during May 2009 at Roquefort-La Bédoule about 20 km southeast of Marseille. The drill sites are located in abandoned quarries in close proximity to outcrops with an existing stratigraphic framework (Figs. 1, 2). The target interval was successfully encountered below the weathering zone.

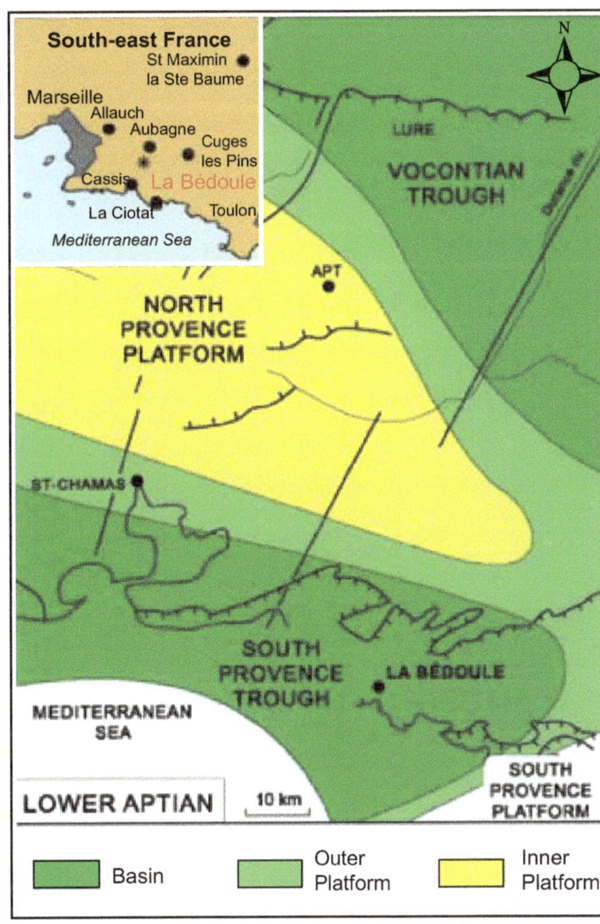


Figure 1. Paleogeographical scheme of the South Provence intrashelf basin during the Lower Aptian (from Renard et al., 2005) with location map of the La Bédoule-Cassis area in southern France (upper left corner).

Figure 2. Drilling operations at section LB1 with a trailer-mounted rig.

The regional and palaeogeographic setting of the stratotype is that of an intrashelf basin—the South Provence Trough—that formed within the Urgonian Carbonate Platform during late Barremian times (Masse et al., 1993). It was isolated from the Vocontian Basin to the north by the North Provence Platform (Monts de Vaucluse – Mont Ventoux) and bounded to the southeast by the South Provence Platform (Mont Faron). Although quite restricted in extent initially, this trough extended westward during Lower Aptian times to join the North Pyrenean Basin of the

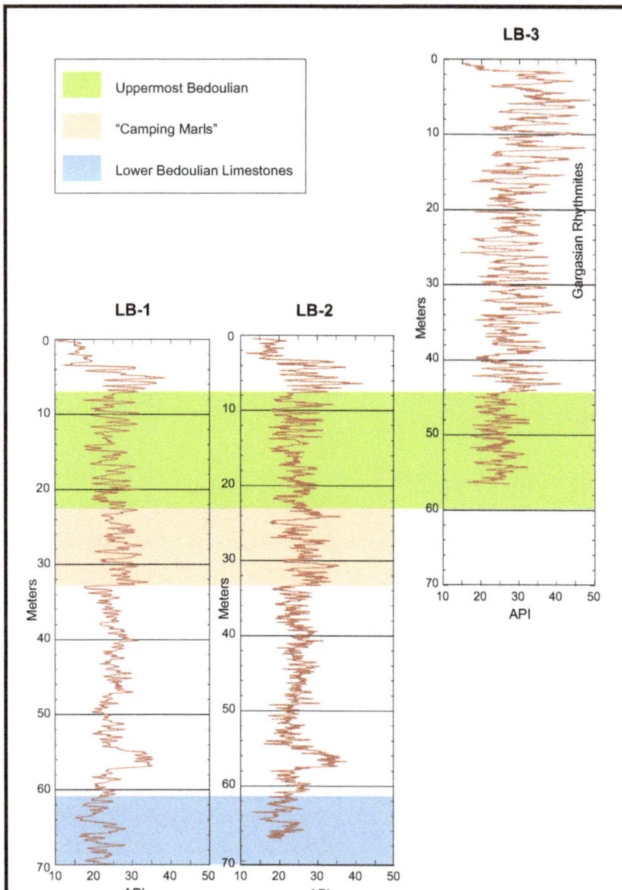


Figure 3. Geological sections and gamma-ray logs of cores drilled in the Aptian of La Bédoule.

'Deshayesites Marls' (Masse et al., 1993). As described by Moullade et al. (1998a, 1998b) and Masse (1998), the section at La Bédoule presents three lithological members:

- Upper Barremian-lower Bedoulian limestone member overlying the Urgonian Platform deposits
- Upper Bedoulian alternating marls and limestones member
- Top Bedoulian - lower Gargasian marly member

The discrimination of these members is of practical value in the field but does not accurately reflect the $CaCO_3$ content which always remains high (80–95%; Masse, 1998). The position of the main OAE1a carbon isotope excursion within the "Camping Marls" in the Upper Bedoulian alternating marls and limestones member is correlated bed by bed to a nearby outcrop section at the former camping ground of La Bédoule (Fig. 3). Three cores at two different sites were drilled with a total recovery of more than 180 meters and a core diameter of 11 cm. Of the three holes, two are covering lower Bedoulian limestones through uppermost Bedoulian sediments (LB1 = 70 m and LB2 = 67 m), while LB3 (56 m) recovered uppermost Bedoulian through middle Gargasian strata. The cores were retrieved continuously with 100% recovery and minimal disturbance at core breaks, which can be bridged by the parallel core in all cases. Cores were cut into working and archive halves and are stored under cool conditions in the core repository at IFM-GEOMAR, Germany.

## Objectives

The drilling campaign was initiated to recover high quality sediment cores in order to determine the timing, succession of events, and feedback mechanism at the onset of OAE1a. Goals include times series analysis of continuously measured proxy data (gamma rays, visible light spectrometry, and XRF scanning) to develop orbital chronologies and to determine leads/lags between different proxy data. Additionally, it is planned to develop ultra-high resolution isotopic ($\delta^{13}$Ccarb and $\delta^{13}$Corg) records of the periods immediately before and during the initial carbon isotope shifts in the early Aptian. These will include proxy records for marine organic carbon accumulation (TOC Rock-Eval records), deep water oxygenation (benthic foraminiferal dissolved oxygen index, redox-sensitive elements such as Mn/Fe, U/Mo, Re/Mo, and Te/Se, and isotope systems such as $\delta^{98/95}$Mo), and SST ($\delta^{18}$O, TEX86, and Ca-isotopes) at intermediate resolution and/or at critical points of the high-resolution data series. A further aim will be determining the amplitude and phasing of $pCO_2$ changes during the OAE as a crucial boundary condition for climate models by using boron isotopes (Pearson and Palmer, 2000).

Initial foraminiferal biostratigraphy indicated that the upper portion of LB2 top is still uppermost Bedoulian (Furcata zone) and that the top of LB3 can be attributed to the very basal *Globigerinelloides barri* zone (middle Gargasian). High-resolution wireline gamma-radiation of the cores was registered to serve for detailed cross-correlation of wells (Fig. 3).

High-resolution analyses of the core are currently underway, including foraminiferal and nannoplankton biostratigraphy (University of Marseille and University of Lyon), and visible light spectrophotometry, XRF scanning, and stable isotope geochemistry (University of Kiel). The advanced analytical phase of this multi-disciplinary project is scheduled for the next three years.

## Acknowledgements

This study is an integral part of the SFB 754 Collaborative Research Center "Climate-Biogeochemistry Interactions in the Tropical Ocean" established by the Deutsche Forschungsgemeinschaft at the University of Kiel and IFM-GEOMAR. We thank Stölben GmbH, especially André Herrlich and Thomas Flohr, for their skills and patience during the drilling. The Ville of Roquefort-La Bédoule (Département des Bouches du Rhone) and the landowner of the main drilling site (Mme. Fiani) generously granted access and facilitated drilling in any possible way. We also want to especially acknowledge the help of Thomas Grelle and Thomas Wonik from the Leibniz Institute for Applied Geophysics in Hannover who kindly provided the wireline logging equipment and support.

## References

Forster, A., Schouten, S., Moriya, K., Wilson, P.A., and Sinninghe Damsté, J.S., 2007. Tropical warming and intermittent cooling during the Cenomanian/Turonian oceanic anoxic event 2: sea surface temperature records from the equatorial Atlantic. *Paleoceanography*, 22:PA1219, doi:10.1029/2006PA001349.

Grötsch, J., Billing, I., and Vahrenkamp, V., 1998. Carbon-isotope stratigraphy in shallow-water carbonates: implications for Cretaceous black-shale deposition. *Sedimentology*, 45(4):623–634.

Masse, J.-P., 1998. Sédimentologie du stratotype historique de l'Aptien inférieur dans la région de Cassis-La Bédoule (SE France). *Géologie Méditerranéenne*, 25 (3–4):31–41.

Masse, J.-P., Bellion, Y., Benkhelil, J., Dercourt, J., Guiraud, R., and Ricou, L.E., 1993. Lower Aptian (114-112 Ma). *In* Dercourt, J., Ricou, L.E., and Vrielynck, B. (Eds.), *Atlas Tethys Paleoenvironmental Maps. Explanatory Notes.* Paris (Gauthier-Villars), 135–152.

Meyer, K.M., and Kump, L.R., 2008. Oceanic euxinia in Earth history: causes and consequences. *Ann. Rev. Earth Planet. Sci.*, 36:251–88.

Moullade, M., Kuhnt, W., Bergen, J.A., Masse, J.-P., and Tronchetti, G., 1998a. Correlation of biostratigraphic and stable isotope events in the Aptian historical stratotype of La Bédoule (SE France). *Comptes Rendus de l'Académie des Sciences - Series IIA - Earth Planet. Sci.*, 327:693–698.

Moullade, M., Tronchetti, G., Busnardo, R., and Masse, J.-P., 1998b. Description lithologique des coupes types du stratotype historique de l'Aptien inférieur dans la région de Cassis-La Bédoule (SE France). *Géologie Méditerranéenne*, 25(3–4):15–29.

Pearson, P.N., and Palmer, M.R., 2000. Atmospheric carbon dioxide concentrations over the past 60 million years. *Nature*, 406:695–699.

Renard, M., de Rafélis, M., Emmanuel, L., Moullade, M., Masse, J.-P., Kuhnt, W., Bergen, J., and Tronchetti, G., 2005. Early Aptian $\delta^{13}C$ and manganese anomalies from the historical Cassis-La Bédoule stratotype sections (S.E. France): relationship with a methane hydrate dissociation event and stratigraphic implications. *Carnets de Géologie/Notebook on Geology*, 2005/04 (CG2005_A04):1–18.

Wagner, T., Wallmann, K., Stüsser, I., Herrle, J.O., and Hofmann, P., 2007. Consequences of moderate 25,000 yr lasting emission of light $CO_2$ into the mid-Cretaceous ocean. *Earth Planet. Sci. Lett.*, 259:200–211, doi: 10.1016/j.epsl.2007.04.045.

## Authors

**Sascha Flögel**, IFM-GEOMAR, Leibniz-Institut für Meereswissenschaften (an der Universität Kiel), Dienstgebäude Ostufer; 8C-110, Wischhofstr. 1-3, 24148 Kiel, Germany, e-mail: sfloegel@ifm-geomar.de.

**Wolfgang Kuhnt**, Christian-Albrechts-Universität zu Kiel (CAU), Institute of Geosciences, Marine Micropaleontology, Ludewig-Meyn-Str. 14, D-24118 Kiel, Germany, e-mail: wk@gpi.uni-kiel.de.

## Photo Credit

Fig. 2: S. Flögel

# 4

# The Great Oxidation Event Recorded in Paleoproterozoic Rocks from Fennoscandia

by Victor A. Melezhik, Aivo Lepland, Alexander E. Romashkin, Dmitry V. Rychanchik, Melanie Mesli, Tor Erik Finne, Ronald Conze and the FAR-DEEP Scientists

With support of the International Continental Scientific Drilling Program (ICDP) and other funding organizations, the Fennoscandia Arctic Russia – Drilling Early Earth Project (FAR-DEEP) operations have been successfully completed during 2007. A total of 3650 meters of core have been recovered from fifteen holes drilled through sedimentary and volcanic formations in Fennoscandia (Fig. 1), recording several global environmental changes spanning the time interval 2500–2000 Ma, including the Great Oxidation Event (GOE) (Holland, 2002). The core was meanwhile curated and archived in Trondheim, Norway, and it has been sampled by an international team of scientists.

## Introduction

The emergence of an aerobic Earth System and a series of interrelated global events (Archean-Paleoproterozoic Transition, APT) during the Late Archean and Early Paleoproterozoic (2500–2000 Ma) led to the irreversible alteration of Earth's surface (Fig. 2). This environmental change poses a fundamental challenge in the geosciences (Melezhik et al., 2005a). The FAR-DEEP Expedition specifically targeted geological formations 2500–2000 Ma old in three different areas (Fig. 1) that recorded most of the events associated with the GOE. The project has three major goals:

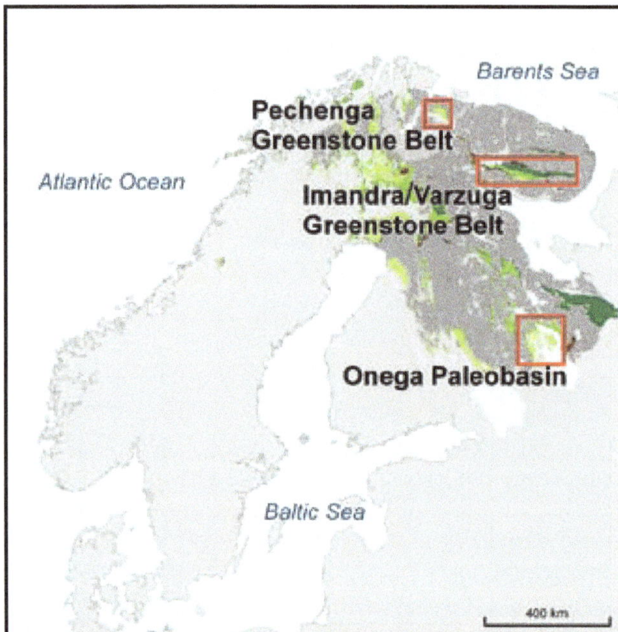


Figure 1. Geographic locations of the FAR-DEEP drilling operations.

- Establish a well-characterized, well-dated, and well-curated succession of rocks for the period of 2500–2000 Ma
- Document the changes in the biosphere and the geosphere associated with the rise in atmospheric oxygen
- Develop a model to explain the genesis and timing of the establishment of the aerobic Earth System.

The samples obtained by the FAR-DEEP provide a representative geological record of the most important global events occurring through the APT (Fig. 2).

## Drilling Operations

The drilling operations were carried out from May to October 2007 on the Russian part of the Fennoscandian Shield. The fifteen holes drilled range in depth from 92m to 503 m, totaling 3650 m of recovered core. To minimize the risk of core contamination for forthcoming biomarker studies, the drilling was performed with clear water and with non-oil-based lubricants. In most cases, the core recovery was close to 100%. Most of the cores were successfully re-oriented, and they are suitable for paleomagnetic studies. The large distances and remoteness of the drilling sites with limited infrastructure required a considerable logistical effort. In addition the coring offered a unique opportunity for students from the FAR-DEEP partner institutions to examine field sites and participate in the recovery of the best available

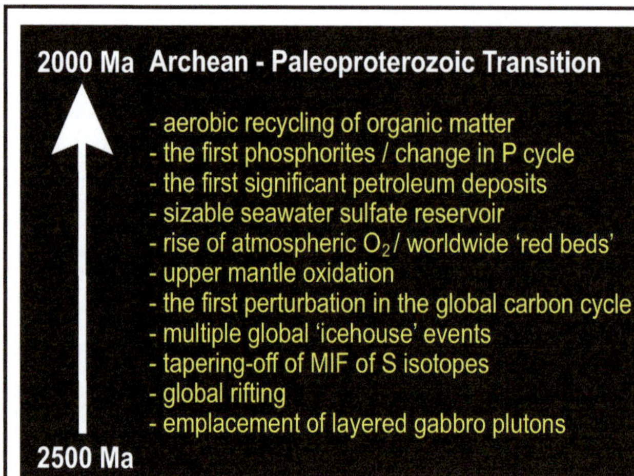


Figure 2. Global environmental events associated with the GOE, and targeted by FAR-DEEP (Holland, 2002).

succession of volcanic-sedimentary rocks recording the time of emergence of an aerobic world.

## Core Archive and Drilling Information System (DIS)

All cores were transported to the Geological Survey of Norway (NGU) in Trondheim for curation and archiving. All technical and geological data gathered during the drilling operations and through the archiving process have been catalogued using the Drilling Information System (DIS, see Web Links) developed by ICDP. During the project, the functions of this software have been continuously updated to meet the specific FAR-DEEP needs. The archive process began in March and was completed by December 2008.

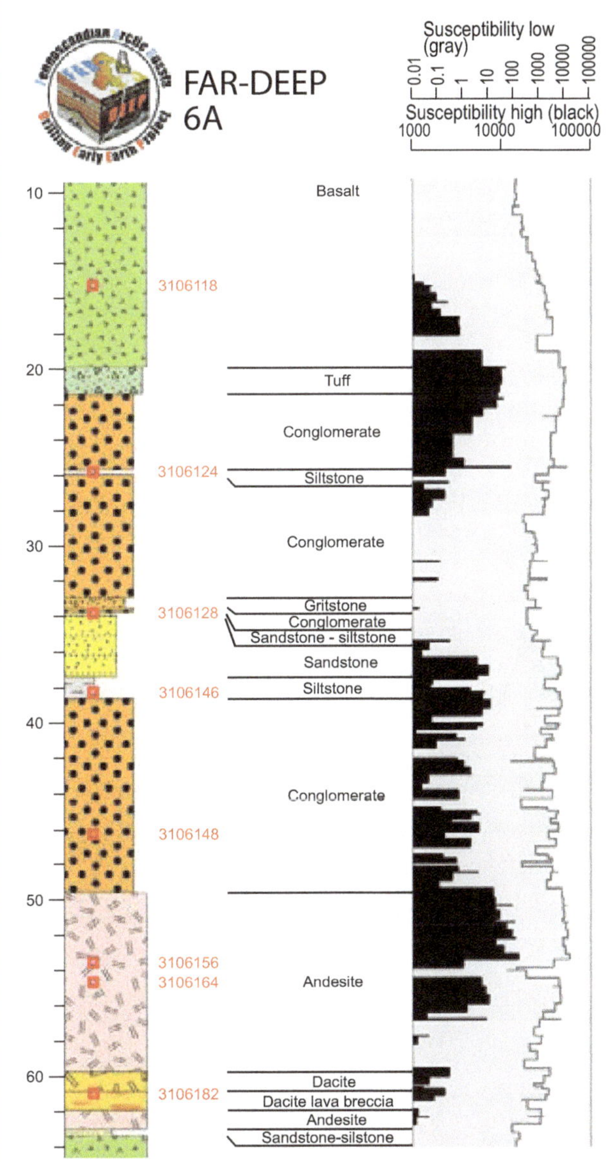


Figure 3. Example of a drillhole log with position of archive samples (red squares) and their numbers. Magnetic susceptibility data are plotted alongside.

At the NGU's core repository the archive process included (1) high-resolution photography of core in boxes, in dry and wet conditions, (2) magnetic susceptibility measurements of full core at ~20-cm intervals, (3) core splitting by sawing, (4) image scanning and photography of split core, (5) detailed lithological description, (6) and routine sampling with ~7-m spacing for general geochemical and petrographic characterizations of the rocks. The archive sample set consists of 554 specimens that have been analyzed for whole-rock major and trace elements, carbonate composition, abundance of C and S, and thin sections have been made from them at the NGU laboratory. Selected specimens from the archive set have been analyzed for S isotopes of sulfides, C- and N-isotopes of the organic matter, and C-, O-, and Sr-isotopes of carbonates by the FAR-DEEP international team of researchers. All documentation obtained during the archiving process and general geochemical profiles of cores (Fig. 3) were made available for the FAR-DEEP partners on the Internet prior to core sampling in Trondheim from March to April 2009.

## "Coring" through Major Paleoproterozoic Events and Paleoproterozoic Reference Sections

The FAR-DEEP cores of sedimentary and volcanic formations in the stratotype areas of the Pechenga and Imandra/Varzuga Greenstone Belts and the Onega Basin (Fig. 1) record the global events of the APT (Figs. 4 and 5).

*Change in fractionation of sulfur isotopes.* Several lines of evidence imply Earth's earliest atmosphere shifted from being anoxic to oxic between 2500 Ma and 2000 Ma (GOE, Fig. 2). The processes driving this shift, as well as their timing and duration, remain enigmatic. Appearance of 'red beds' and sulfates in the stratigraphic record is among the most compelling evidence reflecting the change to oxic conditions. Additional support for the oxygenation has been gained from recent work on S-isotopes that have shown the presence of mass-independently fractionated (MIF) S-isotopes in rocks older than 2360 Ma and disappearance of such a signature thereafter (Bekker et al., 2004; Hannah et al., 2004; Guo et al., 2009). The disappearance of MIF S-isotopes has been typically attributed to the oxygenation of the atmosphere and related change in photochemical reactions, but other processes have also been proposed (Watanabe et al., 2009). FAR-DEEP's drillhole 1A intersected a sequence of marine shales (Fig. 6A) and sandstones deposited prior to 2442 Ma (Fig. 4). These 2442-Ma rocks show a transition from MIF to mass-dependently fractionated sulfur isotopes (Reuschel et al., 2008); thus, they offer new constraints on the S-isotope cycle at the dawn of the GOE. Moreover, isotope geochemistry of marine carbonate rocks interbedded with shales may represent a robust proxy for the global carbon cycle prior to the Huronian-age glaciation (Young et al., 2001).

*Huronian-age global glaciation.* The rapid onset of global glaciation(s) from otherwise climatically invariant conditions at around 2320 Ma (age constraint from South Africa, Hannah et al., 2004) is another significant environmental event during the APT. The triggering mechanisms remain poorly understood (Evans, 2003) and several models have been advanced to explain them (Melezhik, 2006; Claire et al., 2006). In the Imandra/Varzuga Greenstone Belt, drillhole 3A intersected Huronian-age diamictites (Fig. 6B). These are overlain by spinifex-textured komatiites (Fig. 6C) and underlain by Sr-rich limestones, all resting on felsic volcanic rocks. This setting offers unique opportunities for reconstructing paleolatitudinal positions, a geochemical record of seawater Sr- and C-isotopic composition and geochronology.

*Unprecedented perturbation of the global carbon cycle.* The largest positive excursion of $\delta^{13}C$ known in Earth's history as recorded in sedimentary carbonates—termed the Lomagundi-Jatuli Event—is one of a series of prominent Paleoproterozoic environmental events whose interrelationships remain intriguing and only partially resolved (Melezhik et al., 2007). At present, no consensus exists on the causative mechanism(s) responsible for the Lomagundi-Jatuli Event (Bekker et al., 2001, 2003a; Hayes and Waldbauer, 2006). The possible role of local factors in amplification of a global signal remains unresolved (Melezhik et al., 1999; Bekker et al., 2003a, 2003b; Aharon, 2005). Drillholes 4A, 5A, 10A, 10B, and 11A intersected $^{13}C$-rich carbonates of the Lomagundi-Jatuli Event (Fig. 6D) in three different depositional and paleotectonic settings. Drillholes 8A and 8B cored carbonates (Fig. 6E) that post-date the Lomagundi-Jatuli Event. The study of all these cores will help to understand onset, duration, and causes of this unique perturbation of the global carbon cycle.

*Oxidized ocean – abundant Ca-sulfates.* Progressive oxygenation of Earth's surface environments resulted in increased sulfide oxidation during continental weathering and a concomitant increase in the concentration of marine sulfate (Guo et al., 2009). However, to date, available esti-mates of marine sulfate reservoir size remain controversial (Kah et al., 2004; Melezhik et al., 2005b),. Several FAR-DEEP drillholes (5A, 8A, 8B, 10A, 10B, 11A) successfully targeted sedimentary

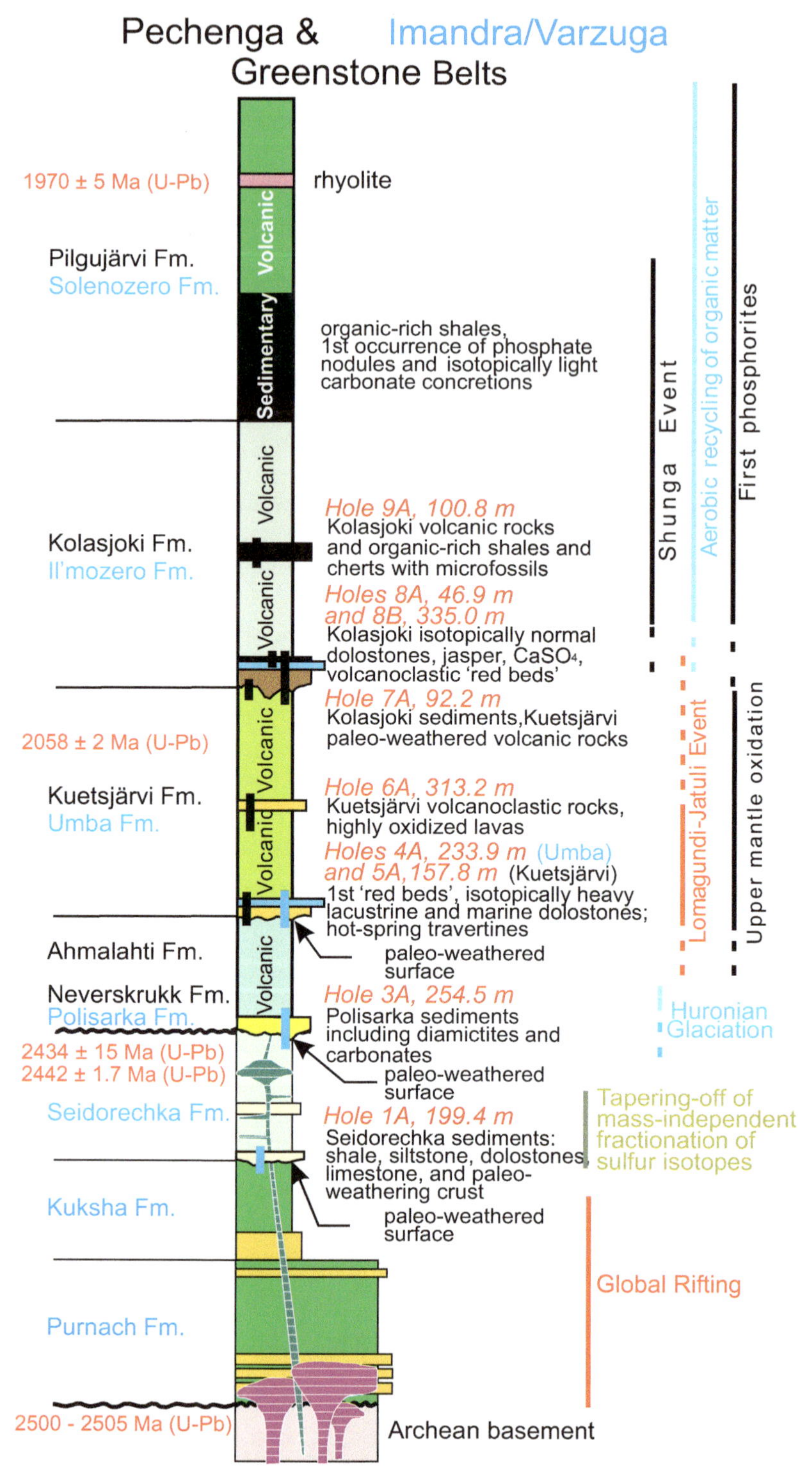


Figure 4. Composite (~10,000-m-thick) and simplified Paleoproterozoic section for the Pechenga and Imandra/Varzuga Greenstone Belts linked to major paleoenvironmental events. Radiometric dates from bottom to top are from Amelin et al. (1995), Melezhik et al. (2007), and Hanski (1992). The Pechenga Belt stratigraphy is shown in black-face letters, and the drilled holes by vertical black lines; features for the Imandra/Varzuga Belt are in blue. The number of the drillhole and its depth are shown in red-face letters; this is followed by text briefly describing rock lithologies intersected by the hole.

formations that contain seawater sulfates. Pseudomorphs after gypsum and anhydrite, with relicts of primary minerals, have been documented in great abundance (Fig. 6F, 6G). Detailed mineralogical and isotopic studies to decipher their environmental significance are in progress and will help constrain the oxygenation history.

*Ferric iron-rich volcanic rocks: upper mantle oxidizing event vs. secondary overprint by oxidized groundwaters.* Was the rise of atmospheric oxygen levels related to increased dioxygen due to evolution of oxygen-producing organisms, or was it controlled by a decrease in oxygen sinks (Kump, 2008)? One hypothesis is that a predominant sink for oxygen in the Archean era was abruptly and permanently diminished during the APT (Kump and Barley, 2007), but the overall cause of the GOE remains unresolved. Approximately 2060-Ma-old volcanic rocks (Fig. 6H, 6I) in the Pechenga Greenstone Belt have high $Fe^{3+}/Fe_{total}$ ratios (average=0.37) which are in sharp contrast to the majority of underlying and overlying volcanic units. The FAR-DEEP drillholes 4A and 6A recovered more than 300 meters of such highly oxidized lavas (Fig. 6I). This anomaly may represent evidence either for relatively oxidized mantle material (e.g., recycled banded iron formations) or a large-scale alteration of Earth's surface by oxidized meteoric and/or groundwaters.

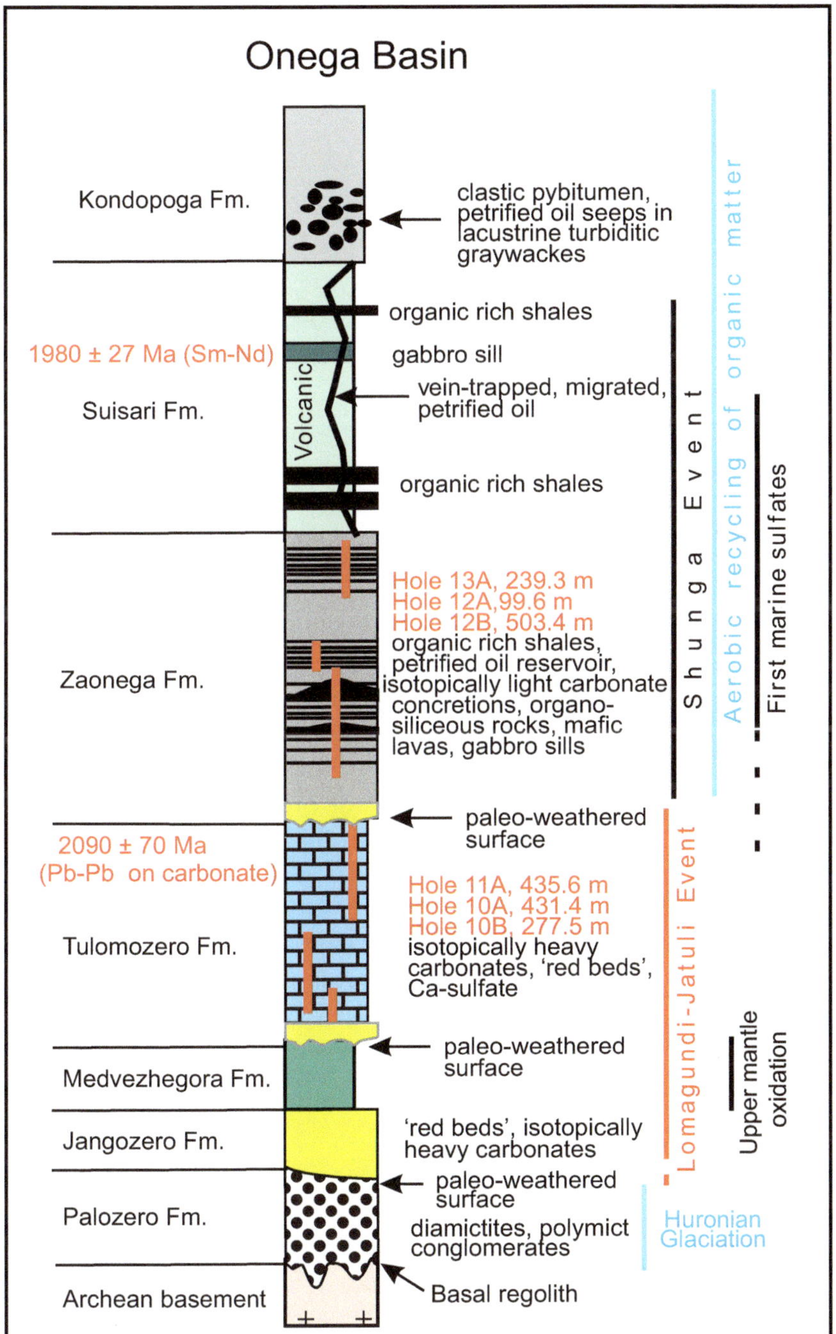


Figure 5. Composite (~5000-m-thick) and simplified Paleoproterozoic section for the Onega Basin linked to major paleoenvironmental events. Radiometric dates from bottom to top are from Ovchinnikova et al. (2007) and Pukhtel et al. (1992). Vertical red lines show the position of the drillholes. The number of the drillhole and its depth are shown in red-face letters; this is followed by text briefly describing rock lithologies intersected by the hole.

*Subaerial and subaqueous hydrothermal systems.* Two types of compositionally different hydrothermal products have been documented in the FAR-DEEP drillcores. The first type is represented by iron oxide-silica-dominated rocks (jasper). This appears in a large variety of occurrences in the FAR-DEEP cores, including numerous amygdales (Fig. 6I), veinlets and veins in volcanic rocks (Fig. 6J), siltstone- and dolostone-hosted layers, beds, and feeder-veinlets in dolostones and siliciclastic sediments (Fig. 6K), and redeposited clasts in fluvial, deltaic and marine sediments. The occurrences of jasper thus represent a complex history of fluid migration and redox alteration in a suite of processes linked to the GOE.

The second type of hydrothermal products is travertine (Fig. 6L), occurring in great abundance as thin crusts and feeder-veinlets in $^{13}C$-rich lacustrine dolostones (Fig. 4, Kuetsjärvi Fm). These oldest ever documented travertines may signify a radical change in Earth surface environment to one allowing the precipitation of hydrothermal carbonates in subaerial conditions. Drillcores from holes 5A, 6A, 7A, 8A, and 8B provide excellent material for studying the complex 'iron story' around the GOE, as well as the earliest known travertine-precipitating hydrothermal systems.

*Revolution in biological cycling of phosphorous and organic matter.* Post-Lomagundi-Jatuli deposits (2000 Ma) record the first known appearance (if siderite in banded iron formations is excluded) of

Figure 6. Major Paleoproterozoic paleoenvironmental events documented in the FAR-DEEP cores and adjacent outcrops. [A] Sandstone-shale from drillhole 1A with tidal bedding formed at the start of the GOE. [B] Diamictite accumulated during the Huronian-global glaciation; drillhole 3A. [C] Spinifex-structured komatiites overlying the diamictites; drillhole 3A. [D] Jatuli-age, red, stromatolitic, $^{13}C$-rich, dolostone recording the unprecedented perturbation of the global carbon cycle during the Lomagundi-Jatuli Event; red color is indicative of oxic environment associated with GOE; drillhole 10A. [E] Post-Lomagundi-Jatuli, pink, isotopically "normal", dolostone recording the recovery from the perturbation in the global carbon cycle; drillhole 8B. [F] Dolomite-quartz pseudomorphs after gypsum rosette and crystals emplaced in red clay; abundant Ca-sulfates suggest a significant ocean sulfate reservoir; drillhole 10B. [G] Dissolution of anhydrite and halite resulted in porous appearance of matrix-supported conglomerate; drillhole 11A. [H] ~2060-Ma weathering surface imprinted in hematite-cemented (black) felsic lava-breccia in vicinity of drillhole 7A; this is indicative of oxic meteoric and ground waters during the GOE. [I] Jasper-filled amygdales (dark red) in oxidized felsic lava from drillhole 6A; this may record an upper mantle oxidizing event. [J] Jasper vein in dacitic lava near drillhole 7A; such veins are truncated by overlying sediments and are the product of subaerial, syn-volcanic hydrothermal activity. [K] Greywacke-hosted jasper layers and beds precipitated from sub-aqueous hydrothermal system; drillhole 8B. [L] The earliest known travertines on Earth from drillhole 5A. [M] $^{13}C$-depleted carbonate concretion incorporating carbon derived from oxic recycling of the organic matter; drillhole 12B. Bar scales in all photos are 1 cm. [N] Pyrobitumen vein representing a 2000-Ma oil migration pathway; drillhole 12A. These units contain both mass-independent and mass-dependent fractionation of sulfur isotopes.

diagenetic carbonate concretions significantly depleted in $^{13}C$ (Fallick et al., 2008). They are varied in size and composition and abundant in sedimentary successions <2000 Ma where they are associated with other diagenetic products such as phosphorites, all of which are seemingly absent in older rocks. Thus, these suggest a major change in the diagenetic mineralization of organic matter, perhaps reflecting increased rates in dissimilatory sulfate reduction (Canfield and Raiswell, 1999). This in turn suggests an elevated concentration of interstitial phosphate. Drillholes 9A, 12A, 12B, and 13A recovered hundreds of meters of cores containing abundant carbonate nodules (Fig. 6M) and carbonate-associated phosphorites, which allow us to address the causes of the emergence of 'modern-style' recycling of organic matter, are and the factors that control the formation of the oldest known phosphorites.

*Supergiant petroleum deposits.* The most remarkable accumulation of organic-matter-rich sediments and generation of petroleum in the Precambrian took place at around 2000 Ma ago in the aftermath of the Lomagundi-Jatuli Event (Melezhik et al., 2009). This is known as the Shunga Event. The causes (e.g., high productivity or anomalous preservation) of such unprecedented global accumulation of organic-matter and pervasive oil generation remain unknown. In the type area at Shunga, three FAR-DEEP drillholes (12A, 12B, 13A) intersected several hundred meters of organic matter-rich source rocks, a uniquely preserved oil migration pathway (Fig. 6N), a petrified supergiant oilfield, and several bodies of enigmatic organo-siliceous rocks containing up to 40 wt% organic carbon. More than 100 archive samples have already been analyzed for major and trace elements as well as for S- and C-isotopes of sulfides, organic matter and associated carbonates, and they indicate a complex geochemical evolutionary pattern. These data, together with the results of forthcoming studies on biomarkers and structural, isotopic, and trace element characteristics of organic matter, will enable several fundamental problems associated with the Shunga Event to be addressed.

## International Sampling and Acknowledgements

The drilling operations were carried out by a Finnish operator, whereas all logistical support was provided by the Russian State Company "Mineral" from St. Petersburg, Russia. Selected specimens from the archive set have been analyzed by partners at the Westfälische Wilhelms-Universität, Münster, Germany (H. Strauss), the Scottish Universities Environmental Research Center, Glasgow, Scotland (A.E. Fallick), the Pennsylvania State University, U.S.A. (L.R. Kump), and the Institute of Precambrian Geology and Geochronology, St. Petersburg, Russia (I.M. Gorokhov). The NGU, Center of Excellence in Geobiology at the University of Bergen in Norway, and the Institute of Geology at the Karelian Research Center in Petrozavodsk, Russia provided resources for the archiving work. In March–April 2009, the NGU provided its facilities and logistical support for an international sampling campaign. Samples have been collected either directly by scientists who have examined the core in Trondheim or by the NGU's staff upon sample requests submitted by FAR-DEEP partners. More than 5500 samples have been collected from cores, and they are being prepared for various analyses and experiments. Scientists from thirteen countries have applied for research grants from various funding agencies to complete studies of the FAR-DEEP cores. Currently, five PhD students from Finland, Germany, and Norway are involved in the FAR-DEEP research program. Four postdoctoral projects in Finland, Great Britain, and Norway have been financed by national funding agencies. A large group of scientists from the U.S.A. has received support from the U.S. National Science Foundation and NASA. Several research groups from Australia, Belgium, Czech Republic, Estonia, and Russia have also been successful in obtaining research grants.

## References

Aharon, P., 2005. Redox stratification and anoxia of the early Precambrian oceans: Implications for carbon isotope excursions and oxidation events. *Precam. Res.*, 137:207–222.

Amelin, Yu.V., Heaman, L.M., and Semenov, V.S., 1995. U-Pb geochronology of layered mafic intrusions in the eastern Baltic Shield: Implications for the timing and duration of Palaeoproterozoic continental rifting. *Precam. Res.*, 75:31–46, doi:10.1016/0301-9268(95)00015-W.

Bekker, A., Holland, H.D., Wang, P.L., Rumble III, D., Stein, H.J., Hannah, J.L., Coetzee, L.L., and Beukes, N.J., 2004. Dating the rise of atmospheric oxygen. *Nature*, 427:117–120, doi:10.1038/nature02260.

Bekker, A., Karhu, J.A., Eriksson, K.A. and Kaufman, A.J., 2003a. Chemostratigraphy of Paleoproterozoic carbonate successions of the Wyoming Craton: Tectonic forcing of biogeochemical change? *Precam. Res.*, 120:279–325, doi:10.1016/S0301-9268(02)00164-X.

Bekker, A., Kaufman, A.J., Karhu, J.A., Beukes, N.J., Swart, Q.D., Coetzee, L.L., and Eriksson, K.A., 2001. Chemostratigraphy of the Paleoproterozoic Duitschland Formation, South Africa: Implications for coupled climate change and carbon cycling. *Am. J. Sci.*, 301:261–285, doi:10.2475/ajs.301.3.261.

Bekker, A., Sial, A.N., Karhu, J.A., Ferreira, V.P., Noce, C.M., Kaufman, A.J., Romano, A.W., and Pimentel, M.M., 2003b. Chemostratigraphy of carbonates from the Minas Supergroup, Quadrilátero ferrífero (Iron Quadrangle), Brazil: a stratigraphic record of Early Proterozoic atmospheric, biogeochemical and climatic change. *Am. J. Sci.*, 303:865–904, doi:10.2475/ajs.303.10.865.

Canfield, D.E., and Raiswell, R., 1999. The evolution of the sulfur cycle. *Am. J. Sci.*, 299: 97–723, doi:10.2475/ajs.299.7-9.697.

Claire, M.W., Catling, D.C., and Zahnle, K.J., 2006. Biogeochemical modelling of the rise in atmospheric oxygen. *Geobiology*, 4:239–269, doi:10.1111/j.1472-4669.2006.00084.x.

Evans, D.A.D., 2003. A fundamental Precambrian-Phanerozoic shift in Earth's glacial style? *Tectonophysics*, 375:353–385, doi:10.1016/S0040-1951(03)00345-7.

Fallick, A.E., Melezhik, V.A., and Simonson, B., 2008. The ancient anoxic biosphere was not as we know it. *In* Dobretsov, N., Kolchanov, N., Rozanov, A., and Zavarzin, G. (Eds.) *Biosphere Origin and Evolution*. New York (Springer Science + Business Media), 169–188.

Guo, Q., Strauss, H., Kaufman, A.J., Schröder, S., Gutzmer, J., Wing, B., Baker, M.A., Bekker, A., Kim, S.-T., Jin, Q., and Farquhar,

J., 2009. Reconstructing Earth's surface oxidation across the Archean-Proterozoic transition. *Geology*, 37:399–402, doi:10.1130/G25423A.1.

Hannah, J.L., Bekker, A., Stein, H.J., Markey, R.J., and Holland, H.D., 2004. Primitive Os and 2316 Ma age for marine shale: Implications for Paleoproterozoic glacial events and the rise of atmospheric oxygen. *Earth Planet. Sci. Lett.*, 225:43–52, doi:10.1016/j.epsl.2004.06.013.

Hanski, E.J., 1992. Petrology of the Pechenga ferropicrites and cogenetic, Ni-bearing gabbro-wehrlite intrusions, Kola Peninsula, Russia. *Geol. Survey Finland Bull.*, 367, 191 pp.

Hayes, J.M., and Waldbauer, J.R., 2006. The carbon cycle and associated redox processes through time. *Philos. Trans. R. Soc. Lond. B Biol. Sci.*, 361:931–950.

Holland, H.D., 2002. Volcanic gases, black smokers, and the great oxidation event. *Geochim. Cosmochim. Acta*, 66:3811–3826, doi:10.1016/S0016-7037(02)00950-X.

Kah, L.C., Lyons, T.W., and Frank, T.D., 2004. Low marine sulfate and protracted oxygenation of the Proterozoic biosphere. *Nature*, 431:834–838, doi:10.1038/nature02974.

Kump, L.R., 2008. The rise of atmospheric oxygen. *Nature*, 451:277–278, doi:10.1038/nature06587.

Kump, L.R., and Barley, M.E., 2007. Increased subaerial volcanism and the rise of atmospheric oxygen 2.5 billion years ago. *Nature*, 448:1033–1036, doi:10.1038/nature06058.

Melezhik, V.A., 2006. Multiple causes of Earth's earliest global glaciation. *Terra Nova*, 18:130–137, doi:10.1111/j.1365-3121.2006.00672.x.

Melezhik, V.A., Fallick, A.E., Filippov, M.M., Lepland, A., Rychanchik, D.V., Deines, Y.E., Medvedev, P.V., Romashkin, A.E., and Strauss, H., 2009. Petroleum surface oil seeps from Palaeoproterozoic petrified giant oilfield. *Terra Nova*, 21:119–126, doi:10.1111/j.1365-3121.2009.00864.x.

Melezhik, V.A., Fallick, A.E., Hanski, E.J., Kump, L.R., Lepland, A., Prave, A.R., and Strauss, H., 2005a. Emergence of the aerobic biosphere during the Archean-Proterozoic transition: Challenges of future research. *Geol. Soc. Am. Today*, 15:4–11.

Melezhik, V.A., Fallick, A.E., Medvedev, P.V., and Makarikhin, V.V., 1999. Extreme $^{13}C_{carb}$ enrichment in ca. 2.0 Ga magnesite-stromatolite-dolomite-'red beds' association in a global context: a case for the world-wide signal enhanced by a local environment. *Earth Sci. Rev.*, 48:71–120, doi:10.1016/S0012-8252(99)00044-6.

Melezhik, V.A., Fallick, A.E., Rychanchik, D.V., and Kuznetsov, A.B., 2005b. Palaeoproterozoic evaporites in Fennoscandia: Implications for seawater sulfate, $\delta^{13}C$ excursions and the rise of atmospheric oxygen. *Terra Nova*, 17:141–148, doi:10.1111/j.1365-3121.2005.00600.x.

Melezhik, V.A., Huhma, H., Condon, D.J., Fallick, A.E., and Whitehouse, M.J., 2007. Temporal constraints on the Paleoproterozoic Lomagundi-Jatuli carbon isotopic event. *Geology*, 35:655–658, doi:10.1130/G23764A.1.

Ovchinnikova, G.V., Kusnetzov, A.B., Melezhik, V.A., Gorokhov, I.M., Vasil'eva, I.M., and Gorokhovskii, B.M., 2007. Pb-Pb age of Jatulian carbonate rocks: the Tulomozero Formation of southeast Karelia. *Strat. Geol. Correl.*, 15(4):359–372.

Pukhtel, I.S., Zhuravlev, D.Z., Ashikhmina, N.A., Kulikov, V.S., and Kulikova, V.V., 1992. Sm-Nd age of the Suisarian suite on the Baltic Shield. *Trans. Russian Acad. Sci.*, 326:706–711. (in Russian).

Reuschel, M., Strauss, H., Fallick, A.E., and Melezhik, V., 2008. FAR-DEEP: Sulphur biogeochemistry in deep time. 33rd International Geological Congress, Oslo, 6–14 August 2008.

Watanabe, Y., Farquhar, J., and Ohmoto, H., 2009. Anomalous fractionations of sulfur isotopes during thermochemical reduction. *Science*, 324:370–373, doi:10.1126/science.1169289.

Young, G.M., Long, D.G.F, Fedo, C.M., and Nesbitt, H.W., 2001. Paleoproterozoic Huronian basin: Product of a Wilson cycle punctuated by glaciations and a meteorite impact. *Sed. Geol.*, 141/142:233–254, doi:10.1016/S0037-0738(01)00076-8.

## Authors

**Victor A. Melezhik**, Geological Survey of Norway, N-4791, Trondheim, Norway; Centre for Geobiology, University of Bergen, N-5020, Bergen, Norway, e-mail: victor.melezhik@ngu.no.

**Aivo Lepland, Melanie Mesli, Tor Erik Finne**, Geological Survey of Norway, N-4791, Trondheim, Norway.

**Alexander E. Romashkin and Dmitry V. Rychanchik**, Institute of Geology of Karelian Research Centre, 185610, Petrozavodsk, Russia.

**Ronald Conze**, Operational Support Group ICDP, GFZ German Research Center for Geosciences (GeoForschungsZentrum Potsdam), D-14473 Potsdam, Germany.

**and the FAR-DEEP Scientists**

## Web Links

http://far-deep.icdp-online.org (Internal Data FARDEEP)
http://www.icdp-online.org/front_content.php?idart=1962

# 5

# New Integrated Data Analyses Software Components

by Ronald Conze, Frank Krysiak, Josh Reed, Yu-Chung Chen, Hans-Joachim Wallrabe-Adams, Colin Graham and the New Jersey Shallow Shelf Science Team, Volker Wennrich and the Lake El'gygytgyn Science Team

## Introduction

Data management in scientific drilling programs such as the Integrated Ocean Drilling Program (IODP), the International Continental Scientific Drilling Program (ICDP), and the Antarctic Drilling Program (ANDRILL) performs two functions: firstly, the capture of drilling and scientific data during an expedition, and secondly, the long-term storage and dissemination of these data. Here we describe the progress in linking data management with stand-alone data capture and visualization applications. This provides a two-way flow of data between the database and the applications, and a more integrated data environment for scientists. The new system has been tested, so far, with cores from the IODP Expedition 313 New Jersey Shallow Shelf and the ICDP Lake El'gygytgyn Drilling Project.

## The Components

1. The Expedition Drilling Information System (ExpeditionDIS) is used for ICDP lake drilling projects and IODP-MSP expeditions in a two-phase data capture process, at the drill site during the drilling operation and in one or more laboratories during the post-drilling phase (Conze et al., 2007). It is based on a relational database using mainly digital forms for the data input.
2. PSICAT, the Paleontological Stratigraphic Interval Construction and Analysis Tool, is a stand-alone, Java-based graphical editing tool for creating and viewing core description diagrams (http://dev.psicat.org/) that can be used in the field as well as in a laboratory environment.
3. Corelyzer as part of CoreWall (http://www.corewall.org/) is a scalable, extensible visualization tool, developed to enhance the study of geological cores. The strength of Corelyzer is the ability to display large sets of core imagery, with multi-sensor logs and annotations. Plugins can be developed to provide additional data visualization functionality.
4. Correlator, another CoreWall product, is an interactive workstation software product for displaying and depth-shifting/merging data from multiple cores and/or downhole geophysical logging data.

## The Concept

The idea is to combine the advantages of all these stand-alone tools into an interoperable configuration.

The link between the ExpeditionDIS, PSICAT, and Corelyzer is realized by a Common Exchange Format (Fig. 1) using XML notation. The general workflow starts with the export of all relevant primary data for core recovered from a drill hole from the ExpeditionDIS in the Common Exchange Format, which can then subsequently be imported by PSICAT and/or Corelyzer. A Common Exchange Format file contains the corresponding driller depths and file paths for core section images as well as any existing lithological visual core description data for each core section, and other corresponding data sets such as petrophysical (e.g., from Multi Sensor Core Logging, MSCL) data.

1. By importing the ExpeditionDIS data as Common Exchange Format, PSICAT generates a visual core description (VCD) form for each core section. Existing VCD data can be modified, and new data can be added using the graphical functionality of PSICAT. The new or modified data can be written back to the

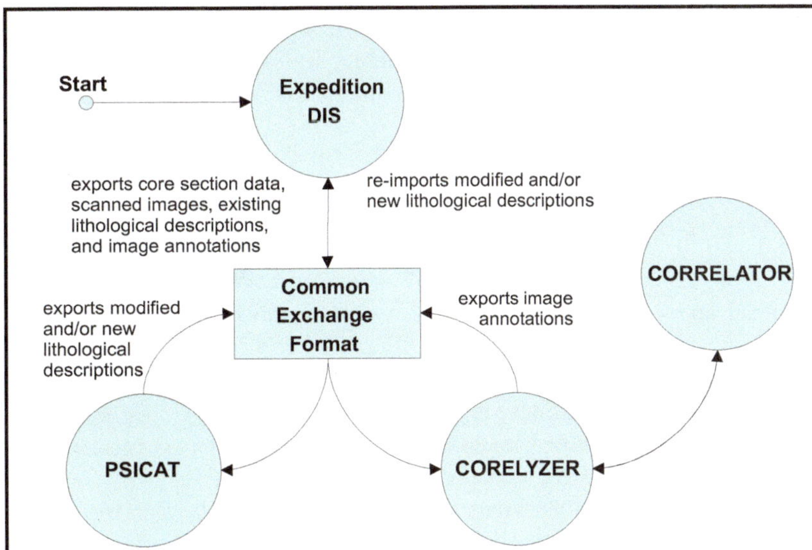


Figure 1. Conceptual diagram showing data exchange between the components (Figure credit by R. Conze).

Figure 3. Corelyzer station at the IODP Exp. 313 New Jersey Shallow Shelf onshore science party at MARUM (University of Bremen). This station was fed continuously with scanned images and MSCL data of new core sections as they were added to the ExpeditionDIS. In conjunction with Correlator, this station was used for correlation between sites using lithological features and/or logging data. (Photo by H. Ando, Ibaraki University)

ExpeditionDIS via the Common Exchange Format file. Upon selecting a subset of core sections or all core sections for a drill hole, a non-editable, continuous sequence, scaled to the driller depth, is displayed.

2. By importing a Common Exchange Format file, Corelyzer reads in all core images and corresponding optional logging data (Fig. 2). The utilization of an existing PSICAT plugin will allow visualization of the lithological units graphically beside these scanned images and log data. The user can add annotations to the images, which are added to the Common Exchange Format file and then exported for use by the other tools.
3. By re-importing the Common Exchange Format data generated by PSICAT, and Corelyzer, the ExpeditionDIS can store all modified or new lithological data.
4. By connecting Correlator, Corelyzer can control visualization and core depth adjustments in runtime using socket commands. It will allow users to utilize rich features in high-resolution core images to generate composite and spliced depth scales. This feature allows more fine-grained interactive stratigraphic correlations.

## Benefits and Outlook

The two test beds demonstrated that the synergetic integration of ExpeditionDIS, PSICAT and Corelyzer is beneficial and enhances the capabilities of the individual stand-alone systems. The ExpeditionDIS and its relational database provide the full documentation of all recovered cores and sections including consistent driller depths and the corresponding scanned images in different grades of resolution. Neither PSICAT nor Corelyzer provides any tools to capture these important primary data, but both need these data as input for generating lithological visual core descriptions and visualization of the scanned images of the core sections and related logging data. In conjunction with Correlator it will be possible to import the depth adjustment tables by ExpeditionDIS using the same Common Exchange Format so that any data can be output on the same composite or spliced depth scales.

Upcoming scientific drilling expeditions of ICDP and ECORD/IODP will provide further test beds for consolidation and development of such features.

## Acknowledgements

This work has been made possible through bilateral memoranda of understanding between ICDP and IODP, ANDRILL, and LacCore as well as the GeoSciNET initiative composed of CoreWall, Geoinformatics for Geochemistry (GfG), System for Earth Sample Registration (SESAR), PaleoStrat, and ICDP.

## References

Conze, R., Wallrabe-Adams, H.J., Graham, C., Krysiak, F., 2007. Joint Data Management on ICDP projects and IODP mission specific platform expeditions. *Sci. Drill.* 4:32–34. doi:10.2204/iodp.sd.4.07.2007

## Authors

**Ronald Conze**, Operational Support Group ICDP, GFZ German Research Centre for Geosciences, Telegrafenberg, D-14473 Potsdam, Germany, e-mail:conze@gfz-potsdam.de.
**Frank Krysiak**, smartcube GmbH Berlin, Puschkinallee 48, D-12435 Berlin, Germany.
**Josh Reed**, Andrill Science Management Office, 126 Bessey Hall, Lincoln, University of Nebraska-Lincoln, P.O. Box 880851, Loncoln, NE 68588-0341, U.S.A.
**Yu-Chung Chen**, University of Illinois at Chicago, Electronic Visualization Laboratory, U.S.A.
**Hans-Joachim Wallrabe-Adams**, MARUM, University of Bremen, Leobener Str., D-28359 Bremen, Germany.
**Colin Graham**, British Geological Survey Edinburgh, ESO – ECORD Science Operator, Murchison House, West Mains Road, Edinburgh EH9 3LA, Scotland, U.K.
**Volker Wennrich**, University of Cologne, Institute of Geology and Mineralogy, Zuelpicher Str. 49a, D-50674 Koeln, Germany.

## Related Web Links

http://www.icdp-online.org
http://www.iodp.org
http://www.ecord.org
http://www.andrill.org
http://lrc.geo.umn.edu
http://www.corewall.org
http://www.geoinfogeochem.org
http://www.geosamples.org
http://www.scientificdrilling.org
http://www.smartcube.de
http://www.wdc-mare.org

# 6

# Automatic Slide-Loader Fluorescence Microscope for Discriminative Enumeration of Subseafloor Life

by Yuki Morono and Fumio Inagaki

## Introduction

The marine subsurface environment is considered the potentially largest ecosystem on Earth, harboring one-tenth of all living biota (Whitman et al., 1998) and comprising diverse microbial components (Inagaki et al., 2003, 2006; Teske, 2006; Inagaki and Nakagawa, 2008). In deep marine sediments, the discrimination of life is significantly more difficult than in surface sediments and terrestrial soils because buried cells generally have extremely low metabolic activities (D'Hondt et al., 2002, 2004), and a highly consolidated sediment matrix produces auto-fluorescence from diatomaceous spicules and other mineral particles (Kallmeyer et al., 2008). The cell abundance in marine subsurface sediments has conventionally been evaluated by acridine orange direct count (AODC; Cragg et al., 1995; Parkes et al., 2000) down to 1613 meters below the seafloor (mbsf) (Roussel et al., 2008). Since the cell-derived AO signals often fade out in a short exposure time, recognizing and counting cells require special training. Hence, such efforts to enumerate AO-stained cells from the subseafloor on photographic images have been difficult, and a verification of counts by other methods has been impossible. In addition, providing mean statistical values from low biomass sedimentary habitats has been complicated by physical and time limitations, yet these habitats are considered critical for understanding the Earth's biosphere close to the limits of habitable zones (Hoehler, 2004; D'Hondt et al., 2007).

Here we report recent developments on the automatic cell count system based on our recently published new cell detection and enumeration method that discriminates the SYBR Green I(SYBR I)-stained cells from the background fluorescent signals, called SYBR-SPAM (SYBR-Stainable PArticulated Matter; Morono et al., 2009). We integrated an automatic slide loader and an LED light-camera monitoring system for high-throughput and high-resolution counting operation.

## Principle of Discriminative Detection

SYBR-I or SYBR-II has been considered a more effective fluorescent dye for cell enumeration in sediments than AO due to its higher fluorescent intensity and sensitivity to nucleic acids (Weinbauer et al., 1998; Lunau et al., 2005; Engelen et al., 2008). However, we found non-autofluorescent SYBR-I-stainable particulate matter (SYBR-SPAM) in heat-sterilized control sediments treated at 450°C for 3 hours (Morono et al., 2009). We also found that when SYBR-I bound to SYBR-SPAM the SYBR-I spectra shifted to longer wavelengths (Fig. 1A), and that the spectra can be distinguished from cell-derived green fluorescence under the observation of a long-pass filter of cut-off wavelength 510 nm (Fig. 1B). To discriminate the cell-derived fluorescent signal more precisely, we obtained microscopic images using band-pass filters at 528/38 and 617/73 nm (center wavelength/bandwidth) that separated the green and red components of SYBR-I fluorescence (Fig. 1C, D). We divided the fluorescent intensity of green by that of the red images to obtain relative intensity profiles of green/red fluorescence, in which cell-derived fluorescence was successfully discriminated as bright signals, whereas SYBR-SPAM and other background signals were entirely eliminated (Fig. 1E).

## Sample Preparation Protocol

A standard sample preparation scheme is shown in Fig. 2. Fifty microliters of sediment slurry (10% [v/v] sediment in ethanol-PBS solution) fixed in 2% paraformaldehyde is mixed with 850 µL of 3% [wt/v] NaCl solution and sonicated at 20 W for 1 min on ice using an ultrasonic homogenizer (Model UH-50, SMT Co. Ltd., Tokyo). The sonicate is then mixed with 100 µL of 10% [wt/v] hydrofluoric acid (HF) and incubated for 20 min at room temperature. This HF treatment has been shown to effectively decrease the number of SYBR-SPAM especially in shallow subseafloor sediments, while intracellular DNA is negligibly affected (Morono et al., 2009). Then, up to 700 µL of the mixture is directly filtered through a 0.22 µm-pore size black polycarbonate membrane without centrifugation. The final sediment volumes on the filter are up to 3.5 x $10^{-3}$ $cm^3$. To eliminate potential carbonate particles and/or precipitates, the membrane is treated with 1 mL of 0.1 M HCl for 5 min on the filtration device. The membrane is then washed with 5 mL of TE buffer (10 mM Tris-HCl, 1.0 mM EDTA, pH 8.0), and roughly 2 x $10^8$ fluorescence microsphere beads (Fluoresbrite Bright Blue Carboxylate Microspheres [BB beads], 0.5 µm, Polysciences, Inc., Warrington, Pa.) are used for focus adjustment (Morono et al., 2009). After air drying, one-fourth of the membrane is placed on the filtration device again, then stained with SYBR-I staining solution (1/40 [v/v] SYBR-I in TE buffer). The stained filter is finally mounted on a glass microscope

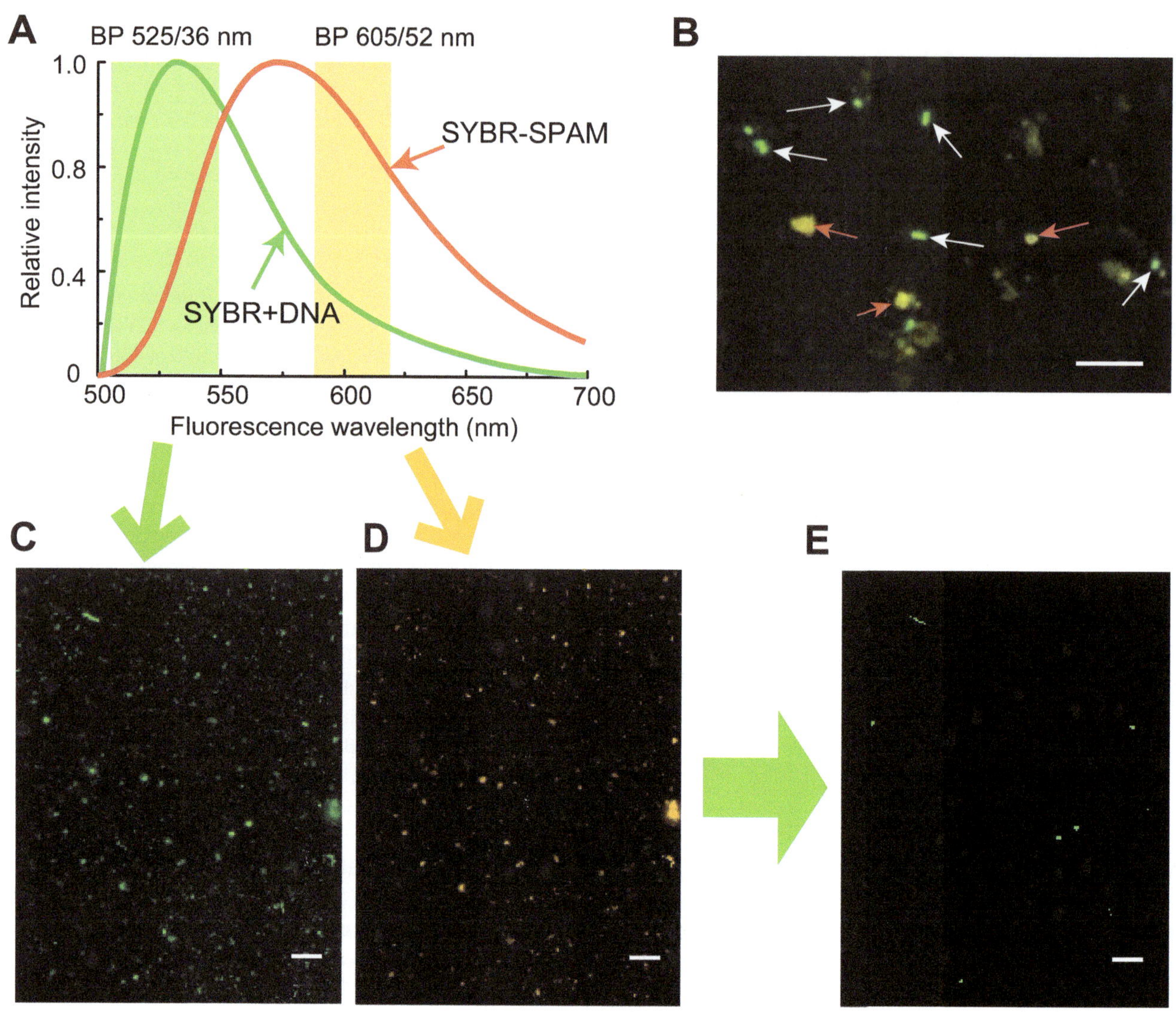


Figure 1. Difference of SYBR-I fluorescence spectrum with intracellular DNA or SYBR-SPAM, and discrimination of cell-derived SYBR Green I fluorescence from background signals using image analysis. [A] Spectrum patterns show "red shift" of SYBR-I fluorescence. When SYBR-I binds to SYBR-SPAM (red line), fluorescence spectra shift to longer wavelengths than SYBR-DNA complex (green line). Green and orange shaded areas show the wavelength range of 525/36 and 605/52 (nm of center wavelength/bandwidth) band-pass filters, respectively. [B] Examples of fluorescence-producing cellular and non-cellular objects stained with SYBR-I. Red arrows, yellowish SYBR-SPAM. White arrows, green *E. coli* cells. The image was obtained using a long-pass filter of cut-off wavelength 510 nm. [C to E] Image analysis to distinguish cell-derived SYBR-I signals from SYBR-SPAM in natural marine sediments (core 1H-1 of Site C0006 Hole E in the Integrated Ocean Drilling Program Exp. 316). Fluorescence microscopic images taken using band-pass filters of 525/36 [C] and 605/52 [D]. Relative intensity profiles of green/red fluorescence [E] show only cell-derived fluorescence signals without background fluorescence. Bars: 10 μm.

slide with 3–5 μL of mounting solution (1:2 mixture of VECTASHIELD mounting medium H-1000 and TE buffer).

## The Slide Handler-Equipped Automatic Fluorescence Microscope System

The images are automatically acquired with a fluorescent microscope system (Fig. 3A) that consists of a basic microscope (BX-51, Olympus), an automatic slide handler (LEP Slide Handler System, Ludl Electric Products, Ltd.), illumination with LED Array Modules (LAMs) of 400 nm and 490 nm (precisExcite, CoolLED, Ltd.), an emission filter wheel (Ludl Electric Products, Ltd.), and a cooled CCD camera (ORCA-AG, Hamamatsu Photonics K.K.) All units are software controlled (MetaMorph 7.5, Molecular Devices). Up to 50 slides of samples can be set in two magazines of the slide handler (Fig. 3B). To keep the sample in good condition for a long time (depending on the scanning area but at least 30 min per sample, i.e., 25 h duration to process all 50 slides), the whole system is put in a dark cooled (15°C) chamber. In addition, we also installed a digital camera to monitor the system without opening the door of the chamber (Fig. 3C, D). LED illumination allows for (i) an instant on/off switch and wavelength change without damage to the lamp source, (ii) a prolonged lifetime in comparison to mercury lamps, and (iii) negligible heating during operation.

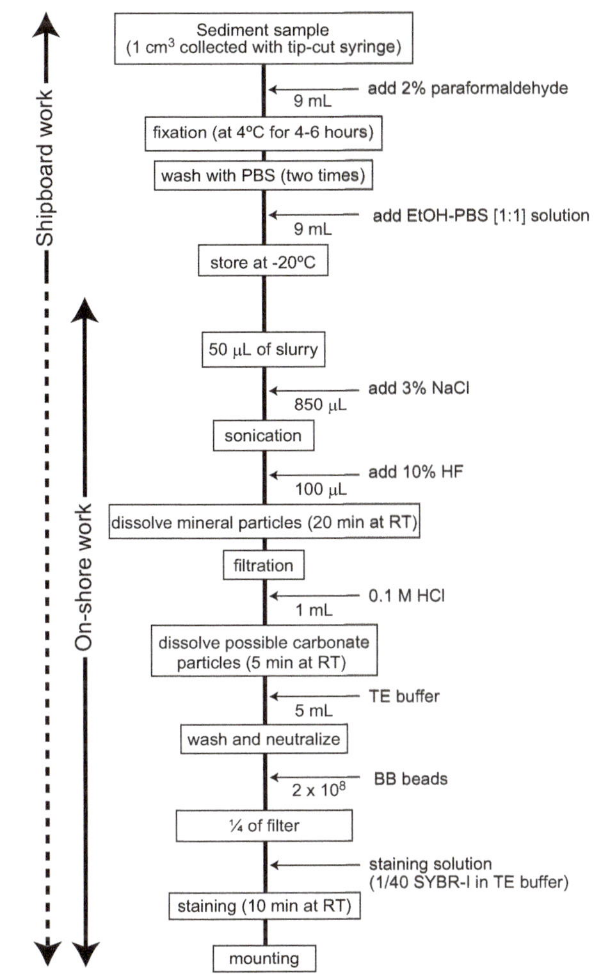


Figure 2. Flow diagram of standard sample preparation protocol. Up to $3.5 \times 10^{-3}$ $cm^3$ of sediment is mounted on the slide

The microscopic imaging is controlled with the software in a "Scan Slide" function that scans a user-defined area of the sample while adjusting the focus on each plane. As the deviation of the slide thickness was comparably large (0.8–1.0 mm), we programmed a wide-focus adjustment macro searching a focus position through a range of 120-μm Z-positions before scanning each slide. After the wide-focus adjustment, the image acquisition sequence with "Scan Slide" proceeds as follows: (i) scan Z-position under 400 nm excitation and 455/50 nm emission filter to focus on BB beads, (ii) fix Z-mortar and acquire bead image, (iii) change excitation to 490 nm and emission to 525/36 nm (green), (iv) acquire cell image, (v) change emission to 605/52 nm (red), and (vi) acquire background image.

The images obtained are analyzed with the macro in the following way. The fluorescence intensity of each pixel in the green image is multiplied by 100 and divided by that of the red image at the same location. The resulting images showing the ratio of the relative intensity of green/red fluorescence were smoothed by median filtering (3 x 3 pixel square), and watershed lines were drawn to separate cells in close proximity to each other. Based on *Escherichia coli* images in control sediments and cellular signals in natural sub-seafloor sediments, we set the threshold value of relative fluorescence at 110 for automatic cell enumeration. Under these threshold conditions, non-specific signals in heat-sterilized sediments as well as on blank filters resulted in a null count. Figure 4 shows the effect of the HF treatment and the image analysis. Counting performed without the modification results in a serious overestimation (100 times higher) of the biomass.

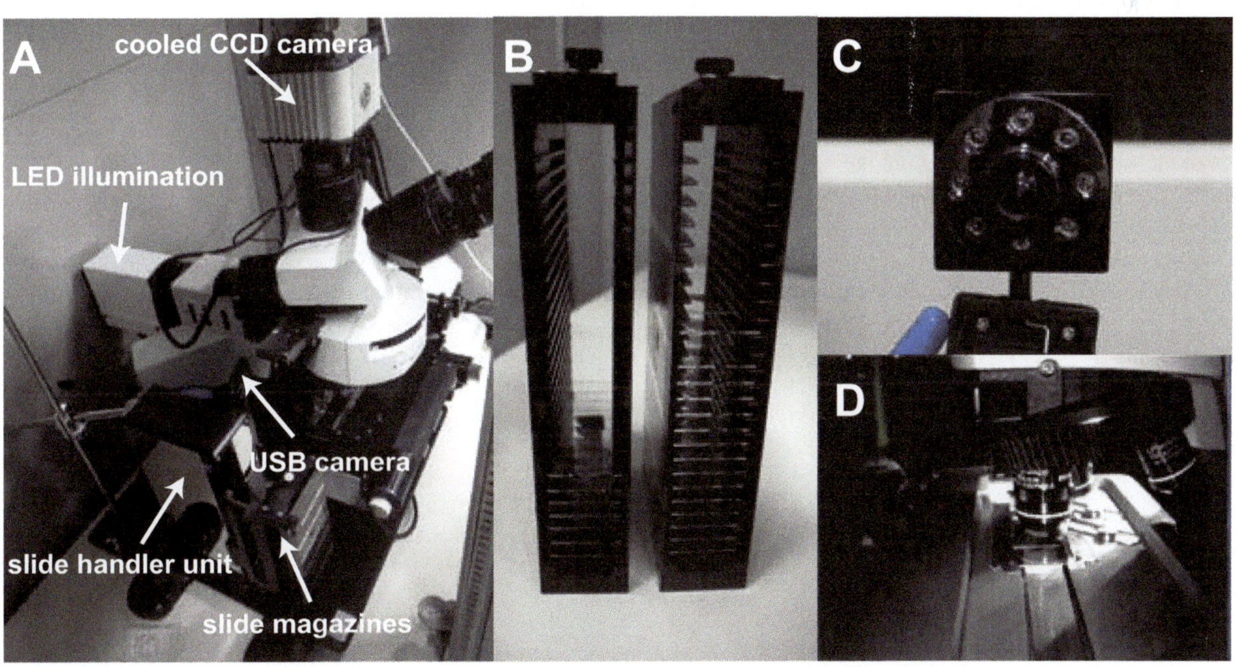


Figure 3. Newly constructed automatic microscopic system with slide handler. [A] Overview of the system. All the units are operated in a dark and cool (15°C) chamber. [B] Slide magazines capable of carrying 25 slides each. [C] USB-connected monitoring camera with infrared LED. [D] The system is monitored without opening the door of the chamber.

## Detection Limit of Life Forms and Future Improvement

The lower limit of detection of DNA-containing lifeform signatures (i.e., Archaea, bacteria, and viral components) can only be determined with a maximum load of samples on the filter and a prolonged time for counting. In our examination, $3.5 \times 10^{-3}$ $cm^3$ was found to be the maximum load on the filter of 2.5 $cm^2$ active filtration area. Although we tried to load more on the filter, sediment particles piled up and resulted in focus failure throughout the image field. The maximum load corresponds to $5.1 \times 10^{-5}$ $cm^3$ of sediment for a field with a 40x objective lens. According to Kallmeyer et al. (2008), three times more cells are needed in source sediment for detecting at least one cell with 95% probability. The detection limit with one image field is therefore $5.9 \times 10^6$ cells $cm^{-3}$. By considering the time (roughly 8 sec) required for taking one field of image with our system, we can obtain a comparison of detection limit vs. required acquisition time as shown in Table 1. Currently we have already tried and succeeded in analyzing a quarter of the filter and counting down to $3.4 \times 10^3$ cells $cm^{-3}$ in a few hours. To lower the limit further, the required time will increase inversely (Table 1). A practical limit would be less than 100 hours of analysis, which corresponds to a detection limit of $1.3 \times 10^2$ cells $cm^{-3}$. For even lower biomass habitats, a brief centrifugation at 100 x *g* (Lunau et al., 2005) or a density gradient separation (Kallmeyer et al., 2008) that reduces mineral particles enables larger amounts of sediment to be placed on the filter and extends the detection limit by 10- to 1000-fold. However, the recovery rate depends on natural cell densities and field conditions (Morono et al., 2009) and may sometimes cause serious loss of the cells. The use of a centrifugation step should be carefully considered, and cell concentration should be described as the minimal value in such a case. The system presented here will be useful for primary microbiological onboard data if deployed on scientific drilling platforms. In addition, such a separation technique will open the door for the application of other analytical tools that detect activities of life in low biomass habitats using nanoscale secondary ion mass spectrometry (NanoSIMS; Kuypers and Jørgensen, 2007). Hence, the direction of future technical developments should be towards both the detection of life components with specific signals (e.g., deep UV excitation, see Bhartia et al., 2008) and the detection of activities and metabolic functions of deep subseafloor life.

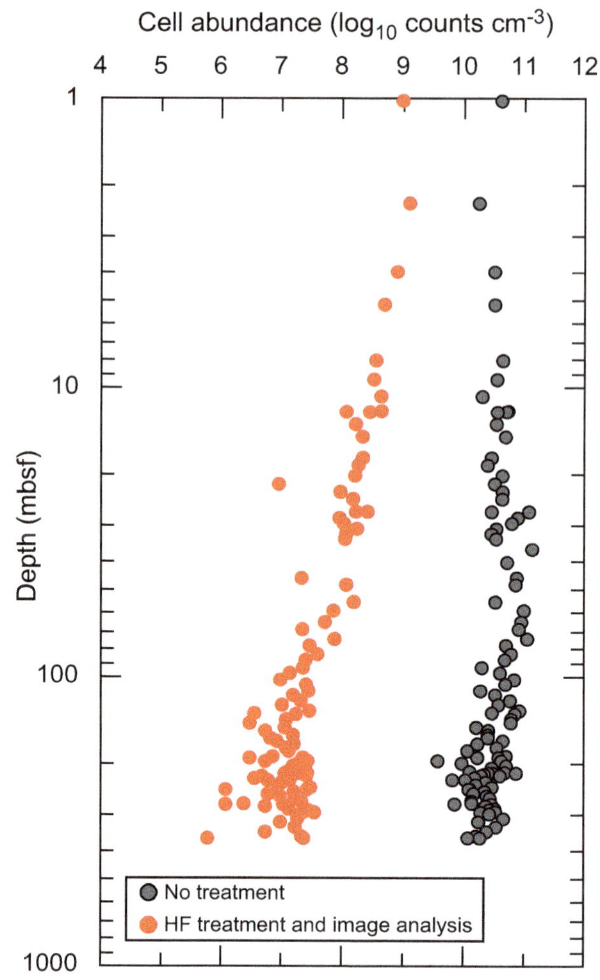


Figure 4. SYBR-I counts of fluorescence signals in sediments collected from offshore of the Shimokita Peninsula of Japan during the *D/V Chikyu* shakedown expedition CK06-06. Gray dots: no acid wash and manual (human-eye) cell counting results without image analysis (seriously overestimated). Red dots: computer-based cell counts in HF-washed sediment samples (modified from Morono et al., 2009).

## Acknowledgements

The authors thank Molecular Devices Co. and Olympus Co. for constructing the automated cell counting system and for developing the software macros. We are grateful to the shipboard scientists and crews of the *D/V Chikyu* Shakedown Expedition CK06-06, and the Integrated Ocean Drilling Program (IODP) Expedition 316 for samples and data. The authors are grateful to N. Masui, T. Terada, S. Tanaka, and S.

Table 1. Correlation of the lower detection limit and the corresponding time required for obtaining images, amount of sediment, number of images, and number of filters.

| Lower limit of detection (cells $cm^{-3}$) | Required time (hrs) | Required sediment volume ($cm^3$) | Number of image fields | Area of analysis ($cm^2$) | Number of required filters |
|---|---|---|---|---|---|
| $1.04 \times 10^4$ | 1 | 0.000287 | 450 | 0.163 | 1 |
| $1.04 \times 10^3$ | 10 | 0.00287 | 4,500 | 1.63 | 1 |
| $1.04 \times 10^2$ | 100 | 0.0287 | 45,000 | 16.3 | 7 |
| $1.04 \times 10^1$ | 1000 | 0.287 | 450,000 | 163 | 66 |
| 1.04 | 10000 | 2.87 | 4,500,000 | 1630 | 652 |

Fukunaga for filter preparation and technical supports. This study was supported in part by JAMSTEC Multidisciplinary Research Promotion Award 2008 (to YM and FI).

## References

Bhartia, R., Hug, W.F., Salas, E.C., Reid, R.D., Sijapati, K.K., Tsapin, A., Abbey, W., Nealson, K.H., Lane, A.L., and Conrad, P.G., 2008. Classification of organic and biological materials with deep ultraviolet excitation. *Appl. Spect.*, 62:1070–1077, doi:10.1366/000370208786049123.

Cragg, B.A., Parkes, R.J., Fry, J.C., Weightman, A.J., Maxwell, J.R., Kastner, M., Hovland, M., Whiticar, M.J., Sample, J.C., and Stein, R., 1995. Bacterial profiles in deep sediments of the Santa Barbara Basin, Site 893. *Proc. ODP Sci. Res.*, 146 (Part 2):139–144.

D'Hondt, S., Inagaki, F., Ferdelman, T., Jørgensen, B.B., Kato, K., Kemp, P., Sobecky, P., Sogin, M., and Takai, K., 2007. Exploring subseafloor life with the Integrated Ocean Drilling Program. *Sci. Drill.*, 5:26–37, doi:10.2204/iodp.sd.5.03.2007.

D'Hondt, S., Jørgensen, B.B., Miller, D.J., Batzke, A., Blake, R., Cragg, B.A., Cypionka, H., Dickens, G.R., Ferdelman, T., Hinrichs, K.-U., Holm, N.G., Mitterer, R., Spivack, A., Wang, G., Bekins, B., Engelen, B., Ford, K., Gettemy, G., Rutherford, S.D., Sass, H., Skilbeck, C.G., Aiello, I.W., Guerin, G., House, C.H., Inagaki, F., Meister, P., Naehr, T., Niitsuma, S., Parkes, R.J., Schippers, A., Smith, D.C., Teske, A., Wiegel, J., Naranjo Padillo, C., and Solis Acosta, J.L., 2004. Distributions of microbial activities in deep subseafloor sediments. *Science*, 306:2216–2221, doi:10.1126/science.1101155.

D'Hondt, S., Rutherford, S., and Spivack, A.J., 2002. Metabolic activity of subsurface life in deep-sea sediments. *Science*, 295:2067–2070, doi:10.1126/science.1064878.

Engelen, B., Ziegelmüller, K., Wolf, L., Köpke, B., Gittel, A., Cypionka, H., Treude, T., Nakagawa, S., Inagaki, F., Lever, M.A., and Steinsbu, B.O., 2008. Fluids from the oceanic crust support microbial activities within the deep biosphere. *Geomicrobiol. J.*, 25:56–66, doi:10.1080/01490450701829006.

Hoehler, T.M., 2004. Biological energy requirements as quantitative boundary conditions for life in the subsurface. *Geobiology*, 2:205–215, doi:10.1111/j.1472-4677.2004.00033.x.

Inagaki, F., and Nakagawa, S., 2008. Spatial distribution of the subseafloor life: Diversity and biogeography. *In* Dilek, Y., Furnes, H., and Muehlenbachs, K. (Eds.), *Links Between Geographical Processes, Microbial Activities and Evolution of Life*. New York (Springer Verlag), 135–158.

Inagaki, F., Nunoura, T., Nakagawa, S., Teske, A., Lever, M., Lauer, A., Suzuki, M., Takai, K., Delwiche, M., Colwell, F.S., Nealson, K.H., Horikoshi, K., D'Hondt, S., and Jørgensen, B.B., 2006. Biogeographical distribution and diversity of microbes in methane hydrate-bearing deep marine sediments on the Pacific Ocean Margin. *Proc. Natl. Acad. Sci. U.S.A.*, 103:2815–2820, doi:10.1073/pnas.0511033103.

Inagaki, F., Suzuki, M., Takai, K., Oida, H., Sakamoto, T., Aoki, K., Nealson, K.H., and Horikoshi, K., 2003. Microbial communities associated with geological horizons in coastal subseafloor sediments from the Sea of Okhotsk. *Appl. Environ. Microbiol.*, 69:7224–7235, doi:10.1128/AEM.69.12.7224-7235.2003.

Kallmeyer, J., Smith, D.C., Spivack, A.J., and D'Hondt, S., 2008. New cell extraction procedure applied to deep subsurface sediments. *Limnol. Oceanogr. Methods*, 6:236–245.

Kuypers, M.M.M., and Jørgensen, B.B., 2007. The future of single-cell environmental microbiology. *Environ. Microbiol.*, 9:6–7, doi:10.1111/j.1462-2920.2006.01222_5.x.

Lunau, M., Lemke, A., Walther, K., Martens-Habbena, W., and Simon, M., 2005. An improved method for counting bacteria from sediments and turbid environments by epifluorescence microscopy. *Environ. Microbiol.*, 7:961–968, doi:10.1111/j.1462-2920.2005.00767.x.

Morono, Y., Terada, T., Masui, N., and Inagaki, F., 2009. Discriminative detection and enumeration of microbial life in marine subsurface sediments. *ISME J.*, 3:503–511, doi:10.1038/ismej.2009.1.

Parkes, R.J., Cragg, B.A., and Wellsbury, P., 2000. Recent studies on bacterial populations and processes in subseafloor sediments: A review. *Hydrogeol. J.*, 8:11–28, doi:10.1007/PL00010971.

Roussel, E.G., Bonavita, M.C., Querellou, J., Cragg, B.A., Webster, G., Prieur, D., and Parkes, R.J., 2008. Extending the sub-seafloor biosphere. *Science*, 320:1046, doi:10.1126/science.1154545.

Teske, A.P., 2006. Microbial communities of deep marine subsurface sediments: Molecular and cultivation surveys. *Geomicrobiol. J.*, 23:357–368, doi:10.1080/01490450600875613.

Weinbauer, M.G., Beckmann, C., and Höfle, M.G., 1998. Utility of green fluorescent nucleic acid dyes and aluminum oxide membrane filters for rapid epifluorescence enumeration of soil and sediment bacteria. *Appl. Environ. Microbiol.*, 64:5000–5003.

Whitman, W.B., Coleman, D.C., and Wiebe, W.J., 1998. Prokaryotes: The unseen majority. *Proc. Natl. Acad. Sci. U.S.A.*, 95:6578–6583, doi:10.1073/pnas.95.12.6578.

## Authors

**Yuki Morono and Fumio Inagaki**, Geomicrobiology Group, Kochi Institute for Core Sample Research (KOCHI), Japan Agency for Marine-Earth Science and Technology (JAMSTEC), Monobe B200, Nankoku, Kochi 783-8502, Japan, e-mail: morono@jamstec.go.jp.

## Photo Credits

Fig. 1 and 3: Yuki Morono, JAMSTEC

7

# High-Precision Orientation of Three-Component Magnetic Downhole Logs

by Christopher Virgil, Andreas Hördt, Torsten Klein, Jochem Kück, Martin Leven, and Erich Steveling

## Introduction

The possible benefits of measuring the magnetic flux density in three components continuously along a borehole have been recognized a long time ago by researchers who developed models and interpretation schemes for 3-component magnetic borehole data (Parker and Daniell, 1979; Gallet and Courtillot, 1989).

Common borehole methods provide data not allowing for an orientation with respect to a global reference, since this requires a highly accurate orientation system independent of the magnetic measurements. A first attempt to obtain the orientation of the sonde was made by Bosum et al. (1988) using a mechanical gyro and accelerometers. However, at that time the data quality of the gyro did not allow for a continuous 3-component measurement. Steveling et al. (2003) provide an example from the Hawaii Scientific Drilling Project (HSDP) drill hole, where directional information of magnetization was used to separate massive lavas from hyaloclastites. However, their directional analysis was limited to the inclination because information on the tool rotation around the vertical axis was not available.

Here, we describe the successful development of an orientation procedure with very high resolution independent of magnetic data. Test data were acquired in the 2.5-km-deep ICDP Outokumpu Research Hole in eastern Finland (Kukkonen, 2007) with the so-called Göttinger Borehole Magnetometer (GBM). The sonde uses three fiber optic gyros (FOGs) exhibiting a small drift of $1.5^{\circ}h^{-1}$ and a high resolution of $9\times10^{-5}$ degrees. In combination with a built-in Förster magnetometer triplet, the GBM can record the magnetic field in three components as well as the tool orientation continuously. In the Outokumpu drill hole, errors (root mean square) were 0.14° for the inclination and 1.4° for the declination of the magnetic flux density.

## Technical Details and Data Preparation

The GBM sonde (3.25 m long, 68 kg) has a diameter of ~86 mm and approximately 140 mm including the centralizer (Fig. 1). The magnetic sensor group of the GBM consists of three Förster magnetometers with a maximum amplitude range of ±50,000 nT for the horizontal components ($B_x$, $B_y$) and ±70,000 nT for the vertical component ($B_z$). The resolution is 6.1 nT for $B_x$ and $B_y$ and 8.5 nT for $B_z$. The fiber optic gyros have a temperature-dependent drift which is minimal for temperatures above 32°C. Furthermore, the Gaussian noise of the FOG data decreases with increasing temperature to an average of $2\times10^{-3}$ degrees in all components. In order to provide optimal operating conditions, the FOGs are mounted in an aluminum cylinder which is thermally isolated against the housing and actively heated by heater elements. To determine the temperature dependent characteristics for posterior recalibration, we conducted several calibration measurements. Due to the delicate electronics inside the FOGs, the maximum ambient operating temperature is 71°C.

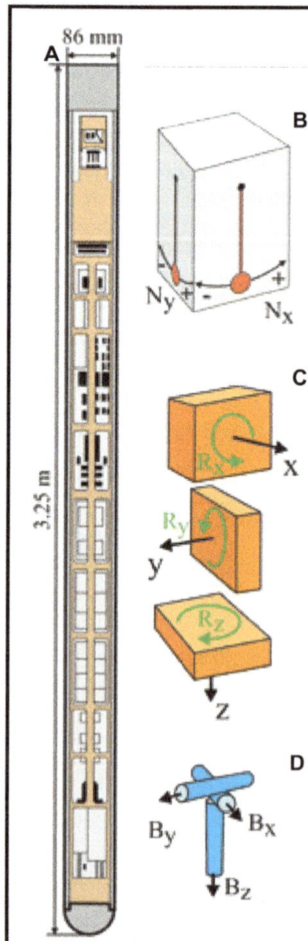

Figure 1. [A] Schematic sketch of the GBM [B] inclinometer, [C] fiber-optic gyros, [D] Förster-magnetometer triplet.

High demands on accuracy require extensive data processing. The first step is the calibration of the Förster probe triplet. The calibration measurements have been done in a magnetic laboratory equipped with a Braunbek coil system with active field compensation (Glassmeier et al., 2007). The quasi-zero-field region inside the Braunbek sys-tem has a residual field of <1 nT. With this system it is possible to determine the sensor offsets, the misalignment between the three sensor axes, and the scaling factors between measured data and genuine field. An alternative method is based on measurements in at least twelve positions while the external field has to be constant (Auster et al., 2002). By minimizing the differ-ence between the external and the measured field, the calibration factors can be computed. Another important step is the drift correction of the FOG data, using the temperature-dependent calibration factors previously determined in the laboratory. Finally, the actual reorientation procedure is carried out, which transforms the magnetic data from tool to geographical coordinates using the orientation information provided by the FOGs.

## Case Study from the Outokumpu Deep Drillhole

The GBM sonde was utilized in a logging campaign in the Outokumpu research borehole in eastern Finland (Fig. 2) in September 2008. This 2516-m-deep well was drilled from 2004 to 2005 and is maintained by the Geological Survey of Finland as a deep geolaboratory. It offers ideal conditions for magnetic measurements as it cuts through a highly magnetized 200-m-thick Cu-Co-Zn bearing ophiolitic layer (1300–1500 m) at moderate temperatures below 40°C (Kukkonen, 2007).

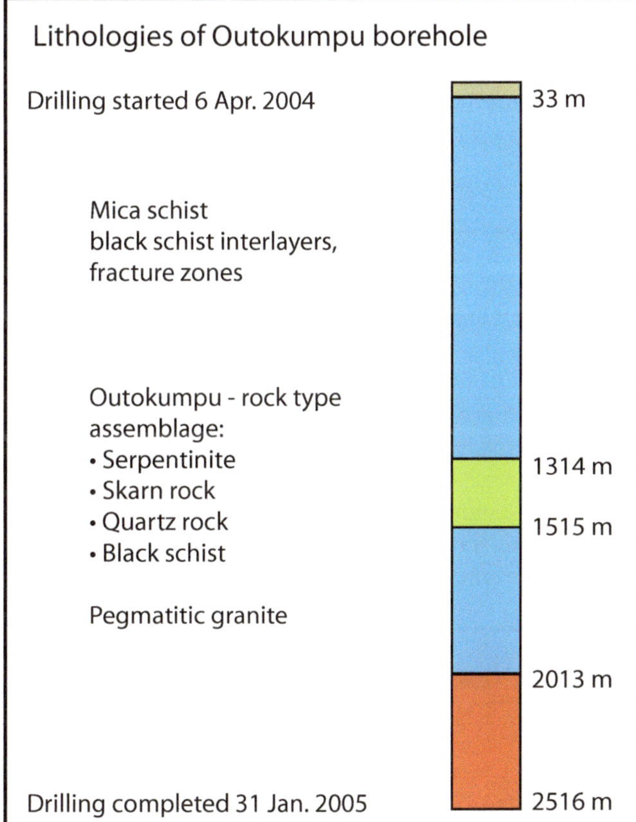

Figure 2. Borehole scheme of the Outokumpu well with simplified lithological units (Kukkonen, 2007).

Figure 3 shows a comparison of downlog and uplog data of sample measurements before and after reorientation. The mean difference between downlog and uplog in the total field (Fig. 3A) is on the order of 30 nT. The horizontal components after recalibration, but before reorientation, are depicted in Figs.3B and 3C. The large variations of ±11,200 nT are due to the rotation around the vertical axis. Depth regions with rapid changes (e.g., at 800 m), which are produced by sudden rotations, can be observed. A possible explanation is that they are caused by borehole wall outbreaks. The cable tension built up by the use of non-rotating centralizers may be released in such breakout sections. This hypothesis is supported by an observed co-occurrence between sections of rapid tool rotation and increasing caliper.

For the reorientation procedure, the magnetic data from the internal tool coordinate system xyz are projected onto the geographical system North, East, and Vertical (downwards). Each magnetic data triplet is multiplied by a rotation matrix computed from the FOG's data. Furthermore, other effects have to be considered, one of which is the Earth's rotation. The fiber optic gyros measure rotation depending on their orientation with respect to Earth's rotational axis, in addition to the tool rotation. Therefore, the Earth's rotational vector has to be projected into the tool system and subtracted from the FOG data in all three components for each depth.

The next processing corrects the misalignment of the FOG system. If the three axes of this system are not exactly perpendicular to each other, rotation around the z-axis will cause a spurious signal in the x- and y-gyros, which would be mistaken for a real rotation. Due to the torque in the cable, the rotation around the z-axis dominates. Thus, we consider only the misalignment between the x- and z-axis, as well as y- and z-axis, which were determined by calibration measurements to be 0.19° and 0.02°, respectively.

Another challenge is the computation of the rotation matrix itself. During the sampling period of 0.5 seconds, rotation occurs continuously around all three axes. However, when computing the rotation matrix, one has to assume rotation in a specific order—for instance, in roll-yaw-pitch convention. Instead of using only one rotation matrix per step, the movement is smoothed by dividing the period into fifty steps. Each matrix corresponds to a rotation by only 1/50 of the measured rotation angles. This reduces the error particularly in regions where fast rotations in all three axes exist.

The final correction takes the tool orientation above ground into account. Before actually starting the logging, the tool is set up in a vertical position above the borehole and is aligned to a north marker by a sighting telescope mounted on the housing. This defines the start orientation in the FOG data with an accuracy of $5 \times 10^{-2}$ degrees. Having finished the uplog this procedure is repeated, so that the orientation at these two time points can be ensured. If the reorientation procedure works perfectly, the two orientations should match exactly after processing. Nevertheless, differences of a few degrees are usually observed in all three components. Based on these deviations, additional drift corrections are computed and applied to the FOG data to minimize the orientation differences. In Figs. 3D–3F, the example logs are shown after the complete reorientation procedure.

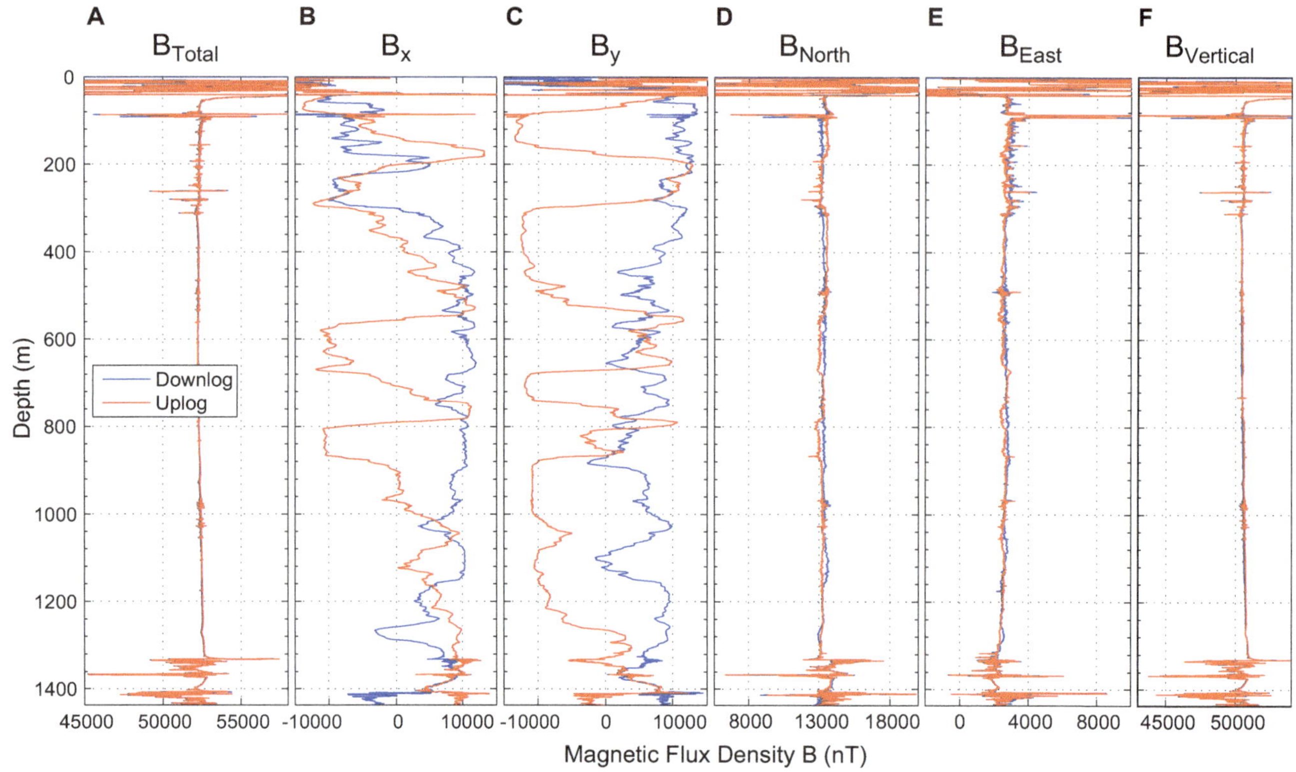


Figure 3. [A] Total magnetic flux density, [B and C] horizontal components before reorientation (internal tool coordinates xyz), [D–F] magnetic flux density after reorientation with geographic coordinates North, East, Vertical.

## Quality Assessment

One possibility to assess the quality of the reorientation procedure is the comparison between the uplog and the downlog. For a quantitative evaluation, we use the root mean square (RMS) of the difference. This results in errors of $RMS_{Total}$ = 50 nT, $RMS_{North}$ = 250 nT, $RMS_{East}$ = 180 nT, and $RMS_{Vertical}$ = 75 nT for the total, north, east, and vertical components, respectively. The deviations are not equally distributed. In regions between two fast rotations (e.g., 800–880 m), the discrepancy reaches up to 500 nT. There are also sections where the deviation is much smaller (less than 45 nT) in all three components between 680 m and 750 m. The errors of inclination and declination of the magnetic flux density result are $RMS_{Inclination}$ = 0.25° and $RMS_{Declination}$ = 0.75°, respectively.

Besides comparing downlog and uplog in the same measurement, the reproducibility between different logs is also an important criterion. In Fig. 4, two logs measured on

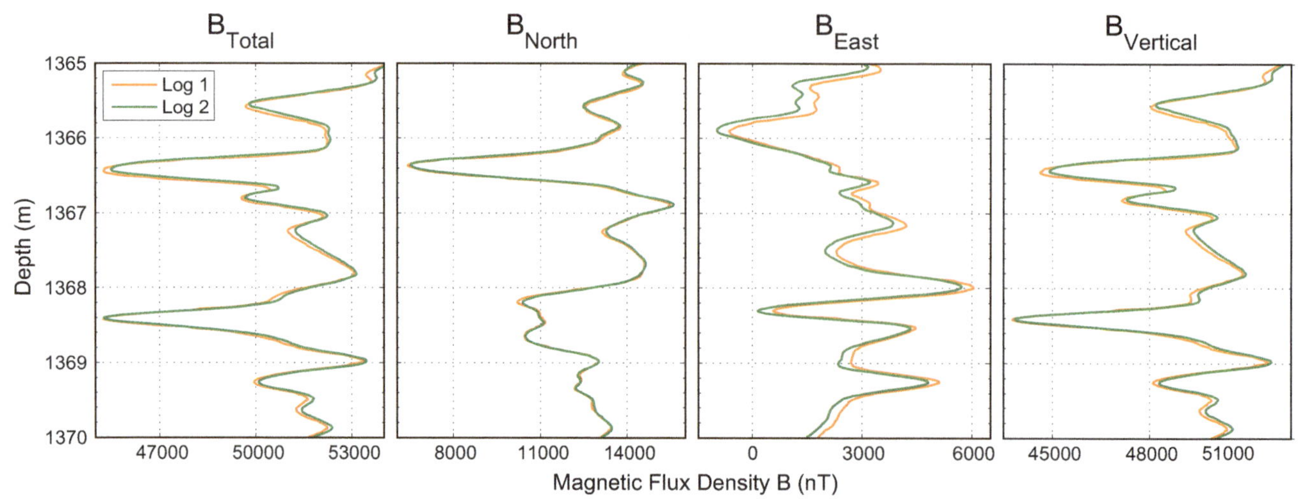


Figure 4. Comparison between two different logs, extract of strongly magnetized region. Each line represents the mean value of downlog and uplog.

two consecutive days under identical conditions are shown. Each log consists of the average value of the corresponding downlog and uplog. To illustrate the spatial resolution, a section of the strongly magnetized Outokumpu rock assemblage is shown in detail (Fig. 4). All features are reproduced very well. The average amplitude deviations between the two logs are 170 nT in the total and vertical components, 10 nT in the east component, and 330 nT in the north component. The deviations in orientation are $RMS_{Inclination} = 0.14^{\circ}$ and $RMS_{Declination} = 1.4^{\circ}$.

## Outlook

The logging results from the Outokumpu drill hole are used to calculate the rock magnetization, based on a model developed by Bosum et al. (1988). Existing susceptibility measurements are used to separate remanent from induced magnetization. Preliminary results indicate that the direction of remanent magnetization is inhomogeneous and scattered. From a statistical analysis we were able to identify two preferred directions of remanent magnetization which might be related to the genesis of the Outokumpu rock assemblage in the depth region of 1300–1440 m. The results of the ongoing detailed analysis including magnetic laboratory measurements will be reported elsewhere. Based on the comparison of our *in situ* direction of remanent magnetization with direction measured in the laboratory, we might also be able to carry out a reorientation of core samples from this section.

We will continue to improve the accuracy of the reorientation procedure. Presently, the main source of error is the fast rotation of up to $10^{\circ}$ around the *z*-axis during the sample period of 0.5 seconds. Since the sampling period is limited by the data transfer rate to the base station, an upgrade of the tool hardware is desirable. We plan to implement a flash memory, which will enable us to increase the sampling frequency by twenty-fold. This will also enhance the spatial resolution of the tool using the same logging speed. Furthermore, we intend to carry out extended calibration measurements to obtain the misalignment angles between all three FOG axes and their scaling factors. We will also re-measure the calibration factors of the magnetic sensors, considering possible temperature dependence. We aim for errors less than $0.1^{\circ}$ for both inclination and declination of the magnetic flux density.

## Acknowledgements

Ilmo T. Kukkonen (GTK, Finland) provided extensive organizational support. Christian Carnein and Martin Töpfer (Operational Support Group OSG/ICDP, GFZ, Germany) helped during the measurements. Johannes Stoll (Metronix, Germany) is thanked for fruitful discussions and advice. This project was supported by the German Science Foundation (DFG, Ho 1506/16-1).

## References

Auster, H.U., Fornacon, K.H., Georgescu, E., Glassmeier, K.H., and Motschmann, U., 2002. Calibration of flux-gate magnetometers using relative motion. *Meas. Sci. Technol.*, 13:1124–1131, doi:10.1088/0957-0233/13/7/321.

Bosum, W., Eberle, D., and Rehli, H.J., 1988. A gyro-orientated 3-component borehole magnetometer for mineral prospecting, with examples of its application. *Geophys. Prospect.*, 36:933–961, doi:10.1111/j.1365-2478.1988.tb02201.x.

Gallet, Y., and Courtillot, V., 1989. Modeling magetostratigraphy in a borehole. *Geophysics*, 54(8):973–983, doi:10.1190/1.1442740.

Glassmeier, K.H., Richter, I., Diedrich, A., Musmann, G., Auster, U., Motschmann, U., Balogh, A., Carr, C., Cupido, E., Coates, A., Rother, M., Schwingenschuh, K., Szegö, K., and Tsurutani, B., 2007. RPC-MAG The fluxgate magnetometer in the ROSETTA plasma consortium. *Space Sci. Rev.*, 128:649–670, doi:10.1007/s11214-006-9114-x.

Kukkonen, I.T., 2007. Outokumpu Deep Drilling Project, Second International Workshop. *Programme and Extended Abstracts*, Espoo, Finland, 21–22 May 2007.

Parker, R.L., and Daniell, G.J., 1979. Interpretation of borehole magnetometer data. *J. Geophys. Res.*, 84:5467–5479, doi:10.1029/JB084iB10p05467.

Steveling, E., Stoll, J. B., and Leven, M., 2003. Quasi-continuous depth profiles of rock magnetization from magnetic logs in the HSDP-2 borehole, Island of Hawaii. *Geochem. Geophs. Geosyst.*, 4(4), 8708, doi:10.1029/2002GC000330.

## Authors

**Christopher Virgil**, **Andreas Hördt,** and **Torsten Klein,** Institut für Geophysik und extraterrestrische Physik, Technische Universität Braunschweig, Mendelssohnstrass 3, D-38106 Braunschweig, Germany, e-mail: c.virgil@tu-bs.de.
**Jochem Kück**, GFZ – German Research Centre for Geosciences, Telegrafenberg, D-14473, Potsdam, Germany.
**Martin Leven and Erich Steveling**, Institut für Geophysik, Georg-August-Universität Göttingen, Friedrich-Hund-Platz 1, D-37077, Göttingen, Germany.

## Related Web Link

http://www.gsf.fi/projects/o_k_deepdrilling/

8

# Seismic Prediction While Drilling (SPWD): Looking Ahead of the Drill Bit by Application of Phased Array Technology

**by Katrin Jaksch, Rüdiger Giese, Matthias Kopf, Andreas Jurczyk, Stefan Mikulla, Stefan Weisheit, Marco Groh, and Kay Krüger**

## Introduction

Geophysical exploration is indispensable for planning deep drilling. Usually 2D- or 3D-seismics investigations are applied and, depending on the resulting geologic model for the underground, the drill site and drilling path are determined. In recent years the focus of exploration has shifted towards small-scale geological structures such as local layers and faults. Depending on the source frequencies and the target depth, 2D- or 3D-seismics from surface cannot always resolve such structures in particular at larger depths. In general, signal frequencies of about 30–70 Hz are typical for surface seismic methods. The deeper and smaller the sought-after structures are, the worse will be the resolution. Therefore, borehole seismic measurements like Vertical Seismic Profile (VSP) or Seismic While Drilling (SWD) have been developed (Fig. 1). For the VSP method geophones are normally integrated in the borehole, while the seismic source generates seismic waves at the surface. The SWD method uses the drill bit as the seismic source. Hence, the quality of the seismic signals is highly dependent on the drilled rock and the type of drill bit, but even well-suited rock conditions and adequate drilling may not provide sufficient data quality.

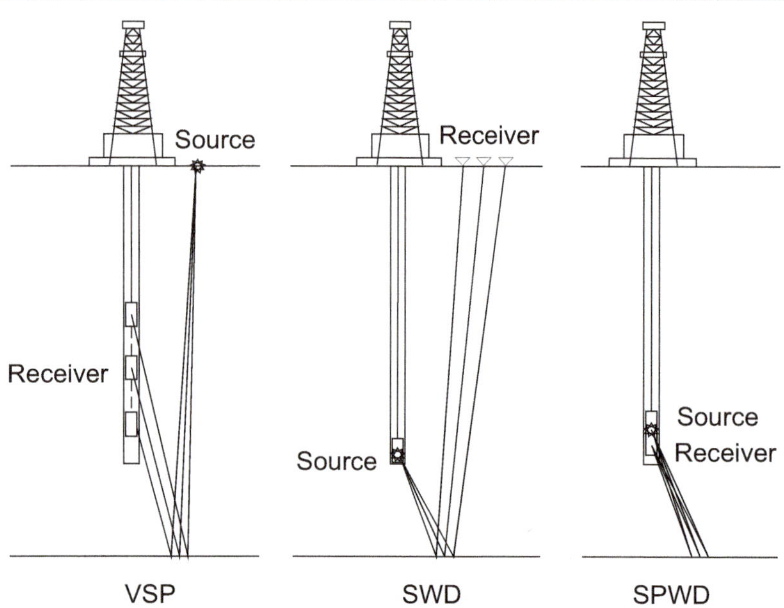


Figure 1. Comparison of the different seismic borehole exploration methods VSP, SWD, and SPWD

Compared to 2D- and 3D-seismics, the distances between source and receiver to the target are shorter for VSP, and for SWD in particular. Signal frequencies up to about 100 Hz are observed yielding a slightly better resolution compared to surface seismics. However, subtle yet important features (e.g., fault zone) often cannot be identified with certainty. Accordingly, a method or tool that would improve the resolution is of high importance since this would allow adjusting the drilling path to minimize risks and costs of drilling.

A device combining source and receiver immediately behind the drill bit would allow seismic exploration of a range of about 50–100 m ahead of the drill bit with a resolution of one meter. The first prototype of such a Seismic Prediction While Drilling (SPWD) device has been designed and manufactured within this project (Wenke et al., 2010). The tool uses signal frequencies between 500 Hz and 5000 Hz, which are significantly higher than the usual VSP and SWD applications.

## A Laboratory Prototype for Seismic Prediction in Boreholes

A SPWD prototype was designed for laboratory conditions without the impact of borehole fluids, high temperatures, and pressures and only for tests in horizontal boreholes. It consists of different units for seismic sources, receivers, and data logger. Figure 2 and Table 1 provide details of the instrumentation and dimensions. Four standard magnetostrictive actuators (AK1 to AK4; Fig. 2) are aligned along the borehole axis to serve as a cascading source and act as seismic sources. Every single actuator unit can be rotated in steps of 15° around the longitudinal axis of the prototype to allow for a variety of source configurations trials.

Table 1. Parameters of the prototype

| | |
|---|---|
| Length | 1.81 m |
| Diameter | 20–23.5 cm |
| Weight | ~60 kg |
| Pressing | pneumatic, 6–8 bar |
| Contact pressure: | |
| actuators | ~ 2 kN |
| geophones | ~ 200 N |

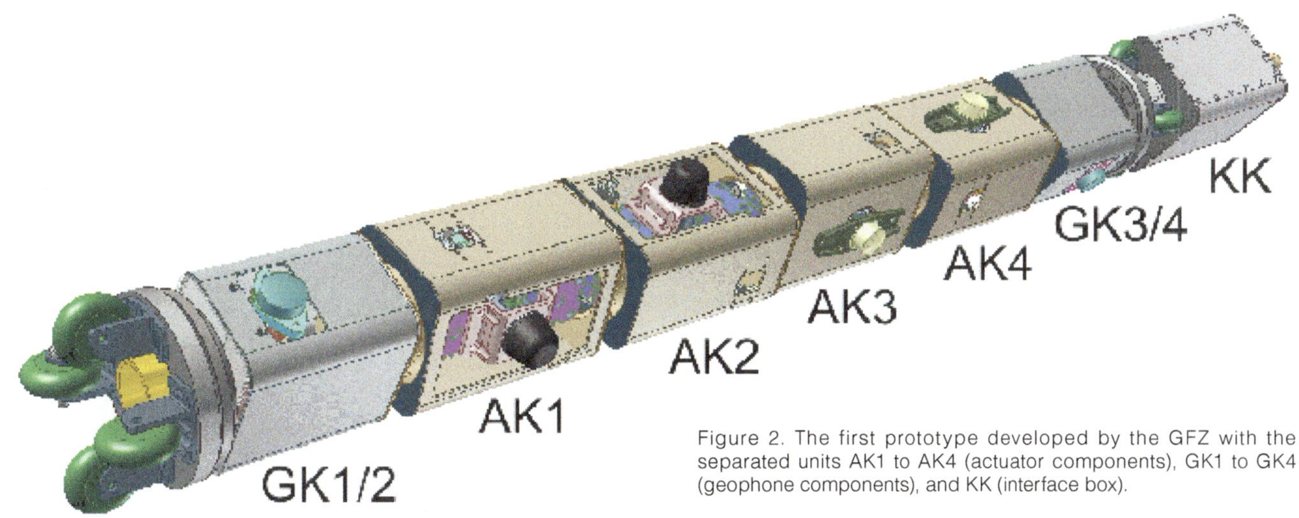


Figure 2. The first prototype developed by the GFZ with the separated units AK1 to AK4 (actuator components), GK1 to GK4 (geophone components), and KK (interface box).

The receivers used are four 3-component geophones integrated in two geophone units (GK1/2 and GK3/4; Fig. 2) with each 3-component geophone mounted in opposite alignment. Again, the geophone units can be rotated in 15° increments perpendicular to the borehole axis. This allows receiving the seismic response from the surrounding rock in defined space directions adjusted to the actuator positions.

The data logger unit (KK; Fig. 2) includes the interface with the external feed cables and the data logger which is used for source activation, triggering, and seismic data acquisition. Additionally, a sensor for inclination, a compass, and an odometer are integrated to determine the position and orientation of the SPWD device in the borehole. An infrared camera at the top monitors the borehole wall.

Several guide rollers are mounted at the top and the end to move the prototype slightly in the borehole. In order to take measurements, the actuators and geophones must be pressured radially to the borehole wall. The prototype is automatically centered in the borehole. The actuators are decoupled from the prototype body so that all seismic energy is initiated exclusively in the borehole wall.

The actuators are controlled in amplitude and phase. Acceleration sensors are integrated in the front stamps of the actuators to measure the emitted seismic signals. With respect to a specific radiation pattern, the signal for each actuator is calculated based on the local coupling conditions at the borehole wall and then applied.

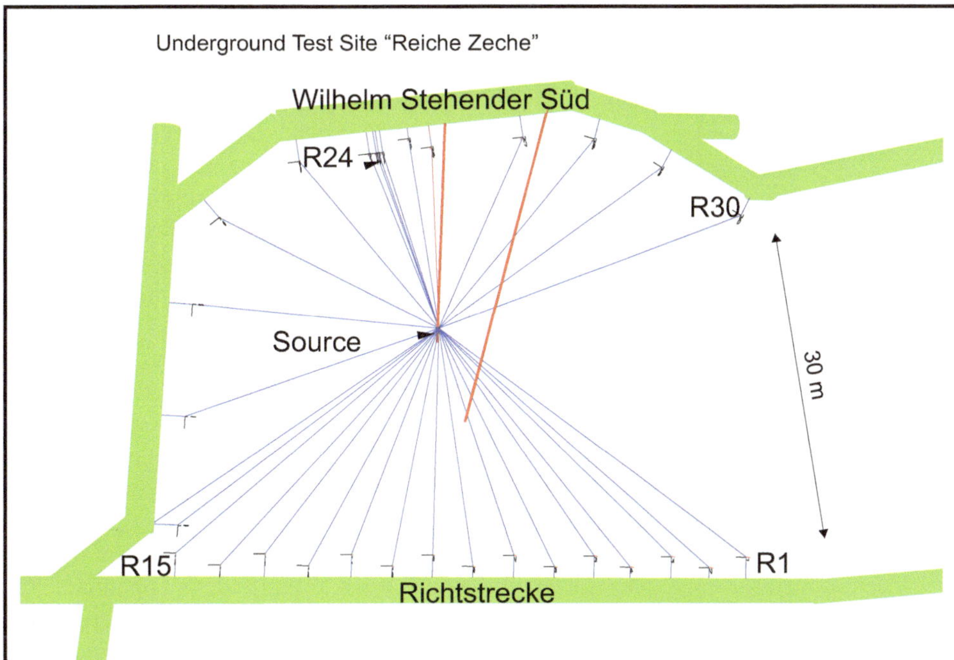


Figure 3. The test site of the GFZ at the education and research mine "Reiche Zeche" in plain view with the three surrounding galleries (green), the boreholes (red), and the geophone anchors (R1-R30, black). The lower horizontal gallery is named Richtstrecke; the upper gallery is called Wilhelm Stehender Süd.

## Test Environment

A test site in the education and research mine "Reiche Zeche" in Freiberg, Germany (Wenke et al., 2010) has been constructed to ensure constant environmental conditions. Surrounded by three galleries, the site comprises a block of homogeneous high-grade gneisses of almost 50 m x 200 m. Previously used as a test site for seismic sources and receivers of an integrated seismic imaging system (Giese et al., 2006), the site has been expanded. Over thirty 3-component geophone anchors one or two meters long are installed around the area about 4–9 m apart (Fig. 3).

Two horizontal boreholes of 20 m and 30 m length were drilled towards a gallery to test the prototype. Diamond coring was applied to yield both high-quality cores and true-to-size caliber. The latter was only disrupted at two centimeter-sized fracture zones which appear to pass through the whole rock block and which are filled with quartz, feldspar, clay, and fluids. The dike-like fault zones and the galleries act as well-defined reflectors for the borehole seismic measurements.

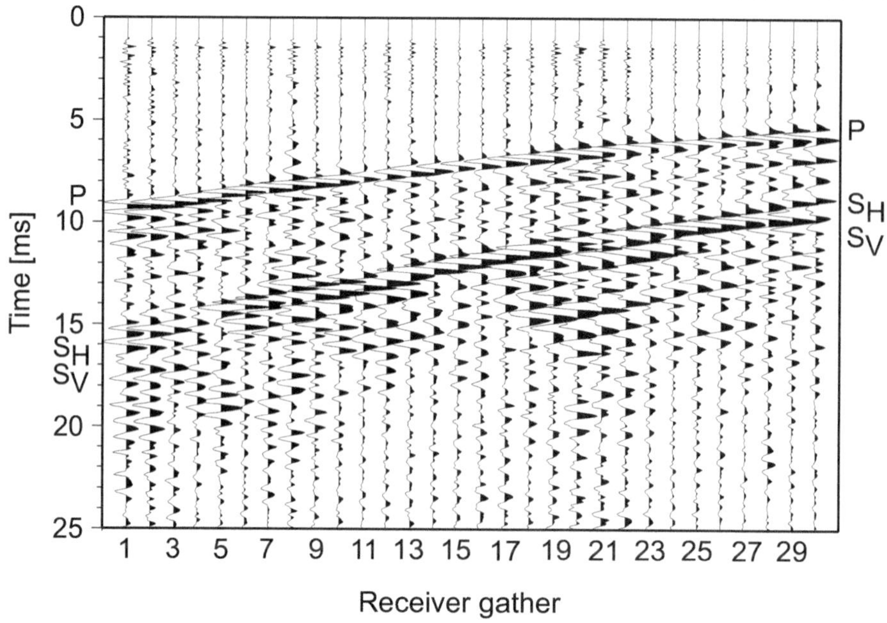


Figure 4. Seismic data of the vertical component at receiver 16 with 30 equidistant shot points in the 30 m borehole. The seismic wave field shows distinct compression (P) and shear (SH, SV) waves and overall a good signal-to-noise ratio.

## Seismic Measurements

Several measurements were implemented during the first year to test the functionality of the different units of the prototype (e.g., the pneumatic contact pressure and the handling of the prototype with rods in the borehole). The performance of the actuators was examined to adapt the parameters of the sweep signals and to maximize the energy of the seismic source in order to reach a sufficient exploration range.

To achieve a directional radiation effect it is essential that all four actuators can be controlled independently of each other. An increased radiation of seismic wave energy in the direction of the borehole axis allows the view in areas to be drilled. Therefore, every actuator related to amplitude and phase of the source signal must be controlled. The next step for focusing the seismic waves is to use the method of phased array, which has been developed in material sciences and for medical diagnostics by sonography (Hedrick and Hykes, 1996). The signal phases for each actuator can be determined depending on the seismic wave velocities of the surrounding rock, the distance of the actuators to each other, and the frequencies used. Furthermore, the influence of the seismic velocity on the radiation pattern was studied by applying a sonic log with the actuators with a broad frequency band (1–14 kHz).

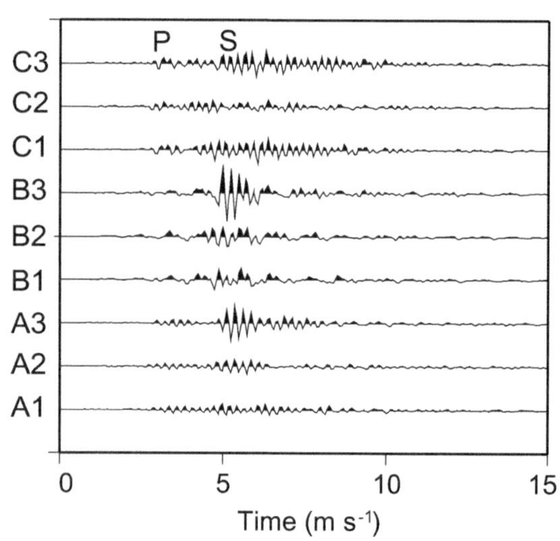


Figure 5. Correlated 3-component recordings of the direct compressional (P) and shear (S) waves at receiver 24 (see R24 in Fig. 3). The source-receiver offset is 17 meters. A – vertical component traces, B – horizontal component traces, C – radial component traces; 1 – sources are in phase, 2 – seismic wave field directed towards Richtstrecke, 3 – seismic wave field directed towards Wilhelm Stehender Süd.

The influence of different seismic source arrangements was investigated in order to achieve the directional characteristics. The main goal of the measurements was to focus the initiated seismic waves in the direction of the borehole axis.

## Results

The borehole data overall show a good signal-to-noise ratio. The direct seismic wave field indicates distinct compression and shear waves (Fig. 4). As expected, the anisotropic gneisses cause shear wave splitting of about ten percent. The quality of the seismic data in amplitude and frequency spectra correlates strongly with the fracture density of the rock along the borehole, which was determined with an optical borehole scan and on core. These regions are associated with a lower signal-to-noise ratio. Independent of the local rock conditions, the borehole data show seismic reflections from ahead and rearwards. In particular the reflections from the gallery ahead are used for the calibration of focusing.

Analysis of several seismic measurements with a focus on the direct seismic waves shows that the phased array technology influences explicitly the directional characteristics of the radiated seismic waves. Their amplitudes can be enhanced three times more in the desired direction and can simultaneously be attenuated in the reverse direction. Figure 4 and 5 present a set of exemplary data recorded at the receiver in geophone anchor R24 (Fig. 3). The source is positioned close to the bottom of the borehole. The frequency content of the source signal is between 500 Hz and 5000 Hz. P-waves arrive at about 3 ms and S-waves at 4.8 ms, corresponding to seismic wave velocities of 5700 m $s^{-1}$ and 3500 m $s^{-1}$, respectively. The amplitudes of the P-wave are highest on the radial component (C in Figs. 4 and 5) directed to the source. The amplitudes of the S-wave are highest perpendic-ular to the P-wave on the vertical (A) and horizontal component (B). If the source wave field is directed towards the gallerie "Wilhelm Stehender Süd" (Fig. 3), the amplitude of the S-wave is clearly amplified (B3 and C3) in comparison to source signals without manipulation of wave field direction (B1 and C1). A slight increase of P-wave amplitudes can also be observed in the radial component in C3 compared to C1. If the wave field is directed towards the gallery "Richtstrecke" (Fig. 3), the amplitude decreases slightly on the radial component C2 compared to original source wave field in C1, whereas no significant reduction of amplitudes is observed on the vertical (A2) and horizontal (B2) components.

## Conclusions

The application of the phased array technology for the directional investigation of borehole surroundings is very promising. However, focusing of the seismic waves has to be improved to maximize the energy in the desired direction. For that purpose several measurements and modeling of different source and receiver configurations must be done to calibrate the initiated seismic signals of the sources. The application of a phased array source system influences the processing and imaging of the data. Methods developed for surface seismic and VSP measurements have to be adapted to take advantage of the improved data quality and known directional information of the recorded wave field.

The next step will be the development of a wireline SPWD prototype for vertical boreholes with depths to 2000 m. Modifications in the arrangement of the different units will be necessary, as well as an adaptation to fluid-fill under the conditions in deep boreholes with respect to pressure and temperature. If the wireline prototype tests successfully, we plan to implement the tool into the downhole assembly of a drill string as a Logging-While-Drilling tool. This would allow a seismic prediction of structures or faults while drilling and will help to reduce risks encountered while drilling.

## Acknowledgements

We are grateful for support from the technical team of the "Reiche Zeche" for the preparation of the test site and from our partners in the SPWD project, the TU Bergakademie Freiberg, GeoThermal Engineering GmbH, and Herrenknecht Vertical GmbH. The project (S)PWD ((Seismic) Prediction While Drilling) is funded by the German Federal Environment Ministry, BMU.

## References

Giese, R., Dickmann, Th., Eppler, Th., and Lüth, S., 2006. Seismic tomography to investigate the tunnel surroundings of the Side Gallery West of the multifunctional station Faido. *In* Löw, S. (Ed.), *Geologie und Geotechnik der Basistunnels am Gotthard und am Lötschberg, Tagungsband zum Symposium Geologie Alptransit*, Zürich (vdf Hochschulverlag an der ETH), 145–154.

Hedrick, W.R., and Hykes, D.L., 1996. Beam steering and focusing with linear phased arrays. *J. Diagn. Med. Sonogr.*, 12:211–215.

Wenke, A., Giese, R., Jaksch, K., Kopf, M., Kreuter, H., Reich, M., Reinhardt, J., and Sohmer, M., 2010. (S)PWD – (Seismic) Prediction While Drilling: Development of a new high-resolution Seismic While Drilling (SWD) concept for geothermal drilling. *Proceedings World Geothermal Congress*, Indonesia, 25–29 April 2010.

## Authors

**Katrin Jaksch**, Helmholtz Centre Potsdam - GFZ German Research Centre for Geosciences, Scientific Drilling, Telegrafenberg, F 358, D-14473, Potsdam, Germany, e-mail: kawi@gfz-potsdam.de.

**Rüdiger Giese, Matthias Kopf, Andreas Jurczyk, Stefan Mikulla, Stefan Weisheit, Marco Groh, Kay Krüger** Helmholtz Centre Potsdam GFZ German Research Centre for Geosciences, Scientific Drilling, Telegrafenberg, D-14473, Potsdam, Germany.

# 9

# Arctic Ocean Scientific Drilling: The Next Frontier

by Bernard Coakley and Ruediger Stein

## Introduction

The modern Arctic Ocean appears to be changing faster than any other region on Earth. To understand the potential extent of high latitude climate change, it is necessary to sample the history stored in the sediments filling the basins and covering the ridges of the Arctic Ocean. These sediments have been imaged with seismic reflection data, but except for the superficial record, which has been piston cored, they have been sampled only on the Lomonosov Ridge in 2004 during the Arctic Coring Expedition (ACEX-IODP Leg 302; Backman et al., 2006) and in 1993 in the ice-free waters in the Fram Strait/Yermak Plateau area (ODP Leg 151; Thiede et al., 1996).

Although major progress in Arctic Ocean research has been made during the last few decades, the short- and long-term paleoceanographic and paleoclimatic history as well as its plate-tectonic evolution are poorly known compared to the other oceans. Despite the importance of the Arctic in the climate system, the database we have from this area is still very weak. Large segments of geologic time have not been sampled in sedimentary sections. The question of regional variations cannot be addressed.

Prior to 2004, the geological sampling in the Arctic Ocean was restricted to obtaining near-surface sediments, i.e., only the upper about 5–15 m could be sampled by means of gravity and piston coring. Thus, more or less, all studies were restricted to the Quaternary, with one exception (Fig. 1; e.g., Clark et al., 1980, 1986; Thiede et al., 1990). In four short sediment cores from Alpha Ridge, upper Cretaceous and lower Tertiary sediments were sampled by gravity coring from ice island T-3. Until recently (Stein, 2008), the absence of technological and logistic solutions for reaching and operating in a permanently ice-covered region thwarted further study of the Arctic Ocean; thus, we have been unable to retrieve long and undisturbed sediment cores.

With the successful completion of IODP Expedition 302 ("Arctic Coring Expedition", ACEX)—the first Mission Specific Platform (MSP) expedition within the Integrated Ocean Drilling Program (IODP)—a new era in Arctic research has begun. For the first time, scientific drilling in the permanently ice-covered Arctic Ocean was carried out, penetrating about 430 m of Quaternary, Neogene, Paleogene, and Campanian sediments on the crest of Lomonosov Ridge close to the North Pole (Backman et al., 2006, 2008; Moran et al., 2006).

ACEX was an outstanding success for two reasons. First, ACEX has proven that with an intensive ice-management strategy (i.e., a three-ship approach with two icebreakers *Sovetskiy Soyuz* and *Oden* protecting the drillship *Vidar Viking* by breaking upstream ice floes into small pieces), successful scientific drilling in the permanently ice-covered central Arctic Ocean is possible. Second, the first scientific results brought new and unexpected insights into the Arctic Ocean climate history and its global significance (Backman and Moran, 2008, and further references therein).

Despite the success of IODP Expedition 302, major questions related to the climate history of the Arctic Ocean and its long- and short-term variability during Mesozoic-Cenozoic times cannot be answered from the ACEX record due to the poor core recovery and, especially, a major mid-Cenozoic hiatus. This hiatus spans the critical time of the transition from the early Cenozoic Greenhouse world to the late Cenozoic Icehouse world (Miller et al., 1987, 1991; Lear et al., 2000; Pearson and Palmer, 2000; Zachos et al., 2001). Nevertheless, the success of ACEX has certainly opened the door for further scientific drilling in the Arctic Ocean. The ACEX results will frame the next round of questions to be answered from new drill holes to be taken by a series of drilling legs during the next decades.

## Workshop on "Arctic Ocean History: From Speculation to Reality"

In order to discuss and plan the future of scientific drilling in the Arctic Ocean, an international workshop was held at the Alfred Wegener Institute in Bremerhaven, Germany, on 3–5 November 2008 (Coakley and Stein, 2008). The coauthors of this article convened that workshop. About ninety-five scientists from Europe, the U.S.A., Canada, Russia, Japan, and Korea as well as observers from oil companies participated in the workshop. All participants were invited to submit abstracts about their experiences, ideas and/or plans of Arctic Ocean research with special emphasis on drilling.

The major targets of the workshop were as follows: (1) to bring together an international group of Arctic scientists, young scientists, and ocean drilling scientists to learn and exchange ideas, experience, and enthusiasm about the Arctic Ocean; (2) to develop a scientific drilling strategy to investigate the tectonic and paleoceanographic history of the Arctic Ocean and its role in influencing the global climate system; (3) to summarize the technical needs, opportunities, and limitations of drilling in the Arctic; and (4) to define scientific and drilling targets for specific IODP-type campaigns in Arctic Ocean key areas to be finalized in the development of drilling proposals.

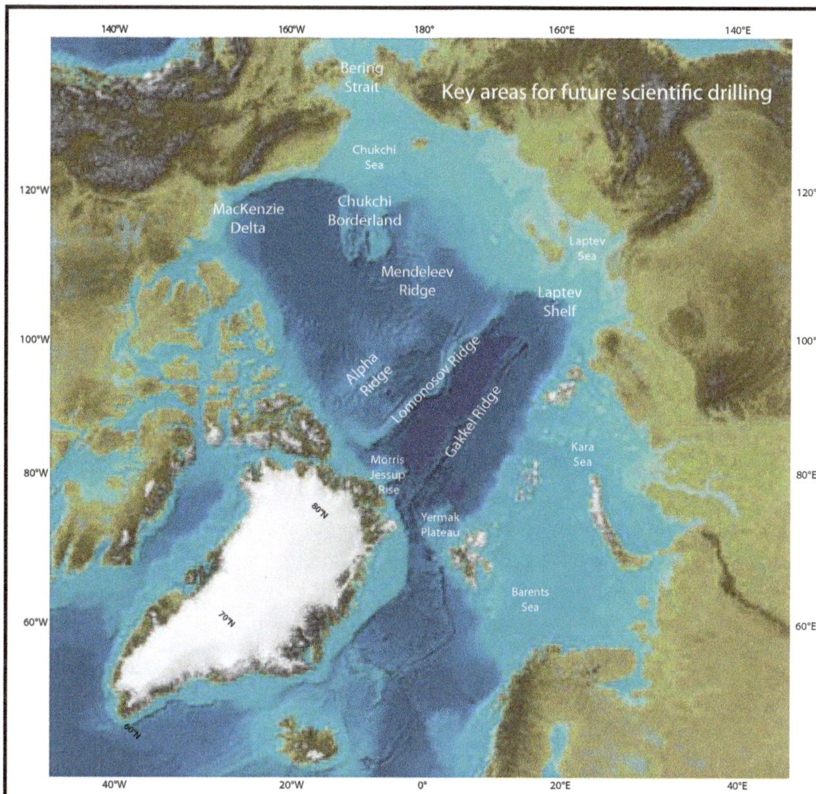


Figure 1. The proposed drilling regions for the Arctic are primarily ridge tops, which will have condensed sections, making it possible to sample relatively long intervals without deep penetration of the seabed. There is also a desire for expanded, high resolution, sections, which can be recovered from the continental shelves. (Background map is from IBCAO, Jakobsson et al., 2008)

The first day of the workshop focused on presentations about the history of the Arctic Ocean, the legacy of high latitude ocean drilling, the existing site survey database, the possibilities of collaboration with industry, and the process of developing ocean-drilling legs through IODP. The next day and a half was spent in thematic and regional break-out groups discussing the particular questions to be addressed by drilling and the particular targets for Arctic scientific drilling. Within the working groups, key scientific questions, site surveys (available and needed), and strategies for reaching the overall goals were discussed, and—as one of the main results—core groups for further developing drilling proposals were formed.

Based on discussions at this meeting, a number of new proposals will be submitted to IODP in 2009/2010, a critical time both for the future of Arctic Ocean science and the future of scientific ocean drilling. As of October 2009, eight active Arctic-related proposals are listed in the IODP system (Table 1). Major themes (hypotheses to be tested by drilling) identified by the workshop participants may be summarized as follows:

*Paleoceanography*:

- Cyclicity between oxic, sub-oxic, and/or euxinic/anxoic conditions during the Cretaceous and Paleocene-Eocene
- Greenhouse vs. icehouse climate
- Polar amplification of greenhouse warming
- Hydrological cycle during greenhouse warming
- Onset of Eocene cooling
- Impact of Eocene-Oligocene transition in global $pCO_2$ and sea level on the Arctic
- Onset and variability of sea-ice cover (seasonal vs. perennial ice cover)
- Circum-Arctic ice-sheet/ice-shelf history and dynamics
- Opening of Bering Strait/Fram Strait and its paleoceanographic consequences
- Causes of extended mid-Cenozoic unconformities
- Nature of the Arctic environment during periods of extreme events (warm/cold)
- Bipolar synchronous vs. asynchronous climate variability?

The varied sedimentary environments of the Arctic Ocean (Stein, 2008) enable two types of studies. Sampling on the tops of the ridges that segment the basin make it possible to collect records that span long intervals of geologic time. Focusing on the shelves and near-shelf areas make it possible to collect expanded, high-resolution records suitable for detailed paleo studies. At the workshop there was little interest in drilling into basinal sediments, given the expectation that these records are largely composed of turbidites and other mass-wasted sediments.

In order to study the long-term Mesozoic-Cenozoic evolution of the Arctic Ocean, we need to obtain undisturbed and complete sedimentary sequences to be drilled along depth transects across the major ocean ridge systems, i.e., the Lomonosov Ridge, the Alpha-Mendeleev Ridge, and the Chukchi Plateau/Northwind Ridge (Fig. 1). High-resolution records will enable detailed studies of climate variability on Milankovich and millennial to sub-millennial time scales. Appropriate sediments can be drilled along the circum-Arctic continental margins characterized by high sedimentation rates. Key areas, for example, are the Kara and Laptev seas and the Mackenzie shelf/slope characterized by large river discharge (Fig. 1). Key locations for studying the history of exchange of the Arctic Ocean with the world's oceans are the Fram Strait/Yermak Plateau and Chukchi Plateau areas (Fig. 1).

*Tectonics*:

- Mode of crustal extension in the Laptev Sea shelf
- Development of the Fram Strait gateway (mode of extension)
- Identification of plate boundaries (Chukchi Plateau)
- Age of magnetic anomalies (Canada Basin)
- Age and evolution of Alpha Ridge, Mendeleev Ridge, Makarov Basin, and Chukchi Plateau
- Correlation of onshore and offshore geology (Paleozoic sediments, Mesozoic magmatism)
- Understanding the 'Amerasia' side of Lomonosov Ridge
- Along-strike geologic variation of Lomonosov Ridge and consequences for Mesozoic evolution

*Petrology*:

- Gakkel Ridge (Fig. 1) mantle melting and geochemistry: western vs. eastern Gakkel Ridge (Global problem: how does continental lithospheric mantle contribute to melting of the asthenosphere? How does extent of melting change as spreading rate goes to zero?)
- Nature and origin of the Chukchi Borderland volcanism
- Origin of Alpha Mendeleev Ridge (hotspot track or segment of a large LIP?) Is the roughly synchronous volcanism recognized in America and Asia somehow related to a High Arctic Large Igneous Province?

*Gas Hydrates*:

The stability of gas hydrates and permafrost below the seafloor are intimately related to climate change (greenhouse gas reservoir). The dissociation of hydrates, which some believe is now underway, could release large quantities of methane to the atmosphere, acting as positive feedback driving global warming and further destabilizing sub-sea permafrost. These processes can destabilize the seafloor, enabling slope failures and thermo-karst as well as influencing biogeochemical processes. Gas hydrates may also be useful energy resources.

A pan-Arctic objective for scientific drilling would be to sample multiple locations that represent different aspects of gas hydrate (GH) evolution and its relationship to climate and geologic history of the Arctic. These could include the MacKenzie shelf (most mature, representative of a deltaic end-member; Fig. 1), the Russian shelf (Laptev Sea, excellent location, wide shelf, but not as mature; Siberia excellent candidate for GH aspect; Fig 1), deep-water locations, particularly Mendeleev Ridge, where pockmarks and other seismic evidence support the presence and venting of subseafloor gases. Integrating the role of GH from these areas will be necessary to understand larger scale phenomena (e.g., carbon cycle).

## Site Surveys

For the precise planning of future drilling campaigns (including site selection, evaluation of proposed drill sites for safety and environmental protection aspects, etc.), compre-

Table 1. Active Arctic-related IODP proposals (as of October, 2009). More details on these proposals including the list of co-proponents and involved institutions can be obtained from the IODP website (http://www.iodp.org/active-proposals).

| Number | Short Title | Contract Proponents | University/ Institute | Country | Platform* | E-mail |
|---|---|---|---|---|---|---|
| 645-Full3 | North Atlantic Gateway | W. Jokat | AWI Bremerhaven | ECORD/ Germany | MSP+NR | Wilfried.Jokat@awi.de |
| 680-Full | Bering Strait Climate Change | S. J. Fowell | University of Alaska Fairbanks | USA | MSP | ffsjf@uaf.edu |
| 708-Pre | Central Arctic Paleoceanography | R. Stein | AWI Bremerhaven | ECORD/ Germany | MSP | Ruediger.Stein@awi.de |
| 746-Pre | Arctic Mesozoic Climate | W. Jokat | AWI Bremerhaven | ECORD/ Germany | MSP | Wilfried.Jokat@awi.de |
| 750-Pre | Beringia Sea Level History | L. Polyak | Ohio State University | USA | MSP+NR | Polyak.1@osu.edu |
| 753-Pre | Beaufort Sea Paleoceanography | M. O'Regan | Stockholm University | ECORD/ Sweden | NR | Matt.oregan@geo.su.se |
| 755-Pre | Arctic Slope Stability | D. Winkelmann | GEOMAR | ECORD/ Germany | | dwinkelmann@ifm-geomar.de |
| 756-Pre | Morris Jesup Rise: Drilling the Arctic Ocean Exit Gateway | M. Jakobsson | Stockholm University | ECORD/ Sweden | | Martin.jakobsson@geo.su.se |

*NR = Non-Riser MSP = Mission Specific Platform

hensive site survey data are needed. The lack of good site survey data and age control for existing seismic reflection records is one of the biggest limitations on the development of Arctic Ocean scientific drilling (see the JEODI Report of Kristoffersen and Mikkelsen, 2004).

For some of the potential study areas, the site survey data base is already quite good. For example, from the Lomonosov Ridge, a large number of deep penetration reflection seismic profiles were acquired on icebreaker-based expeditions in 1991, 1996, 1998, and 2005 (Fütterer, 1992; Kristoffersen et al., 1997; Darby et al., 2005; Jokat, 2005 and further references therein). An intensive PARASOUND survey (in combination with coring) was carried out in 1995 and 1998 (Rachor, 1997; Jokat et al., 1999), and the first high-resolution chirp profiles were collected in 1996 (Jakobsson, 1999). In 1999, the SCICEX program collected high-resolution chirp sub-bottom profiler data, swath bathymetry and side-scan sonar backscatter data on Lomonosov Ridge from an American nuclear submarine (Edwards and Coakley, 2003), contributing significantly to the much improved bathymetric chart of the Arctic Ocean (Jakobsson, 2002; Jakobsson et al., 2008). During the 1995, 1996, and 1998 expeditions, a large number of sediment cores were taken by piston, gravity, and Kastenlot corers in the Lomonosov Ridge area (Backman et al., 1997; Rachor, 1997; Jokat et al., 1999; Stein et al., 2001). That means, in combination with the results from the ACEX drilling campaign (Backman et al., 2006, 2008), future drill areas/sites on Lomonosov Ridge can be identified more accurately. On the other hand, in other key areas for future drilling (e.g., the Alpha-Mendeleev Ridge), site survey expeditions still have to be carried out before a detailed planning and drill site selection can start.

## Outlook

While sampling in the Arctic Ocean is called out as a priority in many of the sections of the IODP Science Plan, these priorities have yet to be realized in a sampling program of commensurate scope and urgency. Concerning the short- and long-term evolution of the Arctic Ocean and its importance for the understanding of the global climate history, most of the key questions mentioned above, as well as the key areas for scientific drilling in the Arctic Ocean, were already identified on several workshops during the last two decades and published in upcoming reports. Several, especially Thiede and the NAD Science Committee (1992), NAD (1997), Hovland (2001), Bowden et al. (2007), and Coakley and Stein (2009), have to be mentioned here. Over the years, however, scientific drilling in the ice-covered Arctic Ocean remained a dream. The ACEX drilling in 2004 (Backman, et al., 2006) was the first major step to transform this dream into reality. Now, further drilling campaigns are needed to follow up in the future. The construction of a new large icebreaker with deep-water drilling capability will certainly be the next milestone in Arctic Ocean research. Such a vessel would guarantee a commitment to Arctic deep drilling, and in combination with a continuous drilling program, could be a potential contribution to the IODP and succeeding programs, as already outlined in the APPG Report (Hovland, 2001). Plans for the development of *Aurora Borealis*, an icebreaker with deep-water drilling capability (Thiede and Egerton, 2004), are pushed forward over the last few years, and it seems possible that it will be completed and available for the international research community within the next decade. Operation of the *Aurora Borealis* would open a new dimension in multidisciplinary Arctic Ocean research.

## References


Backman, J., and Moran, K., 2008. Introduction to special section on Cenozoic Paleoceanography of the Central Arctic Ocean. *Paleoceanogr.*, 23:PA1S01, doi:10.1029/2007PA001516.

Backman, J., Jakobsson, M., Frank, M., Sangiorgi, F., Brinkhuis, H., Stickley, C., O'Regan, M., Løvlie, R., Pälike, H., Spofforth, D., Gattacecca, J., Moran, K., King, J., and Heil, C., 2008. Age model and core-seismic integration for the Cenozoic ACEX sediments from the Lomonosov Ridge. *Paleoceanogr.*, 23:PA1S03, doi:10.1029/2007PA001476.

Backman, J., Jakobsson, M., Knies, J., Knudsen, J., Kristoffersen, Y., Lif, A., Musatov, E., and Stein, R., 1997. Geological coring and high-resolution chirp sonar profiling. *In* Grönlund, E. (Ed.), *Cruise report, Polarforskningssekretaariatets arsbok 1995/96*, Stockholm, 64–66.

Backman, J., Moran, K., Mayer, L.A., McInroy, D.B., and the Expedition 302 Scientists, 2006. *Proc. IODP, 302*, College Station, Texas (Integrated Ocean Drilling Program Management International, Inc.), doi:10.2204/iodp.proc.302.104.2006.

Bowden, S., Corell, R.W., Hassol, S.J., and Symon, C. (Eds.), 2007. *Arctic Research: A global responsibility. ICARP-II* (2nd International Conference on Arctic Research Planning), Copenhagen, Denmark, 10–12 November 2005, Canada (McCallum Printing Group Inc.), 36 pp. and CD-ROM.

Clark, D.L., Byers, C.W., and Pratt, L.M., 1986. Cretaceous black mud from the central Arctic Ocean. *Paleoceanogr.*, 1:265–271, doi:10.1029/PA001i003p00265.

Clark, D.L., Whitman, R.R., Morgan, K.A., and Mackey, S.D., 1980. Stratigraphy and glacial marine sediments of the Amerasian Basin, central Arctic Ocean. *Geol. Soc. Am. Spec. Pap.*, 181, 57 pp.

Coakley, B.J., and Stein, R., 2008. Workshop on "Arctic Ocean History: From Speculation to Reality", Alfred Wegener Institute, Bremerhaven, Germany, 3–5 November 2008. Agenda and Abstract Book, 52 pp., http://www.essac.ecord.org/index.php?mod=workshop&page=workshop.

Coakley, B.J., and Stein, R., 2009. Arctic Ocean History: From Speculation to Reality. Alfred Wegener Institute, Bremerhaven, Germany, 3–5 November 2008. Workshop Report, *ECORD Newsletter*, 12:12 pp.

Darby, D.A., Jakobsson, M., and Polyak, L., 2005. Icebreaker expedition collects key Arctic seafloor and ice data. *EOS, Trans. Am. Geophys. Union*, 86(52):549–552, doi:10.1029/2005EO520001.

Edwards, M.H., and Coakley, B.J., 2003. SCICEX investigations of the Arctic Ocean system. *Chemie der Erde*, 63:281–392,

doi:10.1078/0009-2819-00039.
Fütterer, D.K. (Ed.), 1992. ARCTIC '91: The Expedition ARK-VIII/3 of RV "Polarstern" in 1991. *Rep. Polar Res.*, 107:1–267.
Hovland, M. (Ed.), 2001. *The High-Arctic Drilling Challenge. Final Report of the Arctic's Role in Global Change Program Planning Group (APPG)*, Ocean Drilling Program, 38 pp.
Jakobsson, M., 1999. First high-resolution chirp sonar profiles for the central Arctic Ocean reveal erosion of Lomonosov Ridge sediments. *Mar. Geol.*, 158:111–123, doi:10.1016/S0025-3227(98)00186-8.
Jakobsson, M., 2002. Hypsometry and volume of the Arctic Ocean and its constituent seas. *Geochem., Geophys., Geosyst.*, 3:1–18, doi:10.1029/2001GC000302.
Jakobsson, M., Macnab, R., Mayer, L., Anderson, R., Edwards, M., Hatzky, J., Schenke, H.W., and Johnson, P., 2008. An improved bathymetric portrayal of the Arctic Ocean: Implications for ocean modeling and geological, geophysical and oceanographic analyses. *Geophys. Res. Lett.*, 35:L07602, doi:10.1029/2008GL033520.
Jokat, W., 2005. The sedimentary structure of the Lomonosov Ridge between 88°N and 80°N: Consequences for tectonic and glacial processes. *Geophys. J. Int.*, 163:698–726, doi:10.1111/j.1365-246X.2005.02786.x.
Jokat, W., Stein, R., Rachor, E., and Schewe, I., 1999. Expedition gives fresh view of central Arctic geology. *EOS, Trans. Am. Geophys. Union*, 80(40):465, 472–473.
Kristoffersen, Y., and Mikkelsen, N. (Eds.), 2004. Scientific drilling in the Arctic Ocean and the site survey challenge: Tectonic, paleoceanographic and climatic evolution of the Polar Basin. *Geological Survey of Greenland special publication, JEODI Workshop Report*, Copenhagen, Denmark, 13–14 January 2003, http://www.ecord.org/about/j/wp5-wkshp.pdf.
Kristoffersen, Y., Buravtsev, V., Jokat, W., and Poselov, V., 1997. Seismic reflection surveys during Arctic Ocean-96. *In* Grönlund, E. (Ed.), *Cruise report, Polarforskningssekretaariatets arsbok 1995/96*, Stockholm, 75–77.
Lear, C.H., Elderfield, P.A., and Wilson, P.A., 2000. Cenozoic deep-sea temperatures and global ice volumes from Mg/Ca in benthic foraminiferal calcite. *Science*, 287:269–272, doi:10.1126/science.287.5451.269.
Miller, K.G., Fairbanks, R.G., and Mountain, G.S., 1987. Tertiary oxygen isotope synthesis, sea level history, and continental margin erosion. *Paleoceanogr.*, 2:1–19, doi:10.1029/PA002i001p00001.
Miller, K.G., Wright, J.D., and Fairbanks, R.G., 1991. Unlocking the ice house: Oligocene-Miocene oxygen isotopes, eustasy, and margin erosion. *J. Geophys. Res.*, 96(B4):6829–6848, doi:10.1029/90JB02015.
Moran, K., Backman, J., Brinkhuis, H., Clemens, S.C., Cronin, T., Dickens, G.R., Eynaud, F., Gattacceca, J., Jakobsson, M., Jordan, R.W., Kaminski, M., King, J., Koc, N., Krylov, A., Martinez, N., Matthiessen, J., McInroy, D., Moore, T.C., Onodera, J., O'Regan, A.M., Pälike, H., Rea, B., Rio, D., Sakamoto, T., Smith, D.C., Stein, R., St. John, K., Suto, I., Suzuki, N., Takahashi, K., Watanabe, M., Yamamoto, M., Frank, M., Jokat, W., and Kristoffersen, Y., 2006. The Cenozoic palaeoenvironment of the Arctic Ocean. *Nature*, 441:601–605, doi:10.1038/nature04800.
NAD, 1997. *Nansen Arctic Drilling Implementation Plan*. Report of "Nansen Arctic Drilling Program Implementation Plan Workshop", Arctic and Antarctic Research Institute, St. Petersburg, Russia, 14–15 October 1996, Joint Oceanographic Institutions, Washington, DC, 50 pp.
Pearson, P.N., and Palmer, M.R., 2000. Atmospheric carbon dioxide concentrations over the past 60 million years. *Nature*, 406:695–699, doi:10.1038/35021000.
Rachor, E. (Ed.), 1997. Scientific cruise report of the Arctic Expedition ARK-XI/1 of RV "Polarstern" in 1995. *Rep. Polar Res.*, 226:1–157.
Stein, R., 2008. *Arctic Ocean Sediments: Processes, Proxies, and Paleoenvironment. Developments in Marine Geology*, 2, Amsterdam (Elsevier), 587 pp.
Stein, R., Boucsein, B., Fahl, K., Garcia de Oteyza, T., Knies, J., and Niessen, F., 2001. Accumulation of particulate organic carbon at the Eurasian continental margin during late Quaternary times: Controlling mechanisms and paleoenvironmental significance. *Glob. Plan. Change*, 31(1–4):87–104, doi:10.1016/S0921-8181(01)00114-X.
Thiede, J., and Egerton, P., 2004. *AURORA BOREALIS: A Long-Term European Science Perspective for Deep Arctic Ocean Research 2006-2016*. Strasbourg Cedex, France (European Science Foundation), 80 pp.
Thiede, J., and NAD Science Committee, 1992. The Arctic Ocean record: Key to global change (initial science plan). *Polarforschung*, 61(1):1–102.
Thiede, J., Clark, D.L., and Hermann, Y., 1990. Late Mesozoic and Cenozoic paleoceanography of the northern polar oceans. *In* Grantz, A., Johnson, L., and Sweeney, J.F. (Eds.), *The Geology of North America, Vol. L, The Arctic Ocean Region*, Boulder, Colo. (Geol. Soc. Am.), 427–458.
Thiede, J., Myhre, A.M., Firth, J.V., Johnson, G.L., and Ruddiman, W.F. (Eds.), 1996. *Proc. ODP, Sci. Results, 151*. College Station, Texas (Ocean Drilling Program), 685 pp.
Zachos, J., Pagani, M., Sloan, L., Thomas, E., and Billups, K., 2001. Trends, rhythms, and aberrations in global climate 65 Ma to present. *Science*, 292:868–693, doi:10.1126/science.1059412.

## Authors

**Bernard Coakley**, Geophysical Institute, University of Alaska, Fairbanks, AL, U.S.A.; e-mail: Bernard.Coakley@gi.alaska.edu.
**Ruediger Stein**, Alfred Wegener Institute for Polar and Marine Research, Bremerhaven, Germany; e-mail: Ruediger.Stein@awi.de.

## Related Web Link

http://www.ngdc.noaa.gov/mgg/bathymetry/arctic/arctic.html

# IODP Drilling of the "Shackleton Sites" on the Iberian Margin: A Plio-Pleistocene Marine Reference Section of Millennial-Scale Climate Change

**by Fatima Abrantes, David Hodell, Gabriella Carrara, Luis Batista, and Henrique Duarte**

Few marine sediment cores have played such a pivotal role in paleoclimate research as those recovered from the Portuguese Margin, including MD95-2039 to MD95-2042 (hereafter referred to as the "Shackleton sites") (Fig. 1). These cores preserve a high-fidelity record of millennial-scale climate variability for the last several glacial cycles and can be readily correlated to Greenland ice cores. Moreover, the narrow continental shelf and proximity of the Tagus River results in rapid delivery of terrestrial material to the deep-sea environment, thereby permitting correlation of marine and ice-core records to European terrestrial sequences. Few places exist in the world's ocean where such marine-ice-terrestrial linkages are possible. Consequently, the Iberian Margin cores have become de facto reference sections for the study of abrupt climate change.

The seminal importance of these sections warrants investment by IODP to ensure that the entire Plio-Pleistocene record is continuously cored and properly recovered with multiple holes at several locations. To this end, we held a drilling workshop ("IODP Drilling of the 'Shackleton sites' on the Iberian Margin: In Search of a Plio-Pleistocene Marine Reference Section") in Lisbon on 9–10 November 2009, sponsored by the European Science Foundation Workshops on Marine Research Drilling - MAGELLAN WORKSHOP SERIES, to formulate a drilling proposal to extend these remarkable sediment archives by drilling with the *R/V JOIDES Resolution* (*JR*). The objectives of the workshop were to assemble researchers who have worked on the Portuguese Margin, especially on the Shackleton sites, to provide a forum for invitees to summarize past work, to present new ideas and data, to identify unanswered key questions, and to discuss the best drilling strategy for reaching the overall goals.

We originally envisioned that the workshop would focus solely on paleoceanography, but during its organization Portuguese and Italian colleagues of the tectonics community approached us. Their interest is to establish a borehole observatory (i.e., CORKed and instrumented hole) in the region of the fault, which is believed to have caused the devastating 1755 Lisbon earthquake and tsunami. Given the societal relevance and importance of these geohazards, we invited seven scientists with tectonic interests to attend the workshop and provide input for potential collaborative drilling to achieve both paleoceanographic and tectonic objectives.

Key paleoceanographic questions that emerged from the workshop including the following. (1) Was the bipolar seesaw a persistent feature of Earth's climate system, or did interhemispheric coupling differ for previous glacial periods? (2) How did millennial-scale variability "evolve" as glacial and orbital boundary conditions changed during the Pleistocene? (3) Did millennial-scale climate variability change in frequency or amplitude across the mid-Pleistocene transition (~920 ka and 640 ka), when the average climate state evolved toward generally colder conditions with larger ice sheets, and the spectral character of climate variability shifted from dominantly 41 kyr to 100 kyr? (4) How do millennial and orbital bands of climate variability interact? (5) What role do millennial-scale events play in triggering glacial terminations? Only by integrating marine and ice-core stratigraphies can we begin to address mechanisms of the coupled ocean-atmosphere system, including the causes of Dansgaard-Oeschger variability, glacial-interglacial cycles, and atmospheric $CO_2$ variation.

In an effort to address the paleoceanographic questions in a meaningful way, workshop participants discussed the suggestion by Alley (2003) that "paleoceanographers should consider following the ice-core community's lead and organize a research effort to generate a few internationally coordinated, multiply replicated, multiparameter, high time resolution-type sections of oceanic change." The group

Table 1. Location of proposed drilling site locations.

| Site | Lat (N) | Long (W) | Depth (m) | Distance to land (km) | Objectives |
|---|---|---|---|---|---|
| 1A | 37°52,1 | 10°12,6 | 2978 | 123.4 | Plio- Quaternary Sequence with ±700 m |
| 2A | 37°48,45 | 9°50,35 | 2266 | 91 | Plio-Quaternary Sequence with 750–1000 m |
| 3A | 37°45,5 | 10°02,0 | 2619 | 108 | Plio-Quaternary Sequence with 400–700 m |
| 4A | 37°33,1 | 10°08,8 | 2658 | 118 | Plio-Quaternary Sequence with 400–800 m |
| 5A | 37°32,8 | 10°22,2 | 3559 | 140 | Plio-Quaternary Sequence with 400–800 m |

suported such an approach and proposed that IODP drilling of the Portuguese Margin could serve as a "proof of concept" for adopting an integrated strategy for sampling and analyzing IODP cores that emphasizes a truly multi-proxy approach with attention to resolution, replication, and time control. As a way forward we identified five potential drilling sites in a range of water depths that target sections with the greatest possible sedimentation rates for all time periods of interest and offer the opportunity to obtain continuous sequen-ces that are undisturbedby downslope transport (Fig. 1). In anticipation of the high sample demand for these reference sections, we would plan to drill many holes (up to six) per site to produce multiple copies of composite sections for short- and long-term sampling requests.

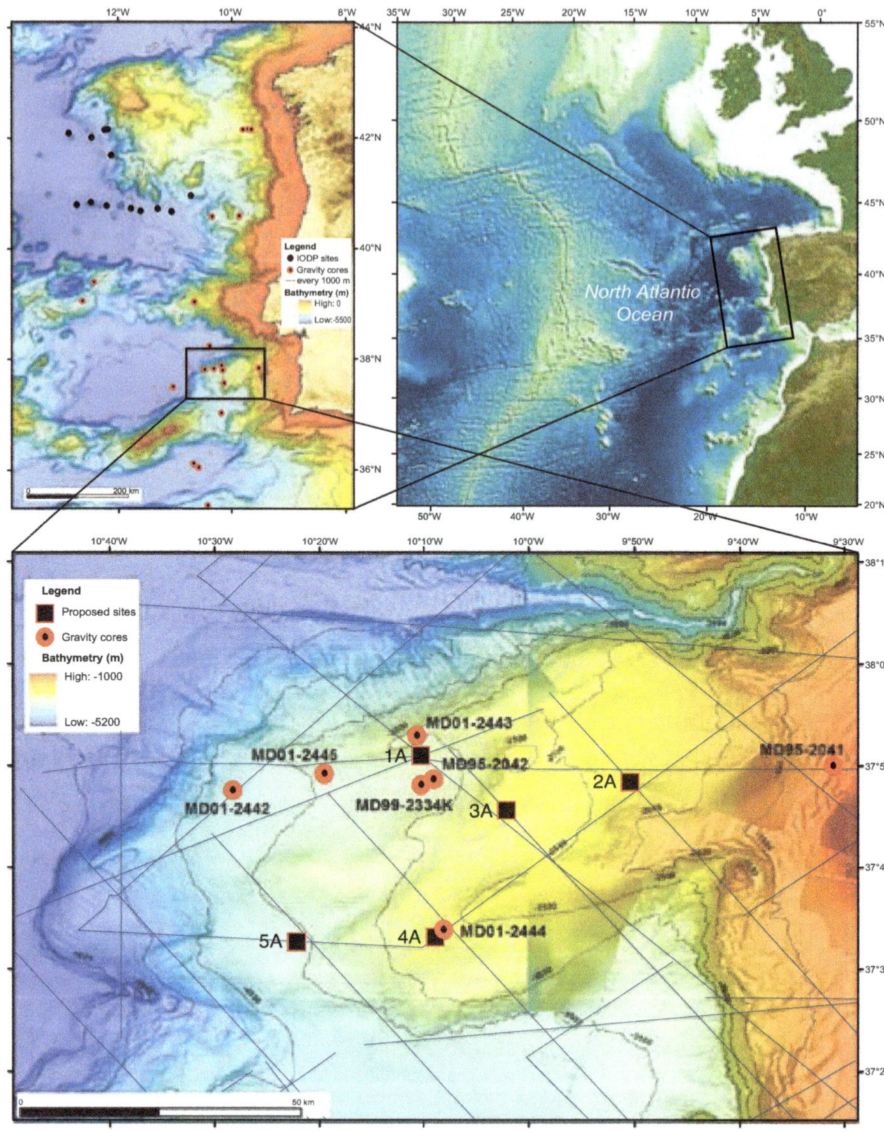


Figure 1. Locations of existing seismic lines, previously collected cores ("Shackleton sites"), and proposed site locations.

In addition to the paleoceanographic questions, a number of tectonics-related questions were generated by the workshop, including: (1) What is the recurrence interval of potential destructive earthquakes and tsunami? (2) What is the rheology of the area? (3) What are the interactions between fluid pressure variation and fault kinematics? Addressing these questions is vital for understanding and mitigating geohazards in this region. As the state of "drilling readiness" for the two sets of objectives is quite different and issues with the borehole observatory site will require additional time to resolve, workshop participants opted to prepare separate but parallel drilling proposals.

The first proposal is to drill the marine-type sections of millennial variability for the last several glacial cycles, known as "Shackleton sites" (to be submitted 1 April 2010). The second is a pre-proposal to address the important geohazards/tectonic concerns of this seismic region (to be submitted in either April or October 2010).

## References


Alley, R.B., 2003. Raising paleoceanography. *Paleoceanogr.*, 18(44):1085, doi 10.1029/2003PA000942.


## Authors


**Fatima Abrantes, Gabriella Carrara, Luis Batista, and Henrique Duarte**, Laboratório Nacional de Energia e Geologia (LNEG), I.P. Estrada da Portela, Bairro do Zambujal – Alfragide Apartado 7586 - 2720-866 Amadora, Portugal, e-mail: fatima.abrantes@ineti.pt.

**David Hodell**, University of Cambridge - Department of Earth Sciences, Cambridge, U.K., e-mail: dhod07@esc.cam.ac.uk.


## Related Web Link

http://www.esf.org > Shackleton > Past Science Meetings

# Workshop on Pliocene Climate

by Johan Etourneau and Nabil Khélifi

The warm Pliocene epoch (5–3 million years ago) is often cited as a good analog for the near future climate because of its striking resemblance to the predictions of the "Intergovernmental Panel on Climate Change" for the next decades. Indeed, relative to today, during the Pliocene epoch, surface temperatures were 3–4°C warmer, sea level was about 5–40 meters higher, atmospheric $CO_2$ concentrations were relatively similar or slightly higher (~400 ± 50 ppmv), and ice sheets were restrained to Antarctica. However, since 3.0 Ma ago, the Earth's climate has undergone a major transition from a warm and relatively stable state towards cold conditions marked by amplified glacial/interglacial cycles and widespread ice sheets in the Northern Hemisphere (NHG), and to a lesser extent over Antarctica. The causes and consequences of this global climate transition—driving warm periods to "icehouse" conditions marked by "Quaternary-style" glacial/interglacial cycles—are still uncertain. Yet, they may include the interaction of several mechanisms tied to oceanic and atmospheric circulations, tectonic-, greenhouse gases-, and biological activity, biogeochemical processes, and changes in Earth's orbit.

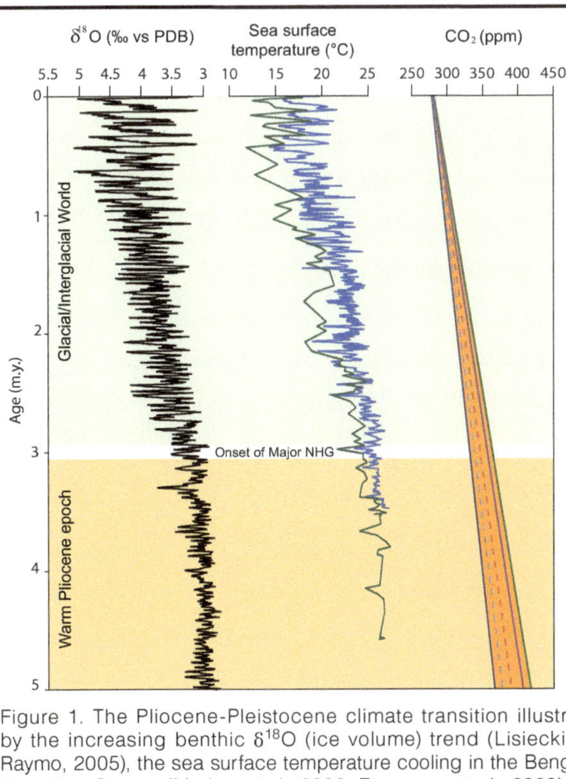


Figure 1. The Pliocene-Pleistocene climate transition illustrated by the increasing benthic $\delta^{18}O$ (ice volume) trend (Lisiecki and Raymo, 2005), the sea surface temperature cooling in the Benguela Upwelling System (Marlow et al., 2000; Etourneau et al., 2009), and the alkenone-based atmospheric $CO_2$ decline (Pagani et al., 2010).

To discuss the different theories explaining this major climate shift, a workshop on Pliocene climate was organized in Bordeaux (France) on 23–25 October 2009. The workshop was sponsored by the European Science Foundation through the "Magellan Workshop Series", the "Université Bordeaux 1", the "Observatoire Aquitain des Sciences de l'Univers", the "Institut des Sciences de l'Univers", and together with considerable financial support from Prof. Gerald Haug, ETH Zürich (Switzerland). In total sixty-three participants from eleven countries participated at the workshop, which was divided into two major parts. Three sessions were dedicated to talks and posters, and three sessions were devoted to group discussions. In addition, Prof. Ralf Tiedemann from Alfred Wegener Institute—Bremerhaven (Germany) and Alan Haywood from the University of Leeds (U.K.) presented keynote lectures on reconstructing and modeling Pliocene climate.

During the three days, a series of thirty-three oral presentations and seventeen posters focused on six major key issues: (1) oceanic gateways, (2) changes in Plio-Pleistocene atmospheric $CO_2$, (3) changes in the North Atlantic Circulation and the impact of Mediterranean outflow water on the formation of North Atlantic deep water, (4) North Pacific and southern oceanic/climatic reorganization, (5) tropical/subtropical Pacific conditions, and (6) modeling the Pliocene climate.

Related to these themes, seven discussion groups were in charge of reviewing the main causes for and consequences of the major Pliocene climate change, and to recommend new directions for future research, including drilling targets to be included in the Integrated Ocean Drilling Program (IODP) future objectives.

More specifically, the discussions addressed the following questions.

- What role could the narrowing of the Indonesian Gateway and the closure of the Panama Seaway have played on the thermohaline circulation and Pliocene climate?
- How varied was the atmospheric $CO_2$ through the Plio-Pleistocene climate transition?
- What may have been the major factor(s) controlling atmospheric $CO_2$ changes, and which impact might

Figure 2. Wind-eroded landforms in the Yadan National Geological Park, Gansu Province, China, resulting from the dryness of a paleolake over the Pliocene-Pleistocene climate transition. Decreasing regional rainfalls are supposed to be closely tied to the Himalayan uplift associated to changes in the atmospheric pattern.

the latter have had on global climate and ice sheet extension?

- Did biogeochemical cycles change in high and low latitudes during the transition from the Pliocene to Pleistocene time, and to what extent might they have impacted on greenhouse gas levels?
- How did the reorganization of the atmospheric circulation affect monsoon system variability and tropical/subtropical/polar precipitation rates over the last 4.0 Ma?
- How can we improve on the currently used models for reconstructing Pliocene climate?

The discussions gave rise to a unanimous consensus: the data spanning the Pliocene are missing in several key sectors of the world ocean, and such data are absolutely needed to further investigate what caused the climate to dramatically cool and generate widespread glaciations in the Northern Hemisphere over the Plio-Pleistocene transition.

A group was specifically tasked with collecting propositions from all participants to define near-future drilling targets. It is clear that the lack of several proxy records from key locations (e. g., bottom and sea surface temperatures, water mass origin, productivity, ocean mixing, $CO_2$ levels, and most importantly a reliable sea-level record covering the last 5.0 Ma) prevents the scientific community involved in Pliocene studies to further explore the causes and consequences of this climate shift. For instance, only one IODP site 806 in the western equatorial Pacific is available to constrain Plio-Pleistocene changes in the entire warm pool conditions. Also, well dated, located, and preserved ocean records for reconstructing African and South Asian climates are too scarce. Oceanic sedimentary archives from East Africa are particularly crucial for a better understanding of the links between hominid evolution and the African climate during the Pliocene.

Another challenge is the poor knowledge of the extent of ice sheets in the northern high latitudes during the Pliocene. Future IODP drilling campaigns need to be planned in the Arctic Ocean. Given the importance of oceanic gateways in controlling the global thermohaline circulation system, it was also concluded that drilling north of the Bering Strait and offshore Indonesia and Panama is also needed. In addition, to better understand the role of the northern and southern polar oceans in the Pliocene cooling, future IODP drilling should be expanded into the Pacific and Indian parts of the Southern Ocean as well, particularly around and within the polar front system.

After three days of discussions, several participants suggested that workshops on Pliocene climate become a regular event (every three years?) addressing, among other issues, the interactions between data and modeling. The workshop also strongly encouraged participants to submit proposals for future IODP expeditions targeting critical locations to understand Pliocene climate changes.

Further information about the program, talks and posters can be obtained from the workshop homepage: http://www.plioclimworkshop.com.

## References

Etourneau, J., Martinez, P., Blanz, T., and Schneider, R., 2009. Pliocene–Pleistocene variability of upwelling activity, productivity, and nutrient cycling in the Benguela region. *Geology*, 37:871–874.

Lisiecki, L.E., and Raymo, M.E., 2005. A Pliocene-Pleistocene stack of 57 globally distributed benthic $\delta^{18}O$ records. *Paleoceanogr.*, 20:PA1003, doi:10.1029/2004PA001071.

Marlow, J.R., Lange, C.B., Wefer, G., and Rosell-Melé, A., 2000. Upwelling intensification as part of the Pliocene-Pleistocene climate transition. *Science*, 290:2288–2291.

Pagani, M., Liu, Z., LaRiviere, J., and Ravelo, A.C., 2010. High Earth-system climate sensitivity determined from Pliocene carbon dioxide concentrations. *Nature Geosci.*, 3:27–30, doi:10.1038/NGEO724.

## Authors

**Johan Etourneau**, Christian-Albrechts-Universität zu Kiel/ Université Bordeaux 1, 24118 Kiel/ 33405 Talence, Germany/ France, e-mail: je@gpi.uni-kiel.de.

**Nabil Khélifi**, Christian-Albrechts-Universität zu Kiel, 24118 Kiel, Germany, e-mail: nk@gpi.uni-kiel.de.

## Photo Credit

Fig. 2: J. Etourneau

# 12

# IODP New Ventures in Exploring Scientific Targets (INVEST): Defining the New Goals of an International Drilling Program

**Wolfgang Bach, Christina Ravelo, Jan Behrmann, Gilbert Camoin, Robert Duncan, Katrina Edwards, Sean Gulick, Fumio Inagaki, Heiko Pälike, and Ryuji Tada**

## Introduction

The INVEST conference, an international meeting to define the scientific goals and required technology for a new ocean drilling program, was held at the University of Bremen on 22–25 September 2009. Based on the large attendance and vigorous engagement of scientists in the discussion of new science/technology ideas, INVEST was extremely successful. Initially 400 participants were expected, but the INVEST steering and organization committees were thrilled to see a much larger number of scientists flock to Bremen to demonstrate their support and enthusiasm for the continuation of an international scientific ocean drilling program. In all, 584 participants, including sixty-four students, from twenty-one nations and >200 institutions and agencies attended the INVEST conference. Contributions to INVEST included 103 submitted white papers that were posted on the INVEST webpage (http://www.marum.de/iodp-invest.html), and breakout discussions in fifty working groups that focused on a range of topics during the course of the conference. In addition, students and early career scientists, as well as national funding agency managers and platform providers, presented a total of eighty-six posters. Interspersed with the working group and plenary sessions were twelve keynote lectures, chosen to highlight overarching themes and new directions in research and technology.

The conference was sponsored by IODP-MI, the Deutsche Forschungsgemeinschaft (DFG) and the MARUM research center. Using input from national workshops that took place in the year prior to the INVEST conference and that provided initial ideas for scientific directions and themes, the INVEST working group sessions were organized within six conference themes: 1) Co-evolution of Life and Planet, 2) Earth's Interior, Crust and Surface Interactions, 3) Climate Change – Records of the Past, Lessons for the Future, 4) Earth System Dynamics, Reservoirs and Fluxes, 5) Earth-Human-Earth Interactions, and 6) Science Implementation.

Each meeting attendee was given the opportunity to participate in three working groups: one within conference themes one through three, one within conference themes four and five, and one within conference theme six. Up to eighteen working groups met in parallel sessions. All working groups within one conference theme met to report to each other the results of the individual working group discussions. The conference theme co-chairs then met with the working group chairs and scribes to prepare a plenary presentation of the conference theme.

The steering committee, with the help of some conference session chairs, used the working group notes and plenary session presentation materials to write the INVEST conference report. The main chapters of the report are "Climate Change Impacts", "The Lithospheric Membrane: The Key Interface and Processing Zone", "Co-evolution of Life and the Planet", and "Earth-Human-Earth Interactions". In addition, several "Cross-disciplinary Research Frontiers" were identified as being important new ocean drilling themes and were highlighted in a separate section of the report. Implementation and outreach aspects are summarized in the chapters "Technology Needs and Developments" and "Outreach, Education, and Branding". "Recommendations for the New Ocean Drilling Program" is the final chapter of the INVEST report and pertains to what the community considers desirable program architecture and science advisory structure. A brief synopsis of each of these chapters is presented below. More details on the INVEST meeting structure, white papers, background material, and the complete report can be found at the INVEST website.

## Climate Change Impacts

Earth's climate results from complex interactions among Earth system components, including the atmosphere, hydrosphere, cryosphere, lithosphere, and biosphere. Variations in these systems over short and long timescales, including the way material cycles among them (e.g., carbon cycle, hydrologic cycle), result in changes to Earth's climate. Understanding how climate has changed in the past and how it will change in the future requires understanding of how the Earth system behaves over a range of conditions. Instrumental records, collected over the last century or so, are insufficient by themselves to study long-term climate change. On the other hand, ocean drilling of sediments and corals provides a unique opportunity to acquire the high-resolution records of past change needed to examine and understand the baseline of natural climate variability against which current climate change can be compared. These records will also allow study of the behavior of the Earth system during different climate states and major climate transitions throughout the Cenozoic, as well as the origin of and mechanisms that lead to abrupt, seasonal- to millennial-scale climate changes.

Fundamental climate change questions that can be addressed by ocean drilling are outlined throughout the INVEST report. This section on "Climate Change Impacts" focuses on scientific questions that, when answered, will underpin our understanding of climate change. It explores the behavior of the climate during abrupt and extreme events, the stability of ice sheets and their relationship to sea level and coastal/shoreline processes, and the impact of regional and global climate on the hydrologic cycle. In the "Earth-Human-Earth Interactions" section, climate change questions are framed in anticipation of future conditions associated with increasing levels of atmospheric carbon dioxide. Climate change questions related to tectonics and hominin evolution are explored in the "Cross-disciplinary Research Frontiers" section.

Ocean drilling can provide unprecedented insight into climatic and oceanic processes through investigation of past rapid and extreme climate events (Fig. 1). These globally significant events represent major deviations from the natural variability of Earth's climate. Extreme events originate from perturbations in a specific component of the Earth system and typically propagate through complicated feedbacks across some or all components, allowing interconnected processes to be examined and understood. Extreme and rapid events stand out in noisy sediment-derived proxy records and are thus easy to quantify. This allows investigation of dynamic system behavior, such as tipping points and thresholds. Ocean drilling remains the only means to reconstruct, at medium to high temporal resolution, the global climate evolution of the Earth system throughout the Cenozoic. Large amplitude perturbations often mark epoch boundaries and provide one of the main avenues to investigate large-scale events, including those that involve extreme climates, ecosystem turnover, and biodiversity evolution. Drilling is also the only means to study how rapid climate phenomena that occur on seasonal to millennial timescales, such as the El Niño Southern Oscillation or Dansgaard-Oeschger cycles, behave under different boundary conditions.

Only ocean drilling can provide the geologic perspective necessary to understand ice-sheet dynamics and the resulting impacts to sea level and shorelines. Over the past 100 m.y., sea-level change reflects global climate evolution from a time characterized by ephemeral Antarctic ice sheets (100 Ma to 33 Ma), to a time when large ice sheets occurred primarily in Antarctica (33 Ma to 2.5 Ma), and finally to a world with large Antarctic and Northern Hemisphere ice sheets (2.5 Ma to present). Over the last ~800 kyr, the cyclic growth and decay of continental ice sheets induced rapid sea-level change with maximum amplitudes of 120–140 m at intervals of ~100 kyr. Ice volume fluctuations can be inferred from eustatic

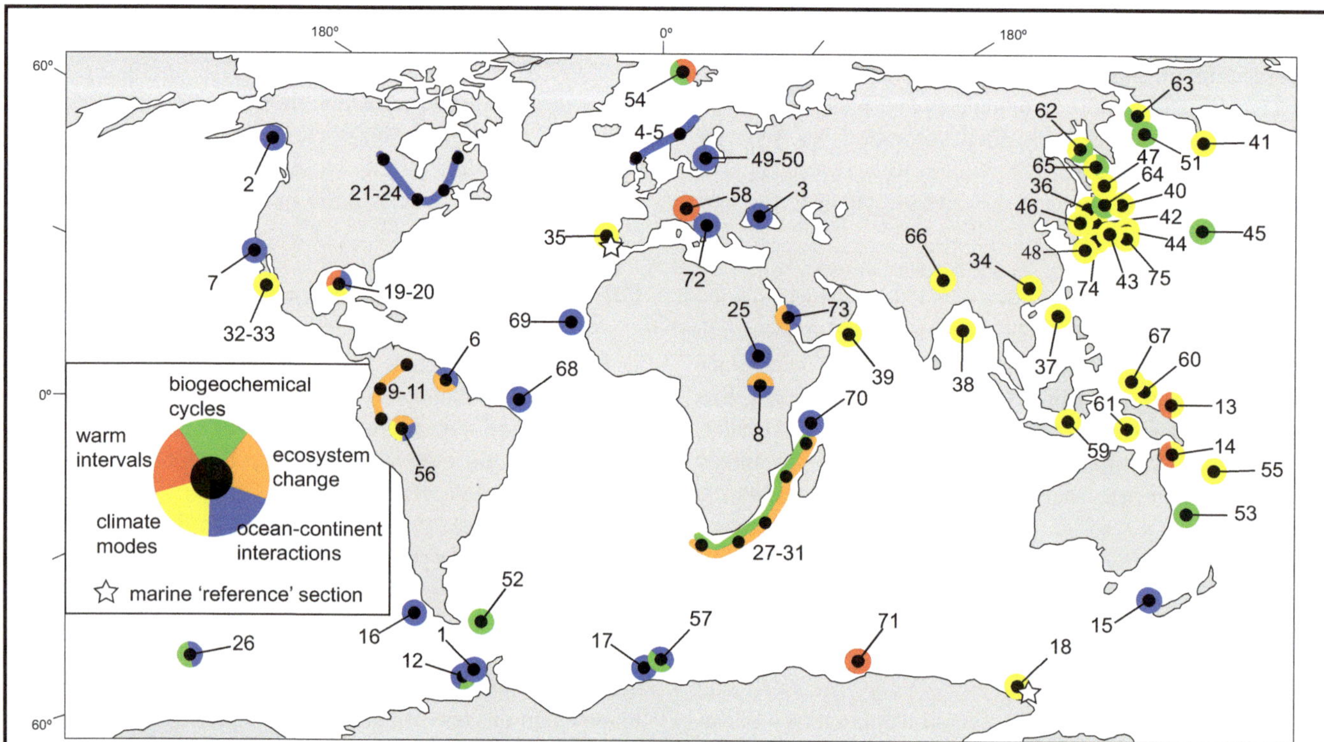


Figure 1. Compilation of potential drilling sites proposed by workshop participants to address the scientific themes 'Warm intervals through time', 'Ocean-continent interactions', 'Biogeochemical cycles', 'Ecosystem change', and 'Climate modes'; from Thurow et al., (2009). Site numbers are linked to information that can be found in Appendix 1 at http://high-resolution.icdp-online.org.

sea-level curves deciphered through comparison of regional sea-level records from various latitudes and tectonic settings. For example, observations from sites distal to glaciated regions (i.e., the "far field") are less affected by isostatic deformation and therefore better suited to constrain glacial eustasy, whereas sea-level data from sites proximal to former ice sheets (i.e., the "near field") provide information on local ice-sheet dynamics.

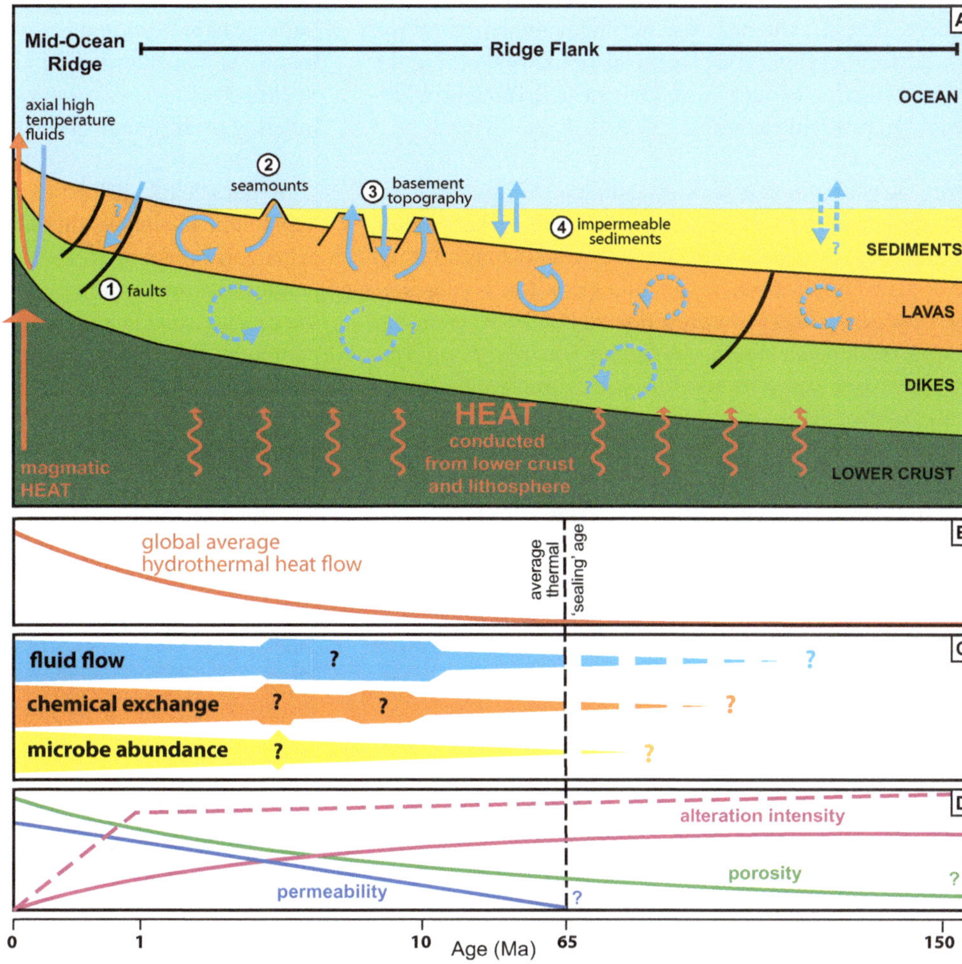


Figure 2. Diagram showing the predicted changes in crustal properties during aging (from Teagle et al., 2009). [A] Parameters that may influence the intensity and style of hydrothermal circulation through the ridge flanks, such as faults, seamounts, basement topography and impermeable sediments. Arrows indicate heat (red) and fluid (blue) flow. [B] The difference between the average measured conductive heat flow and that predicted from conductive cooling plate models. The calculated global hydrothermal heat flow decreases to zero, on average, by 65 Ma. At this age the crust is typically assumed to be 'sealed' to hydrothermal circulation. [C] How parameters such as basement topography and sediment thickness affect fluid-flow, chemical exchange, and microbe abundance in the crust during the aging process remain undetermined. The hydrological, physical, chemical, and biological evolution of hydrothermal circulation through the ridge flank could be investigated across a ridge flank. [D] The controls on the intensity and style of hydrothermal circulation could be investigated by the measurement of multiple parameters such as porosity, permeability, and alteration mineralogy.

Understanding the processes that control changes in the hydrologic cycle is one of the most pressing issues in climate change research because changes in precipitation and evaporation impact the salinity/density distribution in the surface ocean, ecosystems on land, floods, aridification, water resources, and climate-vegetation feedbacks. A comprehensive examination of climatic controls on the hydrologic cycle requires the study of long-term trends and variability over a range of climatic states. In fact, even small global temperature changes impact the energy balance in tropical regions, resulting in changes in evaporation and storm and hurricane activity. In the mid-latitudes, changes in atmospheric thermal gradients can impact atmospheric circulation, the location of storm tracks, and the intensity of storms. There is an urgent need to understand controls on the Intertropical Convergence Zone, climate oscillation modes and their behavior during different mean climate states, wind-driven upper ocean circulation and its coupling to atmospheric forcing, monsoon dynamics, and the relationship between precipitation patterns, density stratification, and biogeochemical processes, to name a few. Future ocean drilling is absolutely necessary to obtain the records needed to constrain past changes in surface ocean conditions associated with major changes in large-scale atmospheric circulation. This must be complemented by continental climate reconstructions obtained from drilling sediments on continental margins and on land. Finally, ocean drilling studies focused on the hydrologic cycle should be used to validate regional climate models used to predict climate change and its associated impact on water resources.

## The Lithospheric Membrane: The Key Interface and Process Zone

Our tectonically active planet evolves by chemical and physical changes from the core and mantle, through the lithosphere to the ocean and atmosphere. Flow of material and energy among these reservoirs drives both gradual changes in the Earth's structure and composition, as well as extremely rapid volcanic events that impact Earth's environment. The lithosphere is the major interface between the solid interior of the Earth and the exterior hydrosphere/atmosphere/biosphere (Fig. 2). Quantifying the fluxes mediated by the lithospheric processing zone is crucial for understanding the state and evolution of our planet, and ocean drilling provides a key tool in this endeavor.

From direct sampling we can examine the fundamental steady-state processes of ocean crust accretion, plate aging, and recycling, as well as intermittent events such as eruptions of large igneous provinces and rifting of continents. These processes affect the presence and state of resources that society depends upon—including sources of energy, fresh water, and nutrients—but they also create deleterious effects on society through earthquakes, tsunamis, and volcanic eruptions.

It is timely to stress that many of the things to be learned about the "lithospheric membrane" of the Earth are of great societal relevance. Understanding initiation of plate boundaries and evolution of plates is fundamental to learning how the Earth works and how the continents we live upon were formed and evolve. Plate boundary formation and related volcanism have impacted past climate and may yield information about agents of climate change. Rifted continental margins host major hydrocarbon reservoirs that are an essential energy resource. Subduction zone hydration-dehydration cycling affects earthquakes and volcanic hazards. Hydrothermal systems form polymetallic sulfide deposits of potential economic interest and also host microorganisms that may be useful for bioengineering. Finally, the ocean crust provides vast opportunities for $CO_2$ capture and storage.

Understanding hydrothermal transfers of heat and mass between the lithosphere and hydrosphere/atmosphere on a global scale has been identified as a high-priority overarching research question. Seawater circulation facilitates microbial growth within the ocean floor and is critical to the transport and distribution of microorganisms; however, the size, activity, and connectivity of the intracrustal oceanic biosphere and its influence on global geochemical and biogeochemical cycles are unknown. Investigation of the coupling between hydrogeological, geochemical, thermal, mechanical, and biological processes and their relationship to the architecture and physical nature of oceanic lithosphere is essential. Of primary importance is addressing the nature of the Mohorovičić discontinuity (Moho), a first-order geophysical interface within our planet, which is uncertain in slow-spreading mid-ocean ridge environments. It could be an igneous boundary or a serpentinization front—a difference with profound consequences for the chemical and rheological properties of the lithosphere. As much as one-quarter of the seafloor exposed at slow-spreading mid-ocean ridges is a heterogeneous assemblage of peridotite and gabbroic lithologies, which is hydrothermally more reactive and undergoes greater changes in physical properties than layered basaltic crust. To understand ocean-crust composition, structure, and evolution, it is essential to drill a complete crustal section across the Moho and into the shallow mantle at a fast-spreading ridge. Complete sampling is also necessary to provide *in situ* confirmation of geophysical imaging of the ocean crust.

When oceanic lithosphere is recycled during subduction and plate collision, sediments may be scraped off, and ocean crust can be accreted to a continent or island arc. Deep in subduction zones, dehydration reactions and melting of subducting sediments and occasionally the upper crust produce continental or oceanic volcanic arcs. Both of these processes can effectively transfer crustal material, fluids, and volatiles from the geologically transient oceanic lithosphere to more permanent continental crust. Ocean drilling is a critical tool necessary to investigate the relative roles of these processes and the magnitudes of element fluxes. Of crucial importance in determination of input fluxes are reliable estimates of the chemical and mineralogical state of the subducting ocean crust. Additionally, subduction zones produce some of the most significant geologic hazards to society including the highest magnitude earthquakes, the greatest tsunamis, and the most explosive eruptions.

Formation, hydrothermal alteration, and subduction of ocean crust generate secular chemical and physical changes throughout the crust and mantle, as well as in the ocean-atmosphere system. Large igneous provinces and hot spot trails are profound examples of massive mantle-crust exchanges and how these vary in time and space. At intermittent intervals, the emplacement of oceanic plateaus and formation of volcanic rifted margins produce significant chemical effects in the ocean and atmosphere. Particularly intense periods of volcanic activity appear to correlate with long-term changes in geodynamic behavior, which could involve interactions between the core and the deep mantle.

Many questions still need to be answered for us to better understand the Earth system. To address these we need to quantify fluxes and interplay within the mantle-crust-ocean system (physical, chemical, and biological) and monitor fault-driven processes. We also need to obtain a better understanding of the components of a subduction zone system that control seismic behavior, which in turn affects the level of hazard posed by the plate boundary. Ocean drilling will continue to significantly contribute to our understanding of these processes.

## Co-Evolution of Life and the Planet

Scientific ocean drilling is poised to offer transformative advances to disciplines within the life sciences and provide insight into how life operates and interacts with Earth processes at and below the seafloor, both today and in the past. In particular, the study of paleoenvironmental controls on marine paleoecosystems through the Cenozoic and of extant life in rock and sediments can best be studied through ocean drilling.

Exploration of deeply seated microbial populations has revealed active, extant populations of microbes down to 1.6 km depth. These sedimentary and crustal habitats are vast in scale, physically and chemically diverse, and together represent the largest biome on Earth. A major challenge in the study of subseafloor life is the need to develop a more comprehensive understanding of the microbes that live there. Future research should reveal the biodiversity of the deep biosphere (Fig. 3), as well as the degree to which it is connected to surface biomes. Metagenomics, observatories, and other emerging technologies will play a significant role in the advancement of our understanding of the function and form of extant life in the subsurface. Advances in our knowledge of subsurface life from past ocean drilling have principally arisen as opportunistic ventures, where a microbiologist would sail on an expedition focused on other scientific objectives. This has resulted in a significantly skewed data set. This view can be changed in the future by utilizing both opportunistic and targeted drilling expeditions for deep biosphere research. A legacy sampling program would also fill critical gaps and lead to a more comprehensive understanding of the largest biome on Earth.

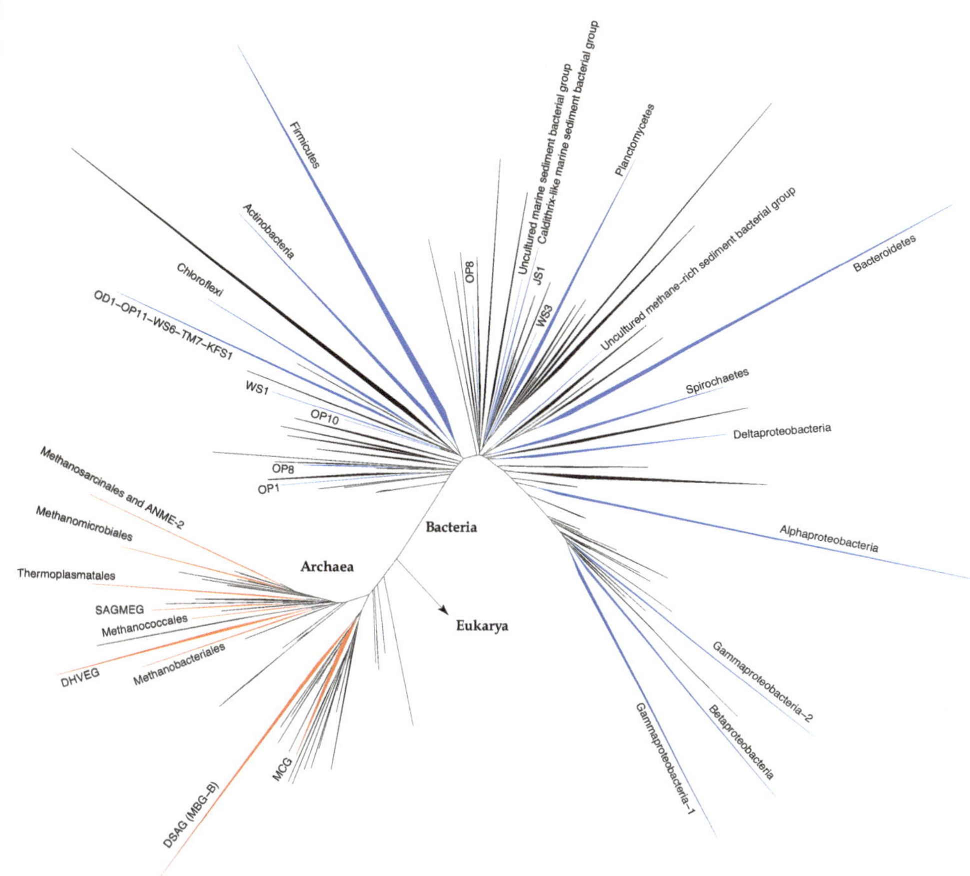


Figure 3. Phylogenetic tree of Archaea and Bacteria based on 16S rRNA gene sequences (from Inagaki, 2010). The red and blue branches represent archaeal and bacterial phyla, respectively, and include sequences frequently detected from the deep subseafloor biosphere.

New approaches and dedicated expeditions will also reveal the physiological function and activity of the deep biosphere and provide a quantitative first-order understanding of the rates and magnitude of the biogeochemical processes that occur there. How these processes impact elemental cycles of carbon, nitrogen, iron, sulfur, etc. are generally not known. This presents an enormous challenge to understanding, at a very basic level, the global implications of the presence of deep ecosystems. Does this biosphere impact Earth processes such as energy and carbon cycling, and can a better understanding of its function help solve global redox and carbon budgets? To meet these challenges new drilling and sampling are required to obtain high-quality fresh samples critical for microbiological and biogeochemical study. Research on the deep subseafloor biosphere has the potential to address urgent questions about energy creation, climate change, and the nature of evolution of life on Earth over geologic time.

Important emerging fields within paleontology are the study of systems ecology and biotic response to major environmental forcing (such as climate change). Drilling is needed to better understand how ecosystems influence and respond to their environment on different timescales in order to predict how populations will respond to present and future global change. High temporal resolution records across paleogeographic gradients and through time can thus be used to study the processes that generate and destroy biodiversity.

## Human-Earth-Human Interactions

### *Geohazards*

A top priority for the new drilling program should be the study of geologic hazards. Geology touches society directly through earthquakes, tsunamis, and landslides, and drilling can provide key constraints. Ocean drilling can be combined with other onshore and offshore national and international geohazards programs to provide knowledge for hazard assessment and mitigation of earthquakes and tsunamis that have the potential to directly impact the majority of the world's population, as well as for submarine slides that are of great concern to subsea and coastal infrastructure.

Research over the past decade indicates that old models of only two types of failure, stick-slip and creep, are incorrect. Slow-slip events, very low frequency earthquakes, and tsunami earthquakes are part of the stress-strain cycle (Fig. 4), and with the recent Haiti earthquake (Mw 7.0) there is recognition that even modest events can potentially cause a devastating loss of life. Specific queries for a drilling-based study of earthquake geohazards include the following: What is the nature of large-slip zones in earthquakes? What controls the size of earthquakes? Are very large destructive earthquakes governed by the same processes as small earthquakes? What are the characteristics of the earthquake repeat cycle? What causes tsunami earthquakes? What controls the range of tsunami efficiency generated by different earthquakes? Are earthquakes on different types of faults (e.g., subduction megathrusts vs. plate-boundary strike-slip faults) fundamentally the same or different? To answer these questions requires drilling using a range of platforms, sampling, and logging strategies, integration with observatories and modeling, and collaboration with onshore studies.

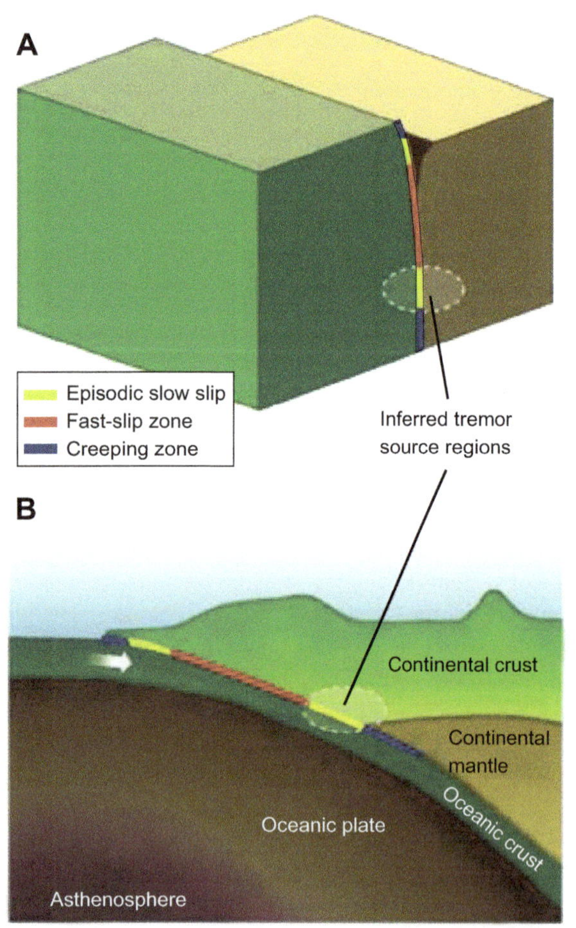


Figure 4. Locations of brittle fast-slip (seismogenic) and slow-slip zones: [A] Vertical strike slip fault; [B] Subduction zone. The new paradigm in earthquakes includes not only classically understood locked zones that break in fast slip producing earthquakes and adjacent creeping zones that nearly continuously deform without generating earthquakes, but also faults with episodic slow-slip zones that produce tremors. Additionally some subduction zones generate tsunami earthquakes that produce inordinately large tsunamis compared to the size of the earthquake (from Kanamori, 2008).

Submarine landslides occur at a wide range of scales and settings. They often comprise distinctive mass-transport deposits recognized on the seafloor or in seismic reflection profiles. Small-scale submarine landslides are relatively frequent. They have displaced oil rigs, damaged pipelines, broken deep-sea communication cables, and devastated segments of coastline. Large- and small-slide events along coastal zones also create local, destructive tsunamis. A variety of triggers have been implicated in the initiation of submarine landslides, including earthquakes, sea-level change, and gas hydrates. There are important questions for examining this geohazard. How safe is the ocean floor? What causes and triggers submarine landslides? What are the frequencies and magnitudes? Is there a size-frequency relationship to submarine landslides? What is the relationship between climate change and submarine landslides? What is the tsunamigenic potential of past and future slides? Can and do gas hydrates cause seafloor instabilities? Can submarine slides cause significant hydrate dissociation? To study these processes requires drilling into past slides and into slide-prone areas to examine rheologic, hydrologic, lithologic, and geotechnical controls on slope stability through sampling, logging, and monitoring.

## *Exploring the Future, Anticipating the Transition to a High $pCO_2$ World*

Increasing atmospheric carbon dioxide content ($pCO_2$) is the main driving force for projected future climate change. One of the primary goals of climate-change research is to quantify the Earth's *equilibrium climate sensitivity*, which is a measure of the climate-system response to sustained radiative forcing caused by changes in atmospheric greenhouse gas contents. Although it is defined as the equilibrium global average surface warming following a doubling of $CO_2$ and greenhouse gas equivalent concentrations, a broader definition would include the sensitivity of the entire Earth system to feedback processes that operate over a wide range of timescales. Phenomena that affect these processes include the carbon cycle, cloud cover, albedo, glacial processes, deep-ocean circulation, weathering, and acidification. Climate sensitivity may be non-linear (feedbacks affected by feedbacks), may differ regionally (such as at high latitudes where sea-ice albedo and other feedbacks may amplify climate change), and may affect the characteristics of climate variability (e.g., the response of climate to perturbations or external forcing in a warm-climate compared to a cold-climate state).

Ocean drilling can deliver unique data necessary to quantify climate sensitivity in the past (Fig. 5) and contribute to understanding the feedback processes that need to be included for successful modeling of this topic of great societal relevance. Ocean sediments contain records of past temperatures, ocean chemistry, and $pCO_2$ from a wide range of boundary conditions and time scales inaccessible by modern, historical, or ice-core records. In particular, continuous high-fidelity records from times with $pCO_2$ levels higher than today are only obtainable through drilling ocean sediments deposited during warm intervals of the past. As such, ocean drilling provides the means to answer some of the key questions related to the quantification of climate sensitivity.

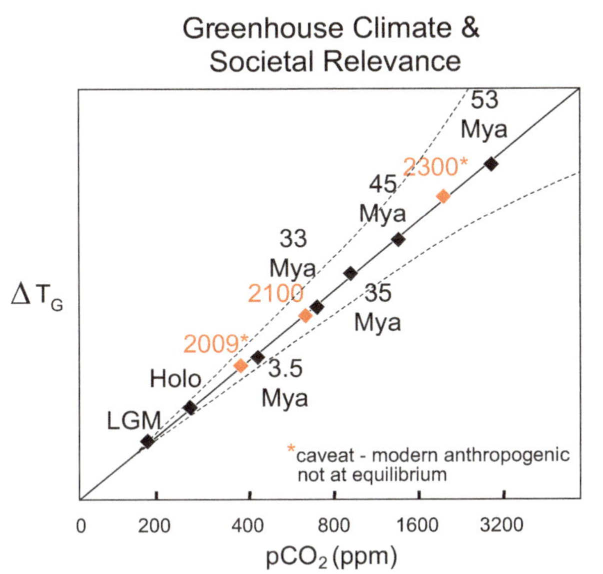


Figure 5. Schematic diagram showing the potential of the geological record to constrain climate sensitivity estimates across a range of $pCO_2$ values. Ocean drilling will provide observations of global temperature change $\Delta T_G$ as a function of atmospheric $pCO_2$ for past climates that can be compared to future predictions (Jim Zachos, personal communication during INVEST meeting; red numbers indicate calendar years now and in the future, whereas the black indicates ages in the past). For progressively higher $pCO_2$, possible geological analogues are found further and further back in time.

Increasing atmospheric $CO_2$ levels will not only cause climate to change, but will also lead to ocean acidification, known as "the other $CO_2$ problem". Acidification will affect bio-calcification, oceanic ecosystems, and carbon-cycle feedbacks. But, in the face of increasing $pCO_2$, carbon-cycle feedbacks are as uncertain as the climatic response. Ocean drilling has already played a major role in framing the current concern about ocean acidification. Although ice-core records indicate that atmospheric $pCO_2$ changes have only varied within a very narrow range during the past ~800,000 years, archives from times of significantly higher $pCO_2$ can only be obtained through ocean drilling. These records provide the means to answer some key questions about the acidification process and the timescales over which the ocean is buffered. Ocean drilling will deliver key records of lysocline shoaling, atmospheric $pCO_2$, carbonate ion concentrations, and biotic responses to changes in ocean acidity in order to quantify the feedbacks and enhance predictions of ocean acidification and its impacts. As such, ocean drilling can make a substantial and unique contribution to this field of immediate societal relevance.

Another urgent avenue of climate-change research is the study of factors that impact rates and amplitudes of sea-level change. The response of ice sheets (and thus sea level) to climate change is difficult to constrain because of the complexity, size, and relatively slow response of ice sheets. Since instrumental records of sea level extend back only about 150 years, refinement of predictions of sea-level rise clearly rely on past high-resolution records of the rates and magnitude of rapid sea-level change.

During glacial-interglacial transitions known as terminations, ice volume decreased, and temperatures, greenhouse gas concentrations, and sea level (magnitude >100 m) increased abruptly. Those periods are therefore regarded as potential analogues for future rapid sea-level rise and coeval abrupt climate change. The reconstruction of rates and magnitudes of sea-level rise during several terminations may help to model ice-sheet dynamics, clarify the mechanisms and sources of catastrophic ice-sheet collapse, understand suborbital climate variability, and determine the timing and volume of meltwater release under varying thermal regimes during deglaciations. In addition, gaining a much better understanding of the dynamics and sensitivity of Greenland, West Antarctic, and East Antarctic ice sheets to climate change during past warm periods is only attainable through ocean drilling, and is critically important to validating and improving ice-sheet and climate models used to predict future sea-level changes.

## Cross-Disciplinary Research Frontiers

### *Extreme Events*

Earth history, on the human and the geologic timescales, is punctuated by extreme events. Increasingly, we have discovered that intermittent, abrupt departures from stable, steady-state conditions can have major impacts on the Earth's environment, the evolution of life, and global biogeochemical cycles. The study of extreme events links many high priority science goals, crosses several broad themes, and will answer fundamental questions about evolutionary processes, and thresholds in the Earth system that, if passed, lead to dramatic ecosystem responses. For example, the processes that dictate changes in biodiversity, the oceanographic and climatic drivers of ecosystem assembly and change, speciation, and extinction can be studied through drilling of extreme events such as mass extinctions and hyperthermals. Bolide impacts and episodes of catastrophic volcanism are examples of abrupt perturbations that have left geologic records whose study will address these issues.

Future drilling can play a vital role in improving our understanding of impacts and their role in Earth's history, particularly through comparison of impacts that caused mass extinction events (e.g., the 65.5 Ma Chicxulub impact) to other large impacts that only caused minor perturbations (e.g., the 35 Ma Chesapeake Bay impact). Identifying the critical factors that lead to global environmental devastation (e.g., the energy of impact, the chemistry of the target rocks, additional environmental stresses, or the vulnerability of life at the time of impact) is possible through ocean drilling of the impact structures and the sediments containing critical fossil and environmental-proxy evidence.

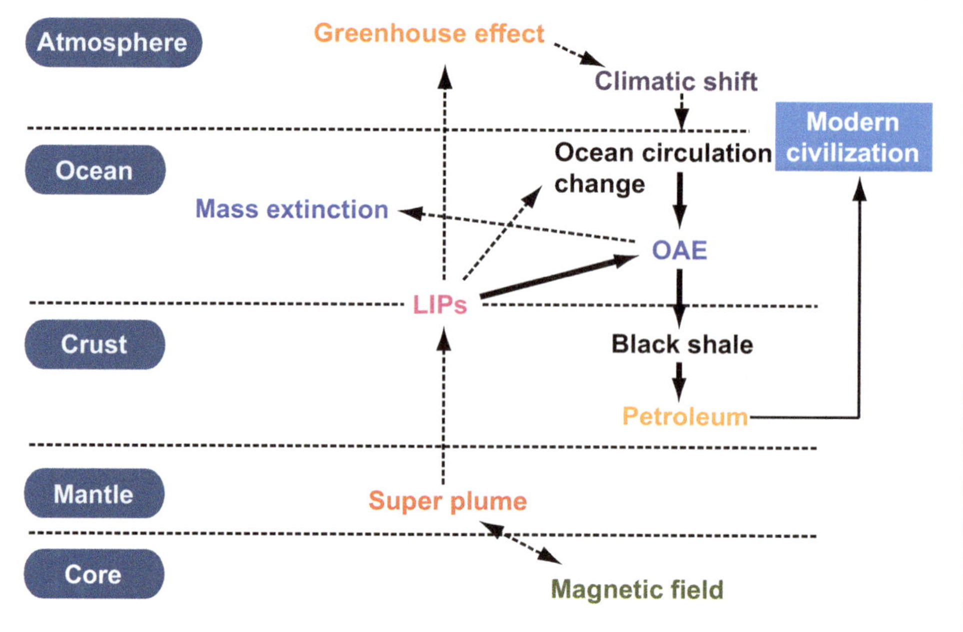


Figure 6. Proposed interconnections between reservoirs and fluxes within the Earth system during extreme volcanic events (Naohiko Ohkouchi, personal communication during INVEST meeting). The emplacement of large igneous provinces (LIPs), an expression of the Earth's interior behavior, could explain episodes of enhanced greenhouse gas input to the atmosphere and subsequent climatic and ocean circulation changes that lead to Oceanic Anoxic Events (OAEs). OAEs are extreme oceanic events marked by enhanced organic carbon burial and related biological impacts.

A fundamental process within the solid Earth is intermittent whole-mantle overturn that results in periods of eruption of large igneous provinces (LIPs) (Fig. 6), high seafloor spreading and associated arc collisions, and sea-level highstands. This mode was last prominent during the Cretaceous to early Tertiary Periods (135–55 Ma) when extraordinary eruption rates produced large igneous systems on both continents and in ocean basins. The rate, volume, and duration of these events need to be constrained to understand the geodynamic mechanisms for their origin and their potential environmental impacts, including mass extinctions, rapid global warming, ocean acidification, and oceanic anoxic events. LIPs are extreme mantle-melting and volcanic events that can be studied to answer questions about geodynamic models whose critical distinctions are magma flux through time, geochemical variability, and internal architecture—all of which are best addressed by drilling to obtain direct volcanic samples and far-field, high-resolution marine sedimentary sections.

## Hominin Evolution

Scientific ocean drilling can transform our understanding of how African climate change affected early human evolution. Key evolutionary events occurred near 3.0–2.5 Ma and 2.0–1.5 Ma that effectively shaped the characteristics that define us as human: bipedalism, exceptionally large brains, and the construction of increasingly sophisticated stone tools. Environmental hypotheses for early human evolution suggest that changing African climate altered the ecological composition of a landscape, which led to specific faunal adaptation or speciation pressures that resulted in genetic selection and innovation. Where hypotheses differ is in the proposed role of climate change in natural selection. The Savannah Hypothesis states that the evolution of African mammalian fauna, including early hominins, was primarily linked to the progressive expansion of more open grassland conditions. The Turnover Pulse Hypothesis is the variant of this idea that focuses on bursts of biotic change initiated by progressive shifts toward greater African aridity at approximately 2.8 Ma and 1.8 Ma. The Variability Selection Hypothesis suggests that changes in the amplitudes of orbitally controlled African climate variability, linked to the eccentricity modulation of precessional monsoonal cycles, may have been an important genetic selection criterion. To test these hypotheses, fundamental questions concerning the timing, nature, and causes of African climate variability must be answered. Drilling targets are margin sediments that reflect the subtropical African geographic domain where hominin fossils are found, including South Africa, Tanzania, Kenya, and Ethiopia. Ocean drilling to obtain records from these regions will likely revolutionize our understanding of the timing and causes of African climate changes and allow us to test the hypothesized role of past climate changes in shaping the course of human evolution.

## Climate-Tectonic Linkages and Feedbacks

Climate and plate tectonics are two forces that shape the Earth in concert; examining the interplay between climatic and tectonic processes will advance our understanding of both. Fundamental questions regarding tectonic-climate linkages can be answered by ocean drilling in conjunction with continental studies. Future studies should focus on how the changing configuration and topography/bathymetry of the continents and oceans influence ocean and atmospheric circulations and biogeochemical cycles, as well as how orogens respond to significant climate shifts such as the onset of Northern Hemisphere glaciation, the Mid-Pleistocene Transition, and the development of the Indian monsoon. Other avenues of research should investigate how tectonics influences discharge of freshwater, nutrients, and sediment from the continents to the ocean and how these affect the biota and biogeochemical cycles on continental margins and in marginal seas. The potential impact of small rivers, groundwater, and eolian transport on global freshwater, nutrient, and sediment budgets also needs to be quantified. Future ocean drilling should focus on capturing a complementary array of continental margin and fan records, including Arctic and Antarctic deposystems; exhumation history; terrestrial climate/vegetation changes; sediment, nutrients, and carbon budgets; and freshwater discharges for the areas of the Earth with the highest fluxes. Such a strategy will undoubtedly advance our understanding of the feedbacks between tectonics and climate and their combined influences on the Earth's surface.

## Technological Needs and Development

To achieve novel scientific ocean drilling objectives will depend on improved drilling capabilities such as enhancing depth penetration, improving core recovery and quality, coring in high-temperature and high-pressure environments, coring in shallow water margins, coral reefs, and sea-ice covered regions (e.g., such as with the planned Research Icebreaker *Aurora Borealis*), and preventing magnetic, chemical, and microbiological contaminations. Overall, the next phase of the scientific drilling program will require significant and even more coordinated engineering efforts.

One of the most significant technological requirements is the measurement of the intrinsic and/or ephemeral properties of cores and boreholes. *In situ* measurements of redox state, chemical compositions, physical parameters, pH, and microbial populations and their activities are absolutely necessary. Newly developed (or improved) logging sensor tools, *in situ* sampling/monitoring devices, and *in situ* microbial colonization systems will be needed to achieve multidisciplinary scientific objectives through borehole observatories and experiments. Real-time hydrocarbon gas monitoring systems that include stable isotope measurements should also be deployed in platform laboratories. Novel and/or improved analytical technologies for quick, high-resolution measurements of temperature-, redox-, and oxygen-sensitive chemical and microbiological components must be developed for use on recently acquired cores, since exploration of high-temperature hydrothermal systems and the deep, hot biosphere has great potential to increase our understanding of the co-evolution of life and the planet. In addition, onboard measurements of physical, chemical, and biological properties of cores are extremely useful for real-time decisions necessary to meet drilling goals.

The development of a high-pressure- (and -temperature) coring system is required for various geochemical and biological reactions because the pressure limit of the currently available high-pressure-coring system is up to 25 MPa, which is not enough for high-pressure gas fields or deep coring (>2500 meters below the sea surface). Furthermore, once core under high pressure and temperature reaches the surface, an onboard high-pressure core transfer system equipped with multiple (micro-) sensors, gas and fluid extraction ports, tracer injection systems, and a mini-core sub-sampling system is needed.

Subseafloor microbes proliferate in narrow niches and vary over local fluid-flow pathways used to transport energy and nutrients, hence high-resolution microbiological sampling is necessary. We need to develop onboard sub-sampling strategies for quick identification of microbially interesting zones and high-resolution sampling capabilities while monitoring and minimizing contamination. Furthermore, the non-destructive identification of core quality and structures (through X-ray CT scanning) is a high priority for future microbiology/biogeochemistry-dedicated drilling expeditions. For high-throughput and high-resolution onboard analyses, computer automated systems, such as the auto-extractor and the automated cell-counting microscope system using fluorescent image analysis, should be deployed.

Penetration through a complete ocean-crust section, the so-called project "Mohole", will require advances in drilling capabilities that include riser drilling capability in 4000 m or more water depth, deep penetration and recovery of crustal rocks of all lithologies, and borehole and drill bit cooling technology for high temperature (>250 °C) environments. Technological issues include improvement of riser-pipe quality and casing strings, the blowout preventer, and the mud-circulation/recovery system.

The newly developed Riserless Mud Recovery (RMR) system has great potential for use in various environments in future drilling, especially for borehole controls (i.e., stability and cooling). Combining the dual-gradient technology with RMR enables environmentally friendly drilling (i.e., clean without mud pollutants to seawater) access to deeper environments and areas previously not drillable by riserless drilling. The RMR technology is directly applicable to all IODP platforms.

Improving core recovery and quality is the fundamental challenge for all drilling environments. Core recovery and quality depend on factors such as depth, temperature, fluid pressure, and lithology, as well as the design and performance of drilling and coring tools. In previous drilling experiences, poor core recovery has plagued (a) chert and/or shales, (b) sand and gravel layers, (c) hydrothermal deposits, (d) rubble basalts and sheeted dyke complexes, and (e) fault and fracture zones. Anticipated improvements include more accurate compensation of drill bit motion, torque, and type of cutting shoe, and operational technologies such as a feedback system of real-time drilling parameters. Other strategies, such as cuttings and side-wall coring, will increase sampling of unrecovered intervals. Borehole management (i.e., stability) requires cuttings removal and compensation of lithostatic and pumping pressures in both riser and riserless drilling modes. In addition, large diameter pipes may provide more opportunities to conduct various geophysical measurements such as pore pressures and resistivity.

Monitoring while drilling using downhole logging tools has greatly expanded our understanding of *in situ* pressure and stress conditions in the borehole. The use of logging tools during drilling or in CORKed boreholes is likely to expand in the new drilling program. Multiple-hole experiments are recommended, including injection tests and cross-borehole communication studies. Broadband and high-sensitivity sensors such as fiber-optic seismo-sensors combined with continuous data recovery should be developed and installed in active subseafloor environments. To study *in situ* conditions of high-temperature and/or high-pressure environments, the durability of logging systems to high temperature and pressure must be improved. Development of new slim-line multi-logging tools and borehole equipment is needed for all platforms. Current shipboard computational and dissemination capabilities should be more effectively integrated between software programs, database mining, and accessibility and be easily interfaced among the multiple drilling platforms and core repositories.

## Outreach and Education, and Branding

Outreach refers to activities that target the general public and funding agencies. Education and educational outreach are aimed at students in primary, secondary, undergraduate, and graduate school realms. Continued coordination through IODP-MI or its successor organization is essential to achieve the best results in the arena of public outreach. A successful branding campaign will be vital to ensure ongoing public recognition of the scientific discoveries and technological achievements of scientific ocean drilling.

Branding the science and accomplishments of ocean drilling should elucidate the linkages to broader objectives or themes, and not necessarily focus on individual expeditions. As visual impression is the key for branding, web sites across the program must have a common layout to promote the impression of a truly integrated program to the scientific community and the public, and to facilitate access to information. A series of bold, clear key messages should be an important element in any branding campaign. For example, the main messages of the drilling program should emphasize the following:

- The program investigates a dynamic Earth; it is a changing, not static, planet.
- Basic science is always valuable to society.
- Scientific ocean drilling is particularly relevant to society, providing knowledge about geohazards and climate change.
- Scientific ocean drilling is on the edge of the science frontier. By exploring Earth through scientific ocean drilling, we make novel and fundamental discoveries.

Outreach and education are vital in raising the profile of the future drilling program. The program needs to employ a full-time science "translator" (scientist or science educator with requisite skills) that can effectively communicate drilling science to non-scientists. Because many ocean drilling scientists are also geoscience educators, there should be mechanisms for them to be closely involved in framing the science to be easily accessible to students and the general public. Consideration should be given to expanding successful current programs to include all international partners and to serve all audiences (undergraduates, graduates, faculty, young scientists, science teachers, etc.). Specific recommendations are (a) the development and dissemination of an archive of basic images documenting the history and goals of the drilling program; (b) admitting science educators, in addition to school teachers, on expeditions; (c) providing communications training for younger scientists on board; (d) offering early career workshops sponsored by the new ocean drilling program; and (e) developing a mentoring plan for young career scientists who go to sea on expeditions and site survey cruises. Generally, the ocean drilling program should make greater use of Google Ocean and GeoMapApp to provide the public with images and video for education and outreach. Finally, in planning the new drilling program, there should be a workshop dedicated to formulating a plan that employs innovative cutting-edge methods of science education, communication, and outreach.

## Recommendations for the Architecture of a New Program

Scientific planning in the new ocean drilling program should be driven from the "bottom up", with scientists playing key roles in defining specific scientific short- and long-term goals and in advising and working directly with management, ship operators, and engineering development to execute the drilling program. The direct involvement of world-class students and scientists will keep the international ocean drilling program fresh and focused on emerging transformative topics that define the frontiers of life and earth sciences. As such, a strong science advisory structure should be a central component of the new program architecture. Furthermore, transformative science is often born from cross-disciplinary perspectives, and the drilling program must have multiple mechanisms to proactively engage scientists and students from disciplines outside of the traditional drilling community, from both academia and industry.

Because of the complex nature of an international drilling program with multiple operators and stakeholders, effective management should focus on fostering stronger international partnerships, well-integrated collaborations with other large geosciences programs, effective fundraising, and creative and efficient coordination amongst the national offices and implementation/ship operators. By working with the science advisory committees and scientific community, management should facilitate the formulation of visionary and innovative scientific goals through the life of the program.

Meeting the scientific goals of the drilling program will require flexibility. Multi-expedition, long-term missions will be required to achieve these ambitious goals. At the same time, many of the highest impact ocean drilling projects may be unanticipated and/or concise and focused. Thus, there must be mechanisms by which the new ocean drilling program can quickly respond to, nurture, and execute brilliant new ideas (identified through a peer-review system) that require ocean drilling. To implement the projects necessary to achieve transformative science will require a new ocean drilling program architecture that, by design, will have the flexibility to react quickly to new opportunities, but also to make decisive commitments to long-term complex, technically challenging projects.

## References


Inagaki, F., 2010. Deep subseafloor microbial communities. In Encyclopedia of Life Sciences, Chichester (John Wiley & Sons, Ltd.), doi:10.1002/9780470015902.a0021894.

Kanamori, H., 2008. Earthquake physics and real-time seismology. *Nature*, 451:271–273, doi:10.1038/nature06585.

Teagle, D., Abe, N., Bach, W., Blackman, D., Coggon, R., Dick, H., Edwards, K., and Ildefonse, B., 2009. Melting, magma, fluids and life: Challenges for the next generation of scientific ocean drilling into the oceanic lithosphere, Workshop Report, 27–29 July 2009. National Oceanography Centre, Southampton, University of Southampton, U.K., 85 pp.

Thurow, J., Peterson, L.C., Harms, U., Hodell, D.A., Cheshire, H., Brumsack, H.J., Irino, T., Schulz, M., Masson-Delmotte, V., and Tada, R., 2009. Acquiring high to ultra-high resolution geological records of past climate change by scientific drilling. *Sci. Drill.*, 8:46–56, doi:10.2204/iodp.sd.8.08.2009.


## Authors and Steering Committee Members


**Wolfgang Bach** (co-chair), Department of Geosciences, University of Bremen, Klagenfurter Strasse, D-28359 Bremen, Germany.

**Christina Ravelo** (co-chair), Ocean Sciences Department, University of California Santa Cruz, CA 95064, U.S.A., e-mail: acr@ucsc.edu (corresponding author).

**Jan Behrmann**, Leibniz-Institut für Meereswissenschaften, IFM-GEOMAR, Gebäude Ostufer, Wischhofstrasse 1-3, 24148 Kiel, Germany.

**Gilbert Camoin**, National Center for Scientific Research (CNRS), CEREGE UMR 6635, CNRS, Europôle Méditerranéen de l'Arbois, BP 80, F-13545 Aix-en-Provence cedex 4, Aix-en-Provence, France.

**Robert Duncan**, College of Oceanic and Atmospheric Sciences, Oregon State University, 104 COAS Administration Building, Corvallis, OR 97331-5503, U.S.A.

**Katrina Edwards**, Department of Biological Sciences, Hancock Auditorium and Museum (AHF) 107F, University of Southern California, Los Angeles, CA 90089-0371, U.S.A.

**Sean Gulick**, The University of Texas at Austin, Department of Geological Sciences, Geology Building 2. 106, 23rd Street at San Jacinto Boulevard., Austin, TX 78705, U.S.A.

**Fumio Inagaki**, Kochi Core Center, JAMSTEC, 200 Monobe Otsu, Nankoku City, Kochi Prefecture, 783-8502, Japan.

**Heiko Pälike**, National Oceanography Centre Southampton, University of Southampton and Natural Environment Research Council, Southampton, University of Southampton Waterfront Campus, European Way, Southampton, SO14 3ZH, U.K.

**Ryuji Tada**, Department of Earth and Planetary Science, University of Tokyo, Sci. Bldg. No.1-734, Hongou 7-3-1, Bunkyo-ku, Tokyo, 113-0033, Japan.

## EuroFORUM 2010:

### *Achievements and Perspectives in Scientific Ocean and Continental Drilling*

*2–7 May 2010, Vienna, Austria*

At the EGU 2010, a special session will be devoted to scientific drilling. The principal goals of this joint Integrated Ocean Drilling Program (IODP) — International Continental Scientific Drilling Program (ICDP) EuroFORUM 2010 are to summarize and review major scientific achievements in ocean and continental drilling with special emphasis on the European contributions to IODP and ICDP. Furthermore, perspectives and visions for drilling projects using a multi-platform approach will be tackled. Co-convenors of EuroFORUM 2010 are R. Stein (AWI Bremerhaven), U. Harms (GFZ Potsdam), H. Brinkhuis (Utrecht University), G. Camoin (CEREGE Aix-en-Provence), J. Erbacher (BGR Hannover), and U. Roehl (MARUM Bremen University). Further details can be downloaded from http://meetingorganizer.copernicus.org/EGU2010/session/2717.

## The ESF Magellan Workshop Series Program

The ESF Magellan Workshop Series was launched in 2006 with the aim of nurturing and coordinating innovative marine scientific drilling proposals for European scientists to strengthen the role of Europe within the international ocean drilling community. More than fifteen workshops, ten short visit grants, and one educational program have been supported. Recent workshops include Drilling of the "Shackleton Sites" and Evolution of the Baltic Sea through the Last Glacial Cycle.

**Next Workrshop:**

Volcanic Basins, Scientific, Economical and Environmental Aspects

*7–10 May 2010, Vienna, Austria*

**A Call for Workshops:**

*Open until 15 June 2010*

All scientists who wish to contribute to the future of scientific ocean research drilling and promote new scientific drilling technologies are encouraged to submit proposals. Priority will be given to proponents from ESF Magellan member countries.

For more information please see http://www.esf.org/ or contact Jochen Erbacher (jochen.erbacher@bgr.de).

## IODP Canada and ICDP Canada at GEOCANADA 2010

*10–14 May 2010, Calgary, Alberta, Canada*

Canadian scientists from IODP and ICDP will host a booth at the GEOCANADA 2010 meeting in Calgary, Alberta (http://www.geocanada2010.ca/). This meeting is jointly sponsored by several scientific societies including the Geological Association of Canada, the Mineralogical Association of Canada, the Canadian Society of Exploration Geophysics, and the International Association of Hydrogeologists - Canadian National Chapter.

GEOCANADA 2010 exhibit hall days are 11 and 12 May. Please visit the IODP Canada (http://www.iodpcanada.ca/) and ICDP Canada (http://www.icdp-canada.ca) booth when you are there! For more information, please contact Hélène Gaonac'h (gaonach.helene@uqam.ca).

## An IODP Canada/ECORD Summer School in 2010

*27 June to 12 July 2010, Quebec, Canada*

Canadian scientists actively involved in IODP activities will hold a 2010 summer school on "Ocean and Climate Changes in Polar and Subpolar Environments" (image below). This event will be held in Quebec (Canada) between Rimouski, Quebec City and Montreal institutions.

The summer school will focus on paleoceanography and paleoclimatology at high latitudes and will include lectures, lab work, field work in the St. Lawrence Estuary, and field trips to the St. Lawrence lowlands.

Ph.D. students and postdoctoral fellows from ECORD countries are welcome to apply. Courses and lectures will cover geophysics, geomorphology, organic, inorganic and isotope geochemistry, physics of sediments, micropaleontology, paleoceanography, modeling of sea ice, geotechnical methods of drilling polar regions, and other topics related to Arctic and subpolar environments.

For information and registration details, please see the http://www.iodpcanada.ca/news_items/open-call-for-ecord-scholarships-2010/ or contact Hélène GAonac'h (coordinator@mail.iodpcanada.ca).

## ECORD Summer School on "Dynamics of Past Climate Changes"

*13–24 September 2010, Bremen, Germany*

The 4th ECORD Summer School in Bremen, to be held at the MARUM-Center for Marine Environmental Sciences at the University of Bremen, Germany, aims to bring Ph.D. students and young postdocs in touch with IODP at an early stage of their career, inform them about research within this international scientific program, and prepare them for future participation on IODP expeditions. Such training will be achieved by taking the summer school participants on a "virtual ship" by exploiting the unique facili-ties linked to the IODP Bremen Core Repository (image below). They will be introduced to a wide spectrum of state-of-the-art analytical technologies and core description methods, including core logging/scanning according to IODP expedition standards. In addition, the topic "Dynamics of Past Climate Changes" will be covered by lectures and discussions with leading paleoceanographers and paleoclimatologists. The latter will include climate modelers, physical oceanographers, and researchers working on polar ice cores and lake records. This comprehensive approach—combining scientific lectures with practicals on IODP-style "shipboard" measurements—is the blueprint for the Bremen ECORD summer school series, which now starts the second three-year-cycle of ECORD summer schools covering the three major topics of the IODP Initial Science Plan. For detailed information visit http://www.glomar.uni-bremen.de/ECORD_Summer_School_2010.html.

## News from the Australian and New Zealand IODP Offices

The Australian and New Zealand IODP Consortium has existed since January 2008. It is a great pleasure for Australia to be back in ocean drilling with colleagues from around the world, and for New Zealand to be a member for the first time. Recently a phase of expeditions in our own region has begun: Canterbury Basin Sea Level and Wilkes Land Glacial History (successfully completed); Great Barrier Reef Environmental Changes (started in February); South Pacific Gyre Microbiology (late 2010); and Louisville Seamount Trail (latest 2010). Port calls by the *JOIDES Resolution* in Townsville and Wellington at the beginning and completion of the Canterbury expedition provided valuable opportunities to raise awareness of IODP (image below that young participants sampling the sediments). We look forward to hosting the ship again in 2010 and 2011.

The scientific communities in both countries are enthusiastically involved in planning for the next phase of ocean drilling. For more information see http://www.iodp.org.au/.

## Scientific Drilling in the Nordic Countries

The researchers network "Scientific Drilling in the Nordic Countries" has recently been funded for three years by NordForsk, the Nordic Council of Ministers' advisory board on research strategy (http://www.nordforsk.org/). The aim of the network is to consolidate knowledge and experience from past and present (and future) scientific drilling projects in Nordic countries and to propagate it to the scientific community. Scientists with experience from scientific drilling enterprises and the few experts in drilling technology within academia are located at various institutions throughout the Nordic countries. For most researchers who encounter a scientific problem that requires drilling, the knowledge resource of managing and executing drilling is not readily available; this network hopes to change that.

Workshops and excursions will be tightly coupled to ongoing coordinated scientific projects and bring together experts, experienced scientists, and novices for knowledge exchange and transfer. For more information visit the network's website at http://www.sdnc.eu/ or contact Henning Lorenz (henning.lorenz@geo.uu.se).

## Canada is Back in ICDP

In 2009 the Natural Sciences and Engineering Research Council of Canada (NSERC) committed itself to supporting Canadian membership in the ICDP for the next five years, giving Canadian scientists the opportunity to continue their involvement in ICDP activities and scientific planning.

ICDP Canadian scientists currently coordinate their activities through the Canadian Continental Drilling Program (CCDP). CCDP is also supported by INRS-ETE and has its coordination office at UQÀM jointly with CCOD, the coordination body for Canadian activities linked to IODP (http://www.iodpcanada.ca/). CCDP is presently chaired by Pierre Francus from INRS-ETE. Further information is available at http://www.icdp-canada.ca/.

## New Scientific Drilling Infrastructure in Sweden

Two years ago, the Swedish geoscientific community established a national program for scientific drilling, the Swedish Deep Drilling Program (SDDP). SDDP was able to successfully promote Swedish membership in ICDP. Furthermore, the Swedish Research Council decided to support an application for a drilling rig. The envisaged drilling platform will be capable of core drilling to at least 2500 m depth in different lithologies. The Department of Engineering Geology at Lund University will organize the management of this infrastructure. The rig is intended primarily for use in SDDP drilling projects but will be rented out to other projects. As it will be capable of drilling to depths of 2000–3000 m, this rig will complement the traditional deep-drilling platform InnovaRig of the GFZ German Research Centre for Geosciences.

## U.S. Implementing Organization Adds New Talent, Restructures

The U.S. Implementing Organization of the Integrated Ocean Drilling Program (USIO) has restructured and added new talent to its team. The USIO operates the *JOIDES Resolution* and consists of the Consortium for Ocean Leadership, Lamont-Doherty Earth Observatory (LDEO) of Columbia University, and Texas A&M University (TAMU).

Ocean Leadership hired Greg Myers as the new Senior Technical Expert (Engineering and Development) and Doug Fils as the new Data Management Technical Expert. Myers will oversee USIO engineering development efforts and coordinate non-IODP funding activities. Fils will interact with IODP-Management International and the USIO to develop and implement approaches to data management needs.

At LDEO, former Scientific Ocean Drilling Vessel project manager Gerardo Iturrino is now the Manager of Engineering Services and will oversee engineering development and maintenance of shipboard and shore-based logging systems. Eric Meissner takes on the newly-created position of Senior Engineering Project Manager. Alberto Malinverno will now super-vise the Science Services group while remaining in his role of Principal Scientist.

Jim Rosser has joined IODP-TAMU as the Manager of Development, Information Technology, and Databases. Rosser brings experience as a U.S. naval commander to oversee the shipboard and shorebased IT infrastructure and the acquisition and management of data collected by the *JOIDES Resolution*. Jay Miller will serve as manager of the newly formed Technical and Analytical Services Department, which is composed of the majority of the shipboard technical support staff.

## ESSAC Office Moves from France to Germany

On 1 October 2009, the European Consortium for Ocean Research Drilling Science Support and Advisory Committee (ESSAC) office moved from Aix-en-Provence, France to Bremerhaven, Germany. The new office, established at the Alfred Wegener Institute (AWI) for Polar and Marine Research, is headed by Ruediger (Rudy) Stein as the new ESSAC Chair. Jeannette (Jenny) Lezius is the new Science Coordinator. More information about ESSAC and its activities is available at http://www.essac.ecord.org/.

## IODP-MI Office in Tokyo

The Integrated Ocean Drilling Program Management International, Inc. (IODP-MI) is integrating most of its activities into one single location. On 1 March 2010 the former Sapporo office moved to new IODP-MI office located at Tokyo University of Marine Science and Technology, Etchujima Campus. A small office of finances and contracts office is maintained in Washington, D.C. Following a few additional staff hires, the planned restructuring of IODP-MI has been completed. The new office structure and contact details can be found at: www.iodp.org.

# 13

# IODP Expedition 333: Return to Nankai Trough Subduction Inputs Sites and Coring of Mass Transport Deposits

by Pierre Henry, Toshiya Kanamatsu, Kyaw Thu Moe, Michael Strasser, and the IODP Expedition 333 Scientific Party

## Abstract

Integrated Ocean Drilling Program (IODP) Expedition 333 returned to two sites drilled during IODP Expedition 322 on the ocean side of the Nankai Trough to pursue the characterization of the inputs to the Nankai subduction and seismogenic zone, as part of the Nankai Trough Seismogenic Experiment (NanTroSEIZE) multi-expedition project. Site C0011 is located at the seaward edge of the trench and Site C0012 on a basement high, Kashinozaki Knoll (Fig. 1). The main objectives of drilling again at these sites were to fill coring gaps in the upper part (<350 m) of the sedimentary sequence, to measure heat flow, and to core the oceanic basement to a greater depth on the Knoll. New results include the observation of a diagenetic boundary within the Shikoku Basin sediments that may be compared to one documented further west by ODP Legs 131, 190 and 196 but occurs here at a lower temperature. Borehole heat flow measurements confirm spatial variations in the Shikoku Basin that were indicated by short probe surveys. Heat flow variations between topographic highs and lows may be related to fluid convection within the basement. This expedition also included the objectives of the Nankai Trough Submarine LandSLIDE history (NanTroSLIDE) Ancillary Project Letter (APL) and cored at Site C0018 a pile of mass transport deposits on the footwall of the megasplay fault, a major out of sequence thrust that presumably slips coseismically during large subduction earthquakes. This brought new insight on the timing of these mass wasting events and on the deformation within the sliding slope sediments.

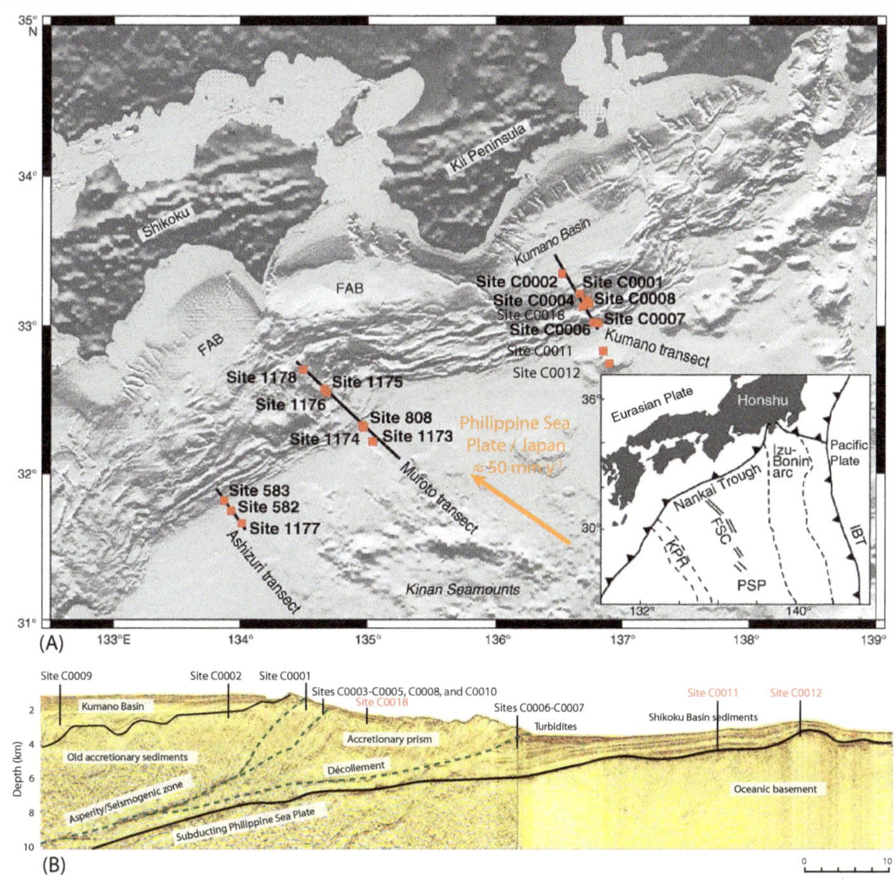


Figure 1. [A] Location map of Kumano, Muroto, and Ashizuri borehole transects across the Nankai Trough. FAB=Forearc Basin. [B] Spliced composite profile of a representative depth section from NanTroSEIZE 3D data volume (Moore et al., 2009) and Line 95 from IFREE mini-3D seismic survey (Park et al., 2008). Projected positions of Stage 1 and 2 drilling sites, including sites C0018, C0011, and C0012, are shown.

## Introduction and Goals

To improve our understanding of earthquakes and tsunamis generated by subduction megathrusts, an ambitious project known as NanTroSEIZE was initiated along the subduction boundary of southwestern Japan in the Kumano transect area (Fig. 1). The overarching goal of NanTroSEIZE is to create a distributed observatory spanning the updip limit of seismogenic and tsunamigenic behavior (Tobin and Kinoshita, 2006a, 2006b). This multi-stage project is in the process of documenting several key components of the subduction margin, including the pre-subduction inputs of sediment and oceanic basement (Underwood et al., 2010), and the plate interface at shallow depths and at depths of 6–7 km where earthquakes occur (Tobin et al., 2009).

Exp. 333 addressed objectives at the NanTroSEIZE input sites seaward of the trench, Sites C0011 and C0012, as well as examined slope instability processes and their relationship with tectonic evolution proposed in the NanTroSLIDE APL. The characterization of the sedimentary column and ocean crust entering the subduction is essential because their evolution at increasing temperature and pressure presumably controls the transition to seismogenic behavior (Hyndman et al., 1997). Furthermore, the distribution of slip during earthquakes on faults affecting the shallow portion of subduction complexes determines locations of tsunami generation and their amplitude (Baba et al., 2006). The depth and frictional properties of the plate boundary fault and of splay faults connected to it may be controlled by the composition of incoming sediments, their diagenetic history, and the release and migration of fluids (Underwood, 2007). Heat flow is a vital measurement for the interpretation of diagenetic history, and it can also aid in identifying vigorous fluid circulation. Heat flow within the oceanic plate is also an important parameter for models calculating temperature along the subduction plate to the depth of the seismogenic zone (Hamamoto et al., 2011; Harris et al., 2011). This was not accurately constrained prior to this expedition.

The NanTroSLIDE APL arose from evidence gathered during earlier NanTroSEIZE drilling. IODP Expedition 316 found slumping near the fault scarp of the mega-splay, and 3D seismic reflection data revealed related mass transport deposits located further downslope (Strasser et al., 2011). Interpretation of 3D reflection seismic data and drilling results document a complex temporal and spatial evolution of the shallow mega-splay fault zone (MSFZ) and anticline structures in the underlying accretionary prism. This is characterized by a general decrease of thrusting motion with time since 2 Ma and alternating periods of high and low structural activity on individual structural segments (Strasser et al., 2009; Kimura et al., 2011). Site C0018 was drilled and cored to recover a Pleistocene-to-Holocene succession of stacked mass-transport deposits (MTDs) within a slope basin on the footwall of the megasplay thrust. Overall, drilling at Site C0018 aimed at (i) establishing a well-dated Quaternary submarine landslide history as it may relate to the tectonic evolution of the accretionary prism and (ii) reconstructing sliding and transport dynamics of submarine landslides as they may relate to the tsunamigenic potential of such event (Henry et al., 2010).

The specific set of questions addressed by Exp. 333 drilling is as follows.

- Is fluid circulation in basement and permeable sedimentary layers influencing heat flow and diagenesis at Sites C0011 and C0012?
- How does lithologic variation and diagenesis impact the physical properties of incoming sediments and guide fault development?
- Was magmatic activity heterogeneous in composition and age on the backarc basin basement high (Kashinosaki Knoll)?
- Is alteration of the upper oceanic basement heterogeneous, and how does such alteration influence geochemical and fluid budgets?
- What is the frequency of submarine landslides, what controls type, size, and magnitude of turbitides and MTDs, and how do related processes change through time?
- How do large MTDs relate to the timing of splay fault activity as inferred from NanTroSEIZE Stage 1 drilling?
- What are the dynamics of large submarine landslides, and can we infer their tsunamigenic potential?

## Geologic Setting and Previous Work

The Nankai Trough is a convergent plate boundary where the Philippine Sea Plate underthrusts the southwestern Japan margin. At the Kumano transect across the Nankai Trough, the velocity between the Philippine Sea Plate and the forearc is 4.5–5.5 cm $y^{-1}$ along an azimuth of $305° \pm 3°$ (Seno et al., 1993; Mazzotti et al., 2001; DeMets et al., 2010; Loveless and Meade, 2010) down an interface dipping 3°–7° (Kodaira et al., 2000b). The subducting lithosphere of the Shikoku Basin was formed by backarc spreading during a time period of 15–25 Ma (Chamot-Rooke et al., 1987; Okino et al., 1994). The three major seismic stratigraphic sequences identified in the northern Shikoku Basin are the lower and upper Shikoku Basin facies and local spill-over of Quaternary trench-wedge turbidites (Ike et al., 2008a). The lower part of the Shikoku Basin displays a complex geometry influenced by basement topography (Ike et al., 2008b). Basement highs are also imaged within the subduction zone (Kodaira et al., 2003; Dessa et al., 2004), where they influence margin structure (Lallemand et al., 1992; Le Pichon et al., 1996; Park et al., 1999; Mazzotti et al., 2002; Bangs et al., 2006; Moore et al., 2009) and seismicity (Kodaira et al., 2000a; Park et al., 2004). The NanTroSEIZE expeditions focused on the Kumano transect, while earlier Ocean Drilling Program (ODP) and Deep Sea Drilling Project (DSDP) sites were drilled on the Muroto transect and the Ashizuri transects, respectively, in the central and western part of Shikoku Basin (Fig. 1; Moore et al., 2001). The combined data provide some information on input sediment variability at basin scale that may relate with margin structure. On the Kumano transect, swath bathymetry and multichannel seismic (MCS) data show a pronounced and continuous outer arc high extending >120 km along strike and marking the limit between the lower forearc slope and the forearc basin. The outer arc-high coincides with a splaying system of thrust faults that branch from a strong seismic reflector interpreted as a major out of sequence thrust, the megasplay fault (Park et al., 2002; Moore et al., 2007). Constraints on the slip history of faults are a major outcome of the previous NanTroSEIZE expeditions. The lower forearc slope of the Nankai Trough consists of thrust imbricates that were accreted to the margin over the last 2 Ma (Kinothis et al., 2009; Moore et al., 2009) but since 400,000–700,000 a, this wedge has been thrust over the trench sediments without accretion of new imbricates (Screaton et al., 2009). NanTroSEIZE expeditions also showed slip on the mega-splay is roughly synchronous with outer arc high uplift and subsequent Kumano forearc basin filling and tilting (Strasser et al., 2009; Gulick et al., 2010); however, in detail, long-term slip rates on individual splay thrust branches vary both in time and space (Kimura et al., 2011). The splay fault branch that was drilled slipped at rates ranging from 9 m $kyr^{-1}$ to 2 m $kyr^{-1}$ in the 2–1.5 Ma interval, but only at a very low rate (<100 m $Ma^{-1}$) over the last 1.3 Ma.

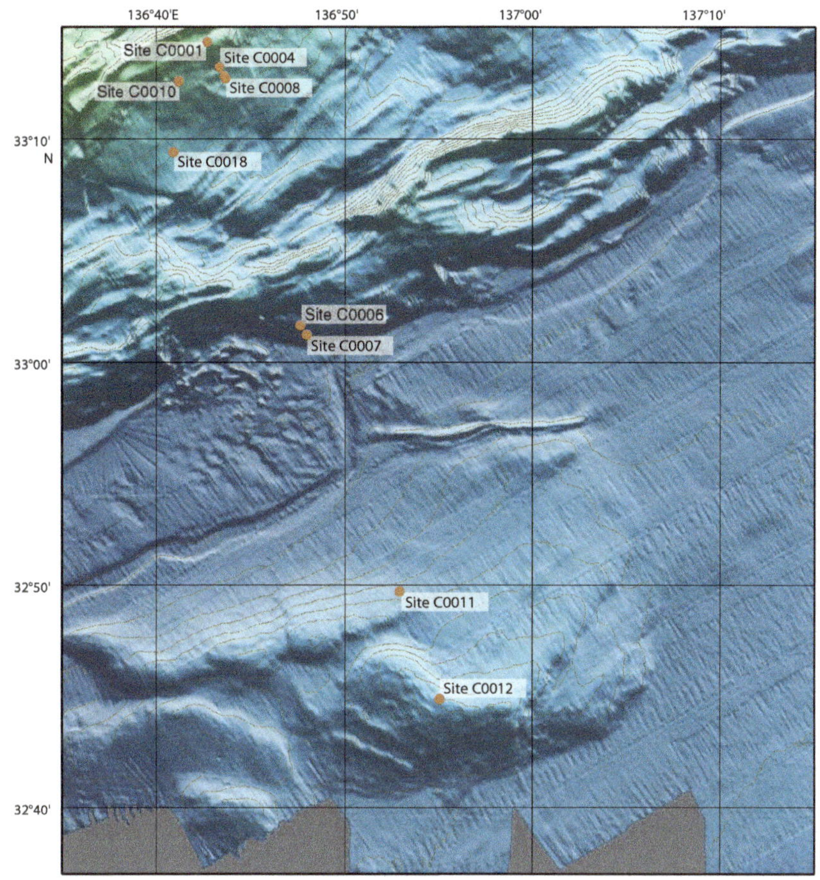


Figure 2. Detailed bathymetric map of Kashinosaki Knoll and Nankai Trough showing location of Sites C0011, C0012, and C0018.

Exp. 322 and 333 focused on two sites on the subducting Philippine Sea Plate. Site C0011 is located on the northwest flank of a prominent bathy-metric high (the Kashinosaki Knoll; Ike et al., 2008a), whereas Site C0012 is located near the crest of the knoll (Fig. 2). Portions of Sites C0011 and C0012 were drilled and cored during Exp. 322 in 2009, but important intervals were not recovered, and heat flow data were not obtained. The upper part of the sedimentary sections was either washed down or cored with rotary core barrel (RCB), which yielded cores of insufficient quality in the shallow unconsolidated sediments. The basement was reached at Site C0012 on top of the Kashinosaki Knoll and cored at 538–576 m below seafloor (mbsf). As a result, IODP Exp. 333 aimed to fill gaps in core data on the upper part of the sedimentary section, acquire much-needed measurements of borehole temperature to assess thermal history and the extent of *in situ* diagenesis, and drill and core to a deeper level in basement at Site C0012 to assess heterogeneity in composition and alteration.

Although the stratigraphic coverage was incomplete, the merger of lithofacies and age-depth models shows how correlative units change from an expanded section at Site C0011 to a condensed section at Site C0012 (Fig. 3). The composite section also captured a previously unrecognized interval of late Miocene volcaniclastic sandstone. An older (early–middle Miocene) turbidite sandstone/siltstone facies with mixed volcaniclastic-siliciclastic detrital provenance occurs in the lower Shikoku Basin facies; this unit may be broadly correlative with superficially similar Miocene turbidites on the western side of the basin (Fergusson, 2003; Underwood, 2007; Underwood et al., 2010). The age of basal sediment (reddish-brown pelagic claystone) at Site C0012 is older than 18.9 Ma.

Geochemical analyses of interstitial water on top of the basement high show clear evidence of upward diffusion of sulfate and other dissolved chemical species from the basement (Underwood et al., 2010). The depth of the sulfate reduction zone is also anomalously deep at Site C0012. Chlorinity values increase toward basement because of hydration reactions in the sediment and diffusional exchange with basement fluids (Underwood et al., 2010). In contrast to Site C0011, where chlorinity decreases with depth, the more saline fluids at Site C0012 are largely unchanged by the effects of focused flow and/or *in situ* dehydration reactions associated with rapid burial beneath the trench wedge and frontal accretionary prism.

Site C0018 is located ~3 km seaward from the splay-fault tip in a slope-basin that (1) represents the depocenter for downslope mass transport from the MSFZ, (2) is characterized by several MTDs identified in 3D seismic data (Strasser et al. 2011), and (3) includes an exceptionally large (as thick as 180 m) MTD. Drilling at Site C0018, which is located where the MTD bodies wedge out and where basal erosion is minimal, allowed for ground truthing and dating of the seismic-stratigraphic interpretation and sampling the MTDs across the most complete and thickest stratigraphic succession.

## Site C0011

A total of 380 meters of sediment was drilled in Holes C0011C and C0011D (Fig. 4). Two lithologic units were identified with very good recovery rates (Expedition 333 Scientists, 2012a). Unit I corresponds to Shikoku Basin hemipelagic/pyroclastic facies, and Unit II corresponds to a volcanic turbidite facies that was originally designated middle Shikoku Basin facies (Underwood et al., 2010). In addition, two lithologic subunits were interpreted within Unit I, Subunit IA (younger) and Subunit IB. Cored lithologies include silty clay, clayey silt, clay, and mudstone interbedded with coarse to fine volcanic ash. The Subunit IA/IB boundary occurs at 251.52 mbsf and separates more indurated dark gray mud and mudstone from overlying sediments that are highly bioturbated. The sediments below the transition are also marked by the disappearance of volcanic ash layers with unaltered glass. Below 347.82 mbsf, which is the Unit I/II boundary, there is an abrupt shift into coarser-grained tuffaceous sandstone and heterolithic gravel and sand.

Four volcanic ash layers comprise provisional stratigraphic markers at Site C0011; they were correlated with dated ash layers on land (Fig. 4). These are Azuki (0.85 Ma), Pink (1.05 Ma), Habutaki (2.8–2.9 Ma) and Ohta (4.0 Ma) (Hayashida, 1996; Satoguchi et al., 2005; Nagahashi and Satoguchi, 2007). Shipboard paleomagnetic interpretations and nannofossil biostratigraphy, done post-cruise, agree well with these preliminary identifications. The shipboard paleomagnetic interpretation also has good continuity with Exp. 322 data and yields an age of 7.6 Ma for the transition between lithologic Units I and II, and the age of the Subunit IA/IB boundary is constrained to 5.32 Ma.

Structural features encountered in Hole C0011C and C0011D are rather sparse. Two sets of conjugate normal faults, striking NNE to SSW and NW to SE can be distinguished. The strikes of these conjugate sets are, respectively, parallel to the N25° maximum horizontal stress direction inferred from borehole breakouts at Site C0011 (Saito et al., 2010), and parallel to Nankai subduction convergence (300°–310°N; Mazzotti et al., 2001; Loveless and Meade, 2010). A NNE to SSW compression would be consistent with the local direction of compression inferred from kinematic modeling of the Zenisu-Izu fault system (20°N; Mazzotti et al., 2001). Furthermore, focal mechanisms of the 2004 earthquakes off the Kii Peninsula, which occurred within the oceanic plate north and northeast of Site C0011, indicate dominantly north–south compression (Ito et al., 2005). It is therefore likely that the state of stress at Site C0011 is influenced by intraplate compressive deformation within the Philippine Sea Plate.

Bulk density and resistivity values generally increase from the surface to 50 mbsf, reflecting normal consolidation of sediment. However, from ~50 mbsf to 80 mbsf density and resistivity decrease and then remain constant from 80 mbsf to 240 mbsf. Below 240 mbsf, density and resistivity increase abruptly across the transition between Subunits IA and IB and then continue along a normal consolidation trend. Two transitions were shown by logging-while-drilling (LWD) data at Site C0011 (Saito et al., 2010): one at 212 mbsf (downward decrease in gamma ray with a small associated decrease in resistivity) and one at 251.5 mbsf (downward increase in resistivity and gamma ray). The transition at 251.5 mbsf correlates with the abrupt porosity and resistivity change observed across the lithologic transition. However, the upper transition is not correlated with any remarkable observations on cores. During hydraulic piston coring system (HPCS) operations, downhole temperature was measured with the third-generation advanced piston corer temperature (APCT-3) tool at ~30-m intervals. We completed one measurement in Hole C0011C at 22.5 mbsf and eight measurements in Hole C0011D from 49 mbsf to 184 mbsf. Data show a nearly linear increase in temperature with depth (0.0913°C m$^{-1}$; Fig. 5), corre-

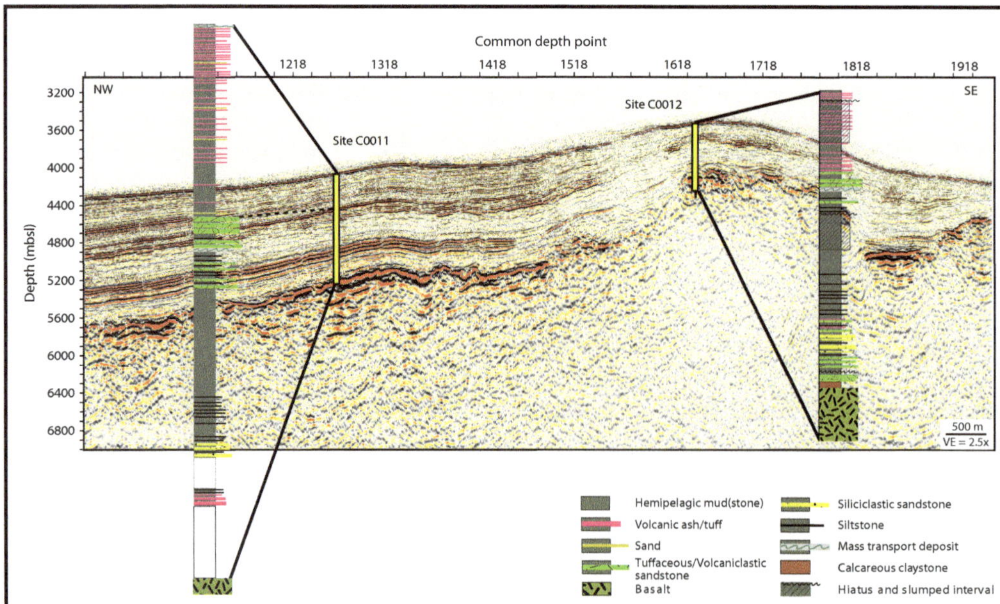


Figure 3. Lithologic columns on seismic background across Kashinosaki Knoll showing sites C0011 and C0012, using results from Expeditions 333 and 322 (Underwood et al., 2010). VE=vertical exaggeration.

sponding to a heat flow value of 89.5 mW m$^{-2}$. Temperature extrapolated at basement (~1050 m, from seismic profile), taking into account heat conductivity variations in the cored intervals, is ~80°C.

Interstitial water chemical composition trends are generally consistent with Exp. 322 results (Underwood et al., 2010). Chlorinity increases rapidly with depth in the upper ~25 m of Holes C0011C and C0011D, stays close to seawater value from 25 mbsf to 250 mbsf, and then gradually decreases. The chlorinity decrease, which continues to the deeper interstitial waters of Hole C0011B taken during Exp. 322 (Underwood et al., 2010), may reflect the updip migration of interstitial water freshened by the smectite-illite reaction at greater depths below the trench and prism toe. Sulfate concentration decreases rapidly in the upper 70 m of Hole C0011C, followed by a less dramatic decrease from 70 mbsf to 183 mbsf, which corresponds to the end of HPCS coring. The scatter of sulfate concentration below this level reflects seawater intrusion resulting from disturbance during coring. Yet sulfate probably remains present at low concentrations down to 280–300 m depth, which could explain why barium concentrations remain low and rise sharply below this level. Methane was the only hydrocarbon gas detected in holes C0011C and C0011D. It occurs only in low concentrations (2 ppmv) in the uppermost 260 m of the cored sequence and then rises to reach the highest values (~900 ppm) at the bottom of Hole C0011D. This indicates that, although sulfate consumption rate may be maximal at 70–80 m, the sulfate-methane transition zone occurs deeper, in the 250–300 m depth range.

Another remarkable feature is a sharp drop in silica concentration from ~800 ppm to ~200 ppm or less at the level of the transition from lithologic Subunit IA to IB. This drop correlates with a decrease in porosity from ~65 % to ~55 % observed over <10 m in moisture and density data, and with a concurrent resistivity increase in LWD (Saito et al., 2010) and core data. This behavior is similar to that observed at ODP Leg 190 Sites 1173 and 1177, where the base of the zone showing retarded compaction was ascribed to dissolution of opal-CT cement and precipitation of quartz (Spinelli et al., 2007). At the level of the tuffaceous sand-stones (350–370 mbsf), a second-ary silica concentration maximum is observed, which may be related to ash alteration.

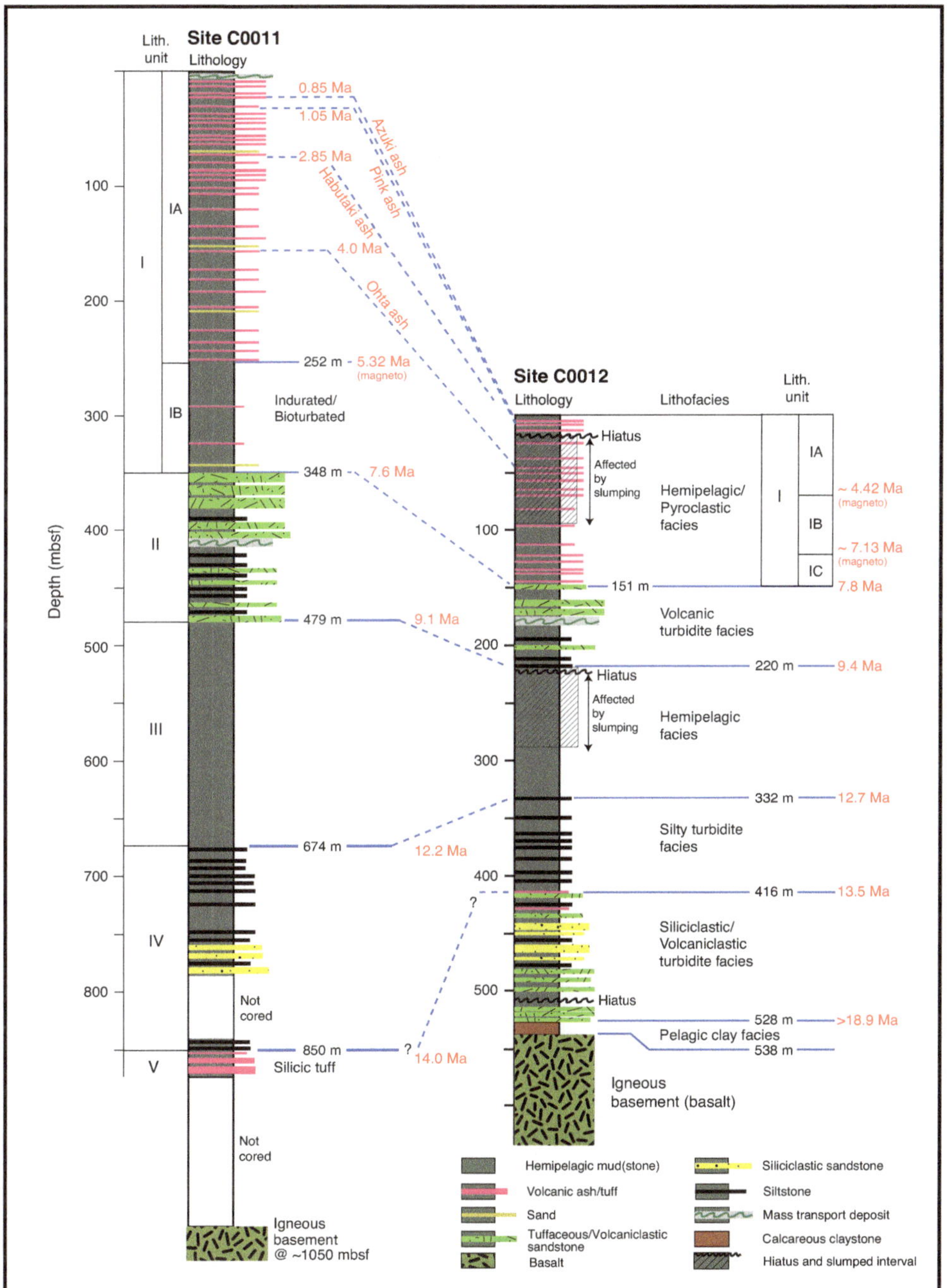


Figure 4. Stratigraphic correlation between Sites C0011 and C0012, based on composite lithology columns and age information from Expeditions 322 and 333. Unit boundary ages taken from integrated age-depth models.

## Site C0012

In Holes C0012C and C0012D, 180 meters of lithologic Unit I (hemipelagic/pyroclastic facies) and the upper part of lithologic Unit II (volcanic turbidite facies) were drilled during Exp. 333 (Figs. 3, 4; Expedition 333 Scientists, 2012b). The remaining

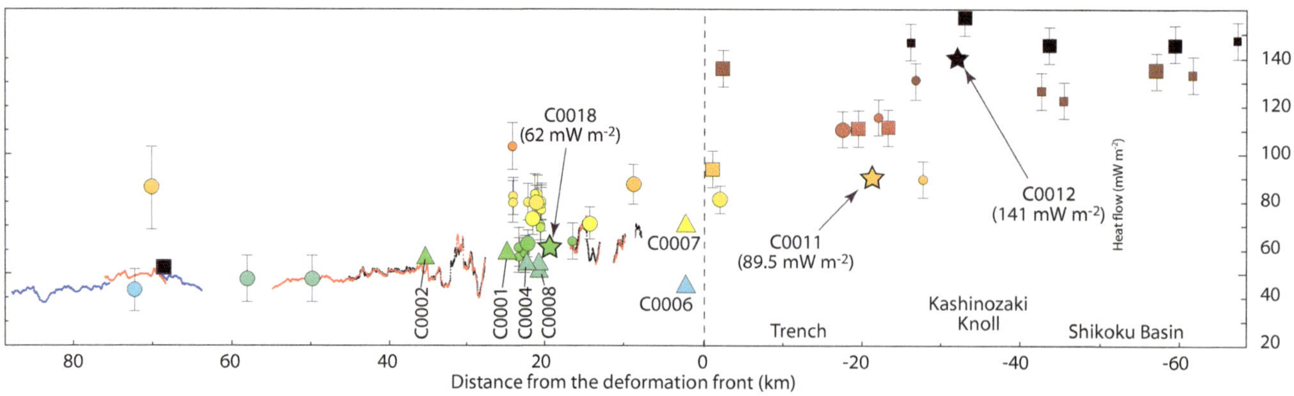


Figure 5. Heat-flow data collection in a 10-km-wide swath along the Kumano transect. Squares are heat flow probe measurements from Kinoshita et al. (2008), circles are long-term measurements from Hamamoto et al. (2011), triangles are borehole measurements from IODP Exp. 315 and 316, stars are from Exp. 333. The symbol sizes indicate data quality as defined by authors (big: good to excellent; small: poor to fair), the symbol color is a function of the heat flow value. Values derived from Bottom Simulator Reflectors (BSRs) along sections of multichannel seismic lines are plotted as lines.

holes (C0012E, C0012F, and C0012G) aimed at drilling red calcareous claystone and basalt at the contact between sediments in the Shikoku Basin and the igneous oceanic crust. Three lithologic subunits were interpreted in Unit I: Subunit IA (youngest), Subunit IB, and Subunit IC. The subunits are distinguished based on the presence, frequency of occurrence, and thickness of volcanic ash layers. The lithologies in Holes C0012C and C0012D include dark greenish gray clay and silty clay and silt interbedded with volcanic ash and minor occurrences of thin sand beds. A major decrease in the frequency of the occurrence of ash layers is recorded at ~71.5 mbsf, thereby defining the Subunit IA/IB boundary. Ash alteration was observed from 91.2 mbsf to the lower part of Subunit IB. At Site C0011, a comparable alteration front occurs at the top of Subunit IB. Ash layers are scarce to ~123.3 mbsf, which defines the base of Subunit IB. Another interval of dark greenish gray clay/silty clay-bearing ash layers corresponding to Subunit IC extends to 149.77 mbsf. This depth for the base of Unit I matches closely with the designation of 150.9 mbsf that was made during Exp. 322 (Underwood et al., 2010). Below 149.77 mbsf, Unit II comprises turbidite sands and sandstones with sharp and well-defined upper and lower boundaries. Commonly, beds have normal grading, but some comprise massive intervals with or without clay clasts. At the base of normally graded beds, pebble and sand clasts are composed of coarse ash and lapilli tuff. The lower part of Unit II comprises several layers of carbonate-cemented sandstones with calcite and barite veins. These beds are separated by mudstone very similar to that of Subunit IC.

Hole C0012E recovered two cores of greenish yellow mudstone intercalated with thin sandstone layers from 500 mbsf, corresponding to the base of lithologic Unit V (volcaniclastic-rich facies) defined during Exp. 322 (Underwood et al., 2010), and one core from 519 mbsf that recovered 6.8 m of reddish brown calcareous claystone with lighter green layers, overlying altered pillow basalts. The interface between the red calcareous claystone and the basaltic basement was also recovered in Holes C0012F and C0012G. The red calcareous claystone corresponds to Exp. 322 lithologic Unit VI (pelagic claystone; Underwood et al., 2010) and contains veins of calcite with traces of barite as well as several layers with accumulations of manganese oxide forming millimeter- to centimeter-sized lumps.

Hole C0012G cored pillows and massive phyric basalts from 525.69 mbsf to the base of the hole at 630.5 mbsf. Two units are defined in the basalt: Unit I is composed of phyric or highly phyric pillow basalt, and Unit II is composed of sheet flows with pillow basalt interlayers. As observed during Exp. 322, the basalt in Unit I is highly altered, and some voids remaining between basalt pillows are filled with analcime. Observations of thin sections showed that all olivine and most glass as well as a fraction of the plagioclases have been replaced by secondary phases—dominantly saponite, celadonite and zeolites—that are also present as vesicle fillings. In Unit II, the massive sheet flows are more crystalline and generally less altered. Basalt experienced localized alteration under iron oxidizing conditions with accumulation of iron hydroxides in veins and alteration halos. Celadonite and saponite are present in the rock mass and, locally, pyrite. This suggests two stages of alteration, under iron oxidizing and iron reducing conditions.

Bedding planes at Site C0012 display a large range of dipping angles from 3° to 70° but are organized in zones of low and high bedding dips. Bedding planes with low dips are characteristically observed between 0 mbsf and 14 mbsf and between 85 mbsf and 145 mbsf. High-dip angle beds are found between 14 mbsf and 85 mbsf where they consistently strike northeast–southwest and dip southeast, and between 145 mbsf and 180 mbsf, where the distribution of strikes and dips appears scattered. Chaotic structures (disrupted beds, folds, and injections of sand or mud) are observed at the bottom of the high-dip angle sections. Faults and shear zones mostly display normal displacement with high-dip angles, striking northwest–southeast, and dipping northeast or southwest, suggesting northeast–southwest extension. Considering the location of Site C0012 on a topographic high southwest of a steep arcuate slope evocative of a slide scar (Fig. 2), the interpretation proposed is that the high-dip angle sections have been affected by sliding events.

The cored portion of Units I and II has an estimated Holocene–Late Miocene age range (~0–8.3 Ma). Three ash beds were correlated to known tephra dated on land, based on visual observations of the shape of the glass shards and dominant associated minerals in smear slides. The Azuki volcanic ash bed, dated on land as 0.85 Ma (Hayashida et al., 1996), was identified at 5.7 mbsf. The Pink ash bed, dated on land as 1.05 Ma (Hayashida et al., 1996), correlates with a characteristic ash bed at 7.7 mbsf. A third major volcaniclastic event, the Ohta ash bed, dated on land as 4.0 Ma (Satoguchi et al., 2005), correlates with an ash bed at 44.95 mbsf. The unusually shallow depths of the presumed Azuki and Pink ash beds imply very slow average sedimentation rates over the last ~1 Ma. This interpretation is, however, consistent with the depth of the Brunhes-Matuyama reversal and of the Jaramillo Chron from natural magnetic remanence data. This interval of condensed sedimentation lies on an angular unconformity at ~14 mbsf, interpreted as the top of a slump. Paleomagnetic and biostratigraphic data suggest that this unconformity correlates with an age hiatus of ~2.0 m.y. (~1 Ma to ~3 Ma). Similar observations within the upper part of Unit III during Exp. 322 suggested that a slumping event associated with the remobilization of the uppermost sedimentary layers also occurred about 9.5 m.y. ago (Underwood et al., 2010). The transition from Subunit IA to IB is constrained at 4.42 Ma from magnetostratigraphy, and thus appears slightly younger at Site C0012 than at Site C0011. The change of the frequency of ash layer occurrence at the Subunit IB/IC boundary is constrained to 7.13 Ma from magnetostratigraphy. The age models at the depth of the transition from Unit I to Unit II are very consistent, and they provide an age of 7.8 Ma, slightly older than at Site C0011.

Porosity and resistivity measured on cores at Site C0012 present, as at Site C0011, an anomalous interval where these parameters remain constant with depth and do not follow a progressive compaction trend, whereas shear strength increases. This interval is found at a shallower depth range (~10 mbsf to 70 mbsf), and it displays a more progressive transition at its base from ~70 mbsf to 100 mbsf across the lithologic Subunit IA/IB boundary. Site C0012 porosity values from below 240 mbsf are generally lower than those from similar depths at Site C0011. A possible explanation for this observation is the removal of about 100 meters of overlying material by erosion or slope failure. This interpretation would be consistent with the observed time gap of ~2 m.y. found between 10 mbsf and 14 mbsf. Near the base of the sediments, calcareous claystones cored below 500 mbsf range in porosity from 0.28 to 0.46 and show P-wave velocities of ~2000 m $s^{-1}$. Within the basalt, measured porosity is extremely variable and ranging from 0.09 to 0.37, and measured P-wave velocities vary between 3000 m $s^{-1}$ and 5000 m $s^{-1}$.

Thermal gradient values are evaluated from the APCT-3 measurements made at ten depths in Holes C0012C and C0012D together, and the mean thermal gradient value determined is 0.135 K $m^{-1}$ (Fig. 5). The estimated heat flow value at this site is 141 mW $m^{-2}$, amounting to ~50% higher than the 89.5 mW $m^{-2}$ determined for the adjacent Hole C0011C that was drilled during this expedition. Based on the determined heat flow value of 141 mW $m^{-2}$, as well as the thermal conductivity values from core measurements (Saito et al., 2010), temperature at the top of basement (at 526 mbsf) is estimated to be ~65°C, which is significantly lower than the estimated value of 80°C at ~1050 mbsf at Site C0011.

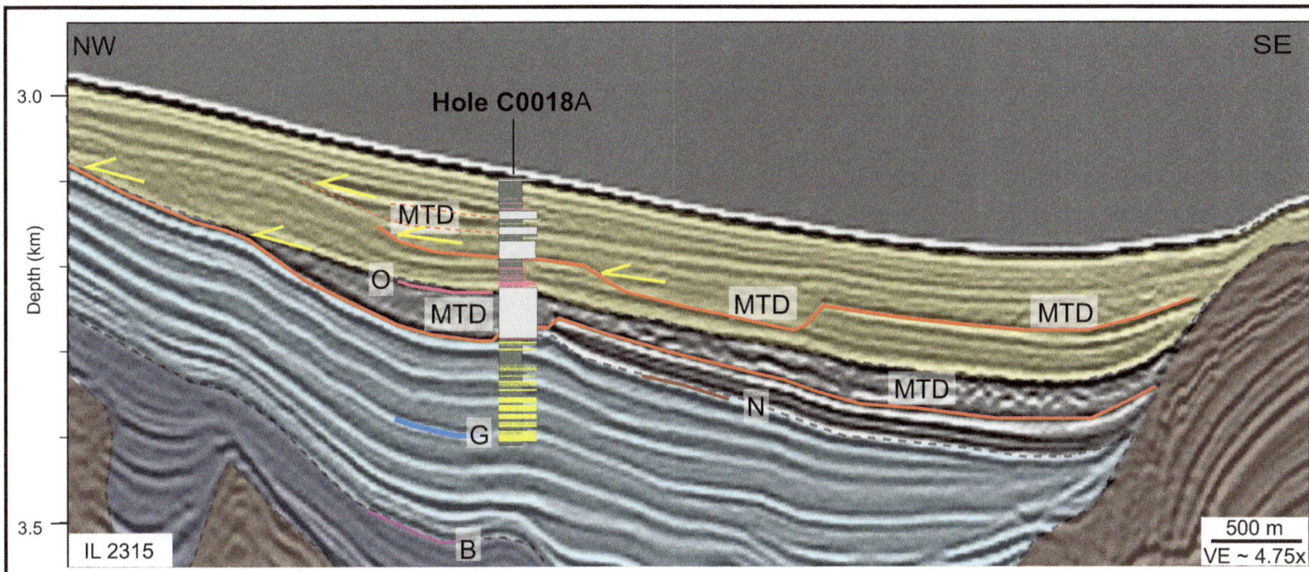


Figure 6. Lithology of Site C0018 on seismic profile. IL=in-line, VE=vertical exaggeration. Key seismic horizons are labeled B, G, N, and O MTD=mass transport deposit. Figure modified after Strasser et al. (2011).

As at Site C0011, results of interstitial water analysis are generally consistent between Exp. 333 and 322. Overall, the combined data sets reflect *in situ* alteration of volcanic ash in the sediment and basalt alteration in the upper igneous crust as well as exchange by diffusion (Underwood et al., 2010). The sulfate profile for Site C0012 indicates that sulfate reduction occurs at a deeper level than at the other sites drilled during this expedition. A minimum in sulfate concentration at ~300 mbsf was observed during Exp. 322, and this was interpreted as being driven locally by anaerobic methane oxidation (Underwood et al., 2010). In Holes C0012E and C0012A, sulfate increases in concentration in the interval below ~450 mbsf, which may indicate diffusional exchange with fluid in basaltic basement that sustain a higher sulfate concentration. A trend of increasing chlorinity, Ca, and Sr concentrations and decreasing Na in the lower part of the boreholes is also interpreted as a consequence of diffusion between the lowermost sediment and basement fluids. At a shallower level, samples from Hole C0012C document a silica concentration drop below the Subunit IA/IB boundary at ~70 m, where a decrease in porosity and an increase in resistivity are also observed. Alteration of volcanic glass shards also becomes more pronounced. As at Site C0011, this suggests a relationship between interstitial water composition, the presence or dissolution of opal-CT cement, and ash diagenesis.

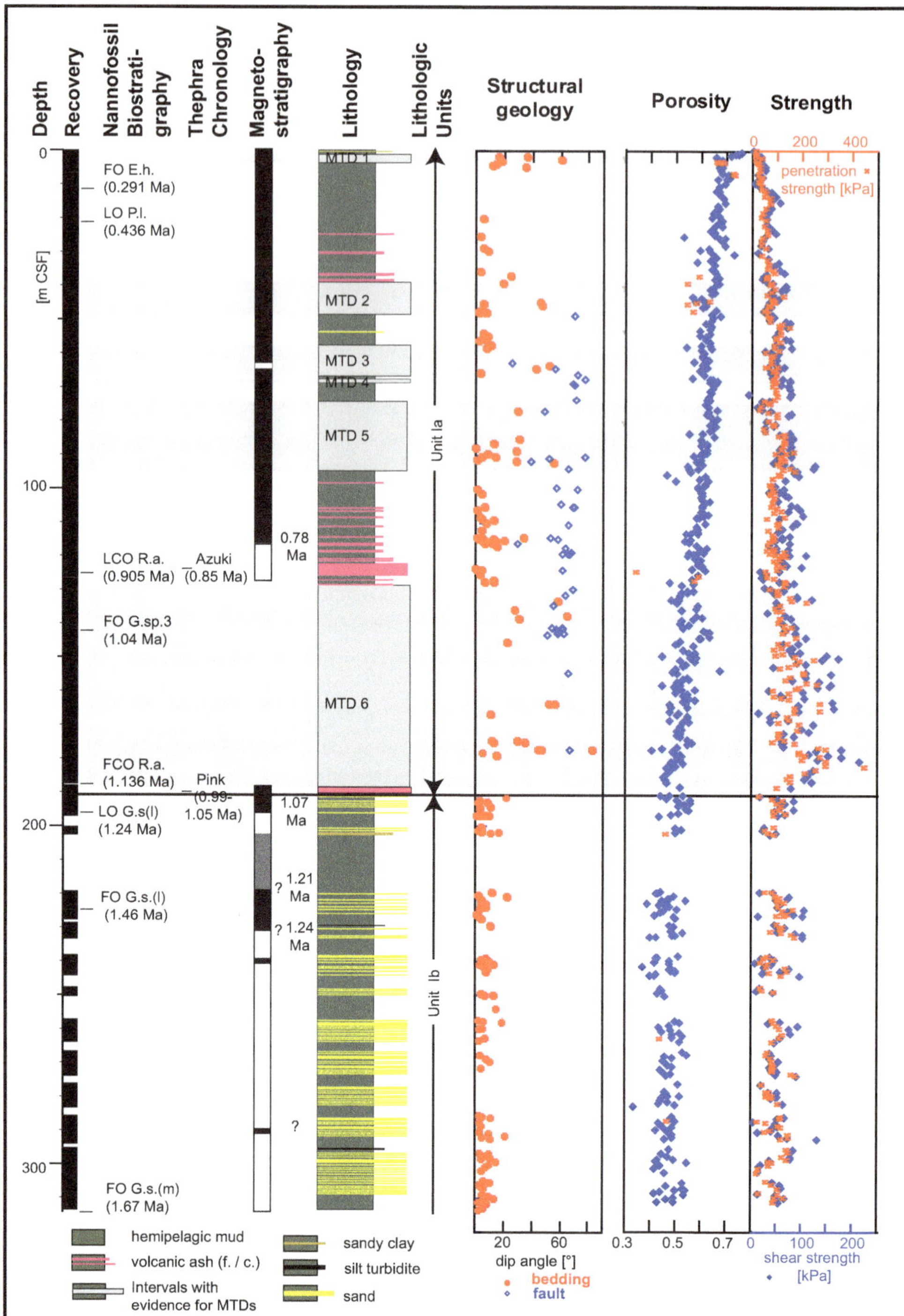


Figure 7. Summary diagram of IODP Expedition 333 Site C0018 shipboard results. Abbreviations for nannofossil data: F(C)O=first (consistent) occurrence, L(C)O=last (consistent) occurrence, E.h.=*Emiliania huxleyi*, P.l.=*Pseudoemiliania lacunosa*, R.a.=*Reticulofenestra asanoi*, G.sp.(l, m)=*Gephyrocapsa* species (large, medium).

In Holes C0012C and C0012D, methane and ethane were either below detection or present at only low concentrations. No heavier hydrocarbon gases (e.g., C3 and C4) were found. The only two samples that contained both methane and ethane were found at depths of ~501 mbsf and ~520 mbsf in Hole C0012E; they had C1/C2 ratios of <100, indicating a possible thermogenic origin of these hydrocarbon gases.

## Site C0018

Sediments cored at Site C0018 (Fig. 6) are subdivided into two lithologic subunits (Expedition 333 Scientists, 2012c), which correlate with seismic Unit 1a and 1b of Kimura et al. (2011) and Strasser et al. (2011). Lithologic Subunit 1a is primarily composed of bioturbated hemipelagic mud intercalated with layers of varying coarse and fine volcanic ash. Six intervals with evidence of MTDs are observed within Subunit 1a (Fig. 7). MTDs at Site C0018 range in thickness from 0.5 m to 61 m and have a cumulative thickness of 98 m, thus accounting for half of the total thickness of Subunit 1a. MTD 6 correlates to the main MTD body identified in seismic data (Fig. 6; Strasser et al., 2011). The upper boundary/contact is well defined for MTDs 1, 2, and 6 and is marked by an MTD-overlying turbidite for two of them (MTDs 2 and 6). Shear zones in fine-grained sediments define the base of MTD 2, 3, and 5. Additionally, physical properties data (Fig. 7) show that MTD intervals display an increased compaction gradient (locally stronger porosity decrease with depth) compared with the linear porosity-depth trend which may be defined over the whole Subunit 1a, and slight reversals (porosity increasing with depth) are observed near the base of MTDs 2, 3, 5, and 6. A thick ash layer, which correlates to the onland tephra "Pink" dated 1.05 Ma (Hayashida et al., 1996), occurs at the base of MTD 6 and also defines the base of Subunit 1a at 190.6 mbsf. This age is supported by magnetostratigraphic constraints correlating the base of a short normal polarity interval immediately below the subunit boundary to the base of the Jaramillo subchron (1.07 Ma), and it is also consistent with the maximum age for this boundary estimated from calcareous nanno-fossils (<1.24 Ma). The Azuki ash layer (Hayashida et al., 1996) was identified above MTD 6, constraining its emplacement between 0.85 Ma and 1.05 Ma. Unit 1a thus spans from 0 Ma to ~1 Ma, and the average sedimentation rate is ~19 cm ka$^{-1}$. Minor normal faults with moderate to high dip angles (50°–80°) are developed in the lower part of Subunit 1a; they reflect vertical maximum principal stress likely in response to the sedimentation and burial processes within the slope basin. Progressive burial and consolidation are indicated by the general downward increase and decrease of undrained shear strength and porosity, respectively (Fig. 7).

Lithologic Subunit 1b comprises a sequence of sandy turbidites interbedded with silty clay. Sub-horizontally dipping 4–20-cm-thick sand layers typically show a sharp and erosional base and normal grading. Bed spacing is 20–30 cm in average, but muddy intervals up to 4 m thick without intercalated turbidites occasionally occur. Turbidite sands have mixed composition with quartz, plagioclase, and abundant lithic fragments of both metamorphic and volcanic origin, indicating a siliciclastic source. Averaged sedimentation rate within Unit 1b is ~40 cm ka$^{-1}$, higher than in Unit 1a. Porosity and undrained shear strength (measured in muddy intervals) are generally constant throughout Unit 1b, and they show values slightly higher and lower, respectively, than what would be expected from downward extrapolating the general trends from Unit 1a. The drop in shear strength below the subunit boundary may be related to the switch from HPCS at this depth to extended punch and extended shoe coring systems (EPCS, and ESCS, respectively), which cause more disturbance to the cores. If real, however, this observation may be interpreted to reflect a general state of slight underconsolidation, potentially resulting from the high sedimentation rates within this interval and the deposition of the thick MTD immediately above.

## Key Results and Future Work

### *Shikoku Basin Stratigraphy and Diagenesis*

Observations on HPCS, EPCS and ESCS cores complemented stratigraphic description in the upper part of the sedimentary columns drilled during Exp. 322. The new observations, however, did not bring major changes in the definition of lithologic units (Saito et al., 2010; Expedition 333 Scientists, 2011). The most remarkable new observations is the correlation between a sharp variation of physical properties (porosity and electrical resistivity; Fig. 8), which was already evident in the LWD data from Exp. 319, a drop in silica concentration, and a concurrent decrease of the abundance of ash layers and increase of their state of alteration. Glass shards from ash layers generally appear fresh above, but are variably altered below, with evidence for glass dissolution and replacement by clay minerals.

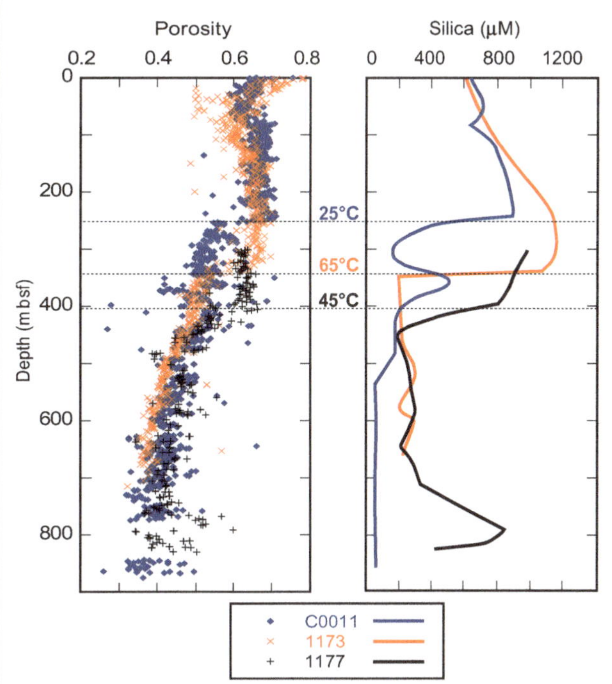


Figure 8. Porosity from sites C0011, 1173, and 1177 from moisture and density measurements, and silica concentration profiles in interstitial water. A concomitant drop of silica concentration and porosity is observed at all sites within the Shikoku Basin hemipelagites. This also correlates with a front of volcanic ash alteration. It was proposed that clay compaction is prevented by the precipitation of opal-CT cements at grain boundaries at the initial stages of diagenesis and that this metastable cement is then dissolved as temperature increases (Spinelli et al., 2007). The temperature estimated from borehole measurements at the depth of the transition is indicated. The diagenetic boundary is found at a different temperature at each site, which may help constraining the kinetics of the reactions.

The transition occurs within 10 m at 250 mbsf at Site C0011 but is found at a shallow level and is more progressive (50–80 m) at Site C0012. However, the sediment at this site has been affected by slumps, and this may preclude an accurate description of the transition. From an analogy with results obtained after Leg 190, it can be proposed that the process responsible for an anomalously high porosity of the sediment above this boundary is cementation by opal-CT and opal dissolution with precipitation of quartz below the boundary (Spinelli et al., 2007). However, the temperature of the reaction would be unusually low at Site C0011 (currently 25°C, as opposed to 65°C at Site 1173, see Fig. 8) and, assuming constant heat flow, it cannot be explained with the reaction kinetics used by Spinelli et al. (2007) because the kinetic time constants calculated at 25°C would be larger than the sediment age. The transition in physical properties is found at an even lower temperature (15°C) at Site C0012, but this may be in part explained by the removal of about 100 meters of sediment by the slump that occurred about 1 Ma ago. Furthermore, the primary silica source is dispersed volcanic glass rather than biogenic opal-A (White et al., 2011), and fresh glass is found again in volcaniclastic sands at a deeper level associated with high silica concentration in the interstitial water. It thus remains an open question whether the kinetics of the opal-A/opal-CT and opal-CT/quartz reactions can accurately model this transition. For instance, other kinetic models could be more appropriate for a volcanic ash silica source (Schacht et al., 2008). It is also possible that initial amorphous silica concentration immediately below the transition was too low to prevent compaction. Answering these questions will require postcruise work (e.g., amorphous silica content determination and modeling work). Chronostratigraphy based on paleomagnetism and biostratigraphy suggests that the age of the transition from opal-cemented hemipelagic mud with numerous ash layers to compacted hemipelagic mudstone with sparse ash layers at Site C0011 (5.25 Ma) is slightly younger at Site 1177 (4.5 Ma), the reference site for the Ashizuri transect. It is even younger at Site 1173 (2.5–3.0 Ma), the reference site for the Muroto transect, where higher heat flow and temperature can account for a more advanced silica diagenesis and result in an upward migration of the transition in the sedimentary column (Spinelli et al., 2007).

Overall, the temperature conditions at correlative stratigraphic levels do not vary by more than ~15°C between the two drill sites. Consequently, differences in fluid composition observed between the two sites (e.g., increasing versus decreasing chlorinity toward basement) cannot result solely from *in situ* reactions in the cored intervals. Temperature and heat flow determinations indicate that the temperature conditions in the uppermost basement at Site C0011 are within the temperature window for the onset of the smectite-illite reaction (~55°C–90°C), which has been proposed as the main cause of freshening at this site (Underwood et al., 2010). Although it is unlikely that the reaction has progressed to a significant extent in any of the cored sections, illitization should thus be expected in the lower part of Site C0011, and it should continue into the trench where the Shikoku Basin sediments are buried to greater depths. Unfortunately, coring during Exp. 322 was aborted almost 200 m above the top of basement at Site C0011.

## Heat Flow and Convection

Exp. 333 provided the first borehole temperature measurements on the incoming plate of the Kumano transect, which constitute an essential constraint for subduction thermal models. Estimation of heat flow was conducted from linear regression of temperature as a function of thermal resistance. Heat flow thus obtained shows a good correspondence with heat flow probe and instrumented corer data (Fig. 5; Kinoshita et al., 2008; Hamamoto et al., 2011). The 90 mW $m^{-2}$ heat flow at Site C0011 is in the lower range of heat flow probe values, while the 140 mW $m^{-2}$ heat flow at Site C0012 is one of the highest. This result confirms the presence of a high heat flow anomaly on the Kashinozaki Knoll. However, constraints on the age of the oldest sediments at the base of C0012 are compatible with Kashinozaki Knoll having the same age (about 20 Ma) as the lithosphere in this part of the Shikoku Basin as indicated from magnetic anomalies (Chamot-Rooke et al., 1987; Okino et al., 1994). When compared with models, heat flow at Site C0011 appears slightly lower that the heat flow expected for a 20-Ma lithosphere (Kinoshita et al., 2008) but can be fit within uncertainties once the effect of sedimentation in the trench is taken into account (Marcaillou et al., 2012). Large-scale heat transfer from depth within the subduction zone to the trench by convection in the crust, as proposed for the Muroto transect to the west (Spinelli and Wang, 2008), appears unlikely on the Kumano transect because the heat flow remains moderate in the trench. However, 1- to 10-km scale heat flow variability could be caused at least in part by thermal convection. Because pore fluid geochemistry profiles from Exp. 322 and 333 do not provide evidence for fluid transfer between Site C0011 and Site C0012 along sedimentary layers, local convection in the basement appears the most likely explanation for the presence of a heat flow maximum in the summit area of Kashinozaki Knoll.

## The Sediment-Basalt Interface and the Basalt

The basalt-sediment interface was cored three times during Exp. 333 with ESCS in Hole C0012E, and with RCB in Holes C0012F and C0012G. Although coring with ESCS was not the original choice, it was used to make the best of a short workable window between two periods of bad weather, and it provided good recovery of altered basalt including the inter-pillow hyaloclastites and analcime filled cavities. Multiple sampling of the red clay above the basement allowed recovery of larger amounts of pore fluid, which will be used for shore-based characterization of fluid isotopic composition (strontium, lithium, dissolved inorganic carbon). From an operational point of view, the use of a polycrystalline diamond compressed (PDC) bit turned out as inappropriate, as the bit was worn down to a point it was no longer functional even before basement coring started. As a consequence of weather-related and technical delays only two days were available for basement coring. One hundred meters were cored with patchy recovery. Pillows and massive flows of phyric basalt were recovered. As observed during Exp. 333, basalts were pervasively altered under iron reducing conditions, most of the glass, olivines, and part of the plagioclases being replaced by saponites and zeolites. Deeper penetration below the seafloor found occasional veins with accumulations of iron hydroxides. The sequence of alteration still needs to be determined. Knowledge of mineralogical composition of the altered basalt will allow quantifying fluid sources available from mineral reactions at the depth of the seismogenic zone, notably the saponite-chlorite reaction.

### *Mass Wasting and Mass Transport Deposits*

The slope-basin stratigraphic succession drilled at Site C0018 documents ~1.5 million years of sedimentation by hemipelagic settling, turbidite, and mass-transport deposition in the footwall of the MGFZ. A lithological transition between a sandy turbidite sequence (below) and ash-bearing hemipelagites that contain MTDs (above) is dated to ~1 Ma (Fig. 7). Six MTDs of thickness 0.5–61 m were identified in the succession overlying this transition, but no positive evidence for MTDs within the turbidite sequence below has been found in the cores. During the same time period, a northward shift in the sediment depocenter is observed within the Kumano forearc basin following ~300 kyr of extensive landward tilting of the outer Kumano basin sediments. This has been interpreted to be related to an episode of motion along the megasplay that formed the modern fault geometry (Gulick et al., 2010). Hence, one possible explanation for the change in sediment delivery and routing pattern inferred from Site C0018 results is that the uplift of the outer arc high confined most of the turbidity currents to the Kumano Basin and thus progressively shut off sand input to the slope-basin seaward of the MSFZ (Strasser et al., 2012). Alternatively, the observed evolution could correspond to a local change of slope and depositional environment from a perched basin trapping sand transported by turbidity currents to a slope environment.

MTD 6 is correlative with this lithological transition and thus potentially may be related to the overall structural evolution of the MSFZ area and/or the change in sedimentation regime. However, all younger MTDs observed at Site C0018 postdate the main structural phase of slip on the shallow MGFZ and may be rather associated with shallow-seated slope failure processes within the post-1-Ma sedimentary cover overlying the MSFZ and anticline structure. The deformation style within these MTDs is heterogeneous; intervals of disturbed sediments are interbedded within intervals inferred to retain original, coherent bedding. In three occurrences the base of the MTD is defined by a shear zone within fine-grained sediments. These observations suggest that slumping and mass-transport are dominated by localized plastic deformation and that slumping only partly evolved into mud flows.

Mass movements are actively shaping the present-day seafloor, as evidenced by numerous surficial slump scars (Kimura et al., 2011; Strasser et al., 2012) and the occurrence of a shallowly buried, presumably Holocene, MTD at Site C0018 (Strasser et al., 2012). Further important findings are that the spacing between individual MTDs suggests submarine slope destabilization does not occur systematically during subduction earthquakes that have a recurrence rate on the order of 100–200 years (Ando, 1975). Hence, yet-to-be-investigated preconditioning factors may exert first-order control on destabilization processes and MTD formation over the last 1 Ma and in the future. These preconditioning factors will be the focus of upcoming studies integrating higher-resolution age constraints, more detailed sedimentological and structural investigations of MTDs, 3D-seismic interpretation and seismic attribute analysis, geotechnical laboratory analysis, as well as data and results from nearby IODP drill sites.

Site C0012 also appears strongly influenced by mass wasting processes. Slumps affecting sediment packages of 70–80 m thickness were identified during Exp. 322 and 333. Their top is marked by hiatuses (from ~1 Ma to 3 Ma and around ~9.5 Ma, respectively), and an even longer hiatus is evident along an unconformity within Unit V near the base of the sedimentary section. Repetition of mass wasting events is very understandable considering the location of Site C0012 on a topographic high immediately above the edge of a major slide scar observed in the bathymetry and seismic data.

## Summary and Conclusions

Borehole heat flow data from Exp. 333 constrain heat flow in the trench and provide a reliable seaward boundary condition for modeling of temperature along the subduction interface (Marcaillou et al., 2012). A sharp contrast in physical properties in the Shikoku Basin hemipelagites is interpreted as a diagenetic boundary, present throughout the basin, and related to the diagenesis of high-silica volcanic ash. This diagenetic boundary is diachronous, and the geometry of turbidite bodies in the lower part of the Shikoku Basin displays considerable variability (Underwood, 2007; Underwood et al., 2010). Nevertheless, an interval of sedimentation deprived of fresh volcanic ash and turbidites is found on the three transects investigated by IODP drilling in the Nankai area. This broadly corresponds to the 4–7 Ma time interval. This interval appears as a favorable interval for the localization of the decolletment of the accretionary wedge but also for the localization of basal shear of slumps at Sites C0011 and C0012. Sampling of cores for geotechnical and mechanical tests will address this problem.

A sequence of cored mass transport deposits remobilizing slope sediments in the shallow MGFZ area documents ~1 million years of submarine landslides in this tectonically active setting (Strasser et al., 2012). Although earthquakes are an obvious potential triggering factor, these mass wasting events appear much less frequent than large earthquakes along this subduction segment. Thus, preconditioning factors such as sedimentation, tectonic loading, and fluid flow must exert key control on slope destabilization, and they are the focus of future investigations. Evidence from Sites C0011 and C0012 shows that large mass wasting events also occurred on topography on the downgoing plate and could explain some of the hiatuses observed within accreted sediment packages.

The basement appears altered by low temperature interaction with pore fluid of seawater origin over at least 100 m. Fluid convection is likely to occur in the fractured basement, but it constitutes a hydrological system largely distinct from the sediment, which comprises clay-dominated intervals that form barriers to flow, at least in the vertical direction. Release of water from mineral reactions will occur both in the sediment and in the basement, as they are buried below the sedimentations in the trench and underthrust beneath the margin. Modeling this evolution will likely be an important objective of post-cruise work.

## The IODP Expedition 333 Scientific Party

Tiago Alves, Thorsten Bauersachs, Hugh Daigle, Koray Ekinci, Shu Gao, Marion Garcon, Kiichiro Kawamura, Yujin Kitamura, Jan Sverre Laberg, Gwangsoo Lee, Youngmin Lee, Yuehan Lu, Boris Marcaillou, Osamu Matsubayashi, Yoshitaka Nagahashi, Beth Novak, Yu Saito, Yasufumi Satoguchi, Elizabeth J. Screaton, Rachel P. Scudder, Bo Ra Song, Asuka Yamaguchi

## Acknowledgements

We are indebted to the captains, operation superintendants, offshore installation managers, shipboard personnel, laboratory manager and officers, curators and laboratory technicians who sailed on the D/V *Chikyu* during Expedition 333 for their dedication and assistance with all aspects of logging, coring, sampling, and shipboard laboratory measurements. We also thank the co-chiefs of Expedition 322 and the Project Management Team and specialty coordinators of NanTroSEIZE for their organizational know-how and guidance with some of the scientific interpretations.

## References

Ando, M., 1975. Source mechanisms and tectonic significance of historical earthquakes along the nankai trough, Japan. *Tectonophysics*, 27:119–140. doi:10.1016/0040-1951(75)90102-X

Baba, T., Cummins, P.R., Hori, T., and Kaneda, Y., 2006. High precision slip distribution of the 1944 Tonankai earthquake inferred from tsunami waveforms: Possible slip on a splay fault. *Tectonophysics*, 426(1–2):119–134. doi:10.1016/j.tecto.2006.02.015

Bangs, N.L.B., Gulick, S.P.S., and Shipley, T.H., 2006. Seamount subduction erosion in the Nankai Trough and its potential impact on the seismogenic zone. *Geology*, 34(8):701–704. doi:10.1130/G22451.1

Chamot-Rooke, N., Renard, V., and Pichon X.-L, 1987. Magnetic anomalies in the Shikoku Basin: A new interpretation. *Earth Planet. Sci. Lett.*, 83:214–228. doi:10.1016/0012-821X(87)90067-7

DeMets, C., Gordon, R.G., and Argus, D.F., 2010. Geologically current plate motions. *Geophys. J. Int.*, 181(1):1–80. doi:10.1111/j.1365-246X.2009.04491.x

Dessa, J.-X., Operto, S., Kodaira, S., Nakanishi, A., Pascal, G., Uhira, K., and Kaneda, Y., 2004. Deep seismic imaging of the eastern Nankai Trough, Japan, from multifold ocean bottom seismometer data by combined travel time tomography and prestack depth migration. *J. Geophys. Res.*, 109:B02111. doi:10.1029/2003JB002689

Expedition 333 Scientists, 2011. NanTroSEIZE Stage 2: Subduction inputs 2 and heat flow. *IODP Prel. Rept.*, 333. doi:10.2204/iodp.pr.333.2011

Expedition 333 Scientists, 2012a. Site C0011. *In* Henry, P., Kanamatsu, T., Moe, K., and the Expedition 333 Scientists, *Proc. IODP*, 333: Washington, DC (Integrated Ocean Drilling Program Management International, Inc.). doi:10.2204/iodp.proc.333.104.2012

Expedition 333 Scientists, 2012b. Site C0012. *In* Henry, P., Kanamatsu, T., Moe, K., and the Expedition 333 Scientists, *Proc. IODP*, 333: Washington, DC (Integrated Ocean Drilling Program Management International, Inc.). doi:10.2204/iodp.proc.333.105.2012

Expedition 333 Scientists, 2012c. Site C0018. *In* Henry, P., Kanamatsu, T., Moe, K., and the Expedition 333 Scientists, *Proc. IODP*, 333: Washington, DC (Integrated Ocean Drilling Program Management International, Inc.). doi:10.2204/iodp.proc.333.103.2012

Fergusson, C.L., 2003. Provenance of Miocene–Pleistocene turbidite sands and sandstones, Nankai Trough, Ocean Drilling Program Leg 190. *In* Mikada, H., Moore, G.F., Taira, A., Becker, K., Moore, J.C., and Klaus, A. (Eds.), *Proc. ODP, Sci. Results*, 190/196: College Station, TX (Ocean Drilling Program), 1–28. doi:10.2973/odp.proc.sr.190196.205.2003

Gulick, S.P.S., Bangs, N.L.B., Moore, G.F., Ashi, J., Martin, K.M., Sawyer, D.S., Tobin, H.J., Kuramoto, S., and Taira, A., 2010. Rapid forearc basin uplift and megasplay fault development from 3D seismic images of Nankai Margin off Kii Peninsula, Japan. *Earth Planet. Sci. Lett.*, 300(1–2):55–62. doi:10.1016/j.epsl.2010.09.034

Hamamoto, H., Yamano, M., Goto, S., Kinoshita, M., Fujino, K., and Wang, K., 2011. Heat flow distribution and thermal structure of the Nankai subduction zone off the Kii Peninsula. *Geochem. Geophys. Geosyst.*, 12:Q0AD20. doi:10.1029/2011GC003623

Harris, R.N., Schmidt-Schierhorn, F., and Spinelli, G., 2011. Heat flow along the NanTroSEIZE transect: Results from IODP Expeditions 315 and 316 offshore the Kii Peninsula, Japan. *Geochem. Geophys. Geosyst.*, 12:Q0AD16. doi:10.1029/2011GC003593

Hayashida, A., Kamata, H., and Danhara, T., 1996. Correlation of widespread tephra deposits based on paleomagnetic directions: Link between a volcanic field and sedimentary sequences in Japan. *Quat. Int.*, 34–36:89–98. doi:10.1016/1040-6182(95)00072-0

Henry, P., Kanamatsu, T., and Moe, K.T., 2010. NanTroSEIZE Stage 2: Subduction inputs 2 and heat flow. *IODP Sci. Prosp.*, 333. doi:10.2204/iodp.sp.333.2010

Hyndman, R.D., Yamano, M., and Oleskevich, D.A., 1997. The seismogenic zone of subduction thrust faults. Island Arc, 6(3):244–260. doi:10.1111/j.1440-1738.1997.tb00175.x

Ike, T., Moore, G.F., Kuramoto, S., Park, J-O., Kaneda, Y., and Taira, A., 2008a. Tectonics and sedimentation around Kashinosaki Knoll: A subducting basement high in the eastern Nankai Trough. *Island Arc*, 17(3):358–375. doi:10.1111/j.1440-1738.2008.00625.x

Ike, T., Moore, G.F., Kuramoto, S., Park, J.-O., Kaneda, Y., and Taira, A., 2008b. Variations in sediment thickness and type along the northern Philippine Sea Plate at the Nankai Trough. *Island Arc*, 17(3):342–357. doi:10.1111/j.1440-1738.2008.00624.x

Ito, Y., Matsumoto, T., Kimura, H., Matsubayashi, H., Obara, K., and Sekiguchi, S., 2005. Spatial distribution of centroid moment tensor solutions for the 2004 off Kii peninsula earthquakes. *Earth Planets Space*, 57:351–356.

Kimura, G., Moore, G.F., Strasser, M., Screaton, E., Curewitz, D., Streiff, C., and Tobin, H., 2011. Spatial and temporal evolution of the megasplay fault in the Nankai Trough. *Geochem. Geophys. Geosyst.*, 12:Q0A008. doi:10.1029/2010GC003335

Kinoshita, M., Kanamatsu, T., Kawamura, K., Sibata, T., Hamamoto, H., and Fujino, K., 2008. Heat flow distribution on the floor of Nankai Trough off Kumano and implications for the geothermal regime of subducting sediments. *JAMSTEC Rep. Res. Dev.*, 8:13–28. http://docsrv.godac.jp/MSV2_DATA/23/JAM_RandD08_02.pdf

Kinoshita, M., Tobin, H., Ashi, J., Kimura, G., Lallemant, S., Screaton, E.J., Curewitz, D., Masago, H., Moe, K.T., and the Expedition 314/315/316 Scientists, 2009. *Proc. IODP*, 314/315/316: Washington, DC (Integrated Ocean Drilling Program Management International, Inc.). doi:10.2204/iodp.proc.314315316.2009

Kodaira, S., Nakanishi, A., Park, J.-O., Ito, A., Tsuru, T., and Kaneda, Y., 2003. Cyclic ridge subduction at an inter-plate locked zone off central Japan. *Geophys. Res. Lett.*, 30:1339–1342. doi:10.1029/2002GL016595

Kodaira, S., Takahashi, N., Nakanishi, A., Miura, S., and Kaneda, Y., 2000a. Subducted seamount imaged in the rupture zone of the 1946 Nankaido earthquake. *Science*, 289(5476):104–106. doi:10.1126/science.289.5476.104

Kodaira, S., Takahashi, N., Park, J.-O., Mochizuki, K., Shinohara, M., and Kimura, S., 2000b. Western Nankai Trough seismogenic zone: Results from a wide-angle ocean bottom seismic survey. *J. Geophys. Res.*, 105(B3):5887–5905. doi:10.1029/1999JB900394

Lallemand, S.E., Malavieille, J., and Calassou, S., 1992. Effects of oceanic ridge subduction on accretionary wedges: Experimental modeling and marine observations. *Tectonics*, 11(6):1301–1313. doi:10.1029/92TC00637

Le Pichon, X., Lallemant, S., Tokuyama, H., Thoué, F., Huchon, P., and Henry, P., 1996. Structure and evolution of the backstop in the Eastern Nankai Trough area (Japan): Implications for the soon-to-come Tokai earthquake. *Island Arc*, 5(4):440–454. doi:10.1111/j.1440-1738.1996.tb00164.x

Loveless, J.P., and Meade, B.J., 2010. Geodetic imaging of plate motions, slip rates, and partitioning of deformation in Japan. *J. Geophys. Res.*, 115:B02410. doi:10.1029/2008JB006248

Marcaillou, B., Henry, P., Kinoshita, M., Kanamatsu, T., Screaton, E.J., Daigle, H., Harcouët, V., et. al., 2012. Seismogenic zone temperatures and heat-flow anomalies in the To-nankai margin segment based on temperature data from IODP expedition 333 and thermal model. *Earth Planet. Sci. Lett.*, 349–350:171–185. doi:10.1016/j.epsl.2012.06.048

Mazzotti, S., Henry, P., and Le Pichon, X., 2001. Transient and permanent deformation of central Japan estimated by GPS2. Strain partitioning and arc-arc collision. *Earth Planet. Sci. Lett.*, 184:455–469.

Mazzotti, S., Lallemant, S.J., Henry, P., Le Pichon, X., Tokuyama, H., and Takahashi, N., 2002. Intraplate shortening and underthrusting of a large basement ridge in the eastern Nankai subduction zone. *Mar. Geol.*, 187(1–2):63–88. doi:10.1016/S0025-3227(02)00245-1

Moore, G.F., Bangs, N.L., Taira, A., Kuramoto, S., Pangborn, E., and Tobin, H.J., 2007. Three-dimensional splay fault geometry and implications for tsunami generation. *Science*, 318(5853):1128–1131. doi:10.1126/science.1147195

Moore, G.F., Park, J.-O., Bangs, N.L., Gulick, S.P., Tobin, H.J., Nakamura, Y., Sato, S., et al., 2009. Structural and seismic stratigraphic framework of the NanTroSEIZE Stage 1 transect. *In* Kinoshita, M., Tobin, H., Ashi, J., Kimura, G., Lallemant, S., Screaton, E.J., Curewitz, D., et al., *Proc. IODP*, 314/315/316: Washington, DC (Integrated Ocean Drilling Program Management International, Inc.). doi:10.2204/iodp.proc.314315316.102.2009

Moore, G.F., Taira, A., Klaus, A., et al., 2001. *Proc. ODP, Init. Repts.*, 190: College Station, TX (Ocean Drilling Program). doi:10.2973/odp.proc.ir.190.2001

Nagahashi, Y., and Satoguchi, Y., 2007. Stratigraphy of the Pliocene to lower Pleistocene marine formations in Japan on the basis of tephra beds correlation. *The Quaternary Research [Daiyonki Kenkyu]*, 46(3):205–213. doi:10.4116/jaqua.46.205

Okino, K., Shimakawa, Y., and Nagaoka, S., 1994. Evolution of the Shikoku Basin. *J. Geomag. Geoelectr.*, 46(6):463–479.

Park, J.-O., Moore, G.F., Tsuru, T., Kodaira, S., and Kaneda, Y., 2004. A subducted oceanic ridge influencing the Nankai megathrust earthquake rupture. *Earth Planet. Sci. Lett.*, 217(1–2):77–84. doi:10.1016/S0012-821X(03)00553-3

Park, J.-O., Tsuru, T., Kaneda, Y., Kono, Y., Kodaira, S., Takahashi, N., and Kinoshita, H., 1999. A subducting seamount beneath the Nankai Accretionary Prism off Shikoku, southwestern Japan. *Geophys. Res. Lett.*, 26(7):931–934. doi:10.1029/1999GL900134

Park, J.-O., Tsuru, T., Kodaira, S., Cummins, P.R., and Kaneda, Y., 2002. Splay fault branching along the Nankai subduction zone. *Science*, 297(5584):1157–1160. doi:10.1126/science.1074111

Park, J.-O., Tsuru, T., No, T., Takizawa, K., Sato, S., and Kaneda, Y., 2008. High-resolution 3D seismic reflection survey and prestack depth imaging in the Nankai Trough off southeast Kii Peninsula. *Butsuri-Tansa [Geophys. Explor.]*, 61:231–241. (in Japanese, with abstract in English)

Saito, S., Underwood, M.B., Kubo, Y., and the Expedition 322 Scientists, 2010. *Proc. IODP*, 322: Washington, DC (Integrated Ocean Drilling Program Management International, Inc.). doi:10.2204/iodp.proc.322.2010

Satoguchi, Y., Higuchi, Y., and Kurokawa, K., 2005. Correlation of the Ohta tephra bed in the Tokai group with a tephra in the Miura group, central Japan. *Chishitsugaku Zasshi [J. Geol. Soc. Japan]*, 111(2):74–86.

Schacht, U., Wallmann, K., Kutterolf, S., and Schmidt, M., 2008. Volcanogenic sediment seawater interactions and the geochemistry of pore waters. *Chem. Geol.*, 249:321–338.

Screaton, E., Kimura, G., Curewitz, D., Moore, G., Chester, F., Fabbri, O., Fergusson, C., et al., 2009. Interactions between deformation and fluids in the frontal thrust region of the NanTroSEIZE transect offshore the Kii Peninsula, Japan: Results from IODP Expedition 316 Sites C0006 and C0007. *Geochem. Geophys. Geosyst.*, 10:Q0AD01. doi: 10.1029/2009GC002713

Seno, T., Stein, S., and Gripp, A.E., 1993. A model for the motion of the Philippine Sea Plate consistent with NUVEL-1 and geological data. *J. Geophys. Res.*, 98(B10):17941–17948. doi:10.1029/93JB00782

Spinelli, G.A., and Wang, K., 2008. Effects of fluid circulation in subduction crust on Nankai margin seismogenic zone temperatures. *Geology*, 36(11):887–890. doi:10.1130/G25145A.1

Spinelli, G.A., Mozley, P.S., Tobin, H.J., Underwood, M.B., Hoffman, N.W., and Bellew, G.M., 2007. Diagenesis, sediment strength, and pore collapse in sediment approaching the Nankai Trough subduction zone. *Geol. Soc. Am. Bull.*, 119(3–4):377–390.

Strasser, M., Henry, P., Kanamatsu, T., Thu, M.K., Moore, G.F., and the IODP Expedition 333 Scientists, 2012. Scientific drilling of mass-transport deposits in the Nankai accretionary wedge: First results from IODP Expedition 333. *In* Yamada, Y., Kawamura, K., Ikehara, K., Ogawa, Y., Urgeles, R., Mosher, D., Chaytor, J., and Strasser, M. (Eds.), *Submarine Mass Movements and Their Consequences, Adv. Nat. Technol. Hazard Res.*, 31:671–681. doi:10.1007/978-94-007-2162-3_60

Strasser, M., Moore, G.F., Kimura, G., Kitamura, Y., Kopf, A.J., Lallemant, S., Park, J.-O., Screaton, E.J., Su, X., Underwood, M.B., and Zhao, X., 2009. Origin and evolution of a splay fault in the Nankai accretionary wedge. *Nat. Geosci.*, 2:648–652. doi:10.1038/ngeo609

Strasser, M., Moore, G.F., Kimura, G., Kopf, A., Underwood, M., Guo, J., and Screaton, E.J., 2011. Slumping and mass transport deposition in the Nankai fore arc: Evidence from IODP drilling and 3-D reflection seismic data. *Geochem. Geophys. Geosyst.*, 12:Q0AD13. doi:10.1029/2010GC003431

Tobin, H.J., and Kinoshita, M., 2006a. Investigations of seismogenesis at the Nankai Trough, Japan. *IODP Sci. Prosp.*, NanTroSEIZE Stage 1. doi:10.2204/iodp.sp.nantroseize1.2006

Tobin, H.J., and Kinoshita, M., 2006b. NanTroSEIZE: The IODP Nankai Trough Seismogenic Zone Experiment. *Sci. Drill.*, 2:23–27. doi:10.2204/iodp.sd.2.06.2006

Tobin, H., Kinoshita, M., Ashi, J., Lallemant, S., Kimura, G., Screaton, E.J., Moe, K.T., Masago, H., Curewitz, D., and the Expedition 314/315/316 Scientists, 2009. NanTroSEIZE Stage 1 expeditions: Introduction and synthesis of key results. *In* Kinoshita, M., Tobin, H., Ashi, J., Kimura, G., Lallemant, S., Screaton, E.J., Curewitz, D., Masago, H., Moe, K.T., and the Expedition 314/315/316 Scientists, *Proc. IODP*, 314/315/316: Washington, DC (Integrated Ocean Drilling Program Management International, Inc.). doi:10.2204/iodp.proc.314315316.101.2009

Underwood, M.B., 2007. Sediment inputs to subduction zones: Why lithostratigraphy and clay mineralogy matter. *In* Dixon, T., and Moore, J.C. (Eds.), *The Seismogenic Zone of Subduction Thrust Faults:* New York (Columbia University Press), 42–85.

Underwood, M.B., Saito, S., Kubo, Y., and the Expedition 322 Scientists, 2010. Expedition 322 summary. *In* Saito, S., Underwood, M.B., Kubo, Y., and the Expedition 322 Scientists, *Proc. IODP*, 322: Washington, DC (Integrated Ocean Drilling Program Management International, Inc.). doi:10.2204/iodp.proc.322.101.2010

White, R., Spinelli, G.A., Mozley, P.S., and Dunbar, N.W., 2011. Importance of volcanic glass alteration to sediment stabilization: Offshore Japan. *Sedimentology*, 58(5):1138–1154. doi:10.1111/j.1365-3091.2010.01198.x

## Authors

**Pierre Henry**, Centre Européen de Recherche et d'Enseignement en Géosciences de l'Environement, Aix-Marseille Université et CNRS, Europôle de l'Arbois, 13545 Aix en Provence Cedex 04, France, e-mail: henry@cerege.fr
**Toshiya Kanamatsu**, Institute for Research on Earth Evolution, Japan Agency for Marine-Earth Science and Technology, 2-15 Natsushima-cho, Yokosuka, Kanagawa 237-0061, Japan
**Kyaw Thu Moe**, Center for Deep Earth Exploration, Japan Agency for Marine-Earth Science and Technology, 3173-25 Showa-machi, Kanazawa-ku, Yokohama, Kanagawa, 236-0001, Japan
**Michael Strasser**, ETH Zürich, Geologisches Institut, Sonneggstrasse 5, 8092 Zürich, Switzerland
**and the IODP Expedition 333 Scientific Party**

# 14

# 500,000 Years of Environmental History in Eastern Anatolia: The PALEOVAN Drilling Project

by Thomas Litt, Flavio S. Anselmetti, Henrike Baumgarten, Jürg Beer, Namik Cagatay, Deniz Cukur, Emre Damci, Clemens Glombitza, Gerald Haug, Georg Heumann, Jens Kallmeyer, Rolf Kipfer, Sebastian Krastel, Ola Kwiecien, A. Feray Meydan, Sefer Orcen, Nadine Pickarski, Marie-Eve Randlett, Hans-Ulrich Schmincke, Carsten J. Schubert, Mike Sturm, Mari Sumita, Mona Stockhecke, Yama Tomonaga, Luigi Vigliotti, Thomas Wonik, and the PALEOVAN Scientific Team

## Abstract

International Continental Scientific Drilling Program (ICDP) drilled a complete succession of the lacustrine sediment sequence deposited during the last ~500,000 years in Lake Van, Eastern Anatolia (Turkey). Based on a detailed seismic site survey, two sites at a water depth of up to 360 m were drilled in summer 2010, and cores were retrieved from sub-lake-floor depths of 140 m (Northern Basin) and 220 m (Ahlat Ridge). To obtain a complete sedimentary section, the two sites were multiple-cored in order to investigate the paleoclimate history of a sensitive semi-arid region between the Black, Caspian, and Mediterranean seas. Further scientific goals of the PALEOVAN project are the reconstruction of earthquake activity, as well as the temporal, spatial, and compositional evolution of volcanism as reflected in the deposition of tephra layers. The sediments host organic matter from different sources and hence composition, which will be unravelled using biomarkers. Pathways for migration of continental and mantle-derived noble gases will be analyzed in pore waters. Preliminary $^{40}Ar/^{39}Ar$ single crystal dating of tephra layers and pollen analyses suggest that the Ahlat Ridge record encompasses more than half a million years of paleoclimate and volcanic/geodynamic history, providing the longest continental record in the entire Near East to date.

## Background and Motivation

The controversial discussion of present and future global warming has demonstrated that it is crucial to increase our knowledge of past climate change to better understand the pattern and dynamics of the global climate system. In continental regions this information can be obtained from lacustrine sediments, where biotic and abiotic parameters provide proxy climate data. In addition, lake sediments constitute unique paleoenvironmental archives storing also information of geologic disasters such as earthquakes or volcanic eruptions. Lake Van in Turkey is an excellent paleoclimate and paleoenvironmental archive, as site survey data indicated that it contains a long continental sedimentary record covering several glacial-interglacial cycles consisting partly of annually-laminated sediments.

Favorable geological conditions in the eastern Mediterranean and the Near East, such as relatively undisturbed accumulation of continuous Quaternary sediments, are rare. Successions such as the pollen records from Tenagi Phillipon and Joannina in Greece (Tzedakis et al., 2001) or from Lake Urmia in NW Iran (Djamali et al., 2008) document climatic changes over the continent on centennial to orbital time scales. These sequences enabled tentative comparisons with the marine isotopic record based on tuning to astronomical parameters, and showed that many stages and substages of the marine isotopic sequence are reflected in continental records. However, the timing of the transitions between them may not be precisely synchronous in the marine and continental realms. Lake Van holds a key position within

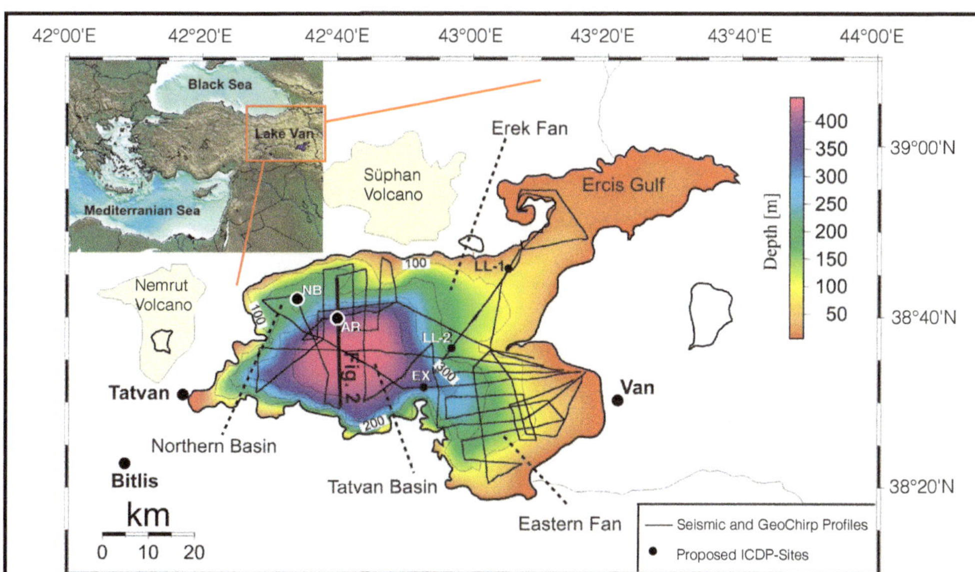


Figure 1. Bathymetric map of Lake Van with location of available seismic profiles and originally proposed drill sites. Ahlat Ridge (AR) and Northern Basin (NB) were finally drilled.

a sensitive climate region between the Mediterranean, Black, Caspian, and Arabian seas allowing for studying Quaternary climate evolution in the Near East during the last ~500 kyr (Litt et al., 2009, 2011). The Lake Van record drilled in 2010 is longer than those of Lake Urmia and Lake Joannina. In addition, the Van record will have a higher resolution and will be better dated than all of these previous records.

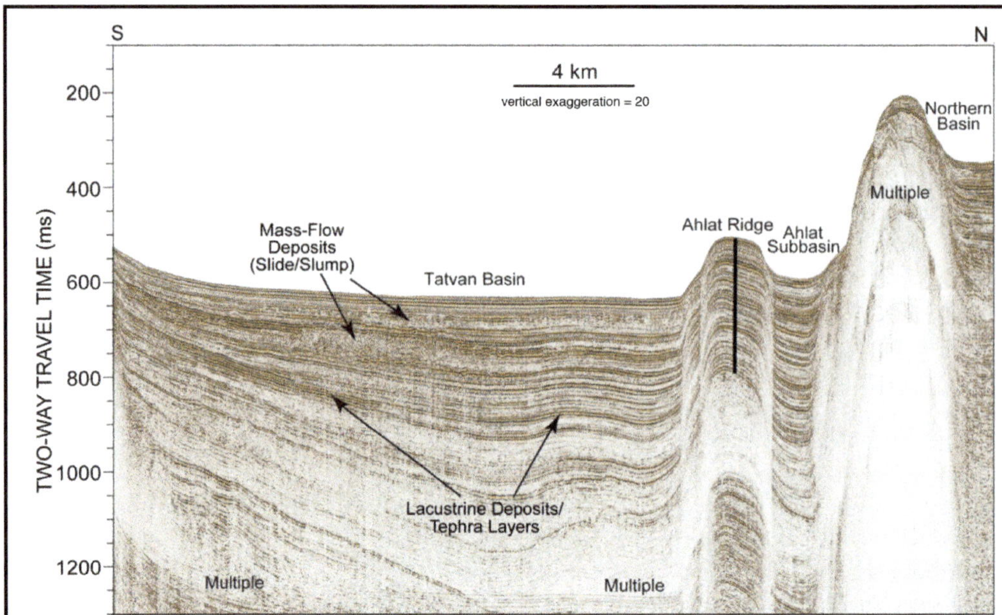


Figure 2. Seismic profile crossing the Tatvan Basin, Ahlat Ridge, and Northern Basin. Ahlat Ridge was selected as key site for the ICDP campaign in order to drill recover a complete sedimentary section for paleoclimatic investigations all the way to basement. See Fig. 1 for location of profile.

Lake Van, situated on a high plateau in eastern Anatolia, extends ~130 km WSW-ENE (Fig. 1). The lake level at present is 1647 m above sea level (a.s.l.) As the fourth largest (by volume) terminal lake in the world and largest soda lake (607 km$^3$, maximum depth 460 m; Kadioglu et al., 1997), it is situated in a region where the Afro/Arabian Plate from the south meets the Eurasian Plate from the north and east. Lake Van fills a tectonic depression within an active fault system that is the cause of regional volcanism and seismic as well as hydrothermal activity (Degens and Kurtman, 1978; Keskin, 2003; Kipfer et al., 1994; Litt et al., 2009; Şengör et al., 2003). Evaporation processes, hydrothermal activity, and chemical weathering of volcanic rocks, together with a supply of alkaline tephra into the lake basin, are responsible for the extreme alkalinity of the lake water (pH 9.8, salinity 21.4‰). Two active volcanoes Nemrut (3050 m a.s.l.) and Süphan (4058 m a.s.l.) rise in the immediate vicinity of the lake (Fig. 1).

The lake's position in a semiarid area at the junction of the atmospheric southwestern jet stream and northern branch of the Subtropical High makes it climatically sensitive. The jet stream steers the cyclone tracks that are responsible for supplying moisture from Mediterranean air masses during winter. The location of the Subtropical High controls the southward extension of the dry continental air masses of northeastern Europe and Asia (Litt et al., 2009).

Terminal and saline, Lake Van reacts very sensitively to lake-level changes caused by any alterations in the hydrological regime in response to climate change. Subaerial paleoshorelines and sedimentological evidence show that the lake level fluctuated up to several hundred meters in the past (Landmann et al., 1996; Lemcke and Sturm, 1997). No erosional unconformity could be detected in the younger lacustrine sections on seismic data below 250 m water depth, limiting the maximum lake-level drop to that depth. These

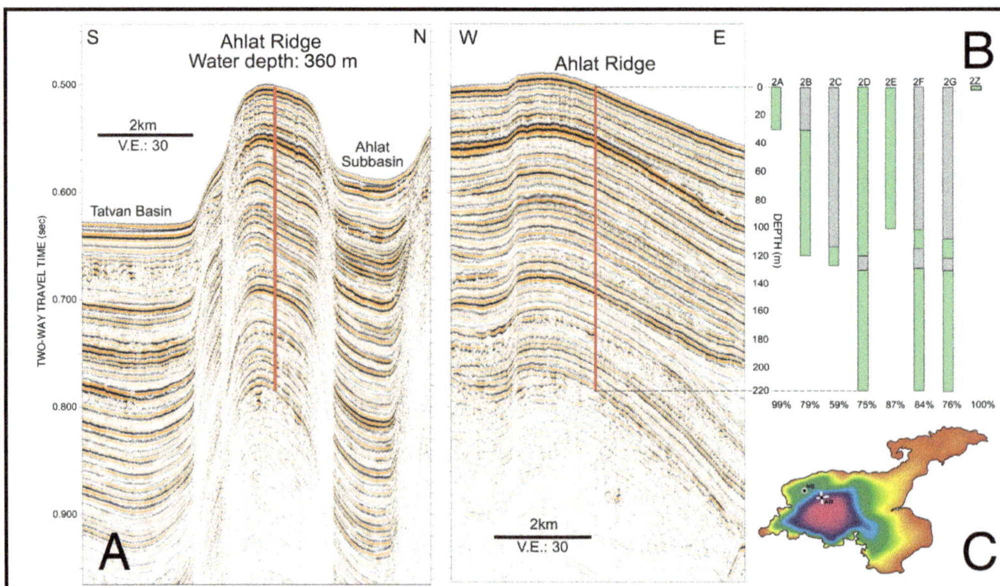


Figure 3. [A] Seismic lines crossing the Ahlat Ridge site. [B] Green bars show the cored sections of the multiple holes 2A–2G; 2Z is the gravity short core. The numbers give the recovery of the individual holes. [C] Location of drill site within lake bathymetry (white).

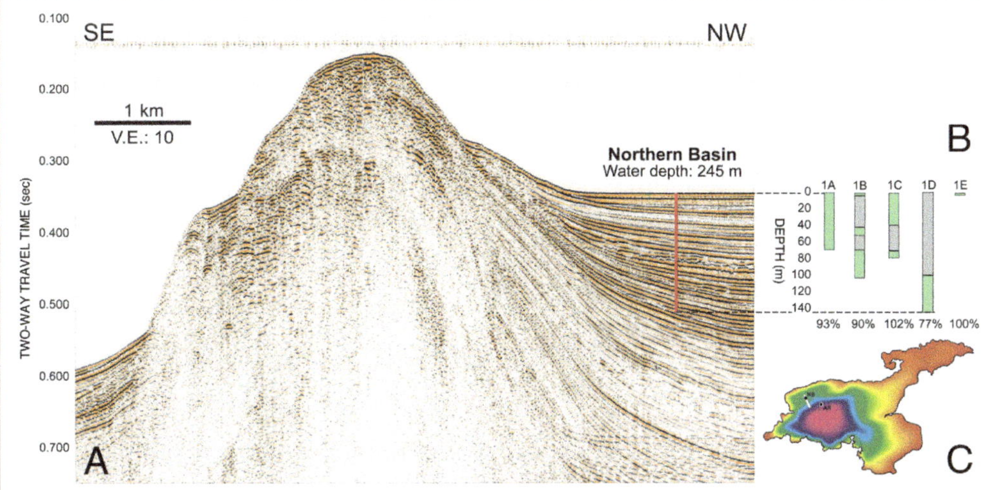


Figure 4. [A] Seismic line crossing the Northern Basin site. [B] Green areas show the cored sections of the multiple holes 1A–1D; 1E is the gravity short core. The numbers give the recovery of the individual holes (values >100 % caused by gas expansion). [C] Location of drill site within lake bathymetry (white).

lake-level changes are sensitively recorded in the lithological and geochemical composition of Lake Van sediments and in the pore-water geochemistry. In particular, the oxygen isotopic composition of the bulk carbonate, consisting of almost pure authigenic aragonite, provides a powerful proxy to track the lake-level fluctuations during the last 15–20 kyr (Lemcke and Sturm, 1997; Litt et al., 2009; Wick et al., 2003). Such a dominance of authigenic carbonate precipitation is also expected from the older sedimentary succession, so that isotope geochemical studies will be the tool of choice for reconstructing the past precipitation/evaporation ratio.

The Lake Van area is strongly affected by earthquakes, a catastrophical natural hazard in the region. An earthquake with a magnitude of 7.2 just north of the city of Van on 23 October 2011 caused more than 600 casualties and major infrastructural damage in the region. In order to investigate recurrence rates of such events and to support seismic hazard studies, long paleoseismic records are needed that document the succession of strong earthquakes and past seismic activities for this tectonically active area. Short sediment cores obtained in 2004 also show strong evidence of earthquake-triggered microfaults, which can be interpreted as seismites (Litt et al., 2009). Consequently, one goal of the PALEOVAN drilling project (http:www.paleovan.info/) was to establish an extended paleoseismic event catalogue on the basis of the deep drill cores.

A major benefit of choosing Lake Van as a drill site was the likelihood of recovering several hundred tephra layers from the active Nemrut and Süphan volcanoes towering Lake Van (Fig. 1), which would provide a tephrostratigraphic framework and allow single-crystal $^{40}Ar/^{39}Ar$ dating. Pre-site work has documented several tens of fallout and flow deposits originating mainly from the alkaline Nemrut volcano, including some eruptions of large magnitude (Sumita et al., 2012). Nemrut volcano has been active in historical times, and future eruptions are likely; therefore, knowledge of its past activity is critical to evaluate the current volcanic hazard.

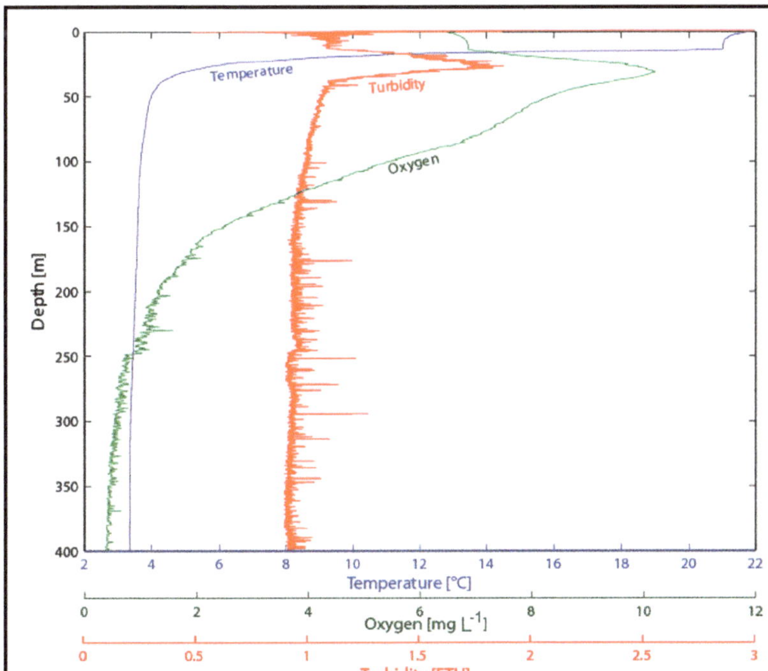


Figure 5. Temperature, turbidity (Formazin Turbidity Units), and oxygen profiles acquired with a CTD probe at the Tatvan Basin mooring site on 28 August 2009 (from Stockhecke et al., 2012). Note that oxygen content is not calibrated, in particular at lower concentrations.

Due to its unique geological and tectonic settings, Lake Van accumulates helium from a depleted Earth mantle source (Kipfer et al., 1994). On regional scales, such a release and transport of He from the solid Earth into the atmosphere is poorly understood. As lakes cover larger areas, they become prime targets to analyze the transport and release mechanism of terrestrial He from the solid Earth. The long profile from Lake Van will allow determination of the *in situ* terrestrial He gradient as a function of depth within a sediment column of several hundred meters. These data will provide the first direct insights into the transport processes of crustal and mantle He through the uppermost layers of the crust and improve the cur-

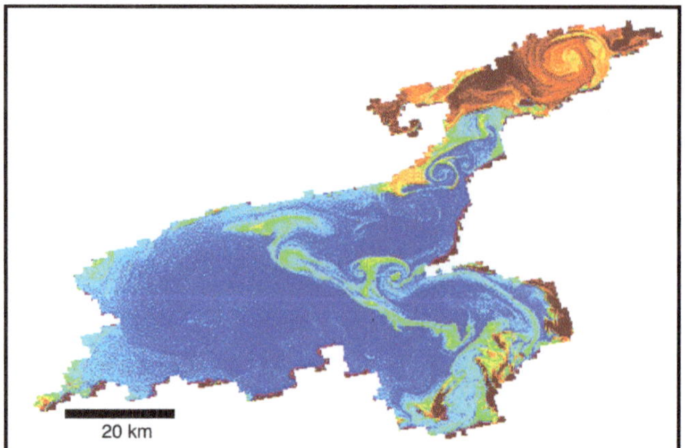


Figure 6. Satellite image (550 nm at-sensor radiance) of Lake Van taken on 20 September 2006. Note the wind-driven surface eddies redistributing suspended surface-water particles mostly consisting of authigenic carbonate ("whitings").

Figure 7. The new large Deep Lake Drilling System (DLDS) operated by DOSECC with the new top-head-drive rotary rig.

rent understanding of terrestrial fluid transport within the continental crust.

## Site Surveys

### *Reflection Seismic Surveys*

A seismic site survey at Lake Van was carried out in 2004 (Litt et al., 2009). In total we collected fifty profiles with a length of ~850 km by means of a high-resolution multi-channel and GeoChirp system (Fig. 1). Three physiographic provinces can be identified based on the seismic data: a lacustrine shelf, a lacustrine slope, and a deep, relatively flat lake basin province (Figs. 1, 2).

The lake has two prominent basins (Tatvan and Northern Basins; Fig. 1), separated by basement rises or ridges. The seismic units in the Tatvan and Northern Basins are largely characterized by an alternating succession of well-stratified and chaotic reflections (Fig. 2). The chaotic seismic facies are interpreted as slump and slide deposits, which were triggered by quick lake-level fluctuations and/or earthquakes. The moderate-to-high-amplitude, well-stratified facies seen in the deep parts of the basins are interpreted as background lacustrine deposits and tephra layers.

The most prominent features of the lacustrine shelf and slope are prograding deltaic sequences, numerous unconformities, submerged channels, as well as closely spaced U- and/or V-shaped depressions (Litt et al., 2009), reflecting the variable lake-level history of Lake Van. Seven clinoform units that are found in the Eastern Fan (Fig. 1) possibly represent relict deltas formed during periods of stationary or slightly rising lake levels. Various volcanic intrusions and extrusions have been identified in the southern part of the lake. Their occurrence closely follows the trend of the northern thrust fault, suggesting a close relationship between the thrust faulting and the volcanic activity.

Based on the high-resolution seismic data collected in 2004, five primary sites were selected (Fig. 1). The Ahlat Ridge site (AR) is the deepest (water depth ~360 m) and most important one (Figs. 1, 3), drilled in order to recover a complete sedimentary section for paleoclimatic investigations. The AR site is located on a sedimentary ridge preventing accumulation of mass-transport deposits, which are widespread in the Tatvan Basin (Fig. 2). On the other hand, the detailed analysis of the seismic data shows that a continuous sedimentary succession without major hiatuses can be drilled at this location down to at least 220 m subsurface depth (Fig. 3).The Northern Basin site (NB) is close to the northern shore of Lake Van (Figs. 1, 2, 4). The proximity to the Süphan and Nemrud volcanoes allows for studying major volcanic eruptions and associated volcanogenic hazards.

Only these two sites were drilled in 2010. Three additional sites were proposed, but their drilling was canceled due to financial limitations. Two sites in different water depths at the Erek Fan (LL-1 and LL-2) were planned to investigate lake-level fluctuations and the evolution of Lake Van. Finally, one site (EX) was proposed to examine the origin of widespread extrusions and intrusions, which are important for the tec-tonic setting of the lake.

### *Monitoring Modern Limnology*

Next to the geophysical site surveys, several limnologic and short-coring campaigns as well as remote-sensing data analysis were undertaken. Numerous conductivity, temperature, and depth (CTD) profiles documented the stratification and mixing behavior of the lake (Fig. 5). The turbidity profile from a late summer profile (Fig. 5) shows a thick layer of suspended particles in the metalimnion at 15–40 m depth accompanied by an oxygen maximum, both indicative of algal blooms and carbonate precipitation. Below, oxygen concentration decreased with depth, and at ~250 m water depth, a drop in oxygen concentration combined with a rise in turbidity can be interpreted as a transition from oxic to anoxic water conditions (Kaden et al., 2010). Satellite data from the different seasons image the wind-driven surface currents of the lake, distributing particles in several eddies (Fig. 6; Stockhecke et al., 2012). The reflectance of the water mostly indicates differences in concentrations of suspended particles from ongoing authigenic carbonate precipitation ("whitings"). The annual particle cycle has been further documented in three years of sediment-trap data. It was shown that carbonate precipitation usually occurs in spring and fall, when runoff (snowmelt and precipitation) supplies Ca ions, which favor carbonate precipitation (Stockhecke et al., 2012). In winter, the fine-grained suspended material (organic and clay particles) settle and form the dark laminae of the varved layers, so that Lake Van sediments provide an annually layered sediment archive (Landmann et al., 1996; Litt et al., 2009, 2011; Wick et al., 2003). Sediment-trap data also documented the annual cycles in biomarkers as well as in terrestrial supply (Huguet et al., 2011, 2012).

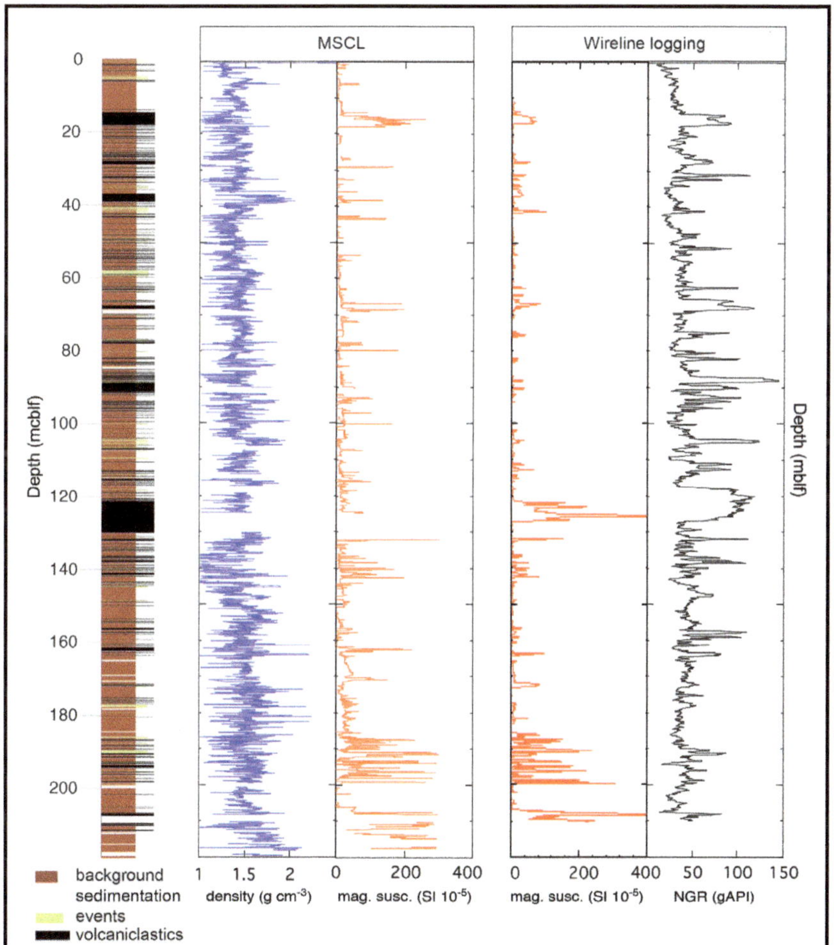


Figure 8. Simplified lithologic log of the Ahlat Ridge composite section (on the basis of visual core descriptions after core opening) distinguishing background sediments (laminated or banded), volcaniclastic layers and event layers (e.g., turbidite deposits or other mass-movement related deposits). Log is juxtaposed to multisensor core logger (MSCL) data (wet bulk density, magnetic susceptibility ) and, on the independent logging depth scale, magnetic susceptibility and natural gamma ray (NGR) as measured by the wireline tools. Note good match between volcaniclastic layers and high magnetic susceptibility and gamma ray values. Note also how wireline logging magnetic susceptibility matches MSCL data and is able to fill gaps of low core recovery. Core-based data plotted in meters composite below lake floor (mcblf), wire-line logging data in meters below lake floor (mblf).

## Drilling Operations

### Core Drilling and Recovery

The ICDP drilling operation was carried out from 2 July to 23 August 2010. As operator of the deep drilling, the U.S. corporation DOSECC (Drilling, Observation and Sampling of the Earth's Continental Crust) developed and assembled a new Deep Lake Drilling System (DLDS). It was specifically designed for coring sediments from deep lakes, and it made its maiden voyage on Lake Van. The DLDS consists of two main parts: the barge and the drilling rig. The barge is a modular system constructed with six separate containers connected in a two-by-three configuration (24.4 m long, 7.3 m wide). The drill rig (Atlas Copco T3WDH) is a top-head-drive rotary rig. The platform also accommodates drilling pipes, mud tanks, a science lab, and a driller's shack. In order to securely set the anchors in water depths of more than 400 m, a 2x2 containers large platform was equipped with a hydraulic winch which served as an additional barge during anchoring operations. During the campaign, two drill and two science teams worked in two 12-hour shifts on the platform, and one science team worked in a shore-based lab. During the shift changes, the cores were transported to the shore and stored in a cooling container at 4°C. The DOSECC drill barge operated in Lake Van at water depths of up to 360 m, and cores from sub-lake-floor depths of 140 m (NB) and 220 m (AR) depth were retrieved, the latter reaching the bedrock at the base of the lacustrine succession as predicted by the seismic data. To obtain a complete sedimentary section, the two sites were multiple-cored (Figs. 3, 4). An average recovery of the cored sections of 91% has been obtained at the AR site and of 71% at the NB site. The length of the total recovered cores is over 800 m.

### Downhole Logging Operations

Downhole logging was performed in Hole 2D at the AR site (to 212 m depth) and in Hole 1D at the NB site (to 127 m depth). Due to borehole instability, the slimhole tools were lowered into the holes through the drill pipes, and several sections of overlapping 15–105-m-long open hole intervals were measured. Continuous data sets of downhole data (spectral gamma ray, magnetic susceptibility [Fig. 8]; dipmeter, resistivity, and temperature as well as partly sonic data) were achieved at both sites. Additionally, vertical seismic profiling at the AR site was carried out by use of an airgun source and a three-component geophone.

The logging-data quality is high, and in particular natural radioactivity (natural gamma ray, K, U, Th-contents), susceptibility (Fig. 8), and resistivity show strong variations, which are promising for further interpretations. Characterization of the physical properties of the lithologies by use of multivariate statistics (cluster analysis) and comparison with results of the core description is ongoing. Of particular interest are variations of physical properties within the tephra deposits, which will be linked to volcanic composition and volcanic source.

## Onsite Laboratory Analyses

### *MSCL Core Scanning*

First core analyses were performed in an onshore laboratory in a 24-hour fashion, following the general concept utilized in similar drilling projects (Melles et al., 2011, 2012; Ohlendorf et al., 2011). A multisensor core logger (MSCL) measured all recovered core sections and yielded wet bulk density, magnetic susceptibility, and p-wave velocity data at a vertical spacing of 1 cm (Fig. 8). These data were used to establish a rough composite section between the different holes at each site, so that drilling operations could be guided in real time in order to have as few gaps as possible in the final composite section. Correlations between the sites were communicated directly to the drilling platform, where presumed overlaps between previously drilled cores were taken into consideration when deciding ideal overlapping drives for the ongoing coring operations. The minimal number of gaps of the resulting composite section (Fig. 8) reflects the usefulness of this critical onsite capability.

### *Core-Catcher Samples and Pore Waters*

The recovered sediment cores were not opened until the core-opening parties at the Integrated Ocean Drilling Program (IODP) Bremen core repository. Core catchers, spaced usually every three meters, provided the available sample material for onsite lithologic descriptions. Smear slides and visual inspections of core-catcher material yielded the first indications of sediment composition, i.e., the distributions of volcanic, detrital, authigenic (carbonate) and biogenic (e.g., diatoms) constituents. The core-catcher samples were also used for initial $^{40}Ar/^{39}Ar$ single-crystal dating of anorthoclase phenocryst analysis focusing on the lower part of the AR site.

Furthermore, core-catcher samples were subsampled for geochemical, pollen, and biomarker analyses and were used for pore-water sampling. Pore water was extracted by hydraulic squeezing, using IODP-style PTFE-titanium squeezers with Teflon disks (Manheim, 1966). Pore-water samples were divided into three aliquots. (1) Salinity and pH values were measured directly in the shore-based lab. (2) One aliquot for major cation measurement was acidified with 2% $HNO_3$ to prevent mineral precipitation. (3) An aliquot for

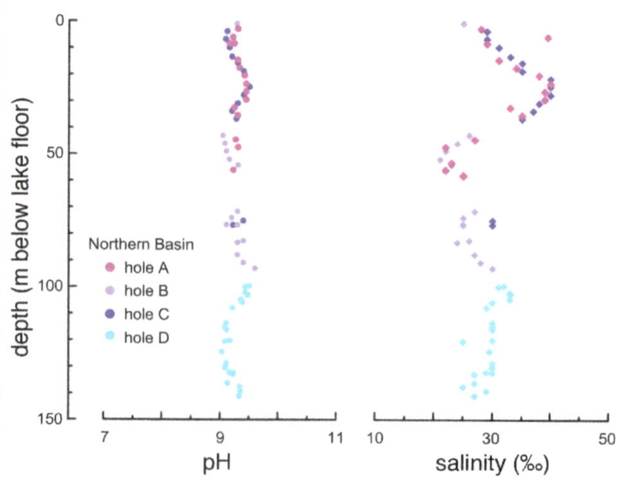


Figure 9. pH and salinity pore water profiles from the Northern Basin analyzed in the on-land lab. Both profiles show distinct "positive" and "negative" fluctuations with regard to recent geochemical state of Lake Van. Observed gradients can only exist over the given temporal (10–100 kyr) and spatial (>20 m) scales if the vertical diffusive transport is suppressed in comparison to diffusive exchange in a continuous and connected pore-water domain. At Ahlat Ridge (not shown), the pore water pH decreases from nearly 10 to 7 in the lowermost 40 m of the cores. According to preliminary results in same depth range, dissolved Ca and Mg concentrations increase continuously—most probably in response to the lower pH—toward the bedrock.

major anions and stable isotope ($\delta^{18}O$, $\delta D$) analyses was not preconditioned. Bulk-water anion and cation quantification as well as isotope analysis was carried out in the respective home laboratories.

The measured pore-water pH and salinity profiles (Fig. 9) are in agreement with the lithological analysis, and they add further evidence to the interpretation that Lake Van evolved from a Ca-carbonate dominated freshwater body with a neutral pH to a high-pH Na-carbonate dominated saline water body. Of particular note is the fact that pH and salinity values change within the sediment cores over a length of 20–50 m. Compared to modern conditions, fluctuations towards higher and lower salinity values are observed. The observed salinity (and pH) gradients in the pore-water profiles indicate that the diffusive transport is strongly attenuated in the sediment column of Lake Van. Such limited exchange allows the

Figure 10. The mobile container laboratory BUGLab operated by GFZ Potsdam located at the shore of Lake Van during the ICDP drilling operation.

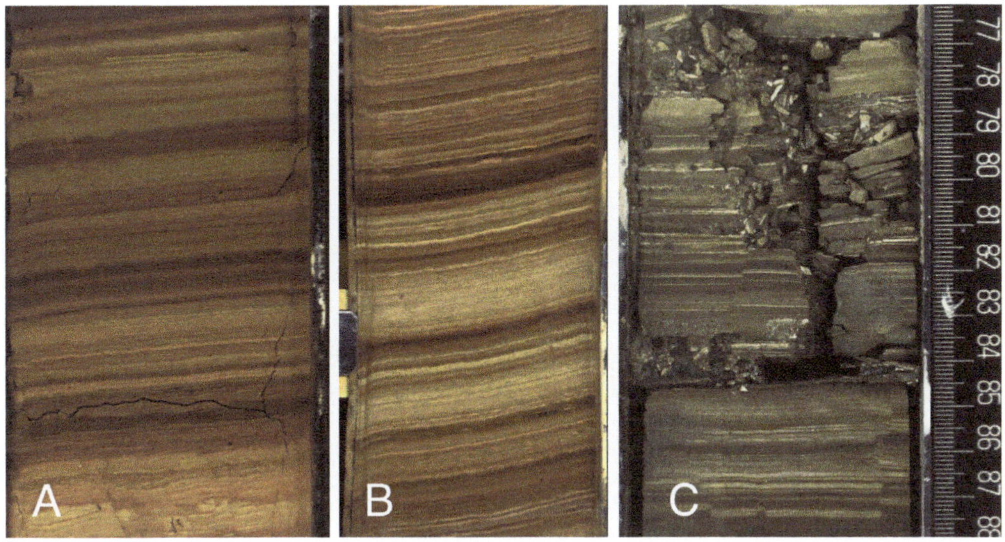


Figure 11. Three examples of finely laminated background deposits from Ahlat Ridge. The example A originates from ~1 m depth and was deposited in the Holocene. Preliminary age data indicates that the example B (40 m depth) can be assigned to the last interglacial (MIS5) and the example C (163 m) to an older interglacial period.

conservation of the composition of the initial co-deposited lake water in the growing sediment column. These favorable "preserving" conditions offer the unique opportunity to apply the pore-water chemistry as a proxy to reconstruct the geochemical evolution of Lake Van (Tomonaga et al., 2011b, 2012).

It is interesting that the pore water of the deeper part of the drilled sections at the NB and AR sites shows quite different salinities: 26‰ at the NB and 20‰ at AR site (Fig. 9). These differences suggest that the shallow Northern Basin of Lake Van may have been disconnected from the deep Tatvan Basin during some time in Lake Van's history. Although caution does not allow us to draw any final conclusions at this point, we tentatively speculate that the observed salinity changes reflect mainly climate-controlled lake level fluctuations.

Moreover, the presence of fresher and more neutral pore waters at the Ahlat Ridge site, together with the basal lithologic succession, indicates that Lake Van did not always host saline and high-pH waters. In contrast, the lake may have been born as a common freshwater body. Only later, the lake evolved towards the present saline and high-pH water mass.

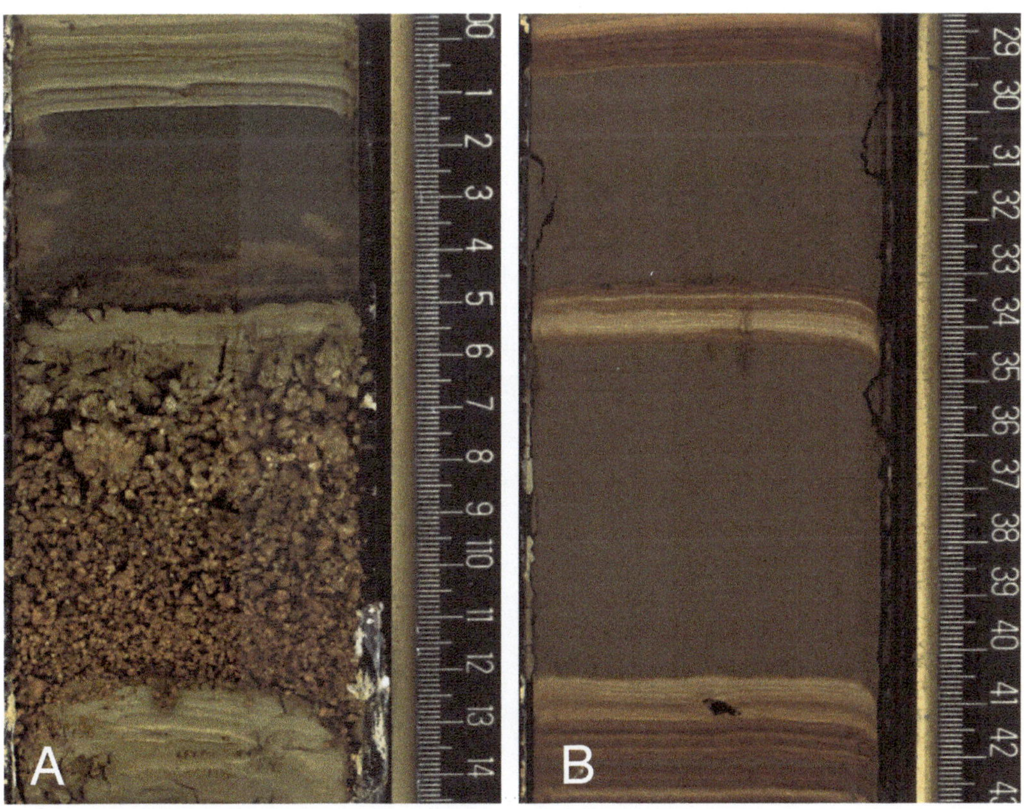


Figure 12. [A] A coarse volcaniclastic layer and a turbidite layer interrupting the continuous, laminated background sedimentation at the Ahlat Ridge site. Note the inverse grading of the pumice particles (deposition from a pumice raft). [B] Two turbidite layers interrupting the laminated sequences (Northern Basin site) consisting mostly of reworked volcaniclastic components.

The indication of limited pore-water exchange means that the dissolved pore-water species, e.g., dissolved noble gases (Blättler et al., 2011; Tomonaga et al., 2011b), may be used to geochemically reconstruct former lake levels and the past ecological state of Lake Van. For this purpose, further pore-water sampling for noble gas analysis was conducted from dedicated core sections of doubled or tripled holes, once completeness of the composite section could be predicted. To achieve these goals, 20–30-cm-long whole-round sections of the cores were sampled immediately after recovery on the drilling platform. At the AR site, ~30 sediment sections were chosen for this purpose and

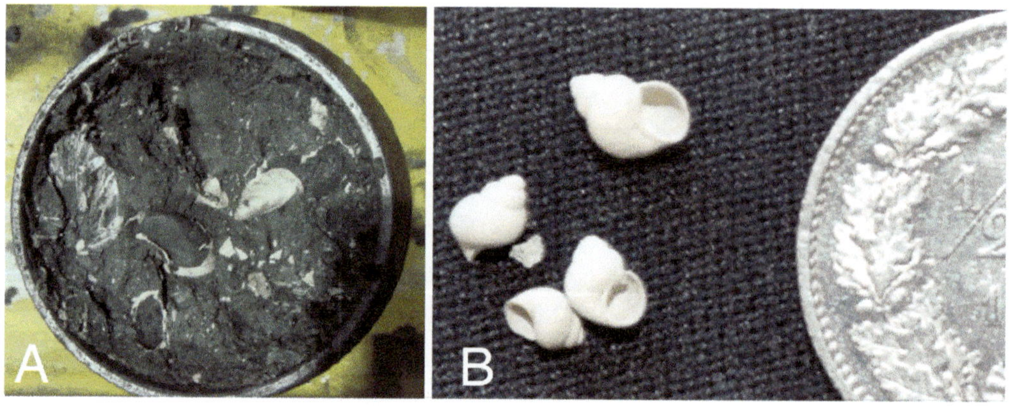


Figure 13. [A] *Dreissena* shells in a core catcher sample just above the sediment-bedrock boundary. [B] Fully conserved and complete shells of *Bithynia*, a freshwater mollusk, found in the sediments at Ahlat Ridge.

transferred into a specially designed sediment press, which allows the bulk sediment to be squeezed without being exposed to atmospheric air into small copper tube containers for noble gas analysis in the pore water of unconsolidated sediments (Brennwald et al., 2003; Tomonaga et al., 2011a). The ongoing noble gas analysis will give insight on the physical transport mechanisms of crustal and mantle derived fluids through the continental crust and on the evolution of the water exchange and salinity in the open water body of Lake Van (Tomonaga et al., 2012).

### *Geomicrobial Analyses*

Onsite geomicrobiological investigations included cell enumeration, quantification of microbial turnover rates, and pore-water analysis. In order to obtain representative results, it is necessary to subsample and process (incubate, fix) the material immediately after retrieval of the core. For this purpose, we used the mobile geomicrobiology laboratory BUGLab of the GFZ German Research Centre for Geosciences (Fig. 10; see also Mangelsdorf and Kallmeyer, 2010). This container hosts a fully air-conditioned laboratory certified for radioisotope use, and it is equipped with fume hood, working bench, nitrogen gas supply, refrigerator, and freezer for incubation experiments and sample storage. The BUGLab is based on a standard sized 20-ft shipping container, and thus provides a functional, mobile laboratory in almost every environment (shore- or ship-based) as long as sufficient power (230–400 V) is available.

Material from undisturbed core catchers from both drilling sites was used for microbiological investigations. Subsamples for cell enumeration as well as for radioisotope incubation experiments were taken immediately when the core material arrived onshore. Samples for cell enumeration were fixed in artificial lake water including 2% formalin (Kallmeyer et al., 2008). Samples for radioisotope incubation experiments were incubated with 35S sulfate for sulfate reduction rate determination (Jørgensen, 1978; Ferdelman et al., 1997). In addition, individual samples were incubated with $^{14}C$-methanol, $^{14}C$-bicarbonate, and $^{14}C$-acetate for quantification of methanogenesis rates and with $^{14}C$-methane for determination of rates of anaerobic oxidation of methane (AOM) (Ferdelman et al., 1997; Treude et al., 2005). All incubations were terminated after 3–5 days, and samples were preserved for analysis in the home lab. Additional samples for quantification of dissolved methane and for pore-water analysis were also taken from the fresh core-catcher material.

## Sediment Lithology

The cores were opened in spring 2011 at the IODP core repository located at the MARUM, University of Bremen. The repository's ideal facilities have been used for splitting, photographic and X-ray fluorescence (XRF) scanning of the core halves, core descriptions, and sampling. The recovered lithologies vary as predicted by the results of the pre-drilling seismic and gravity-coring survey; varved or banded lacustrine sediment sections, termed "background deposits" (Fig. 11), are intercalated by volcanic (tephra) and by event layers such as turbidites (Fig. 12). These events appear to be more abundant and thicker in the NB site than at the AR site. The frequent tephra layers guarantee the lithostratigraphic correlation of both sites and constrain the chronological framework of the recovered sediments. The NB site is characterized by much higher sedimentation rate; in fact, the base of the NB site at 143 m can be correlated to ~42 m at the AR site. A composite section has been established at both sites (Fig. 8), representing the best possible complete sediment succession. Whereas the major parts of both drill sites are characterized by one of the three endmember lithologies (background deposits, event and volcanic layers; Fig. 8), the basal unit at AR is characterized by a marked downcore change in lithology. Just above the contact between the

Figure 14. Washed samples of a core catcher of the basal gravel layers. The rounded nature of the gravel indicates coastal or fluvial reworking related to initial lake transgression.

lacustrine sediments and the bedrock, zebra mussel *(Dreissena)* shells, which can only live in fresh or brackish waters, were found within coarse sand and gravel (Fig. 13a). Moreover, perfectly preserved and intact shells of freshwater gastropods *(Bithynia)* occur at a depth of ~190 m, i.e., ~ 30 m above the basement (Fig. 13b). The presence especially of *Bithynia* shells in the sedimentological record indicates that Lake Van contained freshwater at low pH in the very early days, as confirmed by the pore-water analysis (see above). These basal sediments are underlain by coarse and rounded gravel layers (Fig. 14) indicative of coastal or fluvial origin. Together with the seismic prediction of an acoustic basement at around this depth, these patterns indicate that the drill hole penetrated the initial transgressive phase of Lake Van upon its early creation, and it appears that the retrieved cores hold the entire geological, geochemical, volcanic, seismic, hydrological, and bio-geochemical environmental history of Lake Van starting from its very beginning up to recent. Reaching this basal succession at AR was one of the highlights of the drilling activities as, this boundary marks the birth of Lake Van.

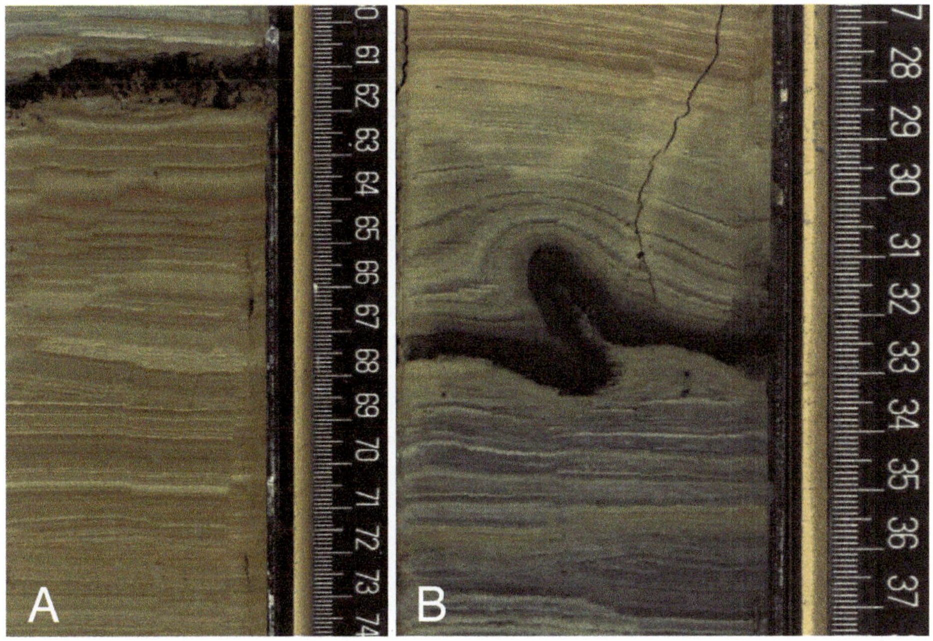


Figure 15. Examples of paleoseismic deformations in the cores. [A] Microfaults dissecting the finely laminated succession. They are only considered of paleoseismic significance if they occur in parallel cores at the same horizons, so that drilling artifacts can be ruled out. [B] Fine tephra layer that has been deformed in a microfold, indicating slumping, potentially induced by seismic shaking.

## Paleoseismic Deformation Structures

The partly annually-laminated sedimentary record down to 220 m also contains an invaluable record of past earthquake activities. Various forms of paleoseismites can be easily identified in the finely-laminated sedimentary sequences. Most common are complex multiple microfaults dissecting the laminations (Fig. 15a). Some of them are lined with coarser, mostly volcanic sand grains, which can be explained by liquefaction processes induced by seismic shaking. In addition, several microfolds can be seen (Fig. 15b), reflecting slumping and sliding processes also indicative of seismic shaking. Up to three parallel cores allow us to rank the observed deformation structures in terms of reliability. Deformation structures are often detected in each of the three parallel cores retrieved per site. Such multiple occurrences are used as criteria to interpret seismic shaking as a deformation trigger, as

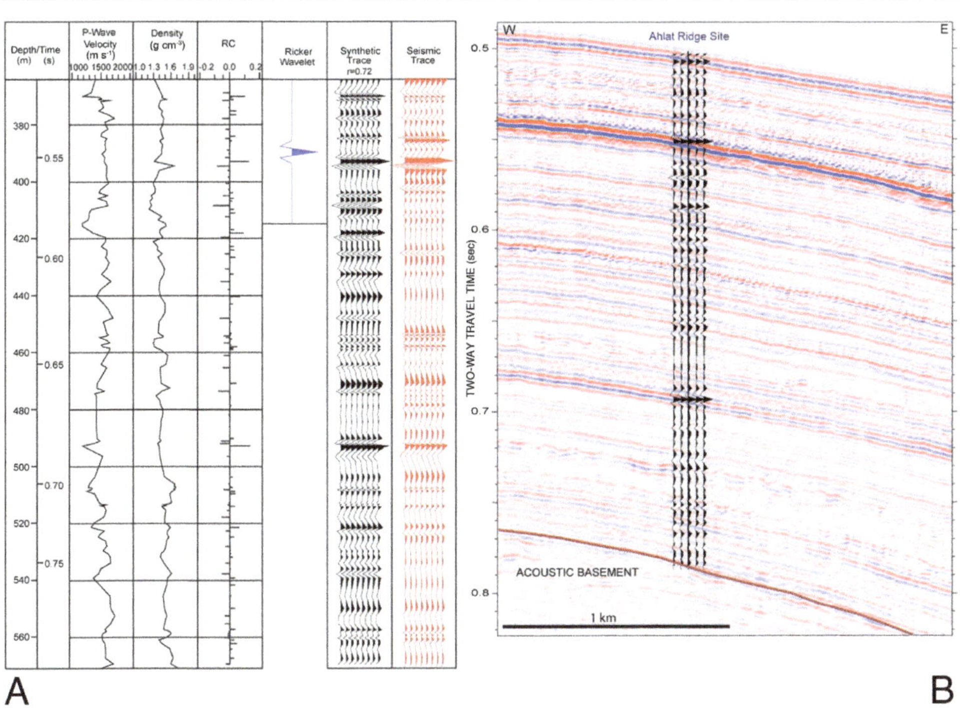


Figure 16. [A] Well to seismic calibration at Ahlat Ridge site. [B] Comparison of the well to seismic tie at Ahlat Ridge site.

single occurrences could also be drilling artifacts. The mapping of these deformation features in the cores will allow us to establish a paleoseismic event catalogue shedding light on recurrence rates and intensities of strong events throughout the past.

In this context, the vulnerability of the area to seismic hazards was dramatically documented by the occurrence of the devastating 2011 M7.2 earthquake near the city of Van.

## Integration of Core and Seismic Data

The integration of core data and high-resolution seismic profiles provides a reliable basis for extrapolation of the stratigraphic information over the lake basin. This allows a detailed analysis of the general sedimentary evolution of Lake Van.

Core logging data was used to generate synthetic seismograms for the synthetic-to-well ties that help to obtain two-way travel time-depth relationships at the well locations. This approach aims to extrapolate the stratigraphy from wells to the 3D space by using the seismic data. The SynPAK module (Kingdom Suite software, http://www.seismicmicro.com) was used to construct synthetic seismograms. The results show good correlations between synthetic and real seismic data in both AR and NB sites (Fig. 16). Strong marker horizons correlate well with prominent and dated tephra layers that were also correlated between the two sites. The geologic age data for the drilled horizons will eventually be assigned to the seismic stratigraphic horizons.

## Chronostratigraphy

The recovered tephra layers are dominated by Süphan tephras in the deeper sections and Nemrut tephras in the shallower sections. Süphan rhyolitic and dacitic tephras carry plagioclase with low radiogenic yield, while Nemrut tephras are alkaline rhyolites and trachytes carrying anorthoclase as felsic phase. Although physical dating of the deeper part of the section is thus more difficult, preliminary single-crystal argon dating of anorthoclase suggests that the AR record encompasses more than 500,000 years of paleoenvironmental and volcanic/geodynamic history (Sumita et al., 2012). Feldspars in the freshwater basement volcaniclastic sediments are as old as ~16 Ma, reflecting earlier post-collisional volcanism in the area. This general age frame is confirmed by initial geochemical results, as well as pollen analyses, which indicate repetitive changes at the glacial-interglacial scale (Fig. 17). Including the current interglacial stage, four to five interglacial stages can be identified by investigating the lithological pattern, total organic carbon concentrations, and pollen data. Laminated sections rich in total organic carbon and tree pollen are indicative of warmer environments, and they represent marine isotope stages 1, 5, 7, 9, and 11 or 13 that dot the last half million years of paleoclimate history. In contrast, non-laminated sections, poor in organic carbon with higher amounts of steppe plants, are indicative of cold conditions and represent glacial periods.

## Outlook on Post-drill Science

A detailed geochronological frame for the long continental record drilled during the PALEOVAN project in 2010 is presently being obtained through varve counting, radiocarbon dating of terrestrial organic matter, $^{40}Ar/^{39}Ar$ single crystal dating of tephra layers, Th/U dating (aragonite laminae), paleomagnetic measurements, $^{10}Be$, OSL/TL and orbital tuning/oxygen-isotope stratigraphy. A precise chronology is a precondition for the analysis of the climate signal in the Lake Van record, which has the potential to reveal signals with frequencies higher than Milankovitch cycles such as North Atlantic Oscillation or Dansgaard-Oeschger cycles. These signals will be documented by high-resolution XRF data. Lower frequency signals (Milankovitch) will be investigated through spectral analysis on the downhole logging data so that cyclicities and sedimentation rates can be investigated. Additional information will be obtained through mineralogical, geochemical, and paleontological analysis as well as the analyses of organic material and its composition using biomarkers such as long-chain alkenones, n-alkanes, alcohols, or fatty acids. Reliable reconstructions of Quaternary climate evolution are essential for understand-ing past climate dynamics and for validating general circulation models used to simulate future climate scenarios. Botanical data based on pollen analysis are well established as proxy data, because plants strongly depend on climate, of which temperature and precipitation are decisive factors for the survival and prospering of plants. Botanical-climatological transfer functions will be used for quantitative paleoclimate reconstructions based on the partly annually-laminated sediments of Lake Van for the last >500 kyr. Data-model comparisons are useful to test these results, as shown by Kaspar et al. (2005). In summary, this successful ICDP drilling campaign yielded a rich, temporally long and

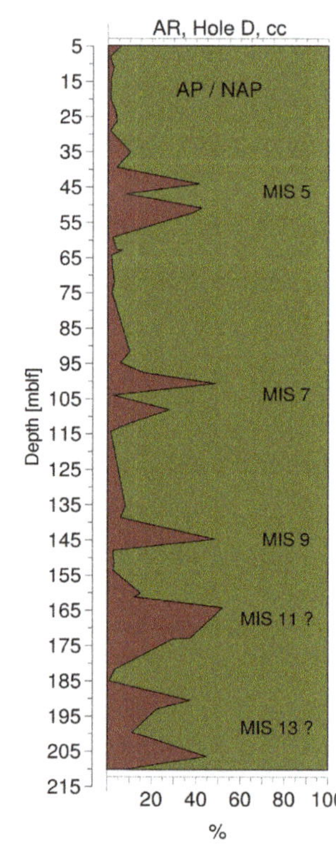


Figure 17. Pollen ratio between trees (arboreal pollen in red: AP) versus herbs (non-arboreal pollen in green: NAP) from Lake Van (AR, hole D, core catcher samples) and tentative correlation with marine isotope stages (MIS).

mostly continuous Lake Van sedimentary succession. This continental record, investigated through a multiproxy approach, has the potential to unravel a large suite of environmental processes occurring in a critical region so far lakking a reliable geological archive of the recent past.

## Acknowledgments

Financial support has been provided by ICDP, the German Research Foundation, the Swiss National Science Foundation, and Turkey's Tübitak. We are grateful to DOSECC and its personnel for recovering the drill cores under partly difficult conditions. We thank the Integrated Ocean Drilling Program and the Bremen Core Repository (MARUM) for logistical support and core curation. Without the tremendous support from the whole Paleovan scientific team (http:// www.paleo van.info/index.php?a=steam), from many local helpers, and from the Turkish authorities, drilling would not have been such a success.

## References

Blättler, R., Tomonaga, Y., Brennwald, M.S., Kwiecien, O., and Kipfer, R., 2011. Are noble gases in the sediment pore water of Lake Van promising proxies for paleoclimate conditions? *Mineralogical Magazine*, 75(3):532 (abstract).

Brennwald, M., Hofer, M., Peeters, F., Aeschbach-Hertig, W., Strassmann, K., Kipfer, R., and Imboden, D., 2003. Analysis of dissolved noble gases in the porewater of lacustrine sediments. *Limnol. Oceanogr. Methods*, 1:51–62. doi:10.4319/lom.2011.1.51

Degens, E.T., and Kurtman, F., 1978. *The Geology of Lake Van*. Ankara (MTA Press).

Djamali, M., de Beaulieu, J.-L., Shah-hosseini, M., Andrieu-Ponel, V., Ponel, P., Amini, A., Akhani, H., et al., 2008. A late Pleistocene long pollen record from Lake Urmia, NW Iran. *Quat. Res.*, 69:413–420.

Ferdelman, T.G., Lee, C., Pantoja, S., Harder, J., Bebout, B.M., and Fossing, H., 1997. Sulfate reduction and methanogenesis in a Thioploca-dominated sediment off the coast of Chile. *Geochim. Cosmochim. Acta*, 61(15):3065–3079. doi:10.1016/S0016-7037(97)00158-0

Huguet, C., Fietz, S., Moraleda, N., Stockhecke, M., Anselmetti, F.S., Sturm, M., Litt, T., Heumann, G., and Rosell-Melé, A., 2012. A seasonal cycle of terrestrial inputs in Lake Van, Turkey. *Environ. Sci. Pollut. Res. Int.*, 19(8):3628–3635. doi:10.1007/s11356-012-0948-3

Huguet, C., Fietz, S., Stockhecke, M., Strum, M., Anselmetti, F.S., and Rosell-Mele, A., 2011. Biomarker seasonality study in Lake Van, Turkey. *Org. Geochem.*, 42(11):1289–1298. doi:10.1016/j.orggeochem.2011.09.007

Jørgensen, B.B., 1978. A comparison of methods for the quantification of bacterial sulfate reduction in coastal marine sediments. *Geomicrobiol.J.*, 1(1):11–27. doi:10.1080/01490457809377721

Kaden, H., Peeters, F., Lorke, A., Kipfer, R., Tomonaga, Y., and Karabiyikoglu, M., 2010. The impact of lake level change on deep-water renewal and oxic conditions in deep saline Lake Van, Turkey. Water Resour. Res., 64:W11508.

Kadioğlu, M., Şen, Z., and Batur, E., 1997. The greatest soda-water lake in the world and how it is influenced by climatic change. *Ann. Geophysicae*, 15:1489–1497. doi:10.1007/s00585-997-1489-9

Kallmeyer, J., Smith, D.C., Spivack, A.J., and D'Hondt, S., 2008. New cell extraction procedure applied to deep subsurface sediments. *Limnol. Oceanogr. Methods*, 6:236–245. doi:10.4319/lom.2008.6.236

Kaspar, F., Kühl, N., Cubasch, U., and Litt, T., 2005. A model-data comparison of European temperatures in the Eemian interglacial. *Geophys. Res. Lett.*, 32:L11703. doi:10.1029/2005GL022456

Keskin, M., 2003. Magma generation by slab steepening and breakoff beneath a subduction-accretion complex: An alternative model for collision-related volcanism in Eastern Anatolia, Turkey. *Geophys. Res. Lett.*, 30:8046. doi:10.1029/2003GL018019

Kipfer, R., Aeschbach-Hertig, W., Baur, H., Hofer, M., Imboden, D.M., and Signer, P., 1994. Injection of mantle type helium into Lake Van (Turkey): The clue for quantifying deep water renewal. *Earth Planet. Sci. Lett.*, 125(1–4):357–370. doi:10.1016/0012-821X(94)90226-7

Landmann, G., Reimer, A., Lemcke, G., and Kempe, S., 1996. Dating Late Glacial abrupt climate changes in the 14,570 yr long continuous varve record of Lake Van, Turkey. *Palaeogeogr. Palaeoclimatol. Palaeoecol.*, 122(1–4):107–118. doi:10.1016/0031-0182(95)00101-8

Lemcke, G., and Sturm, M., 1997. $\delta^{18}O$ and trace element measurements as proxy for the reconstruction of climate changes at Lake Van (Turkey): Preliminary results. *In* Dalfes, H.N., Kukla, G., and Weiss, H. (Eds.), *Third Millennium BC Climate Change and Old World Collapse. NATO ASI Series I, Global Environmental Change 49:* Berlin (Springer), 653–78.

Litt, T., Anselmetti, F.S., Cagatay, M.N., Kipfer, R., Krastel, S., and Schmincke, H.-U., 2011. A 500,000-year-long sedimentary archive drilled in Eastern Anatolia. *Eos*, 92:477–479. doi:10.1029/2011EO510002

Litt, T., Krastel, S., Sturm, M., Kipfer, R., Örcen, S., Heumann, G., Franz, S.O., Ülgen, U.B., and Niessen, F., 2009. 'PALEOVAN', International Continental Scientific Drilling Program, (ICDP): Site survey results and perspectives: *Quat. Sci. Rev.*, 28(15–16):1555–1567. doi:10.1016/j.quascirev.2009.03.002

Mangelsdorf, K., and Kallmeyer, J., 2010. Integration of deep biosphere research into the International Continental Scientific Drilling Program. Sci. Drill., 10:46–55. doi: 10.2204/iodp.sd.10.0.2010

Manheim, F.T., 1966. A hydraulic squeezer for obtaining interstitial water from consolidated and unconsolidated sediments. *U.S. Geol. Surv. Prof. Pap.*, 550-C:256–261.

Melles, M., Brigham-Grette, J., Minyuk, P., Koeberl, C., Andreev, A., Cook, T., Fedorov, G., et al., 2011. The Lake El'gygytgyn Scientific Drilling Project conquering Arctic challenges in continental drilling. *Sci. Drill.*, 11:29–40. doi:10.2204/iodp.sd.11.03.2011

Melles, M., Brigham-Grette, J., Minyuk, P., Nowaczyk, N.R., Wennrich, V., DeConto, R.M., Anderson, P.M., et.al., 2012. 2.8 million years of Arctic climate change from Lake El'gygytgyn, NE Russia. *Science*, 20(337):315–320. doi:10.1126/science.1222135

Ohlendorf, C., Gebhardt, C., Hahn, A., Kliem, P., and Zolitschka, B., 2011. The PASADO core processing strategy — A proposed new protocol for sediment core treatment in multidisciplinary lake drilling projects. *Sed. Geol.*, 239(1-2):104–115. doi:10.1016/j.sedgeo.2011.06.007

Şengör, A.M.C., Özeren, S., Genec, T., and Zor, E., 2003. East Anatolian high plateau as a mantle supported, north-south shortened domal structure. Geophys. Res. Lett., 30(24):8045. doi:10.1029/2003GL017858

Stockhecke, M., Anselmetti, F.S., Meydan, A.F., Odermatt, D., and Sturm, M., 2012. The annual particle cycle in Lake Van (Turkey). *Palaeogeogr. Palaeoclimatol. Palaeoecol.*, 333–334:148–159. doi:10.1016/j.palaeo.2012.03.022

Sumita, M., Schminke, H.-U., and the PaleoVan Scientific Team, 2012. The climatic, volcanic and geodynamic evolution of the Lake Van-Nemrut-Süphan system (Anatolia) over the past 550-600000 years: a progress report based on a study of the products of explosive volcanism on land and in the lake. Annual ICDP/IODP DFG Kolloquium, Kiel, Extended abstract, 157–162.

Tomonaga, Y., Blättler, R., Brennwald, M.S., and Kipfer, R., 2012. Interpreting noble-gas concentrations as proxies for salinity and temperature in the world's largest soda lake (Lake Van, Turkey). *J. Asian Earth Sci.,* in press. doi:10.1016/j.jseaes.2012.05.011

Tomonaga, Y., Brennwald, M.S., and Kipfer, R., 2011a. An improved method for the analysis of dissolved noble gases in the porewater of unconsolidated sediments. *Limnol. Oceanogr. Methods*, 9:42–49.

Tomonaga, Y., Brennwald, M.S., and Kipfer, R., 2011b. Spatial distribution and flux of terrigenic He dissolved in the sediment porewater of Lake Van (Turkey). *Geochim. Cosmochim. Acta*, 75(10):2848–2864. doi:10.1016/j.gca.2011.02.038

Treude, T., Niggemann, J., Kallmeyer, J., Wintersteller, P., Schubert, C.J., Boetius, A., and Jørgensen, B.B., 2005. Anaerobic oxidation of methane and sulfate reduction along the Chilean continental margin. *Geochim. Cosmochim. Acta*, 69(11):2767–2779. doi:10.1016/j.gca.2005.01.002

Tzedakis, P.C., Andieu, V., Birks, H.J.B., de Beaulieu, J-L., Crowhurst, S., Follieri, M., Hooghiemstra, H., et al., 2001. Establishing a terrestrial chronological framework as a basis for biostratigraphical comparisons. *Quat. Sci. Rev.*, 20(16–17):1583–1592. doi:10.1016/S0277-3791(01)00025-7

Wick, L., Lemcke, G., and Sturm, M., 2003. Evidence of Lateglacial and Holocene climatic change and human impact in eastern Anatolia: High-resolution pollen, charcoal, isotopic and geochemical records from the laminated sediments of Lake Van, Turkey. *The Holocene*, 13(5):665–675. doi:10.1191/0959683603hl653rp

## Authors

**Georg Heumann, Thomas Litt, and Nadine Pickarski**, Steinmann Institute of Geology, Mineralogy and Paleontology, Bonn University, Bonn, Germany, e-mail: t.litt@uni-bonn.de

**Deniz Cukur, Sebastian Krastel, Hans-Ulrich Schmincke, and Mari Sumita**, GEOMAR, Helmholtz Centre for Ocean Research Kiel, Wischhofstr. 1-3, 24148 Kiel, Germany

**Flavio S. Anselmetti, Jürg Beer, Gerald Haug, Rolf Kipfer, Ola Kwiecien, Marie-Eve Randlett, Carsten J. Schubert, Mona Stockhecke, Mike Sturm, and Yama Tomonaga**, Eawag, Swiss Federal Institute of Aquatic Science and Technology, Dübendorf and Kastanienbaum, Switzerland, e-mail: flavio.anselmetti@geo.unibe.ch

**A. Feray Meydan, and Sefer Orcen**, Yüzüncü Yıl Üniversitesi, Mühendislik - Mimarlık Fakültesi Jeoloji Mühendisliği Bölümü, Van, Turkey

**Namik Cagatay, and Emre Damci**, Department of Geological Engineering and Eastern Mediterranean Centre for Oceanography and Limnology, Istanbul Technical University, Istanbul, Turkey

**Henrike Baumgarten, and Thomas Wonik**, Leibniz-Institute for Applied Geophysics (LIAG), Hannover, Germany

**Clemens Glombitza, and Jens Kallmeyer**, Potsdam University, Institute for Earth and Environmental Sciences, Geomicrobiology Group, Potsdam, Germany

**Luigi Vigliotti**, Istituto di Scienze Marine, ISMAR-CNR, Bologna, Italy

**and the PALEOVAN Scientific Team**

## Photo Credits

Fig. 7: Flavio Anselmetti, University of Bern
Fig. 10: Jens Kallmeyer, GFZ Potsdam
Fig. 11: PALEOVAN science team
Fig. 12: PALEOVAN science team
Fig. 13: PALEOVAN science team
Fig. 14: PALEOVAN science team
Fig. 15: ALEOVAN science team

## Related Web Link

http://www.paleovan.info/
http://www.dosecc.org/index.php/equipment/deep-lake-drilling-system-dlds

# 15

# IODP Expedition 326 Operations: First Stage of Nankai Trough Plate Boundary Deep Riser Drilling

by Masataka Kinoshita, Harold Tobin, Nobuhisa Eguchi, and Simon Nielsen

## Introduction

Expedition 326 Ultra Deep Riser Top Hole was the first stage of drilling and coring of the Integrated Ocean Drilling Program (IODP) Hole C0002F to the boundary zone between the Philippine Sea and Eurasian Plates in the Nankai accretionary margin, one of the main objectives of the Nankai Trough SEIsmogenic Zone Experiments (NanTroSEIZE) Complex Drilling Program. The expedition objectives were purely operational, with the goal being installation of the wellhead assembly and drilling and casing the uppermost 800 m of the planned 7-km deep hole. Accordingly, no science party was on board during the expedition, and no scientific results are reported. Scientific objectives for the top 1400 m of this Site were previously fulfilled during NanTroSEIZE Stage 1 Expeditions 314 and 315 (Kinoshita et al., 2009).

After a one month operation in July–August 2010, Hole C0002F had been drilled to 872.5 m below sea floor (mbsf), and the hole was lined with cemented-in 20-inch casing. A corrosion cap was set in preparation for return to continue drilling in 2012. We confirm that Hole C0002F is now ready for deep riser drilling to the plate boundary fault zone.

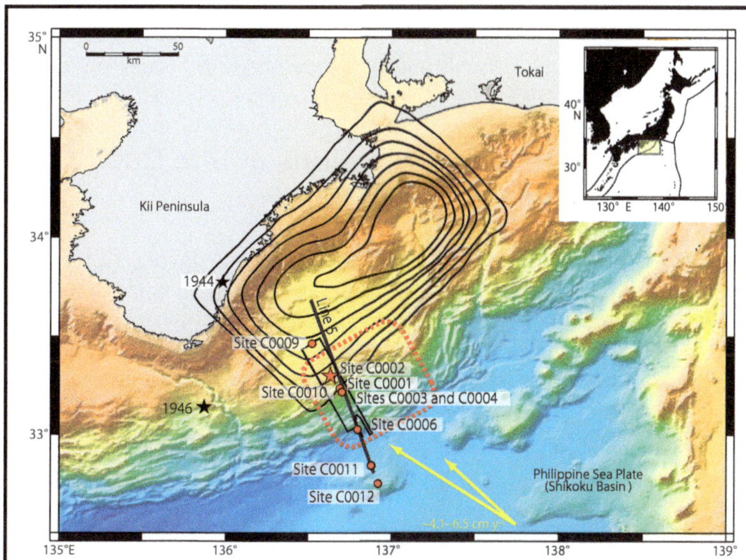


Figure 1. Map of Nankai accretionary complex off Kumano, showing NanTroSEIZE drill sites. Yellow arrows=computed far-field convergence vectors between the Philippine Sea Plate and Japan (Seno et al., 1993; Heki, 2007). Contours indicate estimated slip during the 1944 event (0.5-m intervals). Red box outlines region of recorded very low frequency events (Obara and Ito, 2005). Red circles=NanTroSEIZE drill sites. C0002 is marked with a red star.

## NanTroSEIZE Complex Drilling Project

NanTroSEIZE is a multiexpedition, multistage IODP drilling program focused overall on understanding the mechanics of seismogenesis and rupture propagation along subduction plate boundary faults (Tobin and Kinoshita, 2006). This program includes a coordinated effort to characterize (through integrating core/log with seismic images), sample (core/cuttings), and instrument (borehole observatory) the plate boundary fault system near the updip limit of the locked zone of Tonankai great earthquakes at 5–7 km below seafloor (Figs. 1, 2; Tobin and Kinoshita, 2006).

The main objectives are to understand (1) the mechanisms controlling the updip aseismic–seismic transition along the megathrust fault systems; (2) processes of earthquake and tsunami generation and strain accumulation and release, including the role of recently discovered slow slip and very low frequency earthquake (Ito and Obara, 2006); (3) the absolute mechanical strength of the plate boundary fault and its degree of interseismic locking; and (4) the potential role of a major upper plate fault system (termed the "megasplay" fault) in seismogenesis and tsunamigenesis. This drilling program approaches these objectives through a combination of riser and riserless drilling, long-term observatories, and associated geophysical, laboratory, and numerical modeling efforts.

At Nankai Trough, high-resolution seismic reflection profiles clearly document a large out-of-sequence thrust fault system (the megasplay fault, after Park et al., 2002) that branches from the plate boundary décollement close to the updip limit of inferred coseismic rupture in the 1944 Tonankai M 8.2 earthquake (Fig. 1). Several lines of evidence indicate that the megasplay system is active and may accommodate a significant fraction of plate boundary motion (Moore et al., 2007, 2009). However, the partitioning of strain between the lower plate interface above the oceanic basement and the megasplay system is not understood, and neither are the nature and mechanisms of fault

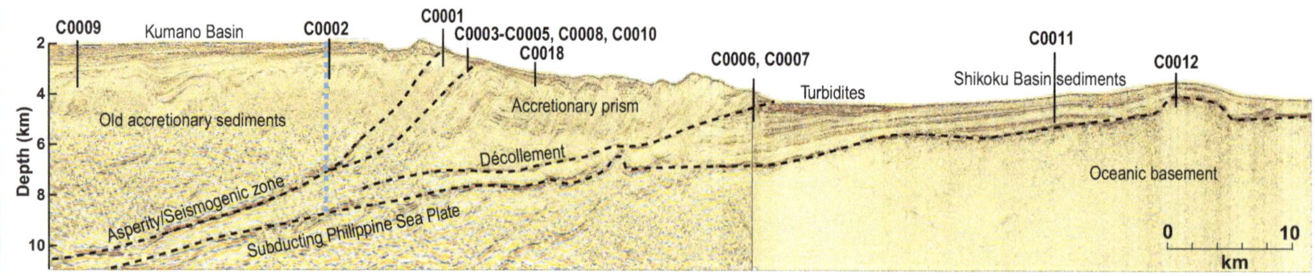


Figure 2. Spliced composite profile of a representative depth section from the NanTroSEIZE 3D seismic data volume (Moore et al., 2009) and Line 95 from IFREE mini-3D seismic survey (Park et al., 2008). Projected positions of Stage 1 and 2 drilling sites, including C0009, C0010, C0011, and C0012.

slip as a function of depth and time on the megasplay. One of the first-order goals in characterizing the seismogenic zone along the Nankai Trough—and which bears on understanding subduction zone megathrust behavior globally and on defining tsunami hazards—is to document the role of the megasplay fault in accommodating plate motion (both seismically and interseismically) and to characterize its mechanical and hydrologic behavior.

IODP Site C0002 is the centerpiece of the NanTroSEIZE project, intended to access the plate interface fault system at a location where it is believed to be capable of seismogenic locking and slip, and to have slipped coseismically during the 1944 Tonankai earthquake. The primary targets include both the basal décollement and the reflector known as the "megasplay fault" (Tobin and Kinoshita, 2006). The megasplay fault zone and the accretionary prism domain are the location of a newly identified class of earthquakes known as very low frequency () earthquakes (Ito and Obara, 2006) as well as the first observation of shallow tectonic tremor (Obana and Kodaira, 2009). The megasplay fault reflector lies at an estimated depth of 5000–5200 mbsf, and the top of the subducting basement is estimated to lie at 6800–7000 mbsf (Fig. 3).

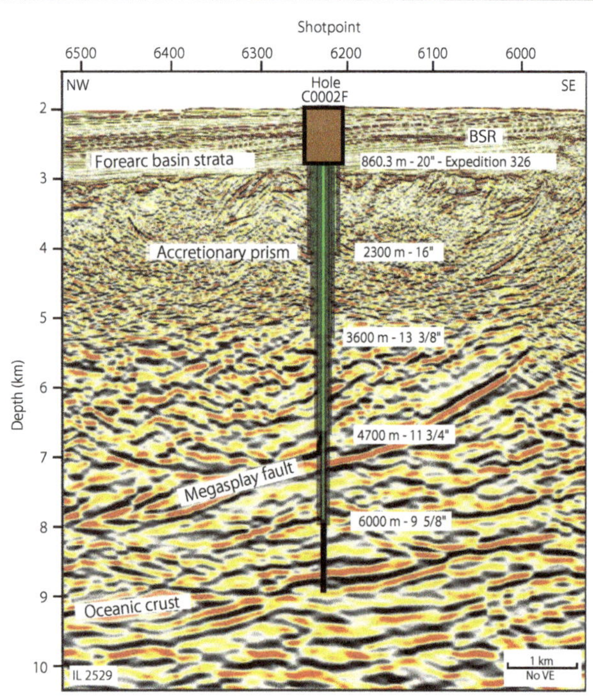


Figure 3. Seismic section of In-line 2529, showing the current status and planned trace for the deep riser Hole C0002F. Brown rectangle denotes the progress made by Expedition 326. Note that the operational options shown here in green are only suggested by the Project Management Team (PMT). BSR=Bottom Simulating Reflector, LWD=Logging-While-Drilling.

During Expedition 314, Site C0002 was drilled to 1401 mbsf with *in situ* measurement of physical properties and borehole imaging through logging while drilling (LWD) but no coring (Expedition 314 Scientists, 2009). Several months later, portions of Site C0002 were cored over the intervals 0–204 mbsf and 475–1057 mbsf on Expedition 315 (Expedition 315 Scientists, 2009). Lithostratigraphy at Site C0002 is characterized by turbiditic sediments to ~830 mbsf, underlain by older rocks of the accretionary prism and/or early slope basin sediments deposited prior to the development of the megasplay fault.

Further background, objectives, and accomplishments to date for the NanTroSEIZE project are discussed in Tobin et al. (2009), Expedition 319 Scientists (2010), and Underwood et al. (2010).

## Summary of Operations at Site C0002 Ultra Deep Riser Top Hole

The Expedition 326 was carried out during 19 July to 20 August 2010. Hole C0002F is located at 33° 18.0507' N, 136° 38.2029' E (Fig. 4) and its water depth is 1968 m. Total drilling depth was 872.5 mbsf.

Our drilling plan for Expedition 326 was to run a mudmat (a steel plate deployed on the seafloor) and a Guidelineless Reentry Assembly (GRA) to the seafloor, and jet in 36-inch conductor casing to 60 mbsf (Fig. 5). The 36-inch conductor pipe and jetting Bore Hole Assembly (BHA) were made up and run, and by 21 July the guide horn was installed, and the vessel started drifting to site. Hole C0002F was spudded on 25 July, jetting the 36-inch conductor pipe in to 54 mbsf. After the casing angle was confirmed by Write Remotely Operated Vehicle (ROV) to be within 1.5° of vertical, the

BHA was exchanged to a 26-inch drill-ahead assembly. Releasing the Drill Ahead Tool (DAT) went smoothly, followed by drilling with 26-inch bit to target depth of 856.5 mbsf on 27 July.

The subsequent wiper trip encountered a few tight spots, including one too close to total depth (TD) that could endanger the cement job for the casing operation. We decided to deepen the hole slightly, to reach a more stable interval at 868.5 mbsf. By 30 July we were able to clear the hole and move upstream to prepare for casing operations.

The lowering of the 20-inch casing began on 31 July, and by 1 August deep sea drilling vessel (D/V) *Chikyu* could drift back towards Hole C0002F with 72 joints of 20-inch casing trailing underneath. However, during the afternoon of 1 August, probably due to a sudden intensifying of the current, the string sheared off just above the Casing Running Tool (CART), and most of the casing was lost on the seafloor. The decision was made to return to the port of Shingu to load up new casing and try again.

The port call lasted less than a day, and by 3 August *Chikyu* returned to sea to stand by, while decisions were made on shore about how to proceed. It was decided to procure a sturdier but untested CART, which was delivered by the supply boat on 7 August. While decisions were still being made onshore, we surveyed current conditions with the help from the tanker ship *Heisei-maru*. Currents sometimes exceeded 4 knots, with an average around 3 knots.

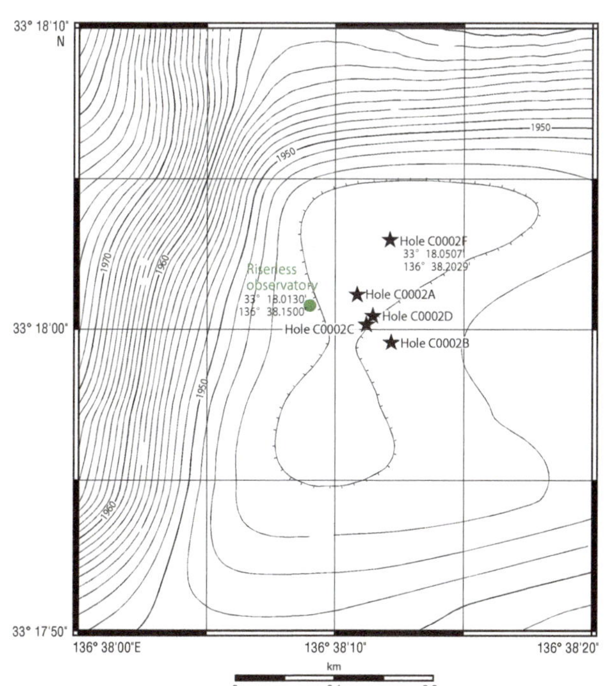


Figure 4. Bathymetric map of Site C0002. Black stars=holes drilled, green circle=proposed hole for Expedition 332 riserless observatory.

Preparations for the second casing operation were completed by 12 August. We reamed down another 4 m (to 872.5 mbsf) and continued wiper trips until 14 August, then returned upstream to a low current area to prepare for setting casing. On 15 August, the new casing was lowered, with four ropes attached to the drill pipe to reduce the Vortex-Induced Vibration (VIV). Use of the rope is proven very effective to suppress VIV (Kitada et al., 2011), thus this option was adopted for the second casing operation.

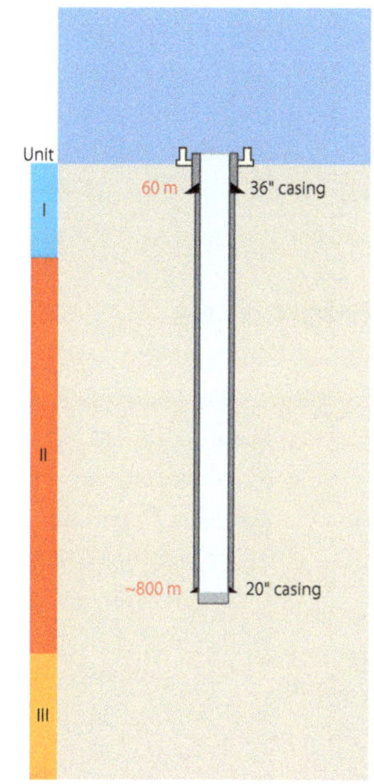


Figure 5. Casing plan for top hole portion of the riser hole, C0002F. The 36-inch conductor casing was set at 54 mbsf, and the 20-inch casing was run to 856.6 mbsf.

Drifting downstream was successfully followed by the casing stabbed into Hole C0002F under the current speeds of >4 knots. On 18 August, the casing was successfully set and cemented, and the drill pipe could be tripped back up. The casing shoe is at 860.3 mbsf. Finally, a steel lid, called the "corrosion cap", was set at the top of the hole by the ROV on 19 August. This completed mission of Expedition 326 to set the wellhead, 36-inch conductor, and 20-inch casing, and to cement to 860.3 mbsf, satisfying the requirement for this tophole portion of the planned riser borehole. The *Chikyu* arrived in Shimizu, Japan, and Expedition 326 concluded on 20 August 2010.

The loss of a substantial length of 20-inch casing in the water column, presumably due to the strong current and VIV, was a costly problem. However, the lesson learned about how to deploy casing and other pipe through the Kuroshio current using the vibration-dampening rope coils proved to be successful in the second attempt.

## Plan Toward the Completion of Hole C0002F

In October 2012, Hole C0002F will be revisited to extend the hole from ~900 mbsf to 3600 mbsf during IODP Expedition 338 "NanTroSEIZE Plate Boundary Deep Riser – 2" (Fig. 3). The proposed extension would access the deep interior of the Miocene accretionary prism. In addition to continuous cutting analyses, mud-gas and logging while drilling (LWD) data will be acquired throughout the interval, whereas the core sampling is planned for a limited interval (2300–2400 mbsf). The hole will be cased with

16-inch casing for the upper 2300 m, then with 13 3/8 inch casing down to 3600 mbsf.

In the following year, this hole is planned to be deepened to ~4700 mbsf for the next casing, followed by the drilling/coring ahead toward the megasplay fault system that is expected at 5000–5200 mbsf.

## References

Expedition 314 Scientists, 2009. Expedition 314 Site C0002. *In* Kinoshita, M., Tobin, H., Ashi, J., Kimura, G., Lallement, S., Screaton, E.J., Curewitz, D., Masago, H., Moe, K.T., and the Expedition 314/315/316 Scientists, *Proc. IODP,* 314/315/316: Washington, DC (Integrated Ocean Drilling Program Management International, Inc.). doi:10.2204/iodp.proc.314315316.114.2009

Expedition 315 Scientists, 2009. Expedition 315 Site C0002. *In* Kinoshita, M., Tobin, H., Ashi, J., Kimura, G., Lallemant, S., Screaton, E.J., Curewitz, D., Masago, H., Moe, K.T., and the Expedition 314/315/316 Scientists, *Proc. IODP,* 314/315/316: Washington, DC (Integrated Ocean Drilling Program Management International, Inc.). doi:10.2204/iodp.proc.314315316.124.2009

Expedition 319 Scientists, 2010. Expedition 319 summary. *In* Saffer, D., McNeill, L., Byrne, T., Araki, E., Toczko, S., Eguchi, N., Takahashi, K., and the Expedition 319 Scientists, *Proc. IODP,* 319: Washington, DC (Integrated Ocean Drilling Program Management International, Inc.). doi:10.2204/iodp.proc.319.101.2010

Heki, K., 2007. Secular, transient and seasonal crustal movements in Japan from a dense GPS array: Implication for plate dynamics in convergent boundaries. *In* Dixon, T., and Moore, C. (Eds.), *The Seismogenic Zone of Subduction Thrust Faults:* New York (Columbia University Press), 512–539.

Ito, Y., and Obara, K., 2006. Dynamic deformation of the accretionary prism excites very low frequency earthquakes. *Geophys. Res. Lett.*, 33(2):LO2311, doi:10.1029/2005GL025270

Kinoshita, M., Tobin, H., Ashi, J., Kimura, G., Lallemant, S., Screaton, E.J., Curewitz, D., Masago, H., Moe, K.T., and the Expedition 314/315/316 Scientists, 2009. *Proc. IODP,* 314/315/316: Washington, DC (Integrated Ocean Drilling Program Management International, Inc.). doi:10.2204/iodp.proc.314315316.2009

Kitada, K., Araki, E., Kimura, T., Kinoshita, M., Kopf, A., Hammerschmidt, S., Toczko, S., et al., 2011. Drill pipe monitoring of vortex-induced vibration during IODP Expedition 332 observatory installations. *In* Kopf, A., Araki, E., Toczko, S., and the Expedition 332 Scientists, *Proc. IODP,* 332: Washington, DC (Integrated Ocean Drilling Program Management International, Inc.). doi:10.2204/iodp.proc.332.106.2011

Moore, G.F., Bangs, N.L., Taira, A., Kuramoto, S., Pangborn, E., and Tobin, H.J., 2007. Three-dimensional splay fault geometry and implications for tsunami generation. *Science,* 318(5853):1128–1131. doi:10.1126/science.1147195

Moore, G.F., Park, J.-O., Bangs, N.L., Gulick, S.P., Tobin, H.J., Nakamura, Y., Sato, S., et al., 2009. Structural and seismic stratigraphic framework of the NanTroSEIZE Stage 1 transect. *In* Kinoshita, M., Tobin, H., Ashi, J., Kimura, G., Lallemant, S., Screaton, E.J., Curewitz, D., Masago, H., Moe, K.T., and the Expedition 314/315/316 Scientists, *Proc. IODP,* 314/315/316: Washington, DC (Integrated Ocean Drilling Program Management International, Inc.). doi:10.2204/iodp.proc.314315316.102.2009

Obana, K., and Kodaira, S., 2009. Low-frequency tremors associated with reverse faults in a shallow accretionary prism. *Earth Planet. Sci. Lett.*, 287(1–2):168–174. doi:10.1016/j.epsl.2009.08.005

Obara, K., and Ito, Y., 2005. Very low frequency earthquake excited by the 2004 off the Kii peninsula earthquake: A dynamic deformation process in the large accretionary prism. *Earth Planets Space*, 57:321–326.

Park, J.-O., Tsuru, T., Kodaira, S., Cummins, P.R., and Kaneda, Y., 2002. Splay fault branching along the Nankai subduction zone. *Science*, 297(5584):1157–1160. doi:10.1126/science.1074111

Park, J.-O., Tsuru, T., No, T., Takizawa, K., Sato, S., and Kaneda, Y., 2008. High-resolution 3D seismic reflection survey and prestack depth imaging in the Nankai Trough off southeast Kii Peninsula. *Butsuri Tansa*, 61:231–241. (in Japanese with English abstract).

Seno, T., Stein, S., and Gripp, A.E., 1993. A model for the motion of the Philippine Sea plate consistent with NUVEL-1 and geological data. *J. Geophys. Res.*, 98:17941–17948.

Tobin, H.J., and Kinoshita, M., 2006. NanTroSEIZE: The IODP Nankai Trough Seismogenic Zone Experiment. *Sci. Drill.*, 2:23–27. doi:10.2204/iodp.sd.2.06.2006

Tobin, H., Kinoshita, M., Ashi, J., Lallemant, S., Kimura, G., Screaton, E., Moe, K.T., Masago, H., Curewitz, D., and the Expedition 314/315/316 Scientists, 2009. NanTroSEIZE Stage 1 expeditions: Introduction and synthesis of key results. *In* Kinoshita, M., Tobin, H., Ashi, J., Kimura, G., Lallement, S., Screaton, E.J., Curewitz, D., Masago, H., Moe, K.T., and the Expedition 314/315/316 Scientists, *Proc. IODP,* 314/315/316: Washington, DC (Integrated Ocean Drilling Program Management International, Inc.). doi:10.2204/iodp.proc.314315316.101.2009

Underwood, M.B., Saito, S., Kubo, Y., and the Expedition 322 Scientists, 2010. Expedition 322 summary. *In* Saito, S., Underwood, M.B., Kubo, Y., and the Expedition 322 Scientists, *Proc. IODP,* 322: Washington, DC (Integrated Ocean Drilling Program Management International, Inc.). doi:10.2204/iodp.proc.322.101.2010

## Authors

**Masa (Masataka) Kinoshita**, Japan Agency for Marine-Earth Science and Technology–Kochi Core Center (JAMSTEC-KCC), Monobe-B200, Nankoku-City, Kochi, 783-8502 Japan, e-mail: masa@jamstec.go.jp

**Harold Tobin**, Department of Geology and Geophysics, University of Wisconsin-Madison, 1215 West Dayton Street, Madison, WI 53706, U.S.A., e-mail: htobin@wisc.edu

**Nobuhisa Eguchi** and **Simon Nielsen**, Center for Deep Earth Exploration–Japan Agency for Marine-Earth Science and Technology (CDEX-JAMSTEC), 3173-25 Showa-machi, Kanazawa-ku, Yokohama, Kanagawa 236-0001, Japan

# 16

# The IODP Expedition 332 : Eyes on the Prism, The NanTroSEIZE Observatories

by Sean T. Toczko, Achim J. Kopf, Eiichiro Araki, and
the IODP Expedition 332 Scientific Party

## Abstract

The Nankai Trough Seismogenic Zone Experiment (NanTroSEIZE) is a major long-term drilling project designed to investigate the seismogenic behavior of subduction zone plate boundaries. Integrated Ocean Drilling Program (IODP) Expedition 332 deployed a long-term borehole monitoring system (LTBMS), an advanced Circulation Obviation Retrofit Kit (CORK)-type observatory. The recovery of pressure and temperature data from a temporary observatory (SmartPlug) deployed during IODP Expedition 319 helped prove the SmartPlug concept. The permanent LTBMS was deployed n the upper 1000 m of Site C0002, while the SmartPlug was recovered from Site C0010 and replaced with a more capable "GeniusPlug", incorporating an extension with a geochem-ical sampler and biological experiment to the original SmartPlug design. SmartPlug pressure and temperature data showed signs of transient pressure events.

## Introduction

Subduction zones account for the majority of global seismic moment release, and slip along subduction megathrusts generate both damaging earthquakes and tsunamis. Understanding the processes that govern the distribution, mechanics, and style of slip along subduction and other plate boundary fault systems is essential to earthquake and tsunami hazard assessment. NanTroSEIZE is a multi-expedition, multistage IODP drilling project focused on understanding the mechanics of subduction plate boundary faults (Tobin and Kinoshita, 2006). The drilling program includes a coordinated effort to characterize, sample, and instrument the plate boundary system at several locations offshore of the Kii Peninsula (Fig. 1), culminating in drilling, sampling, and instrumenting the plate boundary fault system near the updip limit of inferred coseismic slip, at 5~7 km below sea-floor (Fig. 2; Tobin and Kinoshita, 2006) and in installation of a distributed network of integrated borehole observatories to monitor strain and seismological, thermal, and hydrological processes.

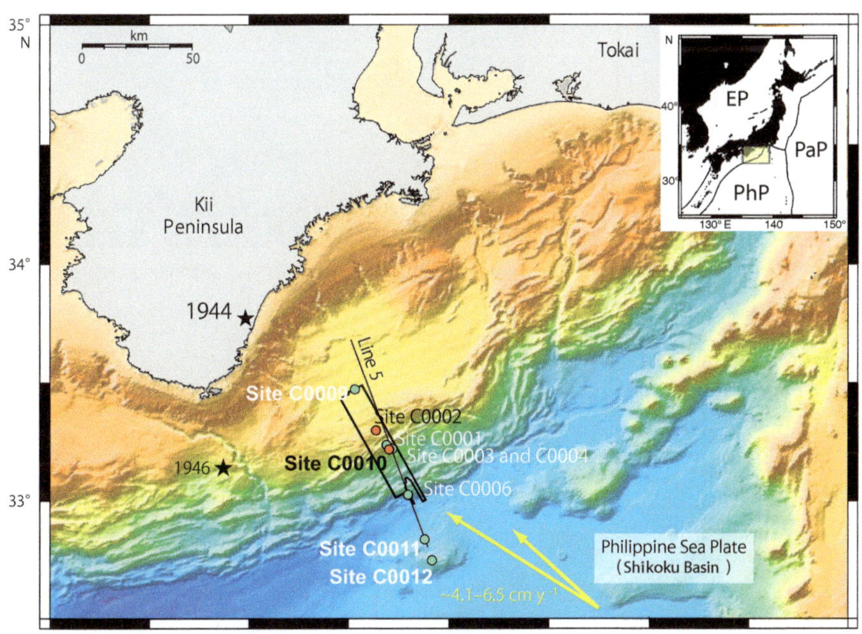


Figure 1. A map of the study area depicting Sites C0002 and C0010 with red circles. The locations of the other NanTroSEIZE sites are shown (green circles), while the inset shows the site location in reference to Japan, and the respective tectonic plates surrounding Japan (EP=Eurasian Plate, PaP=Pacific Plate, PhP=Philippine Plate). The yellow arrows depict the relative movement (4.1–6.5 cm $y^{-1}$) of the Philippine Plate. The elongated box depicts the limits of the 3D survey area, while Line 5 depicts KR0108-5, a 2D seismic survey line.

In the NanTroSEIZE study area, high-resolution seismic reflection profiles across the outer rise clearly document a major out-of-sequence thrust fault system (megasplay fault, after Park et al., 2002) that branches from the décollement close to the updip limit of inferred coseismic rupture in the 1944 Tonankai M 8.2 earthquake (Fig. 2). Several lines of evidence indicate that the megasplay system is active, may accommodate a significant fraction of plate boundary motion, and may slip coseismically (Moore et al., 2007; Strasser et al., 2009). However, the partitioning of strain between the lower plate interface (the décollement zone) and the megasplay system and the nature and mechanisms of fault slip as a function of depth and time on the megasplay are not understood. Thus, the meg-

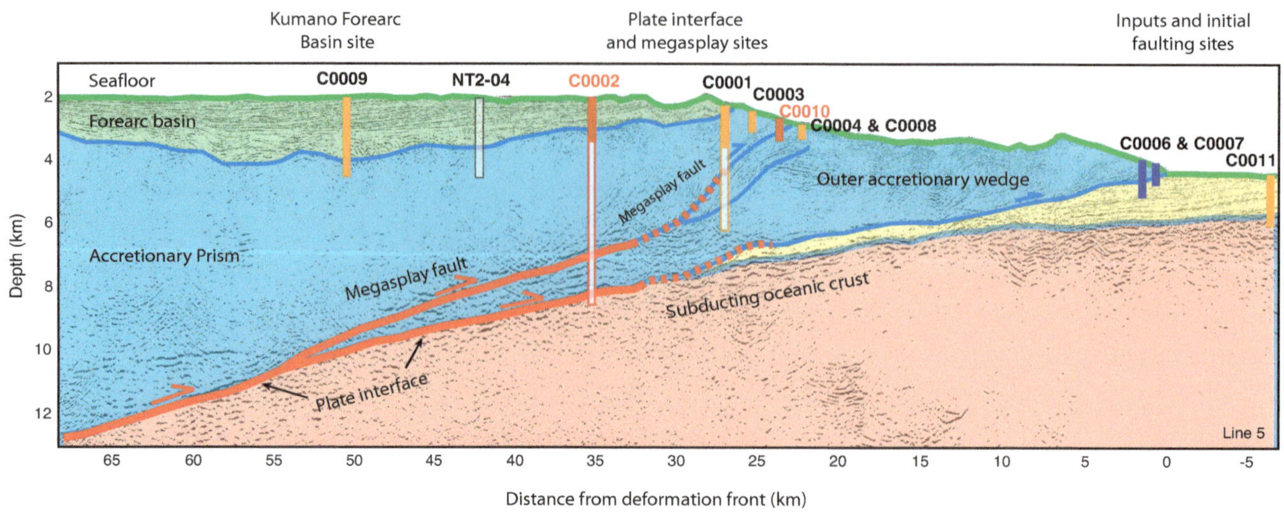


Figure 2. Interpreted seismic profile at regional scale showing interpretation by Park et al. (2002) and NanTroSEIZE drill sites. Sites C0002 and C0010 are indicated in red. All drill sites show current penetration (solid bars) and planned penetration (empty bars).

asplay and the region near its updip terminus comprise one of the primary drilling and monitoring targets for NanTroSEIZE.

NanTroSEIZE Stage 1 included riserless drilling along a transect of eight sites that targeted the frontal thrust region near the trench, the megasplay fault region, and the Kumano forearc basin region (Fig. 2) One of these sites, IODP Site C0002 in the forearc basin, also serves as the ultra-deep riser-drilling target. Expedition 319 installed a temporary borehole observatory (SmartPlug) at Site C0010 in an interval of screened casing across the megasplay fault (McNeill et al., 2010; Saffer et al., 2010).

IODP Expedition 332 continued, expanded, and extended the observatory program, installing the LTBMS and GeniusPlug observatories at Sites C0002 and C0010, respectively.

## The Nankai Margin

The Nankai Trough (Fig. 1) is formed by subduction of the Philippine Sea Plate underneath southwestern Japan at a rate of ~4.1–6.5 cm $y^{-1}$ along an azimuth of 300°–315°N (Seno et al., 1993; Miyazaki and Heki, 2001) down an interface dipping 3°–7° (Kodaira et al., 2000). The Nankai Trough subduction zone forms a sediment-dominated accretionary prism. In the toe region, an incoming sedimentary section ~1–1.5 km thick is accreted to or underthrust below the margin (Moore et al., 2001, 2009).

The megasplay is a major structural boundary to the rear of the accretionary wedge; it forms the boundary between the outer wedge and forearc basin, traverses the entire wedge, and has had a protracted history as shown by the tilted forearc basin sediments trapped behind its leading edge (Moore et al., 2007). The megasplay is also hypothesized to represent a discontinuity in rock physical properties and a mechanical boundary between the inner and outer accretionary wedge and between aseismic and seismogenic fault behavior along the plate boundary (Wang and Hu, 2006). At depth, the megasplay is imaged in seismic reflection data as a high-amplitude reflector (Fig. 2; Bangs et al., 2009), and it branches into a family of smaller splays in the upper few kilo-

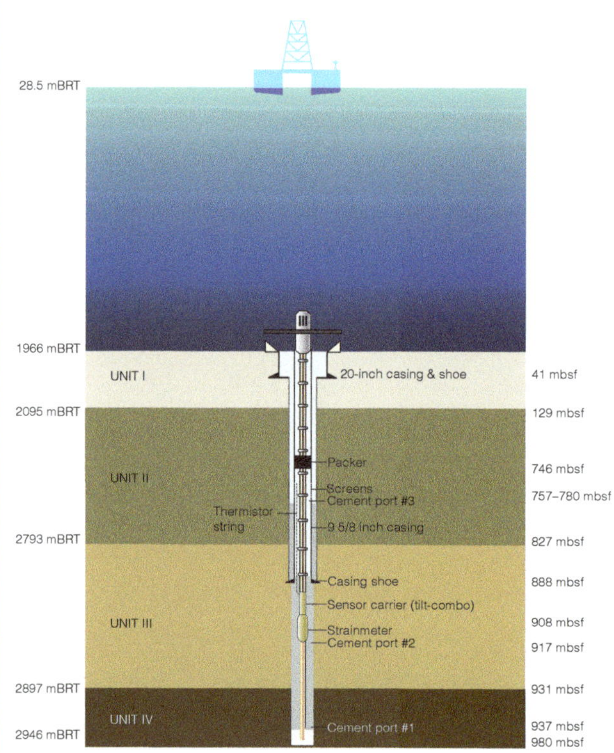


Figure 3. Diagram of the LTBMS installed at Site C0002. The lithological units (Units I–IV) are shown, as well as the depths for different components of the observatory and casing string. Lithological Unit IV is the upper accretionary prism, while Units III to I represents the Kumano forearc basin. (mBRT=meters below rotary table; mbsf=meters below seafloor).

meters below the seafloor, including the fault penetrated at IODP Sites C0004 and C0010. Direct fault intersections to the seafloor are not observed (Moore et al., 2007; Strasser et al., 2009); however, the thrust sheets wedge into these deposits, causing tilt and slumping of even the deposits nearest to the surface. Evidence for mass wasting complexes is consequently found at IODP Sites C0004, C0008, and C0010. Site C0010, as well as Site C0002 further landward in the forearc basin, were the target areas of Expedition 332.

Site C0010, drilled during Expedition 319 (McNeill et al., 2010; Saffer et al., 2010) included Logging While Drilling (LWD) drilling through the megasplay fault zone and into its footwall, setting casing with screens across the fault zone, and installing a simple and temporary borehole observatory (SmartPlug) to monitor fluid pressure and temperature in the shallow megasplay. Major lithologic boundaries as well as the location of the megasplay fault at ~407 mbsf were identified in LWD data and were used to select a depth interval spanning the fault for placement of the two screened casing joints (Saffer et al., 2010).

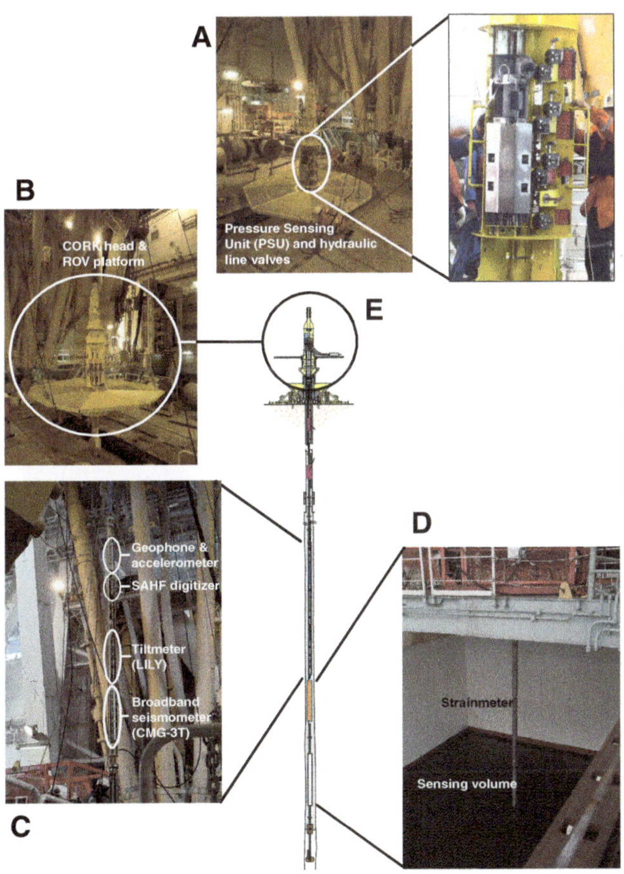


Figure 4. The sensors and placement on the LTBMS CORK observatory at Site C0002. Photos of the different sensor packages and their arrangement on the downhole and CORK head sections were taken during deployment. [A] The Pressure Sensing Unit (PSU), an independent sensor array mounted on the CORK head. [B] CORK head and ROV platform. [C] Sensor carrier, with a geophone and accelerometer assembly, a stand-alone heat flow meter (SAHF) digitizer for the thermistor string, a LILY tiltmeter, and a CMG-3T Guralp seismometer. [D] Strainmeter and sensing volume. [E] Diagram of the Site C0002 CORK observatory.

Site C0002 is located near the southeastern edge of the Kumano forearc basin (Fig. 2). Expedition 332 revisited Site C0002 and drilled with a limited suite of LWD/measurement-while-drilling (MWD) tools for reconnaissance and to identify the most suitable depth intervals to place the sensors of the LTBMS.

## NanTroSEIZE Observatories and IODP Expedition 332

Expedition 332 operations at Site C0010 involved recovering the SmartPlug installed during Expedition 319 and replacing it with an expanded GeniusPlug Observatory. Operations at Site C0002 were the first steps in one of the two final phases of NanTroSEIZE, in this case, installing a permanent observatory in the upper section of the Kumano Basin, which will in the future have a complementary deep observatory installed down to the plate boundary.

### *LTBMS Sensors, Site C0002*

The LTBMS sensor array (Fig. 3) is designed to collect multiparameter observations covering a dynamic range of events, including local microearthquakes, low frequency earthquakes, and large-scale earthquakes related to the Tonankai plate boundary movement approximately 6 km below the sensor array. The LTBMS observatory is comprised of two main components: (1) downhole sensors and (2) a CORK head unit at the seafloor with a pressure and temperature multi-sensor package and data and power connections to the downhole instruments (Fig. 4). The CORK head assembly pressure sensor unit (PSU) is comprised of four pressure transducers, three of which are connected via two-way valves to individual hydraulic lines to one of three pressure ports downhole. The bottom-hole pressure port (#3) was set below the Unit III/IV boundary to sample pore fluid pressure in the accretionary complex (Unit IV) beneath the Kumano forearc basin sediment (Units I–III). Pressure Port #2 was located inside Unit III at 917 mbsf, just below the strainmeter sensing surface.

The downhole sensor suite comprises a strainmeter, a Guralp broadband seismometer, and a tilt combo (tiltmeter, geophone, accelerometer, and thermometer string digitizer)—all attached to a sensor carrier—and a thermistor string (Fig. 4). The strainmeter, along with the tilt combination package, was set into a fracture-free zone (890–917 mbsf) as identified via LWD drilling within the lower Unit III mudstone layer (Kopf et al., 2011a). This provides the strainmeter with the best possible coupling to a relatively homogenous section of the formation. Above the strainmeter is the instrument carrier (890–908 mbsf), hosting the seismometer and the tilt combo package. The sensor carrier also holds a data digitizer for the thermistor string, with the two uppermost nodes positioned in the screened casing interval and two nodes in the cemented section above the instrument carrier.

It was necessary to isolate the seismic sensors from the possibility of signals being transmitted from the casing above by placing the 9-inch casing shoe above the seismic sensor and cemented section. Therefore, the casing shoe was set at 888 mbsf (Fig. 3), above the top of the instrument carrier, in an interval without significant washouts. This ensures a good cementing zone for the bottom of the casing, which also serves to isolate the screened casing section from Unit III. The screened casing (757–780 mbsf) was set in a mudstone layer in Unit II to compare the pore fluid pressure in sedimentary layers with the accretionary prism in Unit IV. Finally, the swellable packer was set near the top of the screened casing interval to minimize the volume associated with the pressure measurement in pressure port #3, and also set to ensure that the packer sealing surface (~1.5 m) does not overlap with any casing joints.

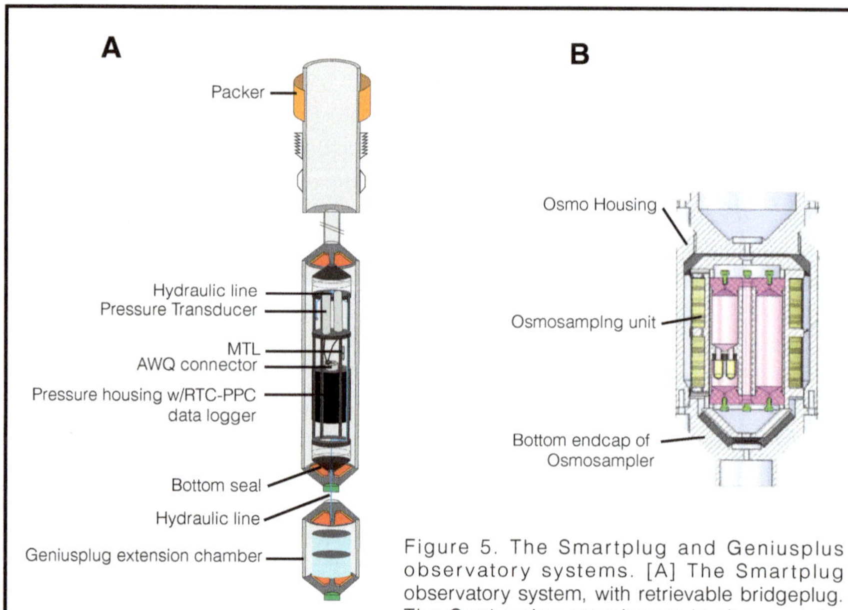


Figure 5. The Smartplug and Geniusplus observatory systems. [A] The Smartplug observatory system, with retrievable bridgeplug. The Geniusplug attachment is shown at the bottom. [B] Close-up and internal diagram of the Geniusplug Osmosampler unit.

The LTBMS CORK in Hole C0002G will provide a wealth of valuable data and observations once it is tied into the DONET seafloor cabled network.

### *SmartPlug Data Set and GeniusPlug upgrade, Site C0010*

The full data set clearly shows a gradual increase in formation pressure and a subtle decrease in the seafloor reference pressure after installation but an overall increase in excess pore pressure. Regardless of this trend, three different types of transient pressure excursions were observed:

1. pressure pulses of up to several hundred pascals in sudden amplitude change relative to background that lasted for 20–60 minutes, which are largely associated with known teleseismic events;
2. tremor-like signals of a fraction of a kilopascal with durations of hours or even days, which can be associated with atmospheric and oceanographic events such as low-pressure systems, storm waves, and tsunamis; and
3. smaller individual pulses of low amplitude (<1 kPa) that lasted for only minutes, which are tentatively associated with either low-magnitude deformational events in the accretionary complex or at the seafloor.

The data collected from the recovered SmartPlug (Fig. 5) prove to be a complete time series data set covering more than fifteen months (since deployment during Expedition 319), and they validate the concept of affordable, durable, replaceable CORK-like observatories (Kopf et al., 2011b). The upgraded version, the GeniusPlug (Fig. 5), is expandable; it uses an attached chamber (30 cm long) which can host other experiments—in the case of Expedition 332, a long-term geochemical and biological experiment.

## Post-Expedition 332 Work

Two papers have been prepared, one focusing on vortex-induced vibration (VIV) mitigation for LTBMS deployment and the other on a preliminary analysis of the SmartPlug pressure and temperature data. These have been added to the proceedings for Expedition 332 (see details in Kitada et al., 2011 and in Kopf et al., 2011a, 2011b).

The LTBMS was due for a service call and hookup to battery source and data storage during the spring of 2011; however, these plans were postponed due to the Tohoku earthquake. The site was revisited during January 2012, and tests were successfully run on all systems. Again, the LTBMS was not hooked to the external power source, this time due to poor weather conditions. Current plans call for the system to be up and running after a remotely operated vehicle (ROV) visit in spring 2013.

## Plans for Future Experiments and Temporary Observatory Recovery

In 2013, the LTBMS (Site C0002) will be hooked to the DONET network for real-time monitoring. The GeniusPlug (Site C0010) will need to be recovered before the geochemical and biological samples cycle through the system, particularly for sampling associated with the 11 March Tohoku earthquake. This is planned for the 2012/2013 NanTroSEIZE operations of D/V *Chikyu*.

## Acknowledgements

We thank all the crew of the D/V *Chikyu* and all the drilling operations and related personnel, particularly the efforts of the Marine Works Japan technicians and the Mantle Quest Japan onboard personnel who worked to make Expedition 332 such a success. We would also like to express our thanks to Tom Pettigrew of Pettigrew Engineering PLLC, who shared his considerable engineering experience with us.

## IODP Expedition 332 Scientific Party

Sebastian Hammerschmidt, Toshinori Kimura, Kazuya Kitada, Rachel Lauer, Demian Saffer, and Geoff Wheat

## References

Bangs, N.L.B., Moore, G.F., Gulick, S.P.S., Pangborn, E.M., Tobin, H.J., Kuramoto, S., and Taira, A., 2009. Broad, weak regions of the Nankai Megathrust and implications for shallow coseismic slip. *Earth Planet. Sci. Lett.*, 284(1–2):44–49. doi:10.1016/j.epsl.2009.04.026

Kitada, K., Araki, E., Kimura, T., Kinoshita, M., Kopf, A., Hammerschmidt, S., Toczko, S., et.al., 2011. Drill pipe monitoring of vortex-induced vibration during IODP Expedition 332 observatory installations. *In* Kopf, A., Araki, E., Toczko, S., and the Expedition 332 Scientists, *Proc. IODP*, 332: Washington, DC (Integrated Ocean Drilling Program Management International, Inc.). doi:10.2204/iodp.proc.332.106.2011

Kodaira, S., Takahashi, N., Park, J.-O., Mochizuki, K., Shinohara, M., and Kimura, S., 2000. Western Nankai Trough seismogenic zone: Results from a wide-angle ocean bottom seismic survey. *J. Geophys. Res., [Solid Earth]*, 105(B3):5887–5905. doi:10.1029/1999JB900394

Kopf, A., Araki, E., Toczko, S., and the Expedition 332 Scientists, 2011a. NanTroSEIZE Stage 2: Riserless Observatory. *Proc. IODP*, 332: Washington, DC (Integrated Ocean Drilling Program Management International, Inc.). doi:10.2204/iodp.proc.332.2011

Kopf, A., Saffer, D.M., Davis, E.E., Hammerschmidt, S., LaBonte, A., Meldrum, R., Toczko, S., et.al., 2011b. The SmartPlug and GeniusPlug: Simple retrievable observatory systems for NanTroSEIZE borehole monitoring. *In* Kopf, A., Araki, E., Toczko, S., and the Expedition 332 Scientists, *Proc. IODP*, 332: Washington, DC (Integrated Ocean Drilling Program Management International, Inc.). doi:10.2204/iodp.proc.332.105.2011

McNeill, L., Saffer, D., Byrne, T., Araki, E., Toczko, S., Eguchi, N., Takahashi, K., and IODP Expedition 319 Scientists, 2010. IODP Expedition 319, NanTroSEIZE Stage 2: First IODP riser drilling operations and observatory installation towards understanding subduction zone seismogenesis. *Sci. Drill.*, 10:4–12, doi: 10.2204/iodp.sd.10.01.2010

Miyazaki, S., and Heki, K., 2001. Crustal velocity field of southwest Japan: Subduction and arc-arc collision. *J. Geophys. Res., [Solid Earth]*, 106(B3):4305–4326. doi:10.1029/2000JB900312

Moore, G.F., Bangs, N.L., Taira, A., Kuramoto, S., Pangborn, E., and Tobin, H.J., 2007. Three-dimensional splay fault geometry and implications for tsunami generation. *Science*, 318(5853):1128–1131. doi:10.1126/science.1147195

Moore, G.F., Park, J.-O., Bangs, N.L., Gulick, S.P., Tobin, H.J., Nakamura, Y., Sato, S., et.al., 2009. Structural and seismic stratigraphic framework of the NanTroSEIZE Stage 1 transect. *In* Kinoshita, M., Tobin, H., Ashi, J., Kimura, G., Lallemant, S., Screaton, E.J., Curewitz, D., Masago, H., Moe, K.T., and the Expedition 314/315/316 Scientists, *Proc. IODP*, 314/315/316: Washington, DC (Integrated Ocean Drilling Program Management International, Inc.). doi:10.2204/iodp.proc.314315316.102.2009

Moore, G.F., Taira, A., Klaus, A., et al., 2001. *Proc. ODP, Init. Repts.*, 190: College Station, TX (Ocean Drilling Program). doi:10.2973/odp.proc.ir.190.2001

Park, J.-O., Tsuru, T., Kodaira, S., Cummins, P.R., and Kaneda, Y., 2002. Splay fault branching along the Nankai subduction zone. *Science*, 297(5584):1157–1160. doi:10.1126/science.1074111

Saffer, D., McNeill, L., Byrne, T., Araki, E., Toczko, S., Eguchi, N., Takahashi, K., and the Expedition 319 Scientists, 2010. NanTroSEIZE Stage 2: NanTroSEIZE riser/riserless observatory. *Proc. IODP*, 319: Washington, DC (Integrated Ocean Drilling Program Management International, Inc.). doi:10.2204/iodp.proc.319.2010

Seno, T., Stein, S., and Gripp, A.E., 1993. A model for the motion of the Philippine Sea Plate consistent with NUVEL-1 and geological data. *J. Geophys. Res., [Solid Earth]*, 98(B10):17941–17948. doi:10.1029/93JB00782

Strasser, M., Moore, G.F., Kimura, G., Kitamura, Y., Kopf, A.J., Lallemant, S., Park, J.-O., et al., 2009. Origin and evolution of a splay fault in the Nankai accretionary wedge. *Nature Geosci.*, 2(9):648–652. doi:10.1038/ngeo609

Tobin, H.J., and Kinoshita, M., 2006. NanTroSEIZE: The IODP Nankai Trough Seismogenic Zone Experiment. *Sci. Drill.*, 2:23–27. doi:10.2204/iodp.sd.2.06.2006

Wang, K., and Hu, Y., 2006. Accretionary prisms in subduction earthquake cycles: The theory of dynamic Coulomb wedge. *J. Geophys. Res.*, 111(B6):B06410. doi:10.1029/2005JB004094

## Authors

**Sean T. Toczko**, Center for Deep Earth Exploration (CDEX), Japan Agency for Marine-Earth Science and Technology (JAMSTEC), Yokohama Institute for Earth Sciences, 3173-25 Showa-machi, Kanazawa-ku, Yokohama, Kanagawa 236-0001 Japan, e-mail: sean.jamstec@gmail.com
**Achim J. Kopf**, MARUM and Department of Geosciences, University of Bremen, 28359 Bremen, Germany, e-mail: akopf@uni-bremen.de
**Eiichiro Araki**, Institute for Research on Earth Evolution, Japan Agency for Marine-Earth Science and Technology (JAMSTEC), 2-15 Natsushima-cho, Yokosuka, Kanagawa 237-0061 Japan
**and the IODP Expedition 332 Scientific Party**

# CORK-Lite: Bringing Legacy Boreholes Back to Life

by C. Geoffrey Wheat, Katrina J. Edwards, Tom Pettigrew, Hans W. Jannasch, Keir Becker, Earl E. Davis, Heiner Villinger, and Wolfgang Bach

## Introduction

An essential aspect of the forty years of deep-sea scientific drilling has been to maximize the scientific return during each expedition while preserving samples for future investigations. This philosophy also extends to borehole design, providing the community with tens of cased legacy boreholes that penetrate into the basaltic crust, each ripe for future investigations of crustal properties and experiments to determine crustal processes (Edwards et al., 2012a). During Integrated Ocean Drilling Program (IODP) Expedition 336 to North Pond on the western flank of the Mid-Atlantic Ridge at 22°N, Hole U1383B (Fig. 1) was planned to be a deep hole, but was abandoned when a 14.75-inch tri-cone bit catastrophically failed at 89.9 meters below the seafloor (mbsf) (Expedition 336 Scientists, 2012). This resulted in about 36 meters of open hole below casing, similar to conditions within tens of legacy boreholes. Because the overall experiment required a return to the "natural" hydrologic state in basaltic basement, it was critical to seal the hole to prevent a hydrologic "short circuit". Thus, a plan emerged at sea to seal Hole U1383B with a simplified Circulation Obviation Retrofit Kit (CORK) termed "CORK-Lite" that could be deployed by a remotely operated vehicle (ROV) on a planned dive series five months later. To prepare for this deployment, a standard ROV platform that is used with CORKs was modified to be self-guiding in the re-entry cone and deployed. The next step was to design a CORK system that could seal the borehole, yet be physically manageable with an ROV, and be ready for shipping and deployment within three months. Several key functional aspects dictated the design of the new CORK-Lite (Table 1).

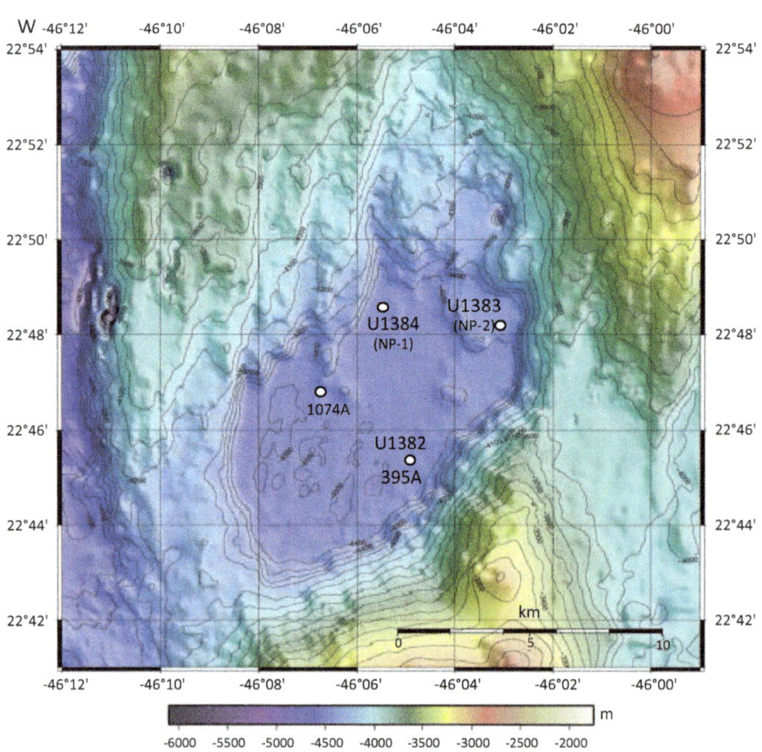


Figure 1. Location of IODP Sites drilled during Exp. 336 (North Pond) and ODP Hole 1074A (duplicated from Expedition 336 Scientists, 2012). The CORK-Lite was deployed at Site U1383. Bathymetry was provided by Schmidt-Schierhorn et al. (2012).

## Design of CORK-Lite

The CORK-Lite has four major components: the body with a seal, a removable cap, a downhole instrument string, and a borehole pressure monitoring instrument. The body is a 4.9-m-long 12-inch pipe with a landing seal ring that has a diameter of 19.5 inches that fits within the 32-inch guide hole in the ROV platform (Figs. 2, 3). The landing seal ring lands on and seals in the 20-inch casing hanger. The body has hooks to hang instruments, two valve bodies that accept hydraulic connectors closed in the horizontal position (Wheat et al., 2011), two flanges (lifting wings) for deployment that also serve to aid in moving the body during ROV operations, and a grooved top ring made of stainless steel to insure a proper seal is achieved with the cap. A re-movable "boot" was designed to fit the bottom of the body to protect it during deployment and to prevent it from penetrating into the sediment during free fall. In addition, a lifting bar assembly was fabricated that connects the body to a float package and eases handling by the ROV.

Table 1. A list of design consideration of CORK-Lite.

| | |
|---|---|
| • | The CORK-Lite has to be installed using an ROV, with weight being balanced by flotation. |
| • | The CORK-Lite body must be self-centering, fit within the 16-inch casing, robust enough to withstand ROV operations, and extend ~2 m above the ROV platform for ease of ROV manipulations. |
| • | The seal must apply for either an over-pressured or under-pressured system. |
| • | Any negative differential pressure across the top seal cap must be vented prior to eventual removal of downhole instruments. |
| • | Two valves and ports (for redundancy) must attach to and penetrate the body for pressure monitoring within the borehole. |
| • | Any instrument string has to fit within the steel pipe used as the CORK-Lite body. |
| • | For safety, the instrument package must reside within the casing; intakes for fluid sampling must extend into the open borehole in an attempt to get "clean" borehole fluids free of possible artifacts from the steel casing and cement above. |

The cap was designed to fit on the top of the body and seal it using a rubber gasket (Fig. 4). In case the borehole formation is over-pressured, four latching dogs are included. These dogs are activated by a mechanical lever system that forces the dogs in place through vertical motion of a floating nut driven by a power screw with attached handle. A two-way valve (closed in the horizontal position) is included in the cap. This valve is necessary to equalize the pressure before removing the cap if the borehole is under-pressured. A pad eye is welded under the cap to attach the downhole instrument string.

The downhole instrument string utilized components from Exp. 336 (osmotic pumps, coils of small bore sample tubing, support rods and various connectors) that were designed for deployment within the 3-inch confines of the Exp. 336 CORKs. With the larger diameter available to the CORK-Lite, seven osmotic packages were coupled into one unit (Fig. 5). These packages include standard, dissolved gas, acid addition, enrichment, BOSS (fluid sampler that is preserved with RNA*later*® for microbial-based analysis), and microbial colonization experiments (Jannasch et al., 2004; Wheat et al., 2011; Orcutt et al., 2010). A frame was designed to hold these packages, protect them during

Figure 2. The CORK-Lite body is lowered over the side of the R/V *Maria S. Merian*. The boot is held in place with a tee handle and secured with a bungie. The lifting bar assemblely is attached to the body with triple-strand polyproplyene rope and floats above. Extra *Alvin* dive weights were attached to the hooks for faster descent.

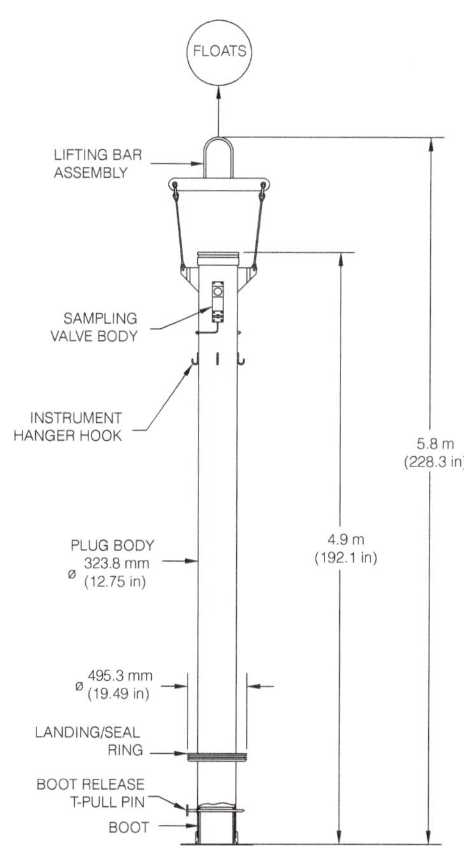


Figure 3. Schematic of the CORK-Lite body during free fall to the seafloor.

deployment and recovery, and guide them into the CORK-Lite body. These frame components have a diameter of 10.75 inches, which easily fit through the 12-inch CORK-Lite body (12-inch I.D.). Because of borehole instability issues within other CORKs, we placed the instrument package near the bottom of the casing yet have the sample intakes extend meters into the open portion of the borehole. Intakes were protected with Tygon tubing and attached to a strength member (rope) and sinker bar. A weak link was positioned just below the instrument package in case the borehole is unstable and traps the sinker bar. Thus, the instrument package can be recovered even if the sinker bar becomes entombed.

The pressure monitoring device was developed at the Pacific Geoscience Centre (Sidney, BC, Canada), identical to most systems deployed on CORKs today, including those deployed on IODP Exp. 336 and previously on Exp. 327 and 328 (see technical descriptions in Davis et al., 2010; Fisher et al., 2011; Edwards et al., 2012b). The instrument includes batteries, electronics, two absolute pressure gauges (Paroscientific Model 8B-7000; one to monitor the formation and the other to monitor the seafloor), a data logger (set to sample pressure and temperature every two minutes like the Exp. 336 CORKs in IODP Holes U1382A and U1383C), an underwater mateable connector (Teledyne ODI), and a stainless steel line that connects to an ROV-deployable hydraulic coupler that fits in the valve package on the CORK-Lite body.

Figure 4. A CORK-Lite was successfully deployed and instrumented in IODP Hole U1383B. The cap (silver material above the gray pipe) is secured to the body utilizing the handle with a short piece of rope that originally was connected to flotation. The pressure logger rests on the ROV platform and is connected to the CORK-Lite body.

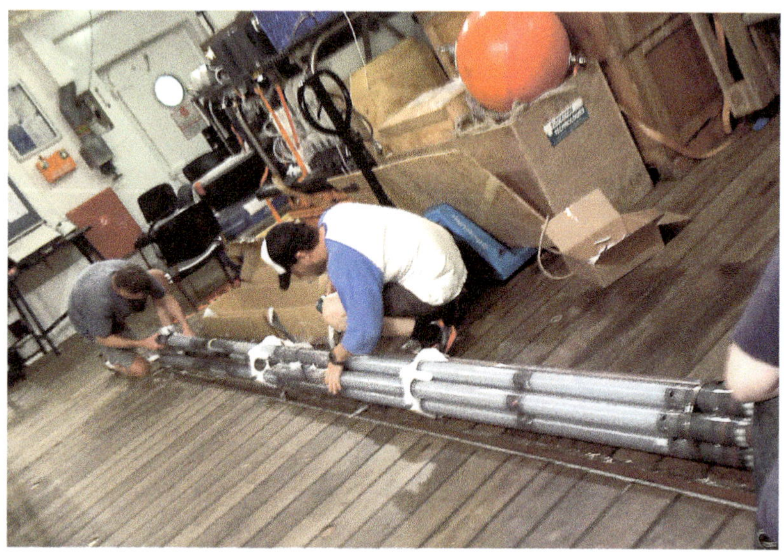

Figure 5. The downhole instrument package is fabricated before connecting it to the intake and ropes that space the package at the proper depth in the borehole. Seven osmotic packages are arranged in a bundle.

## ROV Operations

Operations were conducted with the U.S.-operated ROV *Jason* from the German research vessel R/V *Maria S. Merian* during expedition MSM 20/5. The ROV platform was inspected on *Jason* dive J2-623 (20 April 2012). Before the next dive the CORK-Lite body was deployed with ~770 lbs of flotation. During the second dive the CORK-Lite was located, transported to the borehole, and lowered into place. A black stripe was painted on the body prior to deployment and used as an indicator that the body was in the proper position. After the subsequent dive the downhole instrument string was deployed. The instrument string included (from bottom to top) descent weights, a sinker bar, a 12-m-long three-strand polyproplyene rope with a weak link and intakes that extended 8 m from the instrument package, the instrument package (seven OsmoSampler packages, and three self-contained temperature recorders), a 50-m length of 3/8-inch spectra, the top plug, and flotation. The instrument string was located and installed, and the valves were closed. Later in the dive program the pressure logger was attached, and the data set that was retrieved indicated that the CORK-Lite was sealed (Fig. 6).

## Future Applications

The design, fabrication, and operational effort at Hole U1383B illustrate the engineering potential to seal and instrument any of the tens of legacy boreholes that have been drilled into basement and cased through the sediment, leaving tens to hundreds of meters (in some cases up to 1500 meters) of open borehole (Edwards et al., 2012a).

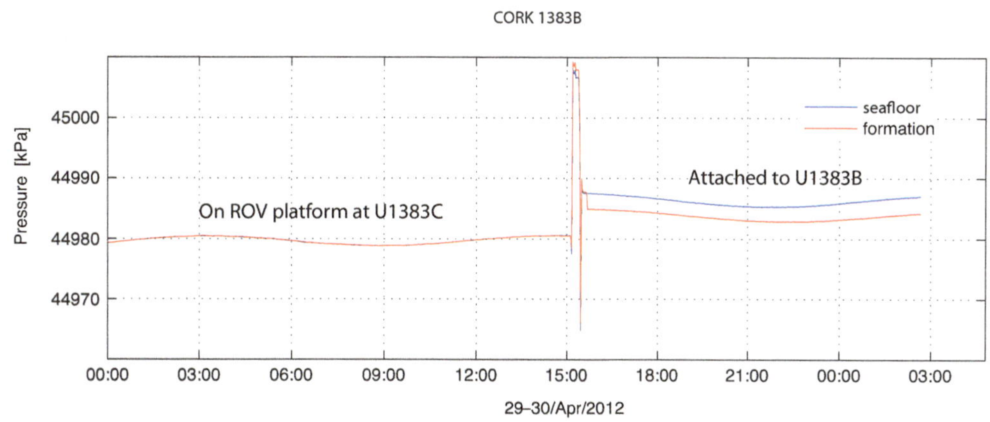


Figure 6. Pressure data from the CORK-Lite at IODP Hole 1383B. Before the instrument was attached, both sensors monitored bottom seawater on the ROV platform at Hole U1383C. After attachment, the two readings are different, signifying that the borehole is sealed and under-pressured relative to bottom seawater.

Placing sensors, samplers, and experiments in legacy boreholes could address a range of fundamental questions about conditions and processes within igneous oceanic crust. For example, the basaltic crustal aquifer plays a substantial role in cooling the Earth, regulating biogeochemical cycles within the oceans, and providing differences in redox potentials (i.e., between oxidizing seawater and reducing basaltic minerals) that offer great potential for abiotic and biologically-mediated electron transfer reactions (Bach and Edwards, 2003; Fisher and Wheat, 2010). With appropriate instrumentation in selected legacy boreholes, we will be able to provide a better measure of the range of biogeochemical processes within the basaltic crust and the global significance of these processes. Legacy boreholes can serve other communities as well, for example, by providing access for a range of sensor suites for geological and geophysical studies.

Because legacy boreholes are typically cased with 10.75-inch pipe or larger, they can accept a range of sensors and instruments that typical CORKs cannot accept (many limit internal instrument diameters to 3.5 inches). Furthermore, there are a number of instrument suites developed by Schlumberger Limited that cannot be used in typical IODP boreholes but could be used in legacy boreholes. Although the initial CORK-Lite did not include electrical cables or an umbilical, future CORK-Lites could be modified to include such connections that penetrate the cap for seafloor interrogation of downhole sensors. Also, it is conceivable to deploy swellable packers, baffles, and other means to eliminate or minimize vertical fluid exchange within the borehole, allowing one to examine specific geologic or hydrologic horizons.

For some applications CORK-Lite provides an inexpensive alternative. It is especially useful for boreholes with single horizons, as it allows for the use of less armored and less expensive umbilicals, and it eliminates the need for an inner casing string or extensive wellhead structure. Furthermore, deploying such systems into legacy boreholes is independent from the drilling schedule. In some ways the CORK-Lite is a rejuvenation of the original CORK concept (Davis et al., 1992), but it is much more versatile. IODP Hole U1383B is just the start! We envision future CORK-Lites that address a range of scientific questions, utilizing a variety of instruments, sensors, and experiments.

## Acknowledgements

We want to thank the engineering and operational staff involved in IODP Exp. 336, the crew of the R/V *Maria S. Merian*, and the team that operates the ROV *Jason*. Funding was awarded from the Gordon and Betty Moore Foundation, the German Science Foundation (DFG), and the National Science Foundation (NSF) through the STC Center for Dark Energy Biosphere Investigations (C-DEBI) (0939564) and individual research grants to CGW (OCE-0939564 and 1030061), KJE (OCE-1060634) and KB (OCE-0946795 and 1060855 for pressure-logging system). This is C-DEBI contribution number 132.

## References

Bach, W., and Edwards, K.J., 2003. Iron and sulfide oxidation within the basaltic ocean crust: Implications for chemolithoautotrophic microbial biomass production. *Geochim. Cosmochim. Acta*, 67(20):3871–3887, doi:10.1016/S0016-7037(03)00304-1

Davis, E.E., Becker, K., Pettigrew, T., Carson, B., and MacDonald, R., 1992. CORK: A hydrologic seal and downhole observatory for deep-ocean boreholes. *In* Davis, E.E., Mottl, M.J., Fisher, A.T., et al., *Proc. ODP, Init. Repts.*, 139: College Station, TX (Ocean Drilling Program), 43–53. doi:10.2973/odp.proc.ir.139.103.1992

Davis, E.E., Malone, M.J., and the Expedition 328 Scientists and Engineers, 2010. Cascadia subduction zone ACORK observatory. *IODP Prelim. Rept.*, 328. doi:10.2204/iodp.pr.328.2010

Edwards, K.J., Becker, K., and Colwell, R., 2012a. The deep, dark energy biosphere: Intraterrestrial life on Earth. *Annu. Rev. Earth Planet. Sci.*, 40:551–568. doi:10.1146/annurev-earth-042711-105500

Edwards, K.J., Wheat, C.G., Orcutt, B.N., Hulme, S., Becker, K., Jannasch, H., Haddad, A., et al., 2012b. Design and deployment of borehole observatories and experiments during IODP Exp. 336. Mid-Atlantic Ridge flank at North Pond.

*In* Edwards, K.J., Bach, W., Klaus, A., and the Expedition 336 Scientists, *Proc. IODP*, 336. Washington, DC (Integrated Ocean Drilling Program Management International, Inc.), in press.

Expedition 336 Scientists, 2012. Integrated Ocean Drilling Program Expedition 336 Preliminary Report. Mid-Atlantic Ridge microbiology, Initiation of long-term coupled microbiological, geochemical, and hydrological experimentation within the seafloor at North Pond, western flank of the Mid-Atlantic Ridge, *IODP Prelim. Rept.*, 336. doi:10.2204/iodp.pr.336.2012

Fisher, A.T., and Wheat, C.G., 2010. Seamounts as conduits for massive fluid, heat, and solute fluxes on ridge flanks. *Oceanography*, 23(1):74–87. doi:10.5670/oceanog.2010.63

Fisher, A.T., Wheat, C.G., Becker, K., Cowen, J., Orcutt, B., Hulme, S., Inderbitzen, K., et al., 2011. Design, deployment, and status of borehole observatory systems used for single-hole and cross-hole experiments, IODP Expedition 327, eastern flank of Juan de Fuca Ridge. *In* Fisher, A.T., Tsuji, T., Petronotis, K., and the Expedition 327 Scientists, *Proc. IODP*, 327: Washington, DC (Integrated Ocean Drilling Program Management International, Inc.). doi:10.2204/iodp.proc.327.107.2011

Jannasch, H.W., Wheat, C.G., Plant, J., Kastner, M., and Stakes, D., 2004. Continuous chemical monitoring with osmotically pumped water samplers: *OsmoSampler design and applications. Limnol.* Oceanogr. Methods, 2:102–113. doi:10.4319/lom.2004.2.102

Orcutt, B., Wheat, C.G., and Edwards, K.J., 2010. Subseafloor ocean crust microbial observatories: Development of FLOCS (FLow-through Osmo Colonization System) and evaluation of borehole construction methods. *Geomicrobiol J.*, 27(2):143–157. doi:10.1080/01490450903456772

Schmidt-Schierhorn, F., Kaul, N., Stephan, S., and Villinger, H., 2012. Geophysical site survey results from North Pond (Mid-Atlantic Ridge). *Proc. IODP,* 336: Washington, DC (Integrated Ocean Drilling Program Management International, Inc.), in press.

Wheat, C.G., Jannasch, H.W., Kastner, M., Hulme, S., Cowen, J., Edwards, K., Orcutt, B.N., and Glazer, B., 2011. Fluid sampling from oceanic borehole observatories: Design and methods for CORK activities (1990–2010). *In* Fisher, A.T., Tsuji, T., Petronotis, K., and the Expedition 327 Scientists, *Proc. IODP,* 327: Washington, DC (Integrated Ocean Drilling Program Management International, Inc.). doi:10.2204/ iodp.proc.327.109.2011

## Authors

**C. Geoffrey Wheat**, Global Undersea Research Unit, University of Alaska Fairbanks, P.O. Box 475, Moss Landing, CA 95039, U.S.A., e-mail: wheat@mbari.org

**Katrina J. Edwards**, Department of Biological Sciences, Marine Environmental Biology Section, University of Southern California, Los Angeles, CA 90089, U.S.A., e-mail: kje@usc.edu

**Tom Pettigrew**, Pettigrew Engineering, 479 Nine Mile Road, Milam, TX 75959, U.S.A., e-mail: pettigrew.engineering@windstream.net

**Hans W. Jannasch**, Monterey Bay Aquarium Research Institute, 7700 Sandholdt Road, Moss Landing, CA 95039, U.S.A., e-mail: jaha@mbari.org

**Keir Becker**, University of Miami, 4600 Rickenbacker Causeway, Miami, FL 33149, U.S.A., e-mail: kbecker@rsmas.miami.edu

**Earl E. Davis**, Pacific Geoscience Centre, Geological Survey of Canada, 9860 West Saanich Road, Sidney, BC V8L 4B2, Canada, e-mail: edavis@nrcan.gc.ca.

**Heiner Villinger and Wolfgang Bach**, Department of Geosciences, University of Bremen, Klagenfurter Strasse, 28359 Bremen, Germany, e-mail: vill@uni-bremen.de, wbach@uni-bremen.de

# 18

# Pressure Core Characterization Tools for Hydrate-Bearing Sediments

by J. Carlos Santamarina, Sheng Dai, Junbong Jang, and Marco Terzariol

## Introduction

Natural gas hydrates form under high fluid pressure and low temperature, where biogenic or thermogenic gases are available. These requirements delimit the distribution of hydrate-bearing sediments to sub-permafrost, deep lakes (>390-m water depth) or ocean sediments (>320 m). Typically, hydrates are found beneath deeper water columns due to thermal fluctuations and diffusion near the sediment surface (Xu and Ruppel, 1999).

The clathrate or cage-like structure formed by water molecules hinders the repulsion between gas molecules allows for very high gas concentration. With the high methane concentration in large areas, natural gas hydrates can become an energy resource and remain a potential source for a potent greenhouse gas. Depressurization and/or warming cause dissociation and volume expansion leading to large-scale sediment destructuration.

A proper characterization of hydrate-bearing sediments requires coring, recovery, manipulation and testing under pressure and temperature (P-T) conditions within the stability field. This report begins with an overview of existing tools, and then describes advances in pressure core technology developed at the Georgia Institute of Technology that have been advanced to address this need.

## Pressure Core Technology: Overview

The development of pressure coring and recovery tools have involved research teams around the world, including initiatives such as the International Ocean Drilling Program and the European Union's Marine Science and Technology Program (Kvenvolden et al., 1983; Pettigrew, 1992; Amann et al., 1997; Dickens et al., 2003; Qin et al., 2005; Schultheiss et al., 2009). A depressurization of cores will cause immediate dissociation of gas hydrates. It is therefore necessary to keep the samples at all times under P-T conditions within the stability field. Pressure core manipulation and transfer technology require a longitudinal positioner/manipulator and ball valves to couple components at equalized pressures (Pressure Core Analysis and Transfer System, PCATS; Schultheiss et al., 2006).

Contact testing tools utilizing P- and S-wave velocities, strength, electrical resistivity profiles and internal core temperature (IPTC; Yun et al., 2006), and non-contact tools utilizing gamma density, X-rays and water-coupled P-waves (Pressure Multi-Sensor Core Logger; Schultheiss et al., 2006; Abegg et al., 2008) are avail-able. Subsampling capabilities have also been developed for biological studies under *in situ* P-T conditions (DeepIsoBUG; Parkes et al., 2009).

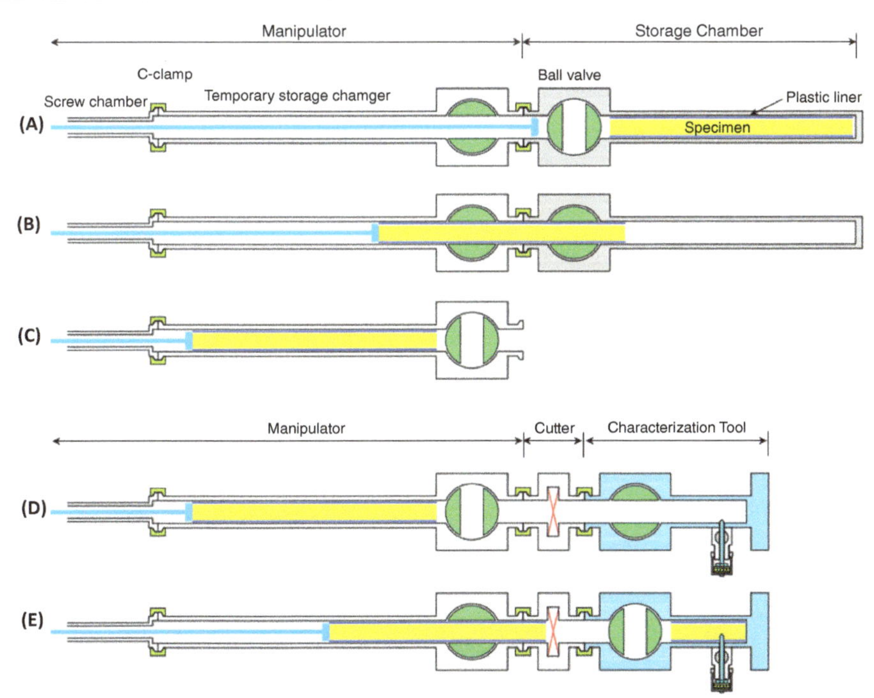


Figure 1. Pressure core manipulation. [A] The manipulator (MAN) couples with the storage chamber, and fluid pressures are equalized at the target pressure p0 before opening the ball valve. [B] The MAN captures the core and transfers it to the temporary storage chamber. [C] Ball valves are closed, and the depressurized storage chamber is separated. [D] The selected characterization tool is coupled to the MAN and is pressurized to $_p0$. [E] Ball valves are opened, and the core is pushed into the characterization tool; stand-alone characterization tools may be detached after retrieving the rest of the core and closing valves. Note: the cutter tool (CUT) is shown in panes [D] & [E]; it is attached in series to cut core to any desired length to meet tool requirements (for stand-alone ESC, DSC, CDC, and BIO tools).

## Pressure Core Characterization Tools (PCCTs)

Our pressure core characterization system includes core manipulation tools and characterization chambers. Tools have been selected to obtain complementary information relevant to science and engineering needs, with emphasis on the measurement of parameters used in hydro-thermo-mechanical analyses.

All tools are designed following key guidelines and objectives: simple and robust systems, portable components for fast deployment, modular design for maximum flexibility, standard dimensions and parts for affordable construction and maintenance, rust-resistance for seawater environment, capability of maintaining and operating at pressure, ability to impose effective stress, and safety for monitoring of hydrate dissociation and gas production during controlled depressurization, heating or fluid exchange (such as with liquid $CO_2$). The modular design allows any two tools/chambers to be coupled through an identical flange-clamp system.

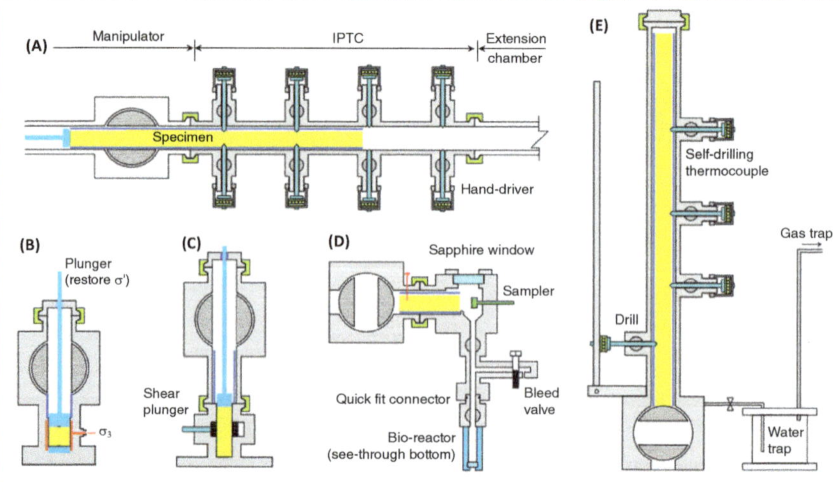


Figure 2. Schematic diagrams of characterization chambers. [A] IPTC with P-T control. [B] ESC with σ'-P-T control. [C] direct shear chamber (DSC) with σ'-τ-P-T control. [D] sampler for multiple bio-reactor chambers (BIO). The outside diameter of the large ball valve shown in all devices is 220 mm. [E] controlled depressurization chamber (CDC) for sediment preservation and gas production.

*Manipulator* (MAN). The manipulator is a longitudinal positioning system that is used to grab and move the core along the interconnected chambers and valves under the required P-T conditions. Figure 1 shows the typical operation sequence used to retrieve a specimen from the storage chamber into the MAN, followed by displacing the core into a test chamber. The geometric analysis of the operation shown in Fig. 1 reveals that the length of the MAN $L_{man}$ (with its "temporary storage chamber") is proportional to the length of the core $L_{core}$ to be manipulated, $L_{man} \approx 3.5 \times L_{core}$. Our system is designed to handle 1.2-m-long cores ($L_{core}$); it uses an internal telescopic screw system (stroke=2.6 m) driven by an external stepper motor, and it can position the specimen with sub-millimeter resolution. It is coupled to the 1.3-m-long temporary storage chamber by means of a dismountable flange-clamp connection. A see-through port is included to confirm the position of the MAN at any time.

*Sub-sampling.* The 1.2-m-long core can be cut into short specimens. Our cutting tool, CUT, houses either a linear or a ring-shaped saw blade within a clamp-type chamber. The saw-based cutting ensures clean surfaces and minimizes specimen disturbance. The CUT is mounted in series between the MAN and any other test or storage chamber as needed (Fig. 1d, 1e).

*Instrumented Pressure Testing Chamber* (IPTC). The chamber was developed to sample fluids and to measure P- and S-wave velocities, undrained strength, electrical conductivity, and internal core temperature (Fig. 2a; details in Yun et al., 2006). Additional tool developments have been implemented by the USGS, within the context of the Golf of Mexico JIP. This cylindrical chamber has two sets of four diametrically opposite port pairs. The first pair drills holes (ID=8 mm) in the plastic liner so that contact probes in successive ports can be pushed into the specimen. In characterization mode, the IPTC is coupled to the MAN on one side and an extension chamber on the other, and measurements can be conducted at any position along the core length. The eight access ports make the IPTC a versatile chamber for conducting well-monitored production studies in view of reservoir calibration models.

*Effective Stress Chamber* (ESC). Pressure cores are recovered and stored at fluid P-T conditions needed to preserve hydrate. However, physical properties such as stiffness and shear strength are functions of both hydrate saturation and effective stress, with the relative effective stress increasing as hydrate saturation decreases. The ESC maintains P-T stability conditions and restores the effective stress (σ′) that the sediment sustains *in situ* (Fig. 2b). It was designed and laboratory-tested at Georgia Tech in 2006 under Joint Oceanographic Institutions (JOI) sponsorship, and it was first deployed in the field by the Korean Institute of Geoscience & Mineral Resource in collaboration with Geotek (Lee et al., 2009).

The original design was based on a zero lateral strain boundary condition. We have updated this chamber to accommodate a stress-controlled boundary condition using a jacket. The resulting triaxial stress configuration consists of $\sigma_3'$ applied with the jacket and $\sigma_1'$ applied by a piston that is advanced through the ball valve and acts directly on the pressure core. The piston and the base pedestal house the sensors needed for the measurements of physical properties, including stiffness (wave velocities), thermal conductivity, and electrical resistivity.

A salient advantage of the flexible wall configuration is the ability to conduct precise fluid conductivity measurements by preventing the preferential flow along the sediment-steel boundaries in rigid wall chambers. This chamber is particularly well suited to monitor production studies under *in situ* effective stress conditions, including assessment of sediment volume change upon dissociation.

*Direct Shear Chamber* (DSC). Two constraints guided the design of the DSC tool. First, the imperfect boundaries that result when cutting heterogeneous cores under pressure cause stress concentration during vertical loading; thus, we selected a "double direct shear" geometry to cut across the specimen away from end effects. Second, overcutting during coring leaves a gap, and the core tends to tilt during shear; therefore, we adopted a double shear plane configuration to avoid bending action. Consequently, the DSC consists of a thick wall stainless steel ring that is pushed to shear the central third of the specimen (Fig. 2c). The DSC includes the piston to restore effective stress (similar to the ESC), a liner trap to capture the plastic liner before the specimen enters the shear chamber, and a small, lateral built-in frame to push the side piston that displaces the ring (Fig. 2c). The maximum shear displacement ($\delta_{max}$) is 15 mm, allowing both peak and residual shear strengths to be determined. The result is strength and volume change data under *in situ* conditions that are necessary for model calibration, production design, and stability analyses.

*Sub-sampling Tool for Bio-Studies* (BIO). Assessment of bioactivity in deep-water sediments without incurring depressurization cycles is crucial to the survival of some barophilic microorganisms. The BIO chamber is loaded with a core segment using the MAN; afterwards, it is detached from the MAN for all successive procedures (Fig. 2d). Its operation involves (1) nitrogen-liquid replacement, (2) core face cleaning and chamber sterilization, (3) sub-sampling using a rotary sampling head, and (4) sample deposition into the bio-reactor that is pre-filled with nurturing solutions (volume=10 mL). All operations can be observed through a sapphire window. Bio-reactors are readily replaced by closing a system of two ball valves and decoupling a quick connect fitting. This device allows the collection of a large number of specimens from a single core segment under *in situ* hydrostatic pressure.

*Controlled Depressurization Chamber* (CDC). Successful pressure coring operations may produce more pressure cores than the available storage. In this case, recovered cores can be selectively depressurized to conduct further studies under atmospheric pressure. The CDC is designed to help preserve the core lithology and to gain valuable in-formation during depressurization, with minimal demand on personnel resources. This stand-alone device has a built-in drilling station to perforate the liner at selected locations in order to reduce the longitudinal expansion of the specimen. A pressure transducer and a thermocouple monitor the gas P-T conditions inside the chamber. In addition, three self-drilling thermocouples are deployed along the CDC; these are driven into the core to monitor the internal sediment temperature during depressurization. Finally, a 2-L water trap and a 55-L gas trap are attached in series to the needle valve that controls the rate of depressurization; these traps allow measurement of the water and gas produced (Fig. 2e).

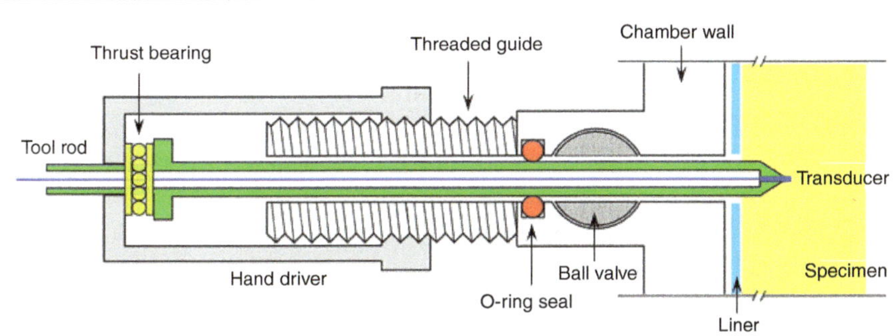


Figure 3. Tool Position. The displacement of sensors, subsampling tools, and drills are controlled under pressure using a screw-based positioning system where the driver advances along the threaded guide while pushing the tool rod (shown in green). Transducers at the tip of the rod are wired through the central hole in the tool rod.

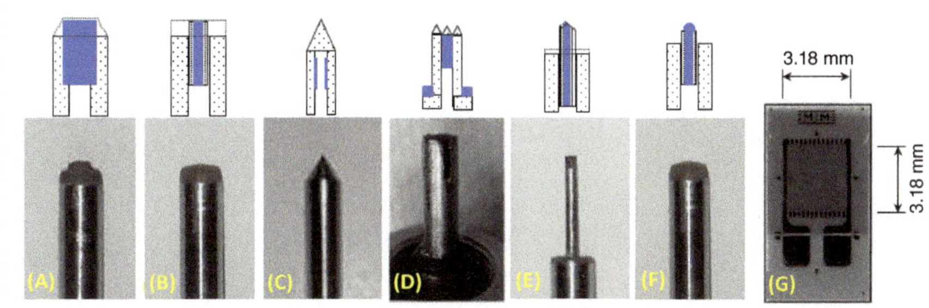


Figure 4. Measurement tools and sensors. [A] Bender elements for S-wave generation and detection. [B] Piezocrystals for P-waves. [C] Penetrometer for strength measurement. [D] Pore fluid sampler. [E] Electrical needle probe for resistivity profiling. [F] Thermocouple instrumented tip. [G] Strain gauge for thermal conductivity determination (TPS–NETL; Rosenbaum et al., 2007).

## Measurement of Physical Properties

Multiple sensing systems have been developed to characterize the sediment and to determine hydrological, thermal, chemical, biological, and mechanical parameters within the chambers, under controlled pressure, temperature, and effective stress conditions as described above. Their deployment in the various devices support the comprehensive characterization of natural hydrate-bearing sediments under *in situ* pressure, temperature, and/or stress conditions, and permit detailed monitoring of gas production tests.

*Tool Position Control.* All contact instruments, sensors, and drills are mounted on polished rods (7.9 mm diameter) that are advanced into the specimen using externally controlled threaded positioning systems to overcome the 1.7 kN force at the maximum working fluid pressure of 35 MPa (Fig. 3). The ball valve between the threaded guide and the chamber permits replacing tools under pressure.

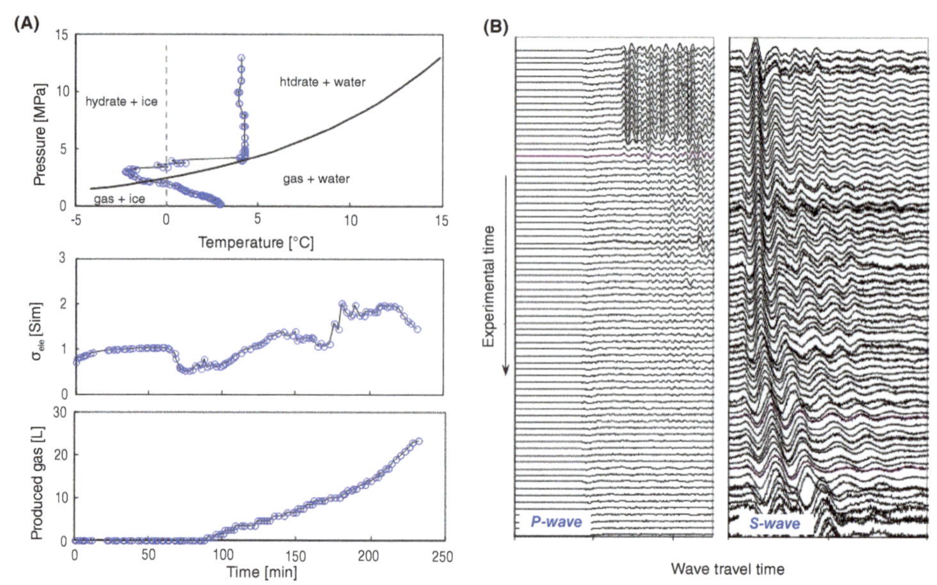


Figure 5. Monitored gas production tests using IPTC: [A] Evolution of pressure, temperature, electrical resistivity, and produced gas (Krishna-Godavari Basin; Yun et al., 2010); [B] Typical wave signatures during gas production: P-wave signatures eventually fade out after gas production; S-waves detect the evolution of the skeleton shear stiffness during hydrate dissociation and gas production (Ulleung Basin; Yun et al., 2011).

*Sensors.* Transducers are mounted at the tip of tool rods and wired through the central bore. Available instruments are shown in Figure 4. Small-strain wave velocity measurements employ bender elements for S-waves and pinducers for P-waves (Fig. 4a and 4b; peripheral electronics and test procedures as described in Lee and Santamarina, 2005a, 2005b).

While large-strain strength data can be gathered using the DSC (Fig. 2c), we have developed a strength-penetration probe as well (Fig. 4c). This device determines the sediment strength using a cone-shaped stud equipped with a full-bridge strain gauge inside. The measured tip resistance during probe penetration reflects the sediment undrained shear strength (Yun et al., 2006).

Fluid conductivity can be determined using the flexible wall system built within the ESC (Figs. 2b), and can be inferred using the fluid sampling tool (Fig. 4d). This is a self-drilling drainage port with a pressure or volume control to drive the interstitial fluids out of hydrate-bearing sediment. The pressure difference can be selected to preserve hydrates within stability conditions.

Electrical resistivity is measured using an electrical needle probe that is gradually inserted into the specimen to determine a radial resistivity profile with millimeter-scale spatial resolution (Fig. 4e; details and measurement procedure in Cho et al., 2004). We have also developed a multiple electrode system at the base of the effective stress cell that allows us to conduct a surface-based electrical resistivity tomography within a specimen.

The thermal probe consists of a thermocouple deployed at the tip of a tool rod. When pushed into the sediment, the thermal probe monitors the temperature inside the core (Fig. 4f). Internal temperature measurements can be used to monitor phase transitions during controlled gas production studies and to determine thermal conductivity. In addition, the transient plane source (TPS) sensor for thermal conductivity measurements—developed at the U.S. Department of Energy National Energy Technology Laboratory (Fig. 4g; Rosenbaum et al., 2007)—can be installed on the tools or on the pedestal of the ESC and DSC.

## Monitoring Dissociation – Gas Production

All PCCT chambers allow core-scale gas production tests by depressurization, heating, or chemical injection (e.g., inhibitors or carbon dioxide). Monitoring data include pressure, temperature, produced gas and water, stiffness (seismic wave velocities), fluid conductivity, and electrical resistivity. Figure 5 shows examples of data gathered during the depressurization of natural hydrate-bearing sediments.

## Conclusions

Pressure core technology is needed for the proper evaluation of natural hydrate bearing sediments. The set of pressure core characterization tools (PCCTs) described in this review allow the manipulation, sub-sampling, and extensive assessment of natural gas hydrate bearing sediments under *in situ* pressure, temperature, and effective stress conditions.

In addition to pressure core testing, comprehensive characterization programs should include sediment index properties analyzed within the framework of available data for natural hydrate bearing sediments, and tests with remolded specimens with synthetic hydrate. Pressure core technology can also be deployed to study other gas rich hydrocarbon formations such as deep-sea sediments, coal bed methane, and gas shales.

## Acknowledgements

Research support provided by the Chevron-managed DOE/NETL Methane Hydrate Project DE-FC26-01NT41330 and Gulf of Mexico Gas Hydrate Joint Industry Project. The Joint Oceanographic Institutions (JOI) supported the initial development of the Effective Stress Chamber (2006 - Tae Sup Yun participated in its design). Additional funding has been provided by the Goiuzeta Foundation. The IPTC chamber has been retrofitted by USGS collaborators W. Winters, D. Mason, W. Waite, and E. Bergeron.

## References

Abegg, F., Hohnberg, H.J., Pape, T., Bohrmann, G., and Freitag, J., 2008. Development and application of pressure-core-sampling systems for the investigation of gas- and gas-hydrate-bearing sediments. *Deep Sea Res. Part I Oceanogr. Res. Pap.*, 55(11):1590–1599. doi:10.1016/j.dsr.2008.06.006

Amann, H., Hohnberg, H.J., and Reinelt, R., 1997. HYACE - a novel autoclave coring equipment for systematic offshore gas hydrate sampling. Deutsche Wissenschaftliche Gesellschaft für Erdgas und Kohle e.V. (DGMK), Report 9706, 37–49.

Cho, G.C., Lee, J.S., and Santamarina, J.C., 2004. Spatial variability in soils: High resolution assessment with electrical needle probe. *J. Geotech. Geoenviron. Eng.*, 130(8):843–850. doi:10.1061/(ASCE)1090-0241(2004)130:8(843)

Dickens, G.R., Schroeder, D.K., Hinrichs, U., and the Leg 201 Scientific Party, 2003. The pressure core sampler (PCS) on ODP Leg201: General operations and gas release. *In* D'Hondt, S.L., Jørgensen, B.B., Miller, D.J., et al., *Proc. ODP, Init. Repts.*, 201: College Station, TX (Ocean Drilling Program), 1–22. doi:10.2973/odp.proc.ir.201.103.2003

Kvenvolden, K.A., Barnard, L.A., and Cameron, D.H., 1983. Pressure core barrel: Application to the study of gas hydrates, Deep Sea Drilling Project Site 533, Leg 76. *In* Sheridan, R.E., Gradstein, F.M., et al., *Init. Repts. DSDP*, 76: Washington, DC (U.S. Govt. Printing Office), 367–375.

Lee, J.S., and Santamarina, J.C., 2005a. Bender elements: Performance and signal interpretation. *J. Geotech. Geoenviron. Eng.*, 131(9):1063–1070. doi:10.1061/(ASCE)1090-0241(2005)131:9(1063)

Lee, J.S., and Santamarina, J.C., 2005b. P-wave reflection imaging. *Geotech. Test. J.*, 28:197–206.

Lee, J.Y., Schultheiss, P.J., Druce, M., and Lee, J., 2009. Pressure core sub sampling for GH production tests at *in situ* effective stress. *Fire in the Ice*, 9(4):16–17.

Parkes, R.J., Sellek, G., Webster, G., Martin, D., Anders, E., Weightman, A.J., and Sass, H., 2009. Culturable prokaryotic diversity of deep, gas hydrate sediments: First use of a continuous high-pressure, anaerobic, enrichment and isolation system for subseafloor sediments (DeepIsoBUG). *Environ. Microbiol.*, 11(12):3140–3153. doi:10.1111/j.1462-2920.2009.02018.x

Pettigrew, T.L., 1992. The design and operation of a wireline pressure core sampler (PCS). *ODP Tech. Note*, 17.

Qin, H., Gu, L., Li, S., Zhu, L., and Chen, Y., 2005. Pressure tight piston corer - A new approach on gas hydrate investigation. *China Ocean Eng.*, 19(1):121–128.

Rosenbaum, E.J., English, N.J., Johnson, J.K., Shaw, D.W., and Warzinski, R.P., 2007. Thermal conductivity of methane hydrate from experiment and molecular simulation. *J. Phys. Chem. B*, 111(46):13194–13205. doi:10.1021/jp074419o

Schultheiss, P.J., Holland, M., and Humphrey, G., 2009. Wireline coring and analysis under pressure: Recent use and future developments of the HYACINTH system. *Sci. Drill.*, 7:44–50.

Schultheiss, P.J., Francis, T.J.G, Holland, M., Roberts, J.A., Amann, H., Thjunjoto, Parke, R.J., et al., 2006. Pressure coring, logging and subsampling with the HYACINTH system. *In* Rothwell, G. (Ed.), *New Techniques in Sediment Core Analysis, Geol. Soc. London, Spec. Pub.*, 267:151–163. doi:10.1144/GSL.SP.2006.267.01.11

Xu, W., and Ruppel, C., 1999. Predicting the occurrence, distribution, and evolution of methane gas hydrate in porous marine sediments. *J. Geophys. Res.*, 104(B3):5081–5095.

Yun, T.S., Narsilio, G.A., Santamarina, J.C., and Ruppel, C., 2006. Instrumented pressure testing chamber for characterizing sediment cores recovered at *in situ* hydrostatic pressure. *Mar. Geol.*, 229(3–4):285–293. doi:10.1016/j.margeo.2006.03.012

Yun, T.S., Fratta, D., and Santamarina, J.C., 2010. Hydrate-bearing sediments from the Krishna-Godavari Basin: Physical characterization, pressure core testing, and scaled production monitoring. *Energy Fuels*, 24:5972–5983. doi:10.1021/ef100821t

Yun, T.S., Lee, C., Lee, J.S., Jang-Jun Bahk, and Santamarina, J.C., 2011. A Pressure Core Based Characterization of Hydrate Bearing Sediments in the Ulleung Basin, East Sea. *J. Geophys. Res.*, 117:151–158.

## Authors

**J. Carlos Santamarina**, School of Civil and Environmental Engineering, Georgia Institute of Technology, Mason Building, 790 Atlantic Drive, Atlanta, GA 30332-0355, U.S.A., e-mail: jcs@gatech.edu

**Sheng Dai, Junbong Jang, and Marco Terzariol**, School of Civil and Environmental Engineering, Georgia Institute of Technology, Mason Building, 790 Atlantic Drive, Atlanta, GA 30332-0355, U.S.A.

# Scientific Drilling in the East African Rift Lakes: A Strategic Planning Workshop

by James M. Russell, Andrew S. Cohen, Thomas C. Johnson, and Christopher A. Scholz

## Introduction and Key Science Themes

The East African Rift lakes offer unparalleled opportunities to investigate fundamental climate, environmental, biological, and geological processes through deep coring into the lake bed. Their sediments hold signals of the evolution of tropical rainfall, temperatures, and winds across 20° of latitude on both sides of the equator from the Miocene to the present. Fossil material in these basins chronicles the development of the East African landscapes in which our own species evolved, and records the explosive evolutionary radiation of literally thousands of species of fish, snails, and other aquatic organisms endemic to East Africa's lakes. Drilling these sediments will also provide insights into the tectonic processes that shape the largest active continental rift system on Earth today.

The past decade has witnessed major advances in our efforts to obtain long climate records from the East African lakes, highlighted by the successful scientific drilling of Lake Malawi in 2005 (Scholz et al., 2007). Since 2005, newly acquired and ongoing geophysical surveys by industrial and academic scientists have provided us with much of the requisite site survey information to propose scientific drilling projects in some of the premier target lakes in East Africa, most notably Lake Turkana, Lake Albert, and Lake Tanganyika (Fig. 1). To strategically plan for the next decade of scientific drilling in East Africa's lakes, the U.S. National Science Foundation (NSF) and Past Global Changes (PAGES) project sponsored "Continental Drilling in the East African Rift Lakes", a workshop attended by thirty-six African, European, and U.S. scientists and hosted by Brown University in Providence, Rhode Island, U.S.A. on 14–16 November 2011.

Recent years have witnessed significant developments in multidisciplinary sciences in the African rift. There are ongoing initiatives to integrate paleoenvironmental and paleoanthropological data to understand the environmental context of hominid evolution. These include the International Continental Scientific Drilling Program (ICDP)- and NSF-sponsored Hominid Sites and Paleolakes Drilling Project (HSPDP, http://www.icdp-online.org/front_content.php?idcat=1225); new initiatives to investigate the fundamental processes underpinning rift initiation and evolution (e.g., GeoPRISMS, http://www.geoprisms.org/); and new observational and climate modeling efforts to understand East African climate history. Workshop participants reviewed past studies and current initiatives to define scientific goals and priorities in the critical areas of East African paleoclimate, rifting processes, and ecosystem evolution. We then reviewed the stratigraphic architecture and environmental history of Lake Turkana, Lake Albert, and Lake Tanganyika to evaluate their potential for addressing key scientific questions in order to develop a strategic plan for drilling projects in the East African lakes.

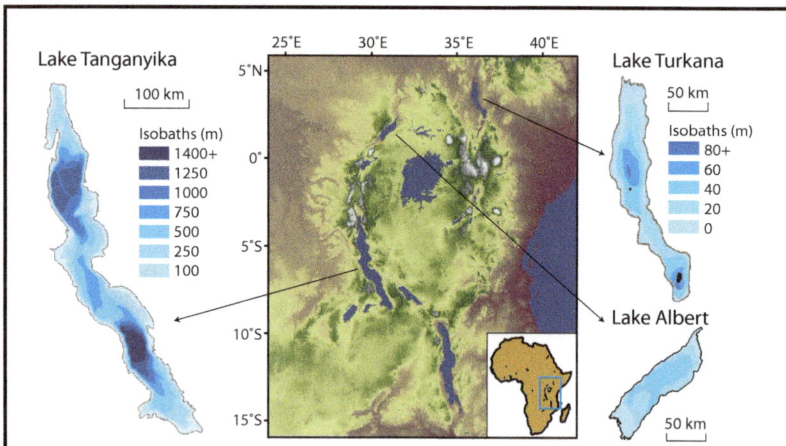


Figure 1. Elevation map of East Africa with bathymetries of Lake Tanganyika, Lake Turkana, and Lake Albert. The inset in the central figure shows the general location of the East African Rift (blue box). Isobaths for Albert and Turkana follow the same color scheme (right).

### Plio-Pleistocene East African Climate

Decades of research have provided us with a relatively coherent picture of climate change at the high latitudes during late Cenozoic time, including iconic marine sedimentary records and ice cores that document glacial-interglacial variability and abrupt, millennial climate change (Fig. 2). Unfortunately, our understanding of tropical climate systems, and particularly continental tropical climate, lags far behind our knowledge of the high latitudes. Only recently we thought that African climate marched largely to the beat of the northern high-latitude ice sheets, yet drill cores recovered by the Lake Malawi drilling project revealed the existence of major droughts in Southeast Africa, far exceeding the aridity experienced during the Last Glacial Maximum, prior to 75,000 years ago (Scholz et al., 2007). These droughts appear to have been forced by eccentricity modulation of orbital precession, and they possibly affected the dispersal of anatomically modern humans both within and out of Africa (Blome et al., 2012). Still, the Malawi drill cores cover only the latter part of the Pleistocene and record climate history at the southernmost end of the African rift—a climate setting that differs considerably from the rest of the rift to the north. Recent modeling studies have suggested strong linkages between Plio-Pleistocene changes in the sea-surface temperature structure of the tropical Pacific and Indian Oceans and changes in East African rainfall (Brierley and Fedorov, 2010), potentially affecting the evolution of our species. These new perspectives call into question many of the fundamental mechanisms and processes causing African climate variability, including the following. Are mega-droughts present at sites to the north, and are they antiphased in their timing relative to those in Malawi as predicted by the precessional insolation forcing model? How consistent is the relationship between precessional orbital forcing and East African climate through the Plio-Pleistocene?

Numerous outcrop and marine sediment records from tropical East Africa suggest that East African climate has responded to a complex and interrelated set of climate forcings. These include orbitally-driven changes in insolation, changes in global ice volume and atmospheric greenhouse gas concentrations, long-term reorganizations of the principal modes of sea-surface temperature variability such as El Niño–Southern Oscillation (ENSO), and tectonically-driven changes in oceanic and atmospheric circulation (Scholz et al., 2007; Tierney et al., 2008; Brierley and Fedorov, 2010). Understanding the governing mechanisms and sensitivity of African climate to these phenomena will require long and continuous climate records extending into at least the Pliocene to observe East African climate responses across multiple events and time scales. Moreover, the Pliocene offers the most recent extended warm period, when atmospheric greenhouse gases approached values similar to those we will witness in the near future, and when continental configurations were similar to those today (Committee on the Earth System Context for Hominin Evolution, 2010). Workshop participants highlighted that such records will allow us, for the first time, to address various questions:

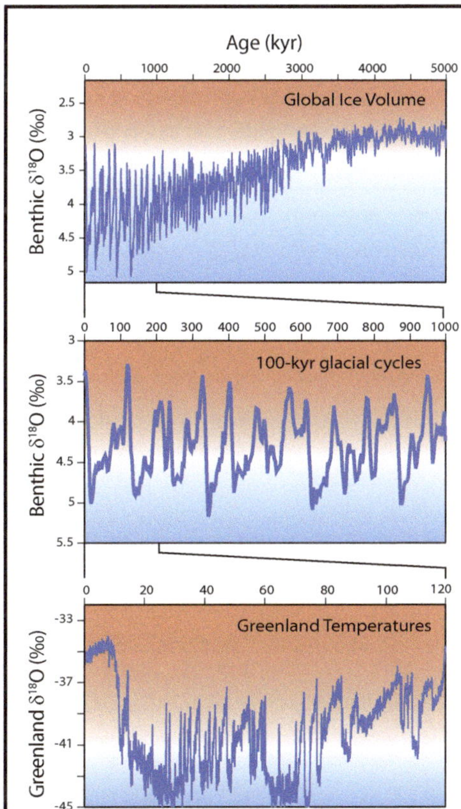

Figure 2. Climate change at different time scales as indicated by marine foraminiferal benthic oxygen isotopic data (Liesecki and Raymo, 2005) and oxygen isotopic reconstructions of Greenland temperatures (Members North Greenland Ice Core Project, 2004).

- What are the dynamics of late Miocene-Pleistocene (last 7 Myr) African climate as a consequence of the mid-Pliocene termination of a permanent El Niño, ocean circulation change with the closure of the Indonesian seaway, the onset, intensification, and changes in the periodicity of Northern Hemisphere Glaciation, and the development of and interactions with East African Rift topography?
- What is the sensitivity and spatial variability of East African hydrology and temperature to orbital-scale climate forcings? Are tropical air temperature and hydrologic change coupled through time, and what are their sensitivities to radiative forcing from greenhouse gases and seasonal insolation?
- How do the rates and amplitudes of East African climate change on millennial to decadal time scales vary as a function of mean climate state? What are the mechanisms underpinning this time scale of variability?

Fundamentally, our goal is to understand the Plio-Pleistocene climate evolution of Africa at a resolution comparable to the records of climate from the high latitudes (Fig. 2), providing key new insights into the dynamics of tropical continental rainfall variability and sensitivity.

## Rift Initiation and Evolution

The general notion is that the East African Rift formed as heat flow from the convecting mantle weakened the lithosphere and allowed the African Plate to stretch and fracture along pre-existing weaknesses in the lithosphere (Ebinger, 1989). This, coupled with the introduction of magma in at least some parts of the rift, led to rapid deformation and localization of strain along pre-existing lithospheric structures. However, these simple conceptual models fail to explain the rates and complexity of rifting processes, and the relative roles of magmatism, volatiles, and pre-existing structures in the rifting process are very poorly known. Both magmatism and reactivation of pre-existing structures can help to explain rift initiation and lithospheric stretching (Dunbar and Sawyer, 1989; Buck, 2004), given the relatively weak forces involved in extensional plate tectonics, whereas volatiles such as $H_2O$ and $CO_2$ can further accelerate fault movement through their effects on the melting point of mantle solids,pore pressures, and metasomatism (Floyd et al., 2001). Magmatism and volatile content also likely play a key role in regulating the movement of faults and associated long-term fault slip rates. Investigation of the volatile contents in drill core pore waters, downhole heat gradients, tephra-stratigraphy, and the frequency of seismically-triggered sedimentary events could all shed light on these issues.

We have a limited understanding of the intertwined processes of rift tectonics, magmatism, erosion, and sedimentation and their evolution in space and time, as in today's rift basins we thus far see only the end products of processes integrated over millions of years. For instance, the evolution of elevated rift flanks has a large impact on regional rainfall and erosion and interacts with the large-scale monsoon circulation over East Africa. Eroded sediment, in turn, may substantially impact the rifting processes and architecture (Spiegel et al., 2007). Addressing these issues will require detailed, multidisciplinary analyses of rift basins and their sediments, and they reflect key initiatives within the GeoPRISMS program (Morgan et al., 2011).

In particular, drill cores could help to evaluate the following three important factors:

- the rates of border fault slip in the rift basins, and how they are constrained by geothermal gradients and volatile concentrations;
- the relationships between long-term fault slip rates and short-term seismic hazards such as earthquake recurrence; and
- long-term erosion rates, sediment budgets and accumulation, and their temporal variability to help to constrain rates of crustal cycling.

Addressing these issues will require investigation of rift evolution in space and time, coupling drill core analyses with detailed seismic stratigraphic surveys, thermochronologic studies of landscape evolution, and studies of erosion and deposition in present-day lakes.

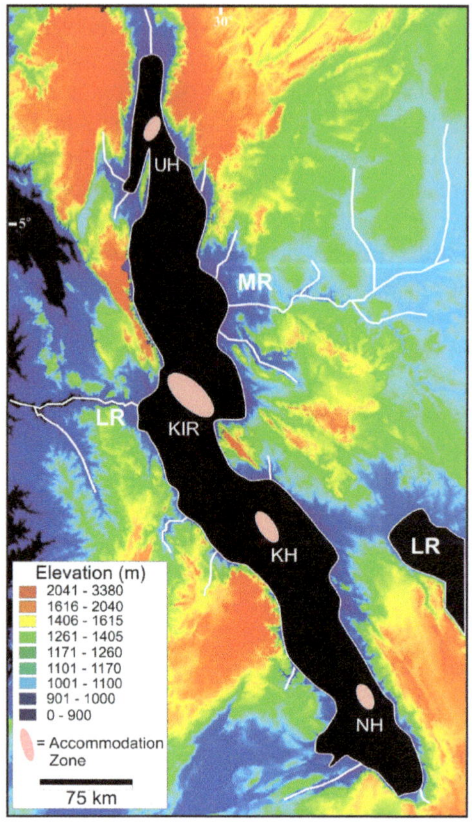


Figure 3. Digital elevation model of the Tanganyika basin showing the location of submerged high-relief accommodation zones within the lake. UH=Ubwari horst, KIR=Kavala Island Ridge, KH=Kalya horst, and NH=Nitiri High. At left, LR=Lukuga River, Tanganyika's outlet. At right, LR=Lake Rukwa, and MR=Malagarasi River. Modified from McGlue et al. (2008).

## *Human, Floral, and Faunal Evolution*

The East African Rift lakes harbor some of the best examples of explosive speciation on the planet (Martens, 1997). Best-known are East Africa's cichlid fishes, which have evolved into groups of literally hundreds of endemic fishes in Lake Tanganyika, Lake Malawi, and Lake Victoria; however, these lakes also contain highly diverse endemic groups of snails, diatoms, ostracodes, and crabs. Tools of molecular genetics now allow for analysis of DNA preserved in fossils of many of these groups, potentially allowing the reconstruction of the genetic history of entire species, from evolution to (in some cases) extinction. Moreover, the East African Rift houses the fossil record of our own lineage and contains some of the most extensive dryland biomes on Earth. Nowhere else can lacustrine scientific drilling provide such fundamentally important insight into the coupling between environmental change and biological evolution (Martens, 1997), including the evolution of *Homo sapiens* (Commiittee on the Earth System Context for Hominin Evolution, 2010).

Lake drill cores can provide more than simply the climate and environmental background in which new species evolve. The rich fossil records preserved in many rift basins could be used to address the rates of change, linearity, and sensitivity of the evolution of species as well as ecosystems and ecosystem processes. Such work would involve a nested approach to identify time scale dependence of ecological and evolutionary responses.

Drilling of extant lakes such as Tanganyika, Turkana, and Albert would dovetail extremely well with the HSPDP, which will provide new drill core from multiple paleolake basins in East Africa, each spanning several hundred thousand years to target key intervals of hominin evolution. These new cores will provide valuable records of environmental variation at hominin sites with direct links to fossil-bearing outcrops; however, these records will be spatially and temporally discontinuous. Long and relatively continuous lacustrine sedimentary records spanning the Plio-Pleistocene, such as what we could obtain from Africa's extant great lakes, would provide a climatic framework in which to interpret the shorter HSPDP records. While marine sedimentary records could further augment this broader context, ongoing piracy in the western Indian Ocean unfortunately constrains marine coring in this region. Drill core from Lake Tanganyika, Lake Turkana, and/or Lake Albert could thus prove instrumental to studies of the climatic context of hominin evolution (Committee on the Earth System Context for Hominin Evolution, 2010).

## Study Sites and Target Selection

Sedimentary records exist in Lake Tanganyika, Lake Turkana, and Lake Albert to answer these questions, but each lake affords very different temporal windows on past climate, different styles of rifting, and different opportunities for understanding evolutionary biology. Moreover, each lake offers very different opportunities in terms of their time scale, resolution, and continuity.

### *Lake Tanganyika*

Tanganyika (32,600 km$^2$, 1471 m deep) is the largest, oldest, and deepest lake in East Africa, with an estimated age of 9–12 Ma (Cohen et al., 1993). Tanganyika is one of the most biodiverse lakes on Earth, hosting endemic cichlid fish, snails, ostracods, crabs, and other organisms, and it lies within the vast but poorly understood Miombo woodland ecosystem of SE tropical Africa. Tanganyika is comprised of four major sub-basins separated by deep-water horsts and provides very clear examples of rift segmentation, fault behavior, and structural growth that could greatly facilitate studies of fault behavior and seismicity. Although the Rungwe volcanic field to the south provides occasional tephras to Tanganyika, there is no extensive volcanism in the basin.

Tanganyika lies within the core path of the Intertropical Convergence Zone's migration over central Africa. Its late Quaternary climate history is largely out of phase with Lake Malawi, indicating that Tanganyika acts as a Northern Hemisphere site despite its location. Drill core data from Tanganyika will therefore complement, not replicate, studies in Lake Malawi. The proximity of both Tanganyika and Malawi to the Rungwe Volcano could facilitate inter-basin tephrstratigraphic studies. Moreover, recent work has demonstrated Tanganyika's potential to record climate processes at orbital, millennial, and even interannual time scales, as the sediments contain a wealth of climate proxies preserved in often laminated sediments (Cohen et al., 2006; Tierney et al., 2008, 2010). Most importantly, recent work has documented deep-water horst blocks (Fig. 3) with condensed sedimentary sequences (Fig. 4), putting early Pliocene-age sediment within relatively easily drilling depths at sites (McGlue et al., 2008). These sites are located at high-relief accommodation zones in deep enough water to have continuous deposition over much, if not the entirety, of this time. The Kalya and Nitiri horsts in Tanganyika's southern basin, in particular, lie in deep water below the depth of maximum lake lowstands and appear to house long and continuous sedimentary sections (McGlue et al., 2008; Fig. 4).

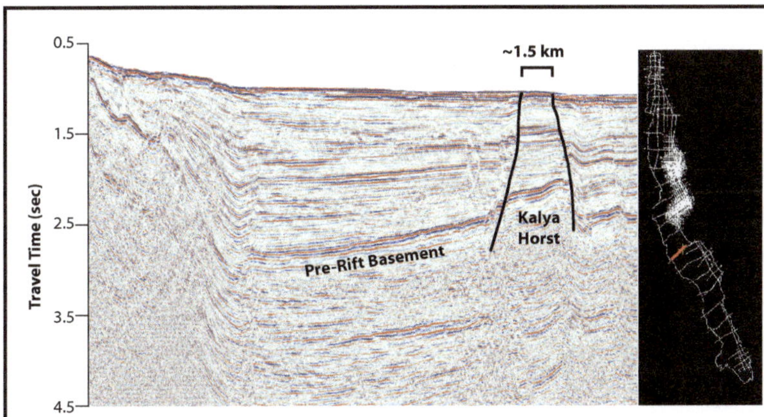

Figure 4. Seismic section from the northern end of the Kalya horst; location of the seismic section within Lake Tanganyika is illustrated by the red line within the seismic grid at right. Note the presence of a continuous and condensed sedimentary sequence above pre-rift basement on the Kalya horst, indicated by the bracket. Data are from Project PROBE (Rosendahl, 1988).

While the challenges and cost of drilling a long hole in deep water (500 m of mud in 700 m of water, for instance) are considerable, the group felt that drilling Tanganyika presented the most exciting scientific opportunities of any African lake. Tanganyika's deep-water horsts are perhaps the only depositional setting in East Africa likely to contain continuous Pliocene-present sediments at easy drilling depth. The potential significance of a continuous, high-resolution, Pliocene-present sediment record for tropical paleoclimatology, paleoanthropology, evolutionary biology, and rifting processes far outweighs the significant time and investment required to move the Tanganyika record forward.

### *Lake Turkana*

Rifting and volcanism have affected the Turkana basin since the Oligocene, with the modern lake developing in the Pliocene. Turkana is well-known for its complex rifting history, and it has extensive and remarkable facies architecture (clastic/carbonate sequences) that house world-famous vertebrate and hominin fossil locales. However, the rifting processes in Turkana are the most complex of the three lakes, and there are extensive and relatively young basalts that may resurface parts of the Turkana basin. The modern aquatic biota is relatively depauperate, but drilling in Turkana could dovetail with planned drilling of early Pleistocene lake beds in west Turkana as part of HSPDP and provide the most direct linkage to paleoanthropology of the three lakes.

Lake Turkana is East Africa's northernmost rift lake, and its sediments could provide important insight into the climate of northeastern Africa that would strongly complement recent drilling efforts on Lake Malawi. Turkana is extremely sensitive to climate variability, and this is reflected in the lake's sedimentary record (Halfman et al., 1994). Moreover, there is a well-established tephrastratigraphy for Turkana that would aid in developing age models for drill cores. However, the modern lake is characterized by very high sedimentation rates, and the lake is certain to have undergone desiccation events that were likely associated with basin deflation. Moreover, on long time scales the complex paleogeography and rift history could complicate paleoclimate reconstruction. To further complicate matters, available seismic reflection data from Turkana are rather poor, requiring significant efforts to properly locate drilling targets. Some aspects of drilling logistics in Turkana are relatively simple, but the remote location, high winds, and lack of facilities will make Turkana a logistically challenging target.

In sum, Turkana presents many exciting and extremely valuable opportunities, particularly the development of a high-resolution climate record spanning the late Pleistocene at the northern end of the East African Rift and in a site of paleoanthropologic importance. Yet, the complications posed by high sedimentation rates, hiatuses, and a complex rifting history make Turkana a lower priority than Tanganyika.

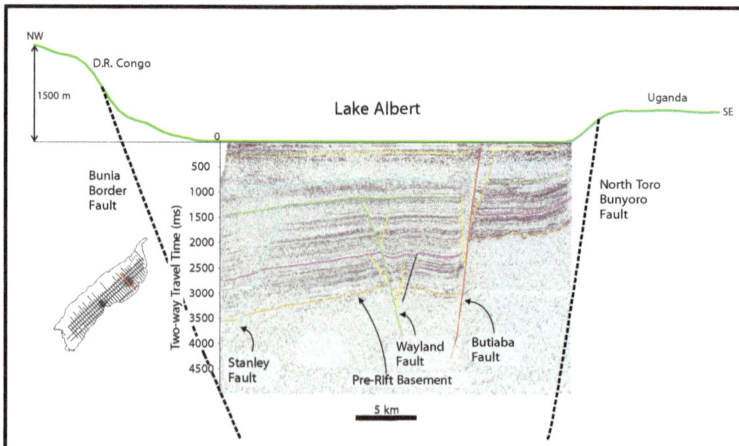


Figure 5. Seismic section from Lake Albert, modified from Karp et al. (2012). The thick green line at top traces the height of the landscape above Lake Albert. Colored lines in the seismic section indicate boundaries between sedimentary units as well as major faults. The location of this section is indicated by the red line within the map of the seismic grid at left. Note the condensed sedimentary section on the southeast side of the basin.

## Lake Albert

Lake Albert is situated near the equator in the western arm of the East African Rift. The basin contains, in places, several kilometers of lacustrine and fluviolacustrine sediments that extend into the Miocene (Fig. 5), with considerable outcrop exposures that have been investigated for their fossil content and paleobiology. Drilling in Lake Albert could elucidate how rift-generated topography and the uplift of the Rwenzori Mountains control sedimentation and rift catchments, and it could further elucidate the history of the source of the White Nile River. Moreover, the Albertine Rift is in many ways simpler than Tanganyika and Turkana, facilitating studies of rifting processes. Drilling in Lake Albert could provide incredibly valuable new records of the late Miocene to present evolution of East African climate. However, discontinuities existing even in late glacial sediments indicate the likelihood of numerous hiatuses in longer drill core records when climate fluctuation may be more extreme.

Perhaps the strongest argument for drilling Albert lies in the extensive infrastructure related to recent oil discoveries in the Albert basin by Tullow Oil, plc (http://www.tullowoil.com). There is an exquisite seismic-stratigraphic framework for Lake Albert, and drilling infrastructure is already available in the basin. Lake Albert is the main depositional sump for much of central equatorial Africa, including Lake Victoria and Lake Edward, and could therefore provide an unparalleled but discontinuous record. Given the likelihood of discontinuities in drill core data, Lake Albert was prioritized beneath Tanganyika, but given Albert's extraordinary record and the near certainty of future industrial drilling activities, the group felt that we should work to partner with industry in drilling activities, perhaps through onshore drilling where condensed sections are present (Fig. 5).

## Summary and Future Work

Scientific drilling in Lake Tanganyika, Lake Turkana, and Lake Albert can answer fundamentally important questions concerning tropical climate variability and change, rifting processes, paleobiology, and human evolution. Among these lakes, Tanganyika stands out as an extraordinary opportunity due to the presence of continuous, deep-water sedimentary records spanning the Pliocene to present, as well as being the nexus of interest in the lake basin for paleoclimatologists, seismologists, and evolutionary biologists. Workshop participants were uniquely enthusiastic about moving forward, and they agreed on several benchmarks to begin a continental drilling project on that lake.

First, and most importantly, existing seismic reflection data from Tanganyika is of very high quality but low spatial resolution. Beach Energy (Adelaide, South Australia) will acquire new deep seismic reflection data from Tanganyika's southern basin, which contains the best drilling targets on deep-water horst environments, and the company could make sections of this data available. These data will provide an excellent seismic-stratigraphic framework for interpreting drill core records but is unlikely to have the resolution needed to detect small-scale unconformities and site drilling sites. The group agreed that our first step should be a proposal to acquire high-resolution data and shallow piston cores over deep-water horsts in Tanganyika's southern basin to determine appropriate drilling targets and identify major sedimentary facies likely to be encountered in scientific drilling.

Secondly, the group expressed interest in accessing extensive existing shallow core collections from Tanganyika, metadata on which is not widely accessible. The group agreed that we would make information about these cores accessible through LacCore (http://lrc.geo.umn.edu), University of Minnesota, to allow for sample analysis, testing, and proxy calibration and to expand the science team involved in Tanganyika drilling.

Finally, there were several more specific studies that could complement a Tanganyika drilling project, ranging from tephrastratigraphic studies of the Rungwe volcanics to limnological studies in the Tanganyika basin. The group agreed to pursue these through PI-driven proposals to relevant science agencies, with coordination and data sharing to build the Tanganyika drilling program.

## References


Blome, M., Cohen, A.S., Tryon, C.A., Brooks, A.S., and Russell, J., 2012. The environmental context for the origins of modern human diversity: A synthesis of regional variability in African climate 150,000–30,000 years ago. *J. Hum. Evol.*, 62(5):563–592. doi: 10.1016/j.jhevol.2012.01.011

Brierley, C.M., and Fedorov, A.V., 2010. Relative importance of meridional and zonal sea surface temperature gradients for the onset of the ice ages and Plio-Pleistocene climate evolution. *Paleoceanography*, 25: PA2214:. doi:10.1029/2009PA001809

Buck, W.R., 2004. Consequences of aesthenosperic variability on continental rifting. *In* Karner, G.D., Taylor, B., Driscoll, N.W., and Kohlstedt, D.L. (Eds.), *Rheology and Deformation of the Lithosphere at Continental Margins*: New York (Columbia University Press), 1–30.

Cohen, A.S., Lezzar, K.E., Cole, J., Dettman, D.L., Ellis, G.S., Eagle, M., Chorokoa, K., et al., 2006. Late Holocene linkages between decade-scale climate variability and productivity at Lake Tanganyika, Africa. *J. Paleolimnol.*, 36(2):189–209. doi:10.1007/s10933-006-9004-y

Cohen, A.S., Soreghan, M., and Scholz, C., 1993. Estimating the age of formation of lakes: An example from Lake Tanganyika, East African rift system. *Geology*, 21:511–514. doi:10.1130/0091-7613(1993)021<0511:ETAOFO>2.3.CO;2

Committee on the Earth System Context for Hominin Evolution, 2010. *Understanding Climate's Influence on Human Evolution*, Washington, DC (The National Academies Press).

Dunbar, J.A., and Sawyer, D.S., 1989. How preexisting weaknesses control the style of continental breakup. *J. Geophys. Res.*, 94:527–530. doi:10.1029/JB094iB06p07278

Ebinger, C.J., 1989. Tectonic development of the western branch of the East African rift system. *GSA Bull.*, 101:117–133. doi:10.1130/0016-7606(1989)101<0885:TDOTWB>2.3.CO;2

Floyd, J.S., Mutter, J.C., Goodliffe, A.M., and Taylor B., 2001. Evidence for fault weakness and fluid flow within an active low-angle normal fault. *Nature*, 411:779–783. doi:10.1038/35081040

Halfman, J.D., Johnson, T.C., and Finney, B., 1994. New AMS dates, stratigraphic correlations and decadal climatic cycles for the past 4 ka at Lake Turkana, Kenya. *Palaeogeogr. Palaeoclimatol. Palaeoecol.*, 111:83–98. doi:10.1016/0031-0182(94)90349-2

Karp, T., Scholz, C.A., and McGlue, M.M., 2012. Structure and stratigraphy of the Lake Albert Rift, East Africa: Observations from seismic reflection and gravity data. *In* Bartov, Y., and Nummedal, D. (Eds.), *Sandstones in Lacustrine Depositional Settings, AAPG Memoir 95*, in press.

Liesecki, L., and Raymo, M., 2005. A Plio-Pleistocene stack of 57 globally distributed benthic $\delta^{18}O$ records. *Paleoceanography*, 20:PA1003. doi:10.1029/2004PA001071

Martens, K., 1997. Speciation in ancient lakes. *Trends Ecol. Evol.*, 12:177–182. doi:10.1016/S0169-5347(97)01039-2

McGlue, M.M., Lezzar, K.E., Cohen, A.S., Russell, J.M., Tiercelin, J.J., Felton, A., Mbede, E., and Nkotagu, H., 2008. Seismic records of late Pleistocene aridity in Lake Tanganyika, tropical East Africa. *J. Paleolimnol.*, 40:635–653. doi:10.1007/s10933-007-9187-x

Members North Greenland Ice Core Project, 2004. High-resolution record of the Northern Hemisphere climate extending into the last interglacial period. *Nature*, 431:147–151. 10.1038/nature02805

Morgan, J., Arrowsmith, R., Benoit, M., Ebinger, C.J., Fischer, T., Flemings, P., Keranen, K., et al., 2011. GeoPRISMS Rift Initiation and Evolution Plan. http://www.geoprisms.org/rie.html

Rosendahl, B.R., 1988. *Seismic Atlas of Lake Tanganyika, East Africa:* Durham, North Carolina, USA (Project PROBE Geophysical Atlas Series, Duke University).

Scholz, C., Johnson, T.C., Cohen, A.S., King, J.W., Peck, J.A., Overpeck, J.T., Talbot, M.R., et al., 2007. East African megadroughts between 135 and 75 thousand years ago and bearings on early-modern human origins. *Proc. Natl. Acad. Sci. U.S.A.*, 42:16416–16421. doi:10.1073/pnas.0703874104

Spiegel, C., Kohn, B.P., Belton, B.X., and Gleadow, A.J.W., 2007. Morphotectonic evolution of the central Kenya rift flanks: Implications for late Cenozoic environmental change in East Africa. *Geology*, 35:427–430. doi:10.1130/G23108A.1

Tierney, J.E., Mayes, M., Meyer, N., Johnson, C., Swarzenski, P.W., Cohen, A.S., and Russell, J.M., 2010. Late-twentieth-century warming in Lake Tanganyika unprecedented since AD 500. *Nature Geoscience*, 3:422–425. doi:10.1038/NGEO1865.

Tierney, J.E., Russell, J.M., Huang, Y., Sinninghe-Damste, J.S., Hopmans, E.C., and Cohen, A.S., 2008. Northern Hemisphere controls on tropical southeast African climate during the past 60,000 years. *Science*, 322(5899):252–255. doi:10.1126/science.1160485


## Authors


**James M. Russell**, Department of Geological Sciences, Brown University, 324 Brook Street, Box 1846, Providence, RI 02912-9019, U.S.A., e-mail: James_Russell@Brown.edu

**Andrew S. Cohen**, Department of Geosciences, University of Arizona, Gould-Simpson Building #77, 1040 East 4th Street, Room 326, Tucson, AZ 85721-0001, U.S.A.

**Thomas C. Johnson**, Large Lakes Observatory, University of Minnesota Duluth, 2205 East 5th Street, Research Laboratory Building 204, Duluth, MN 55812-3024, U.S.A.

**Christopher A. Scholz**, The College of Arts and Sciences, Department of Earth Sciences, Syracuse University, 204 Heroy Geology Laboratory, Syracuse, NY 13244-1070, U.S.A.

# 20

# Unlocking the Opening Processes of the South China Sea

**by Chun-Feng Li, Pinxian Wang, Dieter Franke, Jian Lin, and Jun Tian**

## Introduction

The South China Sea (SCS) is a classical representative of western Pacific marginal seas (Figs. 1, 2). It developed from continental margin rifting, and its central portion is floored with oceanic crust. Despite its relatively short evolutionary history, compelling research opportunities abound for a multitude of key tectonic processes ranging from rifting through seafloor spreading to subduction, as well as for the cyclical climatic changes with broad regional impact. Its relatively small size and young age compared to major ocean basins facilitate tectonic comparisons between conjugate continental margins and make the entire basin accessible through a single expedition of scientific ocean drilling. These attributes make the SCS an exemplary natural laboratory for studying continental break-up, sedimentary basin formation, mantle and lithosphere evolution, and land-ocean interactions. However, lack of any deep scientific borehole in the central SCS basin leaves a gap in understanding of the opening mechanisms, timing of the onset and termination of seafloor spreading, and environmental history in the region.

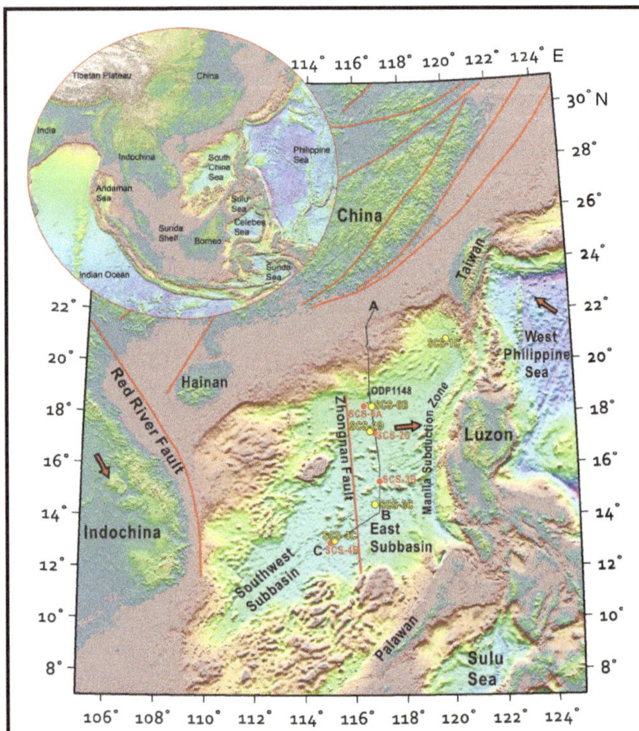


Figure 1. Regional topography and tectonic framework of Southeast Asia (from Shi and Li, 2012). Data based on Smith and Sandwell (1997). Red lines mark major faults. Red and yellow solid points are locations of the proposed drilling sites (red-primary sites; yellow-alternate sites). The black lines labeled with AB and BC are seismic sections shown in Fig. 3. Red arrows indicate the direction of plate movement.

To advance research of Southeast Asia and the SCS through scientific ocean drilling, an international workshop entitled "Unlocking the Opening Processes of the South China Sea" was held on 31 January–1 February 2012 in Shanghai, China. It was the first international workshop devoted to scientific ocean drilling in the SCS in the recent decade. Building on an existing Integrated Ocean Drilling Program (IODP) drilling proposal (Li et al., 2012), this workshop sought contributions from the broad international community, including scientists from non-IODP member states that have a strong scientific interest in the SCS. The overarching goal of the workshop was to develop the most efficient research and drilling strategy to address key questions in tectonics and environmental history of the SCS within a single, initial IODP expedition.

The workshop hosted sixty-five participants from fifteen countries and regions, including scientists from several non-IODP member states surrounding the SCS (Indonesia, Malaysia, Philippines, Singapore, Thailand, and Vietnam). The workshop was jointly sponsored by the IODP-MI, IODP-China, the South China Sea Deep (SCSD) Program of the National Natural Science Foundation of China (NSF-C), and the State Key Laboratory of Marine Geology at Tongji University, China. The lead conveners were Pinxian Wang and Chun-Feng Li of Tongji University, and Dieter Franke of the Federal Institute for Geosciences and Natural Resources, Germany.

The workshop was charged with the following three tasks.

1. To identify key scientific questions on tectonics and environmental history of a marginal sea that can be addressed through scientific drilling in the SCS, and to develop an efficient research and drilling strategy to address these questions
2. To recommend ways to improve the IODP Proposal 735-CPP through focusing drilling objectives, refining drilling targets, and augmenting site survey data
3. To foster cooperation in research of the Southeast Asia region through engagement of a broad international scientific community including scientists

from non-IODP member states surrounding the SCS

## Scientific Background

With its area of 3.5 million km$^2$ and maximal depth of 5500 m, the SCS is one of the largest marginal seas in the world (Figs. 1, 2). Since the Late Mesozoic, it has been at the center stage of many first-order tectonic and paleoclimatic events. Mesozoic subduction of the paleo-Pacific plate, a fragment of which developed roughly along the present-day northern SCS continental margin (Taylor and Hayes, 1983), gradually dispelled Paleo-Tethys and built a massive orogen in Southeast Asia. However, during the transition from the Early to Late Cretaceous, all supportive evidence of the subduction zone disappeared, while regional extension started to prevail (Shi and Li, 2012). This was followed by the Cenozoic opening of the SCS, via continental break-up and subsequently seafloor spreading. The early work of Taylor and Hayes (1980, 1983) and Briais et al. (1993) suggested that the SCS opened from ~32 Ma to ~16 Ma during the Oligocene and Early Miocene. Presently located between the Earth's largest ocean and the largest continent and being near the Pacific Warm Pool, the SCS also plays a significant role in affecting the global climate system (Wang and Li, 2009).

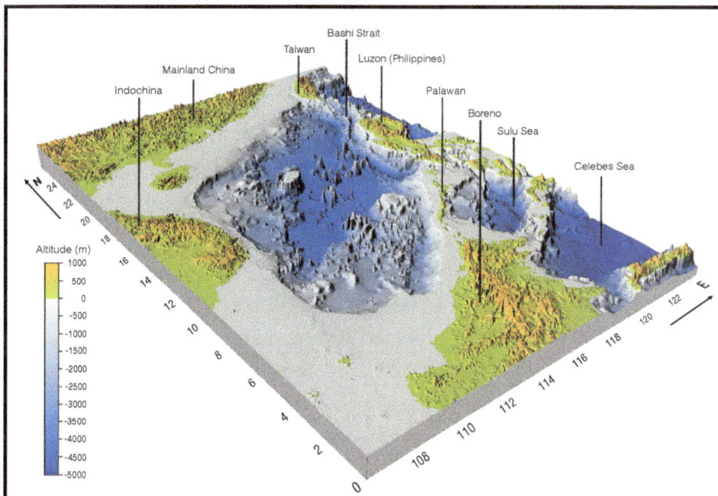


Figure 2. A 3D view of the Southeast Asia topography and South China Sea bathymetry (from Wang and Li, 2009).

The SCS is uniquely situated for studying continental break-up, basin formation, and seafloor spreading processes for several reasons.

1. It is a classical representative of western Pacific marginal seas that developed from continental break-up and has evolved into seafloor spreading.
2. The SCS is located at the junction of the Eurasian, Pacific, and Indo-Australian plates and therefore is sensitive to tectonic and climatic changes associated with the surrounding plates.
3. Its relatively small size and young age facilitates easy tectonic comparisons between the two conjugate continental margins.
4. Despite its short evolutionary history, the SCS has undergone a complete Wilson cycle from continental break-up and seafloor spreading to subduction. Therefore, it is well suited for studying various plate boundary activities, such as oceanic subduction (e.g., the Manila Trench), strike-slip faulting (e.g., the Red River Fault), and active orogenic processes (e.g., Taiwan).
5. The rifting style of the SCS margins may represent an intermediary form of continental extension between the end-member scenarios of highly magma-rich (e.g., North Atlantic margins) and magma-poor or hyperextended (e.g., Iberia and Newfoundland) margins.
6. The SCS is associated with high sedimentation rates and carbonate preservation, providing complementary records to those from the poorly preserved carbonates in the western Pacific Ocean.

All these attributes make the SCS an ideal natural laboratory for studying continental break-up, basin formation, deep mantle evolution, and land-ocean interactions. However, there is not yet one single deep drilling well in the central SCS basin targeting the oceanic crust or its transition from the continental crust. This leaves a large gap in understanding the opening mechanisms and timing of the onset and termination of seafloor spreading in the SCS.

In 1999, ODP conducted the first deep scientific drilling, Leg 184, in the SCS, focusing primarily on Asian monsoon and environmental history of the region (Wang et al., 2000). Leg 184 cored seventeen holes at Sites 1143–1148 in the SCS and recovered 5463 meters of sediment. The suite of sites yields an almost continuous record of the environmental history of the SCS during the last 32 Ma. Leg 184, however, was neither sited for recovering pre-rift and syn-rift strata, nor did it reach the oceanic crust basement. This leaves the SCS basin completely unconstrained in terms of its formation history.

## Scientific Debates on the Tectonics of the SCS

Among the continents of the world, Asia has been subjected to the most significant Cenozoic deformation. Coincident with the large-scale deformation of Asia, many of the marginal seas in the western Pacific stopped spreading during the early Miocene. However, the opening mechanisms and driving forces for the SCS basin are still debated. One model links the opening of the SCS basin with the Red River fault zone, which has at least 500–600 km of left-lateral displacement created during the Oligocene and Miocene

(Tapponnier et al., 1982; Briais et al., 1993). Alternative models relate the opening of the SCS to subduction under Borneo (Taylor and Hayes, 1980, 1983; Holloway, 1982; Hall, 1996) and the influence of subduction beneath the Philippines, driving a back-arc type of extension in the overriding plate (Taylor and Hayes, 1980).

Another major topic of debate is the timing of the onset and termination of seafloor spreading in the SCS. Largely based on interpretation of magnetic anomalies, estimations of the oceanic crust age for the east sub-basin were 32–17 Ma (Taylor and Hayes, 1980, 1983), 32–16 Ma (Briais et al., 1993; Li and Song, 2012), and 31–20.5 Ma (Barkhausen and Roeser, 2004). For the southwest sub-basin, the estimations of the oceanic crust age differ widely from 27–16 Ma (Briais et al., 1993; Li and Song, 2012), to 25–20.5 Ma (Barkhausen and Roeser, 2004), and to 42–35 Ma (Yao et al., 1994). Furthermore, Hsu et al. (2004) proposed the existence of oceanic crust as old as 37 Ma on the northeast part of the SCS. Such large discrepancies in the proposed magnetic ages of the oceanic crust point to the critical importance of direct sampling and dating the oceanic basement rocks through deep drilling in the SCS.

## Workshop Deliberation

The two-day workshop consisted of a mix of presentations, breakout sessions, and plenary discussion. The meeting featured five keynote addresses, thirty oral presentations, and twenty-five posters. These presentations covered a wide range of topics pertinent to tectonics and environmental history of Southeast Asia and the SCS. While the majority of presentations focused on the tectonic and magmatic evolution of the region and comparisons with other examples of continental rifting, basin formation, and seafloor spreading, participants also showcased work on paleoceanography, sedimentology, and biogeochemistry of the region. In addition, the meeting featured a presentation on the IODP proposal process, commentary on the current IODP drilling engineering capability, and a review of the previous panel comments and recommendations on the IODP Proposal 735-CPP.

The workshop devoted a significant amount of time to breakout sessions and plenary discussion. The breakout subgroups were charged with developing recommendations along four thematic lines.

1. What are the "big picture" science questions that are of global significance and can be addressed through scientific drilling in the SCS?
2. How can the drilling objectives, targets, and strategy of the IODP Proposal 735-CPP be prioritized? How can the site survey data be improved?
3. What are other future potential drilling objectives and targets in the SCS?
4. How can we strengthen future international cooperation in research of Southeast Asia and the SCS?

The workshop participants were divided into four subgroups, each of 15–18 members with mixed expertise. Key findings from the four subgroups were remarkably consistent, pointing to a prioritized strategy of scientific drilling in the SCS.

## Workshop Recommendations

Through ample discussion at both plenary discussion and breakout sessions, the workshop reached a consensus on a prioritized ocean drilling strategy in the SCS with the following key recommendations.

**(1) The initial top drilling priority is to constrain the timing of the start and end of seafloor spreading events in the SCS.**

This objective can be achieved through a single expedition drilling into oceanic basement rocks in both the east and southwest sub-basins. It was felt that sampling and dating oceanic basement rocks through deep drilling is the only way to ground truth the hypothesized oceanic crust age, based on interpretation of magnetic anomalies.

**(2) Transect drilling from the oldest visible seafloor-spreading magnetic anomaly to the extinct spreading axis of the SCS.**

All four subgroups suggested that transect drilling should be conducted on both the east and southwest sub-basins.

**(3) The sedimentary sequences obtained while drilling into the oceanic basement should be effectively used for studying environmental history of the region.**

The above recommendations (1) to (3) were later incorporated in the revised IODP Proposal 735-CPP-2 that was submitted to IODP in April 2012.

**(4) The workshop recommended a phased drilling strategy, starting with a leg of drilling the oceanic crust (i.e., IODP Proposal 735-CPP-2) to be followed by future drilling of syn-rifting sequences and Mesozoic strata on both margins of the SCS.**

The participants expressed strong interest in drilling syn-rifting sequences on conjugate margins of the SCS. Drilling targets should enable detailed comparison of extension histories on conjugate sides, aiming to document the full extension history of both margins.

Drilling into Mesozoic sediments also has its merits through providing constraints on the critical period before the SCS formation. Drilling the Mesozoic strata could help to test key questions regarding the connection of the Tethys to paleo-Pacific Ocean. Furthermore, evidence from Mesozoic strata could help to test tectonic models and to explore if there is evidence for pre-SCS extension in the Mesozoic.

**(5) The participants expressed interest in drilling targets associated with submarine hydrology and deep biosphere of the SCS.**

It was felt that hydrology and deep biosphere are important objectives for future drilling in the SCS, but at present information is insufficient to define exact targets. Thus, it would be useful to combine biosphere and microbiological targets with other drilling sites.

Installation of a circulation obviation retrofit kit (CORK) was proposed to study the hydrology, deep biosphere, and carbon cycle. However, it was recognized that the installation of CORK observatories would require intensive effort, and thus careful planning of drilling objectives, sites, and strategy needs to be developed.

Overall, the workshop developed a strong sense of enthusiasm, both from participating scientists and institutions. Participants expressed confidence that if the first phase of drilling (IODP Proposal 735-CPP-2) is approved, subsequent proposals focused on drilling syn-rift sequences and Mesozoic strata and with hydrology/deep biosphere objectives would be developed.

## Progress Forward

In April 2012, a revised IODP proposal entitled "Opening of the South China Sea and its implications for Southeast Asian tectonics, climates, and deep mantle processes since the Late Mesozoic" was submitted to IODP. This revised proposal (Li et al., 2012) has incorporated all of the key recommendations of the workshop. Four primary and five alternate drilling sites in the SCS ocean basin were proposed, all aimed at reaching the oceanic basement in order to recover both oceanic basalts and overlying sediments. The revised proposal has adopted the recommended transect drilling strategy with drilling sites in both the east and southwest sub-basins (Figs. 1, 3).

The proposed drilling sites are designed strategically to determine timing of onset and termination of seafloor spreading in different sub-basins, and timing of a possible ridge jump or a volcanic event amid seafloor spreading. These sites will allow geochemical samplings of basement rocks of different ages, yielding critical information on how the crust and mantle have evolved over the time of basin formation. Together the drilling proposal will address research objectives in three major themes: (1) Cenozoic mechanisms, timings, sequences, and affiliations of seafloor spreading; (2) oceanic crustal accretion and mantle evolution; and (3) paleoceanographic and sedimentary responses to tectonic evolution of the SCS.

## Links to Other Programs

There are currently several ongoing research programs in the region. One major initiative is the SCSD Program funded by the NSF-C. With a total budget of ~$25M USD, this eight-year research program (2011–2018) aims to reveal the life history of a classical marginal sea through research of the interconnection between tectonic/magmatic, sedimentological, and biochemical processes of the SCS (Wang, 2012). The SCSD Program is currently funding 3D Ocean Bottom Seismometers (OBS) experiments and deep-tow magnetics surveys in the east and southwest sub-basins of the SCS. The program will also sponsor additional site surveys and post-cruise research related to the proposed IODP drilling.

Following the Shanghai workshop, new site survey data related to the IODP Proposal 735-CPP-2 were provided by several institutions including the China National Offshore Oil Company (CNOOC), Guangzhou Marine Geological Survey (GMGS), National Central University in Taiwan, Federal Institute for Geosciences and Natural Resources (BGR), Germany, as well as by scien-

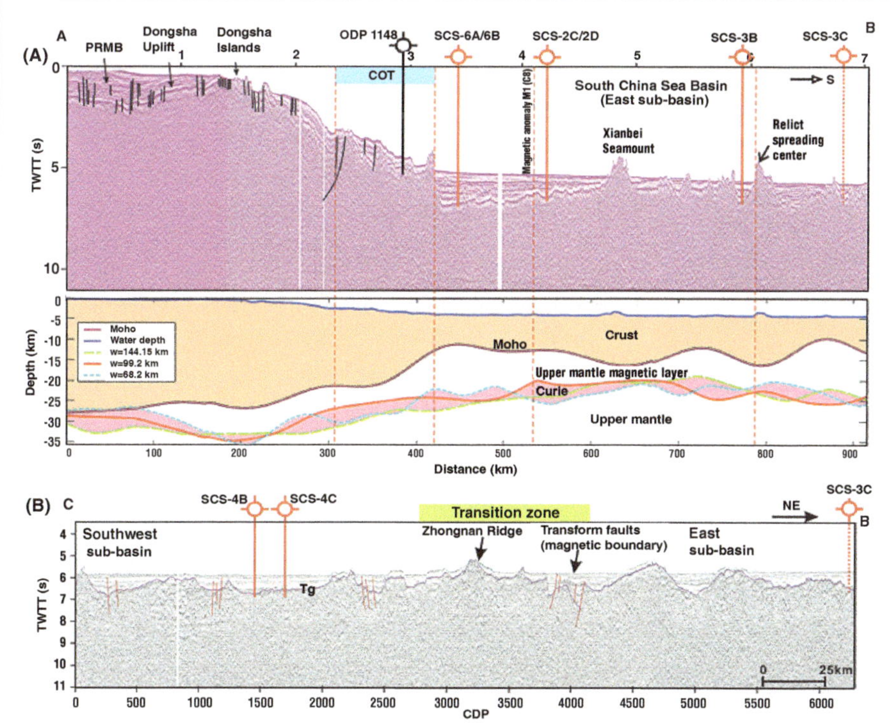


Figure 3. [A] (upper) Proposed sampling transect in the east sub-basin of the SCS (after Li et al., 2010). (lower) The Moho and Curie depths estimated from gravity and magnetic anomalies, respectively. TWTT=two-way travel time. COT=continent-ocean transition zone. PRMB=Pearl River Mouth Basin. w=width of moving windows in estimating Curie depths. [B] Seismic section showing the structures of the southwest sub-basin and the east sub-basin, as well as the transition zone between them. Tg is the Cenozoic basement (after Li et al., 2008).

tists on board a French R/V *Marion Dufresne* cruise in 2012. Acquisition of new site survey data was initiated by the proponents of the IODP Proposal 735-CPP-2 during the summer of 2012.

## Acknowledgements

The workshop was jointly sponsored and funded by the IODP-MI, IODP-China, the South China Sea Deep (SCSD) Program of the National Natural Science Foundation of China (NSF-C), and the State Key Laboratory of Marine Geology at Tongji University. We thank all workshop participants and co-proponents of the IODP Proposal 735-CPP-2 for major scientific input and suggestions.

## References

Barkhausen, U., and Roeser, H.A., 2004. Seafloor spreading anomalies in the South China Sea revisited. *In* Clift, P., Kuhnt, W., Wang, P., and Hayes, D. (Eds.), *Continent-Ocean Interactions within East Asian Marginal Seas, AGU Geophysical Monograph, 149:* Washington, DC (American Geophysical Union), 121–125.

Briais, A., Patriat, P., and Tapponnier, P., 1993. Updated interpretation of magnetic anomalies and seafloor spreading stages in the South China Sea: Implications for the Tertiary tectonics of Southeast Asia. *J. Geophys. Res.*, 98:6299–6328. doi:10.1029/92JB02280

Hall, R., 1996. Reconstructing Cenozoic SE Asia. *In* Hall, R., and Blundell, D.J., (Ed.) *Tectonic Evolution of Southeast Asia. Geological Society of London, Special Publication,* 106:153–184.

Holloway, N.H., 1982. North Palawan Block, Philippines--Its relation to the Asian mainland and role in evolution of South China Sea. *AAPG Bull.*, 66:1355–1383.

Hsu, S.-K., Yeh, Y.-C., Doo, W.-B., and Tsai, C-H., 2004. New bathymetry and magnetic lineations identifications in the northernmost South China Sea and their tectonic implications. *Mar. Geophys. Res.*, 25(1-2):29–44. doi:10.1007/s11001-005-0731-7

Li, C.-F., and Song, T., 2012. Magnetic recording of the Cenozoic oceanic crustal accretion and evolution of the South China Sea basin. *Chin. Sci. Bull.*, 57(24):3165–3181. doi:10.1007/s11434-012-5063-9

Li, C.-F., Shi, X., Zhou, Z., Li, J., Geng, J., and Chen, B., 2010. Depths to the magnetic layer bottom in the South China Sea area and their tectonic implications. *Geophys. J. Int.*, 182:1229–1247. doi:10.1111/j.1365-246X.2010.04702.x

Li, C.-F., Wang, P.X., Franke, D., et al., 2012. Opening of the South China Sea and its implications for Southeast Asian tectonics, climates, and deep mantle processes since the Late Mesozoic. *IODP Complementary Project Proposal,* #735-CPP-2.

Li, C.-F., Zhou, Z., Li, J., Chen, B., and Geng, J., 2008. Magnetic zoning and seismic structure of the South China Sea ocean basin. *Mar. Geophys. Res.*, 29:223–238. doi:10.1007/s11001-008-9059-4

Shi, H., and Li, C.-F., 2012. Mesozoic and early Cenozoic tectonic convergence-to-rifting transition prior to opening of the South China Sea. *Int. Geol. Rev.*,1:1–28. doi:10.1080/00206814.2012.677136

Smith, W.H.F., and Sandwell, T.D., 1997. Global seafloor topography from satellite altimetry and ship depth soundings. *Science*, 277:1957–1962. doi:10.1126/science.277.5334.1956

Tapponnier, P., Peltzer, G., Le Dain, A.Y., Armijo, R., and Cobbold, P., 1982. Propagating extrusion tectonics in Asia: New insights from simple experiments with plasticine. *Geology*, 10:611–616. doi:10.1130/0091-7613(1982)10%3C611:PETIAN%3E2.0.CO;2

Taylor, B., and Hayes, D.E., 1983. Origin and history of the South China Sea basin. *In* Hayes, D.E. (Ed.) *The Tectonic and Geologic Evolution of Southeast Asian Seas and Islands II, AGU Geophysical Monograph*, 27: Washington, DC (American Geophysical Union), 23–56. doi:10.1029/GM027p0023

Taylor, B., and Hayes, D.E., 1980. The tectonic evolution of the South China Sea basin. *In* Hayes, D.E. (Ed.) *The Tectonic and Geologic Evolution of Southeast Asian Seas and Islands, AGU Geophysical Monograph, 23:* Washington, DC (American Geophysical Union), 89–104. doi:10.1130/0091-7613(1982)10%3C611:PETIAN%3E2.0.CO;2

Wang, P.X., 2012. Tracing the life history of a marginal sea – On the South China Sea Deep Research Program. *Chin. Sci. Bull.*, 57:1–22, doi:10.1007/s11434-012-5087-1

Wang, P.X., and Li, Q.Y., 2009. *The South China Sea: Paleoceanography and Sedimentology:* Springer, pp. 506.

Wang, P.X., Prell, W., Blum, P., et al., 2000. *Proc. ODP, Init. Repts.*, 184: College Station, TX (Ocean Drilling Program). doi:10.2973/odp.proc.ir.184.2000

Yao, B., Zeng, W., Hayes, D., et al. 1994. *The Geological Memoir of South China Sea Surveyed Jointly by China and USA* (in Chinese). Wuhan, (China University of Geoscience Press).

## Authors

**Jian Lin**, Woods Hole Oceanographic Institution, Department of Geology & Geophysics, Woods Hole, MA 02543, U.S.A., e-mail: jlin@whoi.edu (Corresponding Author)

**Chun-Feng Li**, Tongji University, State Key Laboratory of Marine Geology, 1239 Siping Road, Shanghai 200092, China, e-mail: cfl@tongji.edu.cn

**Pinxian Wang**, Tongji University, State Key Laboratory of Marine Geology, 1239 Siping Road, Shanghai 200092, China, e-mail: pxwang@tongji.edu.cn

**Dieter Franke**, Federal Institute for Geosciences and Natural Resources (BGR), Stilleweg 2, 30655 Hannover, Germany, e-mail: Dieter.Franke@bgr.de

**Jun Tian**, Tongji University, State Key Laboratory of Marine Geology, 1239 Siping Road, Shanghai 200092, China, e-mail: tianjun@tongji.edu.cn

## Related Web Links

http://www.iodp-china.org/
http://www.iodp.org/workshops/

# New Frontiers in Scientific Drilling of the Indian Ocean

by Stephen J. Gallagher, Neville Exon, Dhananjai Pandey, S. Rajan, Mike Coffin, and Ken Takai

## Introduction

The Indian Ocean exerts a fundamental control on the Earth's climate and hosts a variety of complex tectonic features. It influences the Indian Monsoon and hosts a major part of the thermohaline conveyor. It has been over a decade since scientific drilling occurred in the Indian Ocean, and as such there are major gaps in geoscientific understanding of this region. Future drilling of the sedimentary archives in the region will yield substantial information on the history of uplift, erosion, deposition and monsoonal history. It will improve our understanding of greenhouse/icehouse and ocean gateway dynamics and reef development. The region also hosts exceptional examples of Earth system processes and products that drilling will play an important role in illuminating. It would answer questions associated with subduction and tectonic plate breakup and reorganization. Major geodynamic issues to be investigated include hotspot/spreading ridge interactions and constraints on the mantle reference frame. There are many deep biosphere mysteries that may be solved by drilling sediment, such as the impact of Himalayan uplift and the monsoon on subseafloor community diversity. Drilling oceanic crust will reveal the nature of poorly known microbial communities at ridge systems, providing insights into the composition and abundance of microbial communities in different crustal provinces.

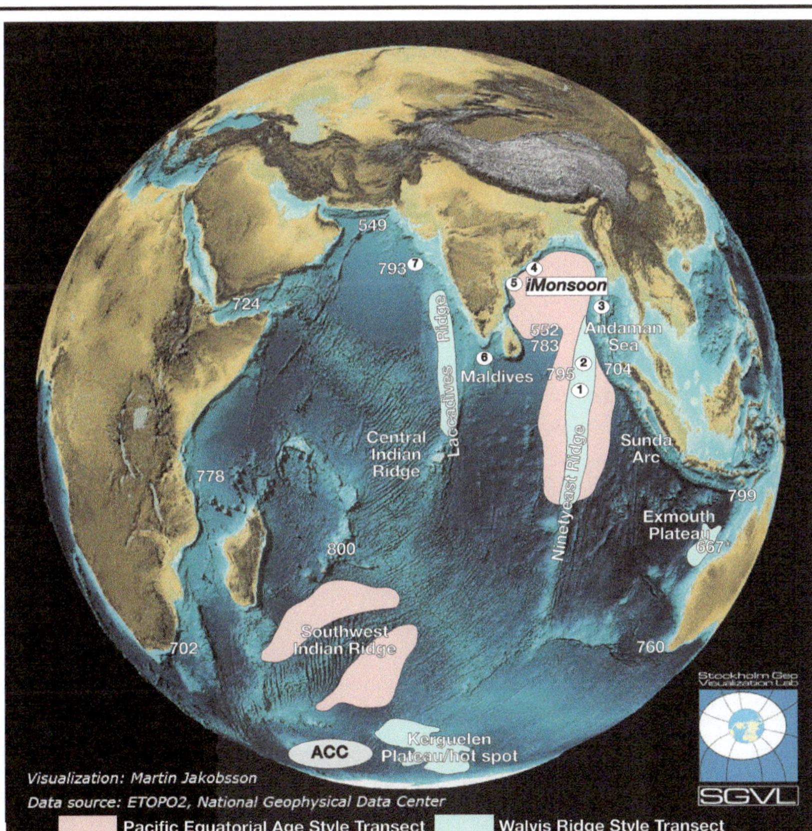


Figure 1. Pre-existing proposals and future drilling prospects in the Indian Ocean. Potential Theme 1 targets include two types of deep ocean transects; ACC=Antarctic Circumpolar Current drifts proposal; Maldives atoll & deep-water proposals. Theme 2 targets include iMonsoon program with locations: 1) DSDP 216, 2) ODP 758, 3) Andaman Sea, 4) Mahanadi Basin, 5) Krishna-Godavari Basin, 6) Cape Cormorin, and 7) Laxmi Basin Kerala-Konkan Basin. Theme 3 targets include most tectonic features shown. Theme 4 could be accommodated in most expeditions. Sourced from: http://www.gebco.net/general_interest/bathymetry_visualisations.html

## Workshop Goals

The Indian Ocean IODP Workshop was hosted in Goa in October 2011 by the National Centre for Antarctic and Ocean Research (NCAOR), India. The workshop was initiated and planned by Australian and Indian scientists to improve existing proposals (Table 1), build new proposals, and initiate alliances essential for drilling proposals.

The workshop was attended by seventy Indian and forty international scientists. Participants stressed the importance of the Indian Ocean in the planned new phase of drilling, the International Ocean Discovery Program (IODP, 2013–2023). The aim is to have several new proposals submitted in 2012 for drilling by 2014 or later, centered around global science problems discussed in four themes.

## Theme 1: Cenozoic Oceanography, Climate Change, Gateways and Reef Development

Drilling Cenozoic marine sequences will lead to an enhanced understanding of Cretaceous to Paleogene ocean oxygenation; greenhouse to icehouse transitions;

the history of the Indian Ocean Dipole; the Antarctic Circum-Polar (ACC) and Agulhas Currents (AC); the timing of the Indonesian Throughflow (ITF) and the origin of reefs. This theme addresses Challenges 1, 2 and 7 of IODP Science Plan 2013–2023. Three subthemes were discussed.

Table 1. A list of the current active proposals in the Indian Ocean.

| Proposal | Short Title | Lead Proponent |
|---|---|---|
| 549-Full-6 | Northern Arabian Sea Monsoon | Lückge, Germany |
| 552-Add | Bengal Fan | France-Lanord, France |
| 595-Full-4 | Indus Fan and Murray Ridge | Clift, U.K. |
| 667-Full | NW Australian Shelf Eustasy | Fulthorpe, U.S.A. |
| 702-Full | Southern African Climates | Zahn, Spain |
| 704-Full-2 | Sumatra Seismogenic Zone | Goldfinger, U.S.A. |
| 724-Full | Gulf of Aden Faunal Evolution | de Menocal, U.S.A. |
| 760-Pre | SW Australia Margin Cretaceous Climate | Grocke, U.K. |
| 778-Full-2 | Tanzania Margin Paleoclimate Transect | Wade, U.K. |
| 783-APL | Indian Monsoon History | Hathorne, U.K. |
| 793-CPP | Arabian Sea Monsoon | Pandey, India |
| 795-Pre | Indian Monsoon Rainfall | Clemens, U.S.A. |
| 799-Pre | West Pacific Warm Pool | Rosenthal, U.S.A. |
| 800-Full | Indian Ridge Moho | Dick, U.S.A. |

*Evolution of the shallow carbonate environment.* Prospective proposals will suggest deep and shallow drill sites in the Maldives Archipelago to better understanding the influence of the monsoon-generated current system and sea-level fluctuations on the Neogene development of the Maldives platform (Betzler et al., 2009). Another potential proposal will address the origin of modern atolls and drowned flat top banks that have developed since the Pliocene.

*Deep ocean record of palaeoceanography/climate.* A series of transects ("Walvis Ridge Transects", Westerhold et al., 2007; "Pacific Equatorial Age Transects", Lyle et al., 2010; Fig. 1) were suggested to reveal the long-term history of palaeo-$CO_2$ levels, sea surface temperatures, ocean anoxic events and the Indian Ocean Dipole. Several proposals focusing on recovering deep oceanic and climatic Cretaceous to Neogene records (proposals 760-Pre, 724-Full and 778-Full) have been submitted.

*Ocean margin boundary currents and gateways.* The ITF and ACC are key drivers of global climate. Understanding the response of these currents to glacio-eustatic and tectonic triggers will allow us to predict their behavior in the future in the context of global climate change. A new version of proposal 667-Full will be submitted in 2012 to investigate ITF and Leeuwin Current history (Gallagher et al., 2009). Another proposal will target ACC history (Barker and Thomas, 2004) by drilling the Conrad Rise. Proposal 702- Full will investigate AC history.

## Theme 2: The History of the Monsoons

The Indian monsoonal circulation is the largest atmospheric transport system on Earth, including latent heat (moisture) sources in the southern Indian Ocean, cross-equatorial transport with large-scale convergence and precipitation over the Indian and Southeast Asian regions, and influencing some three billion people seasonally. This circulation system is complex, driven by cross-equatorial pressure gradients dictated by changes in the Siberian High (winter monsoon), Indo-Asian low (summer monsoon) and both summer- and winter-season changes in the Mascarene High of the southern subtropical Indian Ocean. The magnitude and regularity of the cross-equatorial flow make the Indian monsoon unique. Monsoonal circulation is dynamic on various time scales, from weekly to annual to decadal and linked to short time scale ocean-atmosphere interactions, to millions of years associated with regional and large-scale tectonism. Monsoon runoff from the continent has built the two largest fan systems on Earth, the Indus and Bengal Fans. These host a record of potentially high resolution of the history of uplift, erosion, deposition, and carbon burial. An understanding of the mechanisms driving monsoon climate changes at different time scales has high societal importance. Theme 2 addresses Challenges 1, 3–7, 11 and 13 of the IODP Scientific Plan.

*Monsoonal Paleoclimate and Paleoceanography (MPP).* This focused on the pelagic and hemipelagic archives as recorders of the local and regional responses to monsoon winds and precipitation (Clift and Molnar, 2003). This includes existing proposals 549 Full and 783 APL. The MPP subgroup recommend a workshop to produce a new proposal iMonsoon (Fig. 1) in time for consideration by the program panels and drilling by 2014 or soon after.

*Sedimentary source-to-sink studies, using submarine fan (SF) archives.* Submarine fans are recorders of long-term changes in erosion, transport and deposition linked to tectonics and climate change principally in the Himalayas (Clift et al., 2008; Gupta et al., 2011). These topics are included in proposals 609-Pre, 552-Full and 552-Add, 595-Full and 776- Full targeting the eastern and western river systems. However, amending these proposals with more proximal sites on each fan would allow the transport history to be better constrained. In this regard, the SF sub-committee also recommended a workshop on geohazards in the northern Bay of Bengal, focusing on the frequency of tropical cyclones, Holocene earthquakes, Pleistocene subsidence, and changing sediment supply due to deforestation and agriculture.

## Theme 3: Tectonics and Volcanism

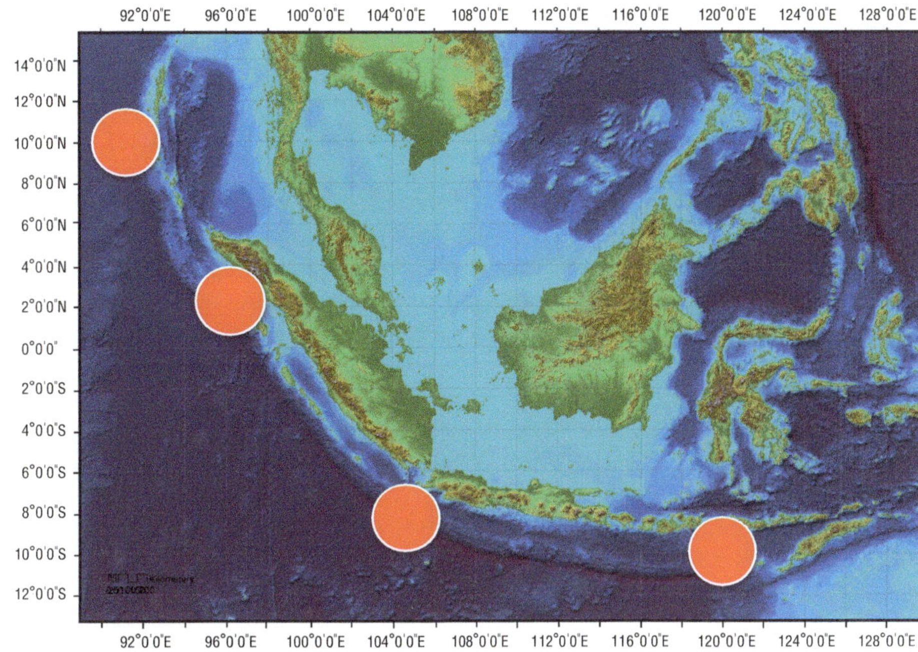


Figure 2. Possible drill sites along the Sunda Arc that represent the transition from collisional tectonics in the east to oblique subduction in the northwest.

The Indian Ocean hosts exceptional examples of Earth system processes and products that ocean drilling will play a key role in illuminating. Tectonic and magmatic objectives address Challenges 12 and 13 of the IODP Science Plan. Tectonic problems include those associated with; 1) subduction—fault properties and slip, sediment effects, role of fluids, segmentation, rupture, tsunamigenesis—along the Sunda arc, site of the great 2004 Indian Ocean earthquake (Gulick et al., 2011) and tsunami (Fig. 2) (proposal 704-Full-2: "Sumatra Seismogenic Zone"); 2) tectonic plate breakup and reorganization associated with Kerguelen and Réunion hotspot activity; 3) India-Madagascar and India Seychelles break-up, origin and evolution of Laxmi and Laccadive ridges and intraplate deformation; and 4) possible incipient plate breakup along the Ninetyeast Ridge. Major geodynamic issues comprise hotspot–spreading ridge interactions and hotspot constraints on the mantle reference frame. Key geochemical questions include those associated with hotspots—Afar, Kerguelen, Réunion—that have sampled nearly the entire scope of mantle source compositions; 1) back-arc spreading magmatism, hydrothermal activity, mineralization, metallogenesis, and biosphere; 2) the testing of existing and development of new models for flood magmatism; 3) the origin and alteration of oceanic core complexes; and 4) the nature of the lower crust and Moho at slower spreading ridges (proposal 800-Full).

## Theme 4: The Deep Biosphere

Scientific drilling has revealed the existence of a deep biosphere in surface and deep subsurface sediment and rocks (Roussel et al., 2008; Edwards et al., 2012). This biosphere is widespread, with some geological and geographical autochthonism, is genetically and geochemically diverse, and comprises a significant fraction of Earth's total living biomass (Orcutt et al., 2011). High temperatures and pressures, variable pH and salinity conditions, and insufficient access to nutrients, carbon, or energy are potential limiting factors for life in the deep subsurface (Takai and Nakamura, 2011). This theme addresses Challenges 5–7 of the IODP Science Plan.

Several geomicrobiological questions were identified that may be answered by drilling of Indian Ocean sediment. How has uplift of the Himalayas and the monsoon impacted the development of the deep biosphere since the Oligocene? How has the drainage from the Himalayan rivers influenced subseafloor community diversity? How has the subseafloor biosphere been inoculated with terrestrial microorganisms? Are there differences between the Bay of Bengal and Arabian Sea? How are the ecosystems related to the formation of ferromanganese nodules?

Other questions may be answered by drilling Indian Ocean Crust. What is the diversity and biogeography of communities in the very heterogeneous ridge systems? Which microbial communities and functions exist in different crustal provinces and structures? How are they related to hydrogeology and to the alteration of the crust?

In the future, these questions may be addressed if deep biosphere research is incorporated in existing IODP proposals. The community is looking for opportunities to join expeditions conducted by themes 1–3 (above) and to expand objectives by requesting coring through Ancillary Project Letters.

## Acknowledgements

We thank IODP-MI, Ministry of Earth Sciences, India (MoES) and NCAOR, U.S. Science Support Program (USSSP), Australian and New Zealand IODP Consortium (ANZIC), and Japan Drilling Earth Science Consortium (J-DESC) for their funding support. We also thank the workshop participants for their contribution to the success of this workshop.

## References

Barker, P.F., and Thomas, E., 2004. Origin, signature and palaeoclimatic influence of the Antarctic Circumpolar Current. *Earth Sci. Rev.*, 66:143–162, doi:10.1016/j.earscirev.2003.10.003

Betzler, C., Hübscher, C., Lindhorst, S., Reijmer, J.J.G., Römer, C.M., Droxler, A.W., Fürstenau, J., and Lüdmann, T., 2009. Monsoon-induced partial carbonate platform drowning (Maldives, Indian Ocean). *Geology*, 37:867–870, doi:10.1130/G25702A.1

Clift, P., and Molnar, P., 2003. Scientific drilling of the Indian Ocean submarine fans: A report on the JOI/USSAC workshop for future IODP drilling. 23–25th July 2003, Boulder, CO (University of Colorado).

Clift, P.D., Hodges, K.V., Heslop, D., Hannigan, R., Long, H.V., and Calvès, G., 2008. Correlation of Himalayan exhumation rates and Asian monsoon intensity. *Nature Geosci.*, 1:875–880.

Edwards, K.J., Becker, K., and Colwell, F., 2012. The deep, dark, energy biosphere. *Ann. Rev. Earth Planet. Sci.*, 40:551–568.

Gallagher, S.J., Wallace, M.W., Li, C.L., Kinna, B., Bye, J.A.T, Akimoto, K., and Torii, M., 2009. Neogene history of the Indo-Pacific Warm Pool, Kuroshio and Leeuwin currents. *Paleoceanography*, 24:PA1206, doi:10.1029/2008PA001660

Gulick, S.P.S., Austin, Jr, J.A., McNeill, L.C., Bangs, N.L.B., Martin, K.M., Henstock, T.J., Bull, J.M., Dean, S., Djajadihardja, Y.S., and Permana, H., 2011. Updip rupture of the 2004 Sumatra earthquake extended by thick indurated sediments. *Nature Geosci.*, 4:453–456.

Gupta, A.K., Mohan, K., Sarkar, S., Clemens, S.C., Ravindra, R., and Uttam, R.K., 2011. East–West similarities and differences in the surface and deep northern Arabian Sea records during the past 21 Kyrs. *Palaeogeogr. Palaeoclimatol. Palaeoecol.*, 301:75–85.

Lyle, M., Pälike, H., Nishi, H., Raffi, I., Gamage, K., Klaus, A., and the IODP Expeditions 320/321 Scientific Party, 2010. The Pacific equatorial age transect, IODP Expeditions 320 and 321: Building a 50-million-year-long environmental record of the equatorial Pacific. *Sci. Drill.*, 9:4–15, doi:10.2204/iodp.sd.9.01.2010.

Orcutt, B.N., Sylvan, J.B., Kanb, N.J., and Edwards, K.J., 2011. Microbial ecology of the dark ocean above, at, and below the seafloor. *Microbiol. Mol. Biol. Rev.*, 75:361–422, doi: 10.1128/MMBR.00039-10.

Roussel, E.G., Bonavita, M.-A.C., Querellou, J., Cragg, B.A., Webster, G., Prieur, D., and Parkes, R.J., 2008. Extending the subsea-floor biosphere. *Science*, 320(5879):1046, doi: 10.1126/science.1154545.

Takai, K., and Nakamura, K., 2011. Limits of life and the biosphere: Lessons from deep subsurface of the Earth. *In* Gargaud, M., Lopez-Garcia, P., and Martin, H. (Eds.), *Origins and Evolution of Life*: New York (Cambridge University Press), 469–486.

Westerhold, T., Rohl, U., Laskar, J., Raffi, I., Bowles, J., Lourens, L.J., and Zachos, J.C., 2007. On the duration of magnetochrons C24r and C25n and the timing of early Eocene global warming events: Implications from the Ocean Drilling Program Leg 208 Walvis Ridge depth transect. *Paleoceanography*, 22:PA2201, doi:10.1029/2006PA001322

## Authors

**Stephen J. Gallagher**, The School of Earth Sciences, The University of Melbourne, Victoria 3010, Australia, e-mail: sjgall@unimelb.edu.au

**Neville Exon**, Research School of Earth Sciences, Australian National University, Canberra 0200, Australia, e-mail: Neville.Exon@anu.edu.au

**Dhananjai Pandey** and **S. Rajan**, National Centre for Antarctic and Ocean Research (NCAOR), Vasco da Gama, Goa, India, e-mail: dhananjai@gmail.com and rajan@ncaor.org

**Mike Coffin**, Institute for Marine and Antarctic Studies, University of Tasmania, Private Bag 129,HOBART, TAS 7001, Australia, e-mail: Mike.Coffin@utas.edu.au

**Ken Takai**, Subsurface Geobiology Advanced Research (SUGAR) Project, Japan Agency for Marine-Earth Science & Technology (JAMSTEC), 2-15 Natsushima-cho, Yokosuka 237-0061, Japan, e-mail: kent@jamstec.go.jp

## Related Web Links

http://www.ncaor.gov.in/IODP/index.html
http://www.iodp.org/workshops/http://www.ncaor.gov.in/IODP/index.html
http://www.iodp.org/Science-Plan-for-2013-2023/

# IODP Workshop on Using Ocean Drilling to Unlock the Secrets of Slow Slip Events

by Laura M. Wallace, Eli A. Silver, Nathan Bangs, Rebecca Bell, Joshu Mountjoy, Stuart Henrys, and Ingo Pecher

## Introduction

Subduction margins produce the largest and most destructive earthquakes and tsunami on Earth. Knowledge of the mechanics of fault slip behavior on subduction thrust interfaces is necessary to understand and mitigate the hazards posed by these major plate boundary features. The observation of episodic slow slip events (SSEs) occurring at many subduction zones worldwide (Fig. 1) is arguably one of the most exciting discoveries in the Earth sciences in the last decade. SSEs involve transient, aseismic slip on a fault, lasting days to months, at a rate intermediate between steady plate boundary displacement rate and seismic slip. Prior to the discovery of slow slip, most studies of fault behavior assumed that slip on faults occurs either as steady creep or suddenly in an earthquake. Geodetic and seismological detection of slow slip and its associated slow seismic phenomena (such as non-volcanic tremor, low-frequency and very-low frequency earthquakes) has transformed our understanding of the spectrum of fault slip behavior (Ide et al., 2007) and has fundamental implications for deformation mechanisms and rheology on subduction megathrusts. Despite the fact that there is now abundant evidence for SSEs at many of the Earth's subduction zones, the physical mechanisms leading to slow slip behavior and the relationship of SSEs to destructive seismic slip on subduction thrusts remain unknown.

A variety of theories regarding the origin of SSEs have been proposed; many consider episodic slow slip as a consequence of high fluid pressures within a conditionally stable frictional regime. These proposed mechanisms for episodic SSE behavior arise largely from theoretical and modeling studies (Liu and Rice, 2005, 2007; Shibazaki and Iio, 2003) and interpretations of physical properties from seismic attributes (Audet et al., 2009; Bell et al., 2010; Kodaira et al., 2004; Song et al., 2009). In many ways, our understanding of the origins of slow slip events has been hampered by the fact that many of the best-studied SSEs worldwide (e.g., Cascadia, Nankai) occur at depths of 30–40 km below the Earth's surface (Dragert et al., 2001; Obara et al., 2004), and scientists must rely on indirect evidence to investigate these fascinating phenomena. However, the discovery of shallow SSEs (at <10 km depth) at subduction margins in New Zealand, central Japan and Costa Rica (Douglas et al., 2005; Outerbridge et al., 2010 Ozawa et al., 2007; Sagiya, 2004; Wallace and Beavan, 2010) highlights the exciting potential for IODP drilling to gain direct access to the SSE source area, and thus to play a key role in unlocking the secrets of SSEs.

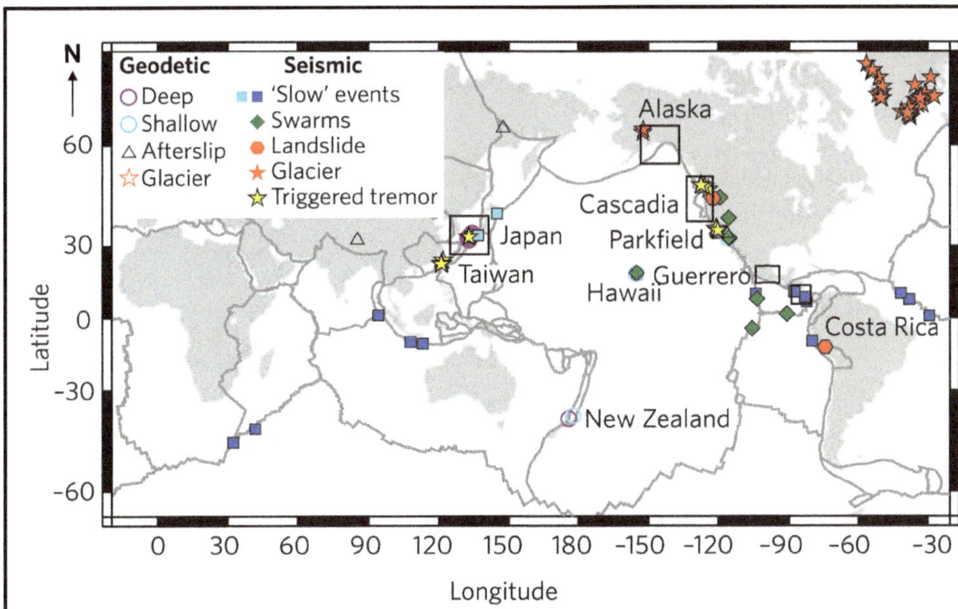


Figure 1. Map showing the global distribution of seismically and geodetically observed slow slip phenomena, from Peng and Gomberg (2010). Note that this map includes slow earthquakes and/or slow slip behavior from all fault types (not just restricted to subduction zones), and slow slip-type phenomena observed in landsliding and glacial settings.

## Workshop Goals and Overview

To discuss and develop a plan to use scientific ocean drilling to help advance the emerging and highly topical field of slow slip events, Integrated Ocean Drilling Program (IODP) sponsored a workshop in Gisborne, New Zealand on 1–3 August 2011. The workshop attracted seventy geoscientists from a

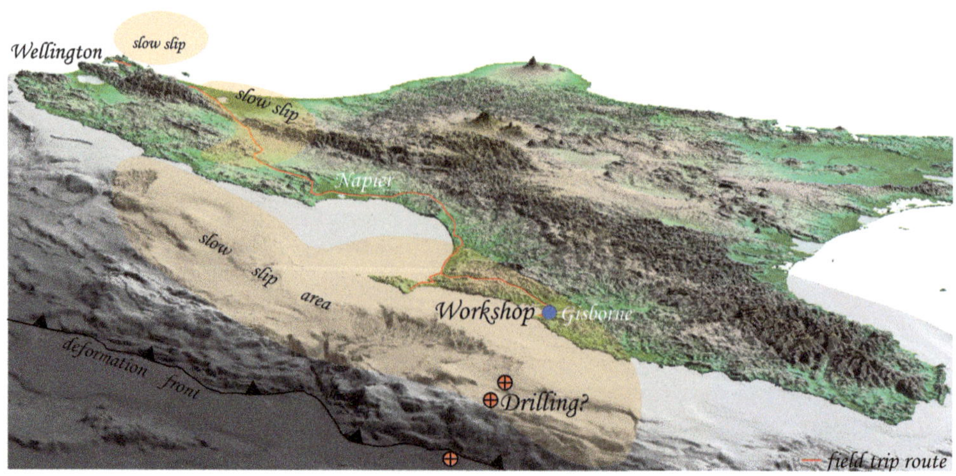

Figure 2. Oblique view of the Hikurangi subduction margin, including locations of slow slip (orange shaded areas), the location of the workshop and field trip route (orange solid line), and some of the proposed slow slip event drilling targets offshore Gisborne.

dozen countries around the world and was also sponsored by the Consortium for Ocean Leadership, the New Zealand Ministry of Science and Innovation, and the U.S. National Science Foundation's (NSF's) GeoPrisms program. Gisborne was chosen as a venue due to its proximity to the source area of shallow slow slip (<5–15 km depth) that occurs at the northern Hikurangi subduction margin in New Zealand. The main three-day workshop was followed by a two-day field trip (sponsored by the New Zealand Ministry of Science and Innovation) from Gisborne to Wellington (Fig. 2), designed to expose participants to the onshore, uplifted components of the Hikurangi forearc and to give participants insights into the geological and tectonic context of slow slip in New Zealand.

The primary goals of the workshop were to summarize critical requirements of a drilling program to discern the physical mechanisms responsible for SSE behavior, to develop strategies to achieve the scientific goals, to determine what types of data are needed to develop an effective drilling program, and to identify the expertise and technologies needed for drilling a SSE source area successfully. Oral presentations at the workshop centered around thematic sessions, including (1) observations of and theories for SSE occurrence, (2) lessons learned from previous IODP subduction zone drilling, and (3) potential slow slip drilling targets in New Zealand, Costa Rica, and central Japan. The talks were interspersed each day with breakout discussion sessions and broader group discussions. On the final day, breakout groups developed implementation plans for each location.

## Recommendations and Outcomes of the Workshop

Overall, the workshop participants agreed that slow slip events offer a number of compelling scientific questions that can be realistically addressed by IODP drilling. Some of these questions are listed as follows. (1) What does a slow slip zone look like? Is the slip localized to one or two sharp, thin discontinuities, or is slip distributed continuously throughout a zone of finite thickness? (2) Do slow slip events occur in rocks consistent with a conditionally stable frictional regime, as suggested by theoretical studies? (3) Are slow slip events associated with low effective stress due to high pore fluid pressure? (4) Can a single fault patch host both SSEs and "normal" earthquakes? (5) Are SSEs restricted to a specific pressure or temperature range and can they propagate all the way to the trench? Participants concluded that these fundamental questions can be addressed by a strategy that includes conducting geophysical experiments, monitoring shallow instrumented bore holes above SSE source regions, logging and coring of the region above the SSE source and incoming sedimentary section, and, ultimately, by direct sampling and monitoring of the slow slip patch by deep riser drilling. During the workshop the participants discussed generic methodologies to address these questions and developed plans which could be implemented in specific locations around the world.

Auxiliary studies (e.g., non-drilling) are required for any successful SSE drilling program, and these were a major topic of discussion. In summary, participants recommended a broad range of studies with particular emphasis on collocated, integrated studies using geological and geophysical techniques that are designed and constrained with laboratory and numerical modeling. Of particular importance to determining the spatial distribution of SSEs and related seismicity are onshore geodetic and seismic experiments combined with offshore, long-term deployment of ocean bottom seismographs equipped with absolute pressure sensors to monitor seismicity and vertical seafloor deformation during SSEs. Other critical data sets include active source 2D and 3D seismic imaging and wide-angle refraction, passive source studies, heat flow surveys, and multibeam seafloor mapping. These auxiliary data are required to help identify drilling targets and to complement borehole monitoring, core, and log data.

Shallow level borehole monitoring (to detect tilt, strain, seismicity, geochemical, and hydraulic transients) along a transect (Fig. 3) is key to addressing questions related to the spatial distribution of slow slip beneath offshore subduction margins, and to reveal the possible relationship between

SSEs and normal seismicity, as well as to discern changes in fluid flow and geochemistry within the upper plate during the SSE cycle. Depending on the monitoring needs and available resources, instrumentation in the boreholes could range from sophisticated, Circulation Obviation Retrofit Kits (CORK) type installations to simpler Simple Cabled Instrument for Measuring Parameters In Situ (SCIMPI) type devices that do not require standard casing installation. Coring and logging along the same transect are required to characterize in detail the lithology, stratigraphy, structure, fluids, and physical properties within the upper plate above the SSE source regions. Of particular importance is comprehensive coring and logging of the entire input section (e.g., undeformed seafloor sediments on the incoming plate), which will reveal the lithologies likely to be found at depth within the SSE source area. Participants developed an idealized shallow (<1–2 km depth) riserless transect design that should achieve the shallow monitoring and coring/logging objectives (Fig. 3). Riser drilling, logging, sampling, and monitoring of the SSE source area would constitute the ultimate, final stage of an SSE drilling program, and would provide the most direct information on the physical conditions (frictional properties, mineralogical composition, fluid pressure conditions, temperature, among others) that lead to and control slow slip event behavior. Targets to intersect SSEs at 5–7 km depth have been identified at some of the shallow SSE locations that were discussed at the workshop. Participants agreed that deep drilling of an SSE source area is within reach and is the ultimate way to solve the mystery of why SSEs occur.

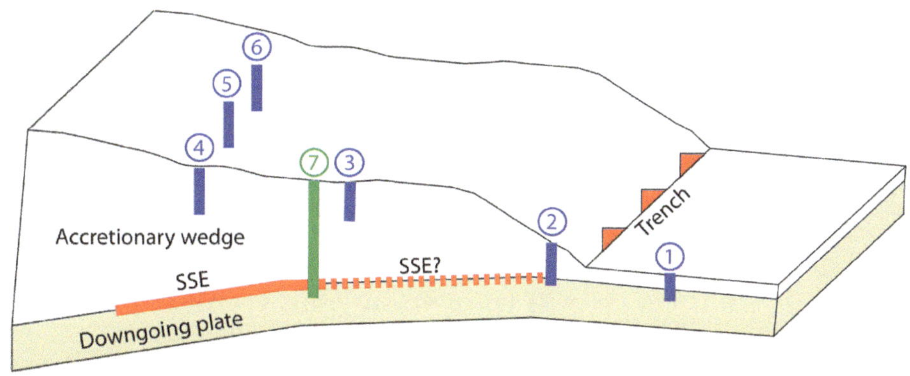


Figure 3. Schematic illustration (not to scale) of a drilling and borehole monitoring scheme which could be applied to address primary SSE questions. See workshop report for further details on the transect design. Note that holes 5 and 6 are positioned off the main transect and are intended for basic monitoring to constrain off-transect variation in SSE behavior. Red patches on the subduction interface show where SSE slip is constrained from onshore instruments. The position of the deep riser hole (#7) should be constrained by current knowledge of the shallowest part of the interface undergoing SSE. In addition to shallow monitoring and sampling of the upper plate, Site 3 would serve as a pilot hole for the deep riser Site 7. Site 1 is primarily intended to core and log the input section on the subducting plate. The dashed red line shows a possible trenchward continuation of SSE behavior. Offshore instrumentation (borehole monitoring and seafloor geodesy) will be needed to constrain the updip limit of SSE.

Participants also outlined the characteristics of an "ideal" SSE drilling target. The ideal location must have slow slip occurring somewhat predictably, at depths shallow enough for the source area to be accessible by riser drilling. SSE drilling locations should have a relatively high frequency of SSEs so that multiple cycles can be monitored over a realistic observation period, and the SSEs must be large enough to be measurable and resolvable. Close proximity of the SSE source to the coastline is also advantageous, in order to maximize potential to link with and complement onshore data gathering infrastructure. An SSE target should also have existing 2D seismic reflection profiles to characterize the structure, interface properties, and geological context for the SSEs.

An interesting discovery during the workshop was the realization by participants that the world's best-documented shallow SSEs in Costa Rica, central Japan, and New Zealand bear some striking similarities to each other. All three locations have relatively cold temperatures on the interface in the SSE source (~100°C–150°C), similar SSE durations (generally ~2–3 weeks), comparable equivalent moment magnitudes per event ($M_W$ ~6.5), and a rough, irregular subducting plate (e.g., subduction of seamounts, ridges, or arc volcanoes). We expect that continued comparisons of these three subduction zones begun at the workshop will lead to new insights into the mechanisms behind shallow SSEs.

Following the workshop, some new and/or revised proposals have been submitted to IODP to conduct ocean drilling related to SSE processes for some of the candidate SSE locations. A detailed workshop report is available to view on the IODP website.

## Acknowledgements

We appreciate generous support for the workshop by IODP-MI, the New Zealand Ministry of Science and Innovation, the Consortium for Ocean Leadership, and NSF/GeoPrisms. We also thank the workshop participants for their contributions to the vibrant, exciting discussions held there, as well as for the excellent oral and poster presentations. Kat Hammond (GNS Science), Joan Gerritse (the Portside Hotel, Gisborne), and Kim Rampling (the Gisborne Conference Center) provided logistical support that was crucial to the success of the workshop.

## References

Audet, P., Bostock, M.G., Christensen, N.I., and Peacock, S.M., 2009. Seismic evidence for overpressured subducted oceanic crust and megathrust fault sealing. *Nature*, 457:76–78. doi:10.1038/nature07650

Bell, R.E., Sutherland, R., Barker, D., Henrys, S., Bannister, S., Wallace, L.M., and Beavan, J., 2010. Seismic reflection character of the Hikurangi interface, New Zealand, in the region of repeated Gisborne slow slip events. *Geophys. J. Int.*, 180:34–48. doi:10.1111/j.1365-246X.2009.04401.x

Douglas, A., Beavan, J., Wallace, L., and Townend, J., 2005. Slow slip on the northern Hikurangi subduction interface, New Zealand. *Geophys. Res. Lett.*, 32: L16305. doi:10.1029/2005GL023607

Dragert, H., Wang, K., and James, T.S., 2001. A silent slip event on the deeper Cascadia subduction interface. *Science*, 292:1525–1528. doi:10.1126/science.1060152

Ide, S., Beroza, G.C., Shelly, D.R., and Uchide, T., 2007. A scaling law for slow earthquakes. *Nature*, 447:76–79. doi:10.1038/nature05780

Kodaira, S., Iidaka, T., Kato, A., Park, J.-O., Iwasaki, T., and Kaneda, Y., 2004. High pore fluid pressure may cause silent slip in the Nankai Trough. *Science*, 304:1295–1298. doi:10.1126/science.1096535

Liu, Y., and Rice, J.R., 2005. Aseismic slip transients emerge spontaneously in three-dimensional rate and state modeling of subduction earthquake sequences. *J. Geophys. Res.*, 110:B08307. doi:10.1029/2004JB003424

Liu, Y., and Rice, R., 2007. Spontaneous and triggered aseismic deformation transients in a subduction fault model. *J. Geophys. Res.*, 112:B09404. doi:10.1029/2007JB004930

Obara, K., Hirose, H., Yamamizu, F., and Kasahara, K., 2004. Episodic slow slip events accompanied by non-volcanic tremors in southwest Japan subduction zone. *Geophys. Res. Lett.*, 31:L23602. doi:10.1029/2004GL020848

Outerbridge, K.C., Dixon, T.H., Schwartz, S.Y., Walter, J.I., Protti, M., Gonzalez, V., Biggs, J., Thorwart, M., and Rabbel, W., 2010. A tremor and slip event on the Cocos-Caribbean subduction zone as measured by a global positioning system (GPS) and seismic network on the Nicoya Peninsula, Costa Rica. *J. Geophys. Res.*, 115:B10408. doi:10.1029/2009JB006845

Ozawa, S., Suito, H., and Tobita, M., 2007. Occurrence of quasi-periodic slow slip off the east coast of the Boso peninsula, Central Japan. *Earth Planets Space*, 59:1241–1245.

Peng, Z., and Gomberg, J., 2010. An integrated perspective of the continuum between earthquakes and slow-slip phenomena. *Nature Geosci.*, 3:599–607. doi:10.1038/ngeo940

Sagiya, T., 2004. Interplate coupling in the Kanto district, central Japan, and the Boso Peninsula silent earthquake in May 1996. *Pure Appl. Geophys.*, 161:2327–2342, doi:10.1007/s00024-004-2566-6

Shibazaki, B., and Iio, Y., 2003. On the physical mechanism of silent slip events along the deeper part of the seismogenic zone. *Geophys. Res. Lett.*, 30:GL017047. doi:10.1029/2003GL017047

Song, T.-R.A., Helmberger, D.V., Brudzinski, M.R., Clayton, R.W., Davis, P., Perez-Campos, X., and Singh, S.K., 2009. Subducting slab ultra-slow velocity layer coincident with silent earthquakes in southern Mexico. *Science*, 324:502–506.

Wallace, L.M., and Beavan, J., 2010. Diverse slow slip behavior at the Hikurangi subduction margin, New Zealand. *J. Geophys. Res.*, 115:B12402. doi:10.1029/2010JB007717

## Authors

**Laura M. Wallace** (co-chair and local host), GNS Science, P.O. Box 30-368, Lower Hutt, New Zealand; *now at* University of Texas, Institute for Geophysics, J.J. Pickle Research Campus, 10100 Burnet Road (R2200) Austin, TX 78758-4445, U.S.A., e-mail: lwallace@ig.utexas.edu

**Eli A. Silver** (co-chair), Earth and Planetary Sciences, University of California Santa Cruz, 1156 High Street, Santa Cruz, CA 95064-1077, U.S.A.

**Nathan Bangs** (co-chair), University of Texas, Institute for Geophysics, J.J. Pickle Research Campus, 10100 Burnet Road (R2200) Austin, TX 78758-4445, U.S.A.

**Rebecca Bell,** Department of Earth Science and Engineering, Imperial College London, South Kensington Campus London SW7 2AZ, London, U.K.

**Joshu Mountjoy,** National Institute of Water and Atmospheric Research (NIWA), 301 Evans Bay Parade, Hataitai, Wellington 6021, Private Bag 14901, Wellington, New Zealand

**Stuart Henrys,** GNS Science, 1 Fairway Drive, Avalon 5010, P.O. Box 30-368, Lower Hutt 5040, Lower Hutt, New Zealand

**Ingo Pecher,** GNS Science, 1 Fairway Drive, Avalon 5010, P.O. Box 30-368, Lower Hutt 5040, Lower Hutt, New Zealand; *now at* University of Auckland, School of Environment, Private Bag 92019, Auckland, New Zealand

## Related Web Links

http://www.iodp.org/doc_download/3480-iodpreportfinal

http: drill.gns.cri.nz/DrillNZ/Latest-News/Archive/DrillNZ-Archive-2011/2011-DrillNZ-Events/Slow-Slip-IODP-Workshop

# The Towuti Drilling Project: Paleoenvironments, Biological Evolution, and Geomicrobiology of a Tropical Pacific Lake

by James Russell and Satria Bijaksana

## Introduction

Three zones of deep atmospheric convection energize the Earth's heat and moisture budgets: tropical Africa, Amazonia, and the Indo-Pacific. The Indo-Pacific region is by far the largest and most important of these (Fig. 1), and it exerts an enormous influence on global climate (Pierrehumbert, 1999). Lacustrine scientific drilling through International Continental Scientific Drillig Program (ICDP) has provided a suite of lake records from the northern and southern tropics (Fig. 1), providing key insights into the mechanisms responsible for long-term variations in the monsoons at the modern-day limits of the seasonal migration of the Intertropical Convergence Zone (ITCZ) in South America, Africa, and Asia (Fritz et al., 2007; Hodell et al., 2007; Scholz et al., 2007). However, ICDP has not yet provided a long climate record from an equatorial zone of deep convection, nor from the tropical western Pacific.

Lake Towuti (2.75° S, 121.5° E, 561 $km^2$, 203 m maximum depth) is the largest tectonic lake in Indonesia, and it is part of the Malili lake system, a set of five, ancient (1–2 Myr) tectonic lakes in central Sulawesi (Fig. 2). Lake Towuti's location in central Indonesia provides a unique opportunity to reconstruct paleoclimates and paleoenvironments in the crucially important yet understudied Indo-Pacific Warm Pool (IPWP). Moreover, the Malili lakes have extremely high rates of floral and faunal endemism and are surrounded by one of the most diverse tropical forests on Earth. Drilling in Lake Towuti will document the environmental and climatic context that shaped the evolution of Malili's unique lacustrine and terrestrial ecosystems and their resilience to long-term environmental change. Lake Towuti is surrounded by ultrabasic (ophiolitic) rocks that release high levels of iron, nickel, chromium, and other metals, catalyzing biogeochemical reactions within a diverse, exotic microbial community. Drill core will provide insight into long-term changes in the geomicrobiology of this system, including coupled changes in climate, microbial biogeochemical cycling, carbon storage, metal deposition, and microbial processes operating at depth in the sediment.

## Site Description

The clear potential of Lake Towuti to the scientific drilling community was recognized already in 1995 by the Past Global Changes (PAGES) workshop on Continental Drilling for Paleoclimate Records, which designated the Malili Lakes a high-priority drilling target (Colman, 1996). However, only now do we have the requisite site survey information to begin planning a drilling program on Lake Towuti. Over the course of four field seasons we have acquired ~150 meters of sediment core and over a thousand kilometers of CHIRP and airgun seismic reflection data, such that we can now propose the first lake drilling project in Southeast Asia.

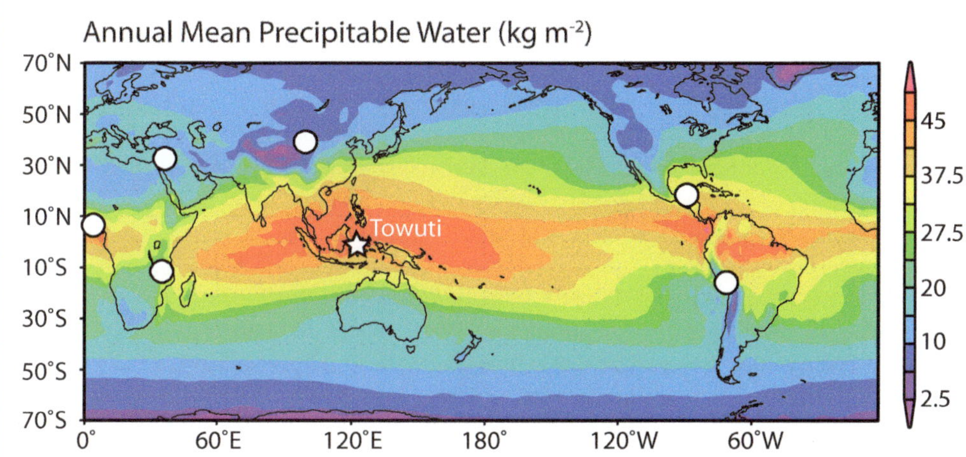


Figure 1. Annual mean precipitable water content of the atmosphere from NCEP Reanalysis (Kalnay et al., 1996), highlighting the importance of the Indo-Pacific to the global atmospheric water cycle. The star indicates the location of Lake Towuti. Circles indicate locations of previous ICDP lake drilling projects in the tropics and subtropics.

Existing sediment cores (10–15 m long) from Lake Towuti reveal large variations in sedimentation at millennial and longer time scales during the past 60 kyr. Variations in sediment metal content are associated with large variations in paleohydrology. These are measured by isotopic indicators, changes in vege-

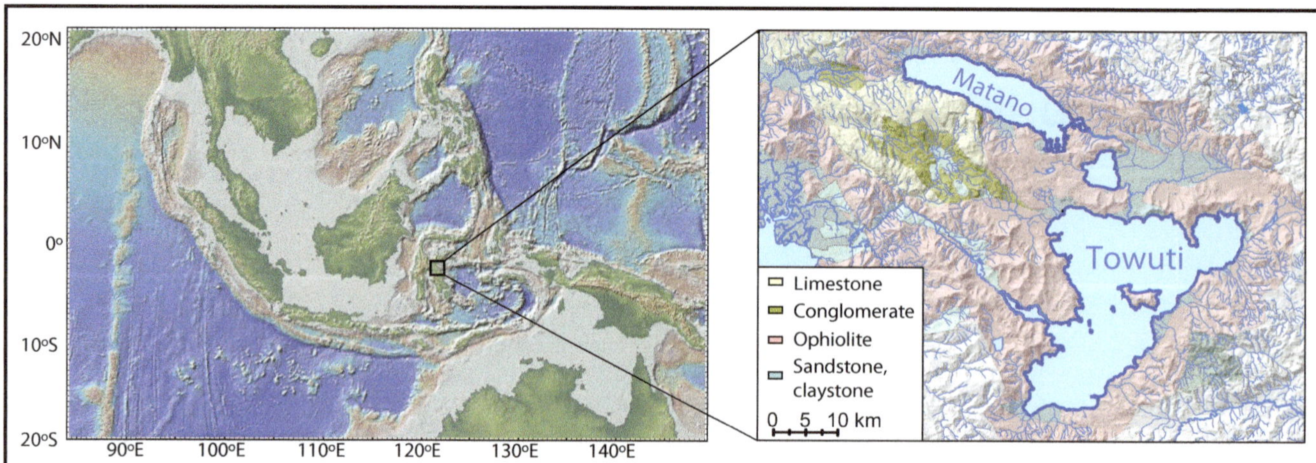


Figure 2. Topographic map of Indonesia and surrounding countries showing the location of the Malili Lakes region within Sulawesi. At right is a simplified geologic map of the Malili Lakes, courtesy of PT Vale-Inco Tbk.

tation observed through fossil pollen, as well as substantial variations in sedimentary organic carbon content and other indicators of lacustrine carbon cycling. All of these indicate coherent linkages between climate, ecosystems, and lacustrine geomicrobiology.

Seismic reflection data document up to 150 meters of stratified lacustrine fill in the basin (Fig. 3), likely spanning the last ~700,000 years. With this abundance of lacustrine sediment, we can test the response of IPWP hydrology across glacial-interglacial variations since the mid-Pleistocene transition. This lacustrine unit is underlain by fluviolacustrine material of a similar thickness that records the history of initial infilling and structural evolution of the Towuti basin. There is no evidence for widespread gas that could pose a drilling hazard, nor extensive tectonic deformation of these sediments; these conditions allow us to optimize our coring locations for critical science objectives. This dynamic setting provides the context in which our science groups developed their hypotheses and questions.

Lake Towuti's climate is influenced by the Australasian monsoons, changes in the Walker Circulation driven from both the Indian and Pacific Oceans, and the location and strength of convection within the ITCZ. Lake level monitoring shows a strong influence of the El Niño-Southern Oscillation (ENSO), and while sediments from Towuti cannot resolve individual El Niño events, the long-term climate variability of Sulawesi is governed by the interplay between meridional monsoon variations and zonally-oriented anomalies that originate in the Walker circulation. Lake Towuti is hosted within a fault-bounded extensional basin along the Matano-Palu Fault, a strike-slip fault system formed in association with the collision between Australia and Eurasia. Bedrock is dominated by the large East Sulawesi Ophiolite (Kadarusman et al., 2004). Lateritic nickel ores that developed on these ultrabasic rocks have attracted the mining industry, with whom we will partner in our proposed work.

## Workshop Structure and Findings

To develop the Towuti Coring Project, sixty-two people from nine countries participated in a workshop in Bandung, Indonesia during 26–29 March 2012 to define scientific priorities and logistical demands for deep scientific drilling in Lake Towuti. Presentations on the first day focused on IPWP climate dynamics and history, the structural geological evolution of Sulawesi, and the limnology, geomicrobiology, and evolutionary biology of the lake. Participants spent the following three days in breakout groups to focus on developing the scientific hypotheses and strategies, with breakout groups divided into five themes: paleoclimate, geochronology, geomicrobiology, evolutionary biology, and structural geology. Discussions during day 4 focused on drilling project logistics, planning, and timelines.

The paleoclimate group reviewed extensive data sets from mainland Asia documenting the long-term history of the Asian monsoon, highlighted by loess and speleothem records spanning hundreds of millennia (Wang et al., 2008). These records show a dominant response of the Asian monsoon to orbital precession, yet existing sediment cores from Lake Towuti highlight substantial changes in regional hydrology associated with the Last Glacial Maximum. Drilling long, 700,000-year sequences would allow us to test whether hydrological changes in central Indonesia are driven by tropical insolation or glacial forcing across multiple glacial-interglacial cycles, including changes in the Walker Circulation forced by exposure of the Sunda Shelf and greenhouse gas variations (DiNezio et al., 2011). We could also investigate the amplitude of land surface temperature changes across multiple glacial cycles, its relation to temperature, and the extent to which high-latitude events such as Heinrich and Dansgaard/Oeschger variability influence the hydrology of the maritime continent.

The Malili lakes are ultraoligotrophic systems that contain a unique and fascinating microbial community (Crowe

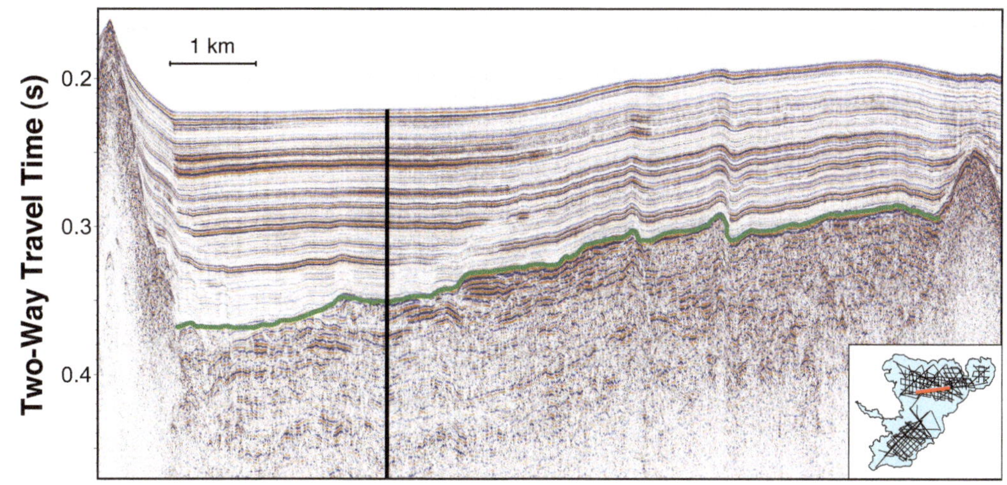


Figure 3. Example of seismic section from northern Lake Towuti (Wattrus et al, unpublished). Inset at bottom right indicates the location of this section (thick red line) within our seismic reflection grid (thin black lines). The boundary between stratified lacustrine sediment (upper unit) and fluviolacustrine sediment (lower unit) is marked by a green line. One hypothetical drilling target is illustrated by the black line.

et al., 2008) that dominates the lakes' biology. At present, the high metal and low sulfur content of lake water and sediments fuel a geomicrobiological community in these lakes driven largely by metal cycling. Existing sediment cores suggest large changes in metal deposition through time, with possible impacts on microbial composition, activity, and the lacustrine carbon cycle. Through drilling, we will investigate long-term rates of microbial evolution and ecosystem change in a ferruginous basin. We will further investigate how external forcing—including changes in chemical and physical weathering of the landscape, terrestrial geomorphology, and climate-driven changes in lake mixing—affects Towuti's geomicro-biological community and sediment geochemistry. Indeed, this work will provide important constraints on the behavior of sediment climate proxies in metal-rich, organic carbon-poor sediments

Lake Towuti is the largest of the Malili Lakes complex, and is situated at the downstream end, with flow from Lake Matano through Lake Mahalona into Towuti. Sarasin and Sarasin (1895) were among the first Western scientists to explore the Malili Lakes, and they noted the lakes' unique fauna. Subsequent work during the twentieth century confirmed the presence of diverse, highly endemic flora and fauna in the Malili Lakes (Rintelen et al., 2011), and data have shown that each of the Malili Lakes contain a locally endemic fauna despite the hydrologic connections between the basin. This could hint at long-term changes in the hydrologic connectivity among these lakes related to movement along the region's numerous and active faults. Drilling in Lake Towuti will allow us to investigate whether tectonic evolution of the Malili Lakes' drainages influenced the timing, rate, and mode of speciation in these basins. Drilling will also elucidate the sensitivity and resilience of central Sulawesi's rainforests to climate, fire, and environmental variability.

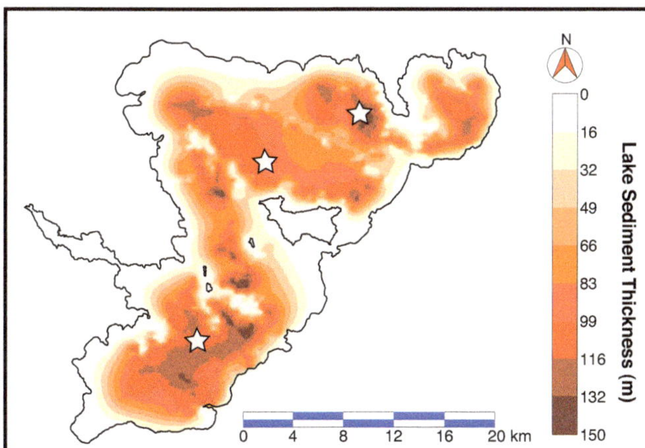


Figure 4. Coring targets (indicated by stars) overlain on a map of the thickness of stratified lacustrine sediment (Wattrus and Russell, unpublished).

Investigations of these areas will require a robust geochronology. The high iron and magnetic mineral content, coupled with little reductive diagenesis, makes Towuti an ideal site for dating sediment cores through paleomagnetic variations. Existing paleointensity data show robust correlation to reference sections from the Sulu Sea (Schneider and Mello, 1996), placing the Towuti chronology on the marine oxygen isotope time scale. Although Towuti is far from major volcanoes, ashes likely derived from the North Sulawesi volcanic field are present in sediment cores and appear common in seismic reflection data. These could provide additional constraint and may also provide a history of large-magnitude eruptions in a poorly studied region.

Participants discussed possible drilling targets to address these issues and whether to drill both the lacustrine and fluviolacustrine sections in Towuti. Participants agreed on the importance of recovering a complete lacustrine and fluviolacustrine section to improve our understanding of the biological and environmental evolution of the Towuti basin. Through these discussions, three primary drilling targets (Fig. 4) were identified:

- a primary site in central Towuti where existing sediment cores and seismic data indicate the most quiescent depositional setting,
- a site situated in northern Towuti to record variations in sediment input from the Mahalona River, to reconstruct the hydrological connectivity of the Malili system, and
- a site in southern Towuti, which is completely dominated by ultramafic rocks and where some of the oldest lacustrine sediments are found.

We are now pursuing pre-drilling scientific activities including improved sedimentological surveys of the Towuti basin, limnological and climate monitoring, and educational and outreach activities within the local community.

## References


Colman, S.M., 1996. Continental drilling for paleoclimatic records: Recommendations from an international workshop. *PAGES Workshop Report Series,* 96–4, 104 pp.

Crowe, S.A., O'Neill, A.H., Katsev, S., Hehanussa, P.E., Haffner, G.D., Sundby, B., Mucci, A., and Fowle, D.A., 2008. The biogeochemistry of tropical lakes: A case study from Lake Matano, Indonesia. *Limnol. Oceanogr.*, 53:319–331, doi:10.4319/lo.2008.53.1.0319

DiNezio, P.N., Clement, A.C., Vecchi, G.A., Soden, B.J., Broccoli, A.J., Otto-Bliesner, B., and Braconnot, P., 2011. The response of the Walker Circulation to LGM forcing: Implications for detection in proxies. *Paleoceanography,* 26:PA3217.

Fritz, S.C., Baker, P.A., Seltzer, G.O., Ballantyne, A., Tapia, P.M., Cheng, H., and Edwards, R. L., 2007. Quaternary glaciation and hydrologic variation in the South American tropics as reconstructed from the Lake Titicaca drilling project. *Quat. Res.*, 68:410–420, doi:10.1016/j.yqres.2007.07.008

Hodell, D.A., Anselmetti, F.S., Ariztegui, D., Brenner, M., Curtis, J.H., Gilli, A., Grzesik, D., et al., 2007. An 85-ka record of climate change in lowland central America. *Quat. Sci. Rev.*, 27:1152–1165, doi:10.1016/j.quascirev.2008.02.008

Kalnay, E., Kanamitsu, M., Kistler, R., Collins, W., Deaven, D., Gandin, L., Iredell, M., et al., 1996. The NCEP/NCAR 40-Year Reanalysis Project. *Bull. Amer. Meteor. Soc.*, 77:437–471. doi:10.1175/1520-0477(1996)077<0437:TNYRP>2.0.CO;2

Kadarusman, A., Miyashita, S., Maruyama, S., Parkinson, C.D., and Ishikawa, A., 2004. Petrology, geochemistry, and paleogeographic reconstruction of the East Sulawesi Ophiolite, Indonesia. *Tectonostratigraphy,* 392:55–83, doi:10.1016.j.tecto.2004.04.008

Pierrehumbert, R.T., 1999. Subtropical water vapor as a mediator of rapid global climate changes. *In* Clark, P.U., Webb, R.S., and Keigwin, L.D. (Eds.), *Mechanisms of Global Climate Change at Millennial Time Scales. Geophysical Monograph Series,* 112: Washington, DC (American Geophysical Union), 339–361, doi:10.1029/GM112p0339

Rintelen, T.v., Rintelen, K.v., Glaubrecht, M., Schubart, C., and Herder, F., 2011. Aquatic biodiversity hotspots in Wallacea – the species flocks in the ancient lakes of Sulawesi, Indonesia. *In* Gower, D.J., Johnson, K.G., Richardson, J.E., Rosen, B.R., Rüber, L., and Williams, S.T. (Eds.), *Biotic Evolution and Environmental Change in Southeast Asia:* Cambridge (Cambridge University Press).

Sarasin, P., and Sarasin, F., 1895. Reisebericht aus Celebes. IV. Reise durch Central-Celebes vom Golf von Boni nach dem Golf von Tomieni. *Verhandlungen der Gesellschaft für Erdkunde,* 30:312–352.

Schneider, D.A., and Mello, G.A., 1996. A high-resolution marine sedimentary record of geomagnetic intensity during the Brunhes Chron. *Earth Planet. Sci. Lett.*, 144:297–314, doi:10.1016/0012-821X(96)00164-1

Scholz, C., Johnson, T.C., Cohen, A.S., King, J.W., Peck, J.A., Overpeck, J.T., Talbot, M.R., et al., 2007. East African megadroughts between 135 and 75 thousand years ago and bearings on early-modern human origins. *Proc. Natl. Acad. Sci. U.S.A.*, 42:16416–16421, doi:10.1073/pnas.0703874104

Wang, Y., Cheng, H., Edwards, R.L., Kong, X., Shao, X., Chen, S., Wu, J., Jiang, X., Wang, X., and An, Z., 2008. Millennial- and orbital-scale changes in the East Asian monsoon over the past 224,000 years. *Nature,* 451:1909–1093, doi:10.1038/nature06692


## Authors


**James Russell**, Department of Geological Sciences, Brown University, Box 1846, Providence, RI 02912, U.S.A., e-mail: James_Russell@Brown.edu

**Satria Bijaksana**, Global Geophysics Research Group, Faculty of Mining and Petroleum Engineering, Institut Teknologi Bandung, Jalan Ganesa 10, Bandung, 40132, Indonesia, e-mail: satria@fi.itb.ac.id

# Scientific Drilling in the Basin of Mexico to Evaluate Climate History, Hydrological Resources, and Seismic and Volcanic Hazards

by Erik T. Brown, Josef P. Werne, Socorro Lozano-García, Margarita Caballero, Beatriz Ortega-Guerrero, Enrique Cabral-Cano, Blas L. Valero-Garces, Antje Schwalb, and Alejandra Arciniega-Ceballos

## Introduction

A group of fifty scientists and students from eight countries met on the campus of UNAM (Universidad Nacional Autónoma de México) in Mexico City on 4–8 March 2012 to plan for a program of continental drilling that will address a wide range of ongoing issues and hazards facing the Mexico City region. The initial impetus for the workshop was investigation of a long and continuous climate and ecological record preserved in the lake sediments underlying the city. Workshop attendees included participants with scientific interests in sediment core and borehole instrumentation in this region, including experts in volcanology, seismology, hydrology, geodesy and the associated geological hazards. The workshop was supported by the International Continental Scientific Drilling Program (ICDP) and UNAM.

The Basin of Mexico (Fig. 1; 9600 km$^2$, 2240 m asl) is a hydrologically closed basin in the central-eastern part of the Mexican Volcanic Belt. Active volcanism around this basin dates from the Oligocene to the present. The lake system of the basin may have developed an association with the basin's closure by emplacement of the Chichinautzin volcanic field in the southern sector of the basin after 780 ka (Mooser et al., 1974; Urrutia-Fucugauchi and Martin del Pozzo, 1993). In any case, continuous deposition has allowed accumulation of thick sequences (>400 m) of lacustrine sediment interbedded with sporadic volcanic horizons.

The hydraulic regime of the former lakes of the Basin of Mexico has been modified since pre-Hispanic times for flood control and agricultural practices. Multiple efforts to drain the basin continued during colonial times in an effort to control periodic flooding and to foster urban development. Mexico City is built upon these lakebeds. Lake Chalco is located on the southernmost sector of the Basin of Mexico; due to its proximity to freshwater sources, Chalco has typically been less saline than other lakes within the basin. This lake was reduced to a small marsh during the nineteenth and early twentieth centuries. It is among the largest blocks of undeveloped agricultural land within the Mexico City megalopolis and is thus a prime target for scientific drilling.

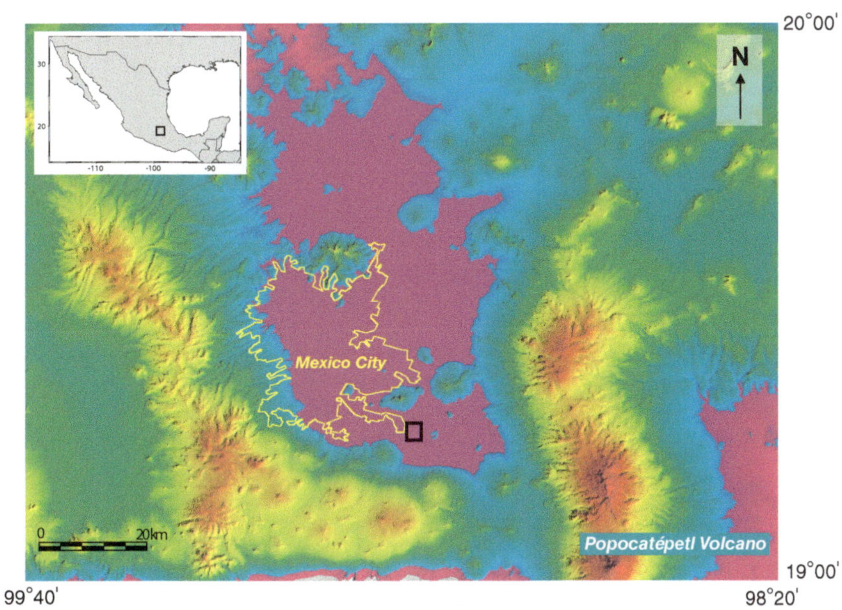


Figure 1. Location map of the Basin of Mexico, showing the location of previous coring at paleolake Chalco.

Sediments from the Chalco Basin have the potential to yield a unique and remarkable record of climate history directly relevant to millions of people. An understanding of past variability in the regional hydrological regime, including variations in monsoonal precipitation, provides a basis for evaluation of ongoing climate change. In addition, the sediments can provide histories of volcanic activity, and their physical properties are relevant to models of seismic wave propagation in the basin as well as to understanding its intense subsidence and regional groundwater resources. Drilling for paleoenvironmental studies can be leveraged and complemented by continuous subsurface geophysical monitoring through instrumented boreholes that will provide additional insights into regional geologic hazards and better groundwater management.

## Climate and Ecological History

A significant scientific objective of the Lake Chalco drilling will be obtaining a continuous, high-resolution record of past climates in the continental neotropics over the past ~500 kyr. Previous studies at this location (Fig. 2) have exploited geochemical, geological, and ecological proxy records (Bradbury, 1989; Lozano-Garcia et al., 1993; Lozano-Garcia and Ortega-Guerrero, 1994; Urrutia-Fucugauchi et al., 1994; Caballero-Miranda, 1997; Lozano-Garcia and Xelhuantzi-Lopez, 1997; Caballero and Ortega Guerrero, 1998). This body of research demonstrates that Chalco sediments have the potential to provide unique knowledge of interannual through orbital-scale variations in the North American monsoon and the hydrological balance of the neotropics. Indeed, the cores we expect to retrieve from Chalco would be the longest continuous climate record from tropical North America, and therefore they could become the "type section" to which other records are compared. Southern North America is projected to become more arid in the coming decades and centuries in response to anthropogenically-driven climate change (Seager et al., 2007). In this already dry region, the increasing population pressures water supplies, so more intense drought may pose major societal challenges. Knowledge of past climate fluctuations will add to our understanding of ongoing climate change.

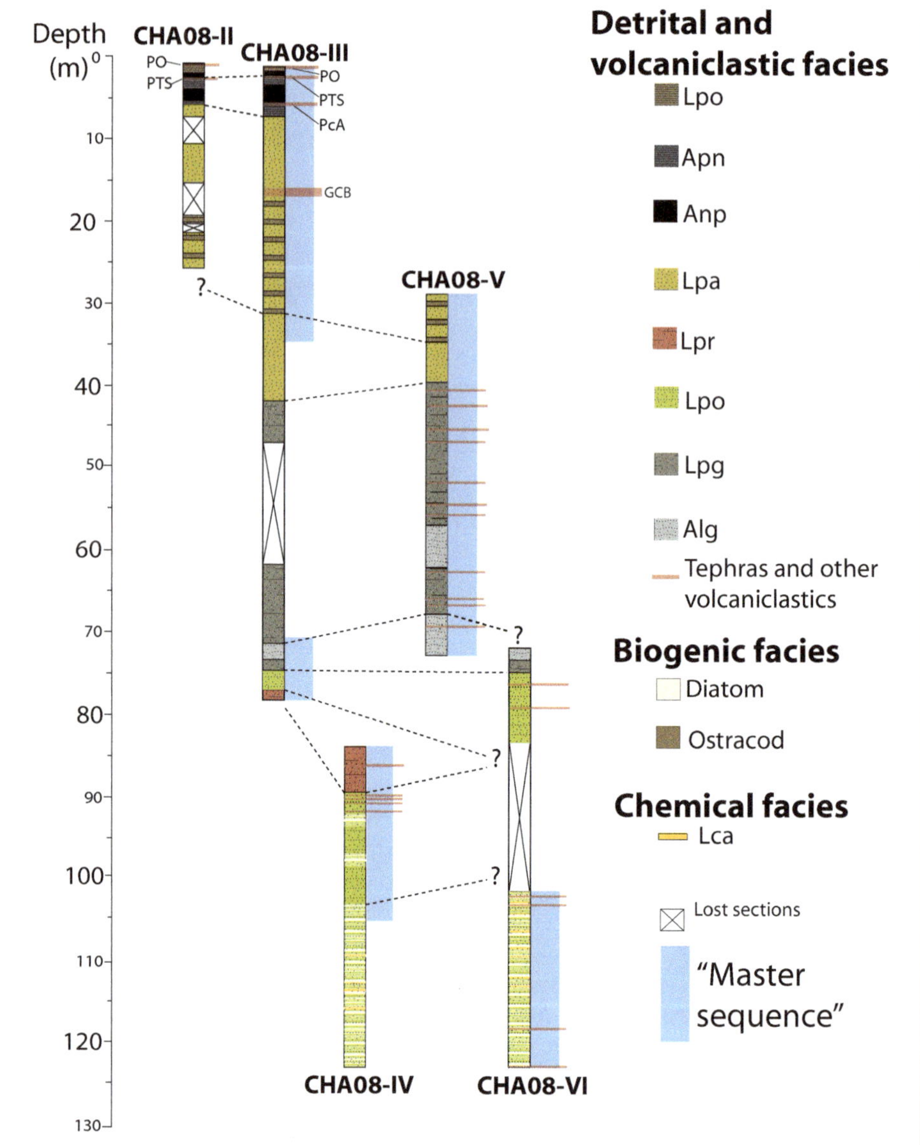


Figure 2. Stratigraphy of core recovered from Chalco in 2008 by UNAM researchers supported by UNAM funding. The base of this section (122 m) has a preliminary age assignment of ~220 ka, suggesting that the full 350–400-m sequence could represent 800 ky of sediment accumulation. Modified from Herrera-Hernandez (2011).

## Volcanic History and Hazard

Volcanic activity has occurred throughout the Mexican Volcanic Belt since the Oligocene. For example, Popocatépetl, located at the southeastern end of the Basin of Mexico (35 km from Chalco), is regarded as one of Mexico's most dangerous volcanoes due to the massive eruptions (e.g., 23 ka and 14 ka) preserved in the geological record (Siebe et al., 2004a, 2004b, 2005; Siebe and Macías, 2006; Arana-Salinas et al., 2010). Popocatépetl resumed activity in 1994 after some sixty-five years of dormancy, and it presents a clear hazard to the region.

The drilling program serves needs in two areas.

1. Tephra layers preserved in sediments of the Chalco Basin will provide a unique record of eruptive histories. Precise dating of large and small eruptive events will help to refine the eruptive history from adjacent large stratovolcanoes and will aid in volcanic hazard prediction.

2. Installation of seismic motion sensors at depth will enable monitoring of seismic activity related to Popocatépetl's magma chamber and to aid in evaluating potential future eruptions.

## *Hydrological Resources and Subsidence*

Mexico City utilizes groundwater from its underlying lacustrine aquitard and aquifer. The fine-grained, organic-rich Quaternary lacustrine sequences (Ortega-Guerrero et al., 1997) are highly susceptible to non-recoverable consolidation. Subsidence due to groundwater extraction in the Basin of Mexico was first recognized in the 1890s. In the subsequent 120 years this led to over ten meters of accumulated subsidence in downtown Mexico City, with accompanying damage to housing and urban infrastructure (Ovando-Shelley et al., 2007). During the 1950s subsidence was so widespread in the historic downtown area that extraction wells were relocated to outlying areas, including Chalco (Joint Academies Committee on the Mexico City Water Supply, 1995). In the Chalco region this led to the significant subsidence mentioned above, with current rates of 150–200 mm $yr^1$. Continued withdrawal of groundwater and subsidence has stressed the upper lacustrine sequence, leading to surface fracturing (fracture lengths of ~500 m, with openings up to one meter and vertical displacement as much as two meters), most prevalent at the interface of lacustrine deposits with adjacent volcanic structures. A deep aquifer is also present within the Basin of Mexico; while it remains essentially unexploited, there is ongoing discussion of its potential to meet future needs of the growing city.

The Lake Chalco drilling project will serve multiple needs in this area.

1. physical characterization of sediments associated with aquifers and aquitards to improve understanding and modeling of subsidence, particularly including characterization of the deep (relatively unexploited) aquifer that has not been well-studied;

2. creation of master *in situ* and synthetic logs that can be used for stratigraphic correlation among boreholes within the basin (most of them for groundwater exploitation) and thus allow basin-wide, stratigraphic correlation; and

3. installation of permanent geodetic, hydrologic, and geotechnical monitoring and sampling instruments to characterize changes to groundwater in response to changing well water withdrawal.

## Seismic Risks

Seismic hazards are omnipresent in Mexico City; the 1985 M 8.0 earthquake caused well over 10,000 casualties and considerable damage to city's buildings and infrastructure. Continuing activity (most recently, an M 7.4 event on 20 March 2012) poses an ongoing threat to the city. Recent tectonic research has focused on episodic tremor and slip (ETS) events in the Guerreo and Oaxaca segment of the Mid-America trench (Brudzinksi et al., 2007). The subhorizontal geometry of the subducting slab underneath southern and central Mexico effectively widens the sesimogenic area and extends inland the occurrence of ETS events well into the Basin of Mexico.

This project serves two needs in this area: 1) Physical characterization of recovered core material will provide information needed to improve and refine geotechnical earthquake engineering models. 2) Installation of geodetic and seismic sensors—including collocated borehole strain meters, pore pressure sensors, deep seismic and tilt meters, and a surface GPS receiver—in deep boreholes will provide previously unavailable monitoring of micro-seismic activity isolated from surface noise.

## Workshop Outcomes

We planned a scientific program that addresses the needs of studies of seismic hazards, volcanoes, hydrology, and paleoclimate/ecology. Prior geophysical studies provide a basis for a broad understanding of the geometry and depth of the basin, but they may be insufficient for selection of drill sites for core recovery. To address this, we are planning and coordinating efforts for magnetic, gravity, and passive seismic surveys to be undertaken in June and July 2012.

The plan is to undertake a two-phase drilling program (proposal to be submitted to ICDP in January 2013, so fieldwork could be as soon as 2014). Phase 1 will drill and log four 500-m wells suitable for use as instrumented boreholes for long-term monitoring of groundwater as well as seismic and volcanic activity. Downhole logs from Phase 1 (in combination with geophysical surveys) can be used to develop strategies for site selection for Phase 2 drilling. The second phase will recover a continuous sequence of core for the physical, chemical, biological, and geological characterization of sediments discussed above.

## References


Arana-Salinas, L., Siebe, C., and Macías, J.L., 2010. Dynamics of the ca. 4965 yr 14C BP "Ochre Pumice" Plinian eruption of Popocatépetl volcano, México. *J. Volcanol. Geoth. Res.*, 192:212–231, doi:10.1016/j.jvolgeores.2010.02.022

Bradbury, J.P., 1989. Late Quaternary lacustrine paleoenvironments in the Cuenca de Mexico. *Quat. Sci. Rev.*, 8:75–100.

Brudzinski, M., Cabral-Cano, E., Correa-Mora, F., DeMets, C., and Márquez-Azúa, B., 2007. Slow slip transients along the Oaxaca subduction segment from 1993 to 2007. *Geophys. J. Int.*, 171:523–538. doi: 10.1111/j.1365-246X.2007.03542

Caballero, M., and Ortega Guerrero, B., 1998. Lake levels since about 40,000 years ago at Lake Chalco, near Mexico City. *Quat. Res.*, 50:69–79.

Caballero-Miranda, M., 1997. Reconstrucción paleolimnológica del Lago de Chalco durante el último máximo glaciar – El registro de diatomeas entre 34,000 y 15,000 años A.P. Rev. Mex.. *Cienc. Geol.*, 14:91–100.

Herrera-Hernández D., 2011. Estratigrafía y análisis de facies de los sedimentos lacustres del Cuaternario tardío de la cuenca de Chalco, México. MSc Thesis, Posgrado en Ciencias de la Tierra, Instituto de Geofísica, Universidad Nacional Autónoma de México. 122 pp.

Joint Academies Committee on the Mexico City Water Supply, 1995. Mexico City's Water Supply: *Improving the Outlook for Sustainability*: Washington, DC (National Academy Press), 230 pp.

Lozano-Garcia, M.S., and Ortega-Guerrero, B., 1994. Palynological and magnetic susceptibility records of Lake Chalco, central Mexico. Palaeogeogr. *Palaeoclim. Palaeoecol.*, 109:177–191.

Lozano-Garcia, M.S., and Xelhuantzi-Lopez, M.S., 1997. Some problems in the Late Quaternary pollen records of central Mexico basins of Mexico and Zacapu. *Quat. Int.*, 43:117–123.

Lozano-Garcia, M.S., Ortega-Guerrero, B., Caballero-Miranda, M., and Urrutia-Fucugauchi, J., 1993. Late Pleistocene and Holocene paleoenvironments of Chalco Lake, Central Mexico. *Quat. Res.*, 40:332–342.

Mooser, F., Nairn, A.E.M., and Negendank, J.F.W., 1974. Palaeomagnetic investigations of the tertiary and quaternary igneous rocks: VIII a palaeomagnetic and petrologic study of volcanics of the valley of Mexico. *Geol. Rundsch.*, 63:451–483, doi:10.1007/BF01820824

Ortega-Guerrero, A., Cherry, J., Aravena, R., 1997. Origin of pore water and salinity in the lacustrine aquitard overlying the regional aquifer of Mexico City. *J. Hydrology,* 197:47–69, doi:10.1016/S0022-1694(96)03280-5

Ovando-Shelley, E., Ossa, A., and Romo, M.P., 2007. The sinking of Mexico City: Its effects on soil properties and seismic response. Soil Dynam. *Earthquake Eng.*, 27:333–343, doi:10.1016/j.soildyn.2006.08.005

Seager, R., Ting, M., Held, I., Kushnir, Y., Lu, J., Vecchi, G., Huang, H., et al., 2007. Model projections of an imminent transition to a more arid climate in southwestern North America. *Science*, 316:1181–1184.

Siebe, C., and Macías, J.L., 2006. Volcanic hazards in the Mexico City metropolitan area from eruptions at Popocatépetl, Nevado de Toluca, and Jocotitlán stratovolcanoes and monogenetic scoria cones in the Sierra Chichinautzin volcanic field. In Siebe, C., Macías, J.L., Aguirre, G. (Eds.), Neogene–Quaternary Continental Margin Volcanism. A Perspective from Mexico: *Geol. Soc. Am. Spec. Pub.*, 402:253–329, doi:10.1130/2004.VHITMC.SP402

Siebe, C., Arana-Salinas, L., and Abrams, M., 2005. Geology and radiocarbon ages of Tláloc, Tlacotenco, Cuauhtzin, Hijo del Cuauhtzin, Teuhtli, and Ocusacayo monogenetic volcanoes in the central part of the Sierra Chichinautzin, México. *J. Volcanol. Geoth. Res.*, 141:225–243, doi:10.1130/2004.VHITMC.SP402

Siebe, C., Rodríguez-Lara, V., Schaaf, P., and Abrams, M., 2004a. Geochemistry, Sr-Nd isotope composition, and tectonic setting of Holocene Pelado, Guespalapa and Chichinautzin scoria cones, south of Mexico City. *J. Volcanol. Geoth. Res.*, 130:197–226, doi:10.1016/S0377-0273(03)00289-0

Siebe, C., Rodríguez-Lara, V., Schaaf, P., and Abrams, M., 2004b. Radiocarbon ages of Holocene Pelado, Guespalapa, and Chichinautzin scoria cones, south of Mexico City: Implications for archaeology and future hazards. *Bull. Volcanol.*, 66:203–225, doi:10.1007/s00445-003-0304-z

Urrutia-Fucugauchi, J., and Martin del Pozzo, A.L., 1993. Implicaciones de los datos paleomagneticos sobre la edad de la Sierra de Chichinautzin, Cuenca de Mexico. *Geofisica Internacional*, 32:523–533.

Urrutia-Fucugauchi, J., Lozano, S., Ortega-Guerrero, B., Caballero, M., Hansen, R., Bohnel, 1994. Paleomagnetic and paleoenvironmental studies in the southern basin of Mexico, *Geofís. Int.*, 33:421–444.

## Authors

**Erik T. Brown and Josef P. Werne**, Large Lakes Observatory and Department of Geological Sciences, University of Minnesota Duluth, 2205 East 5th Street, Duluth, MN 55812-3024, U.S.A., e-mail: etbrown@d.umn.edu

**Socorro Lozano-García**, Instituto de Geología, Universidad Nacional Autónoma de México, Cd. Universitaria, 04510 México D.F., México

**Margarita Caballero, Beatriz Ortega-Guerrero, Enrique Cabral-Cano, and Alejandra Arcniega-Ceballos,** Instituto de Geofísica, Universidad Nacional Autónoma de México, Cd. Universitaria, 04510 México D.F., México

**Blas L. Valero-Garces**, Instituto Pirenaico de Ecologia Consejo Superior de Investigaciones Cientificas, E-50080 Zaragoza, Spain

**Antje Schwalb**, Institut für Geosysteme und Bioindikation Technische, Universität Braunschweig, D-38106 Braunschweig, Germany

## The International Ocean Discovery Program—Preparing for Launch

The Integrated Ocean Drilling Program (IODP) international Working Group+ (IWG+) has released a New IODP Framework, establishing the business model for scientific ocean drilling to apply from October 2013 (www.iodp.org/doc_download/3485-new-iodp-framework-17-august-2012). Planning for the transition to the new IODP by the IWG+ (www.iodp.org/international-working-group-plus) has been underway since 2010. The New IODP Framework and a new Science Plan (http://www.iodp.org/science-plan-for-2013-2023) will serve as a foundation for the next decade of international collaboration in scientific ocean drilling.

The U.S. National Science Foundation (NSF) has received approval from the National Science Board for operation of the US operated drilling vessel *JOIDES Resolution* (*JR*) under the new program framework for one year starting October 2013. Based on experiences gained the NSF will then prepare and submit a proposal covering another ten years of internationally co-funded *JR* operations within IODP. Meanwhile, Japan's Ministry of Education, Culture, Sports, Science, and Technology (MEXT) will develop long-term funding models for the riser-drilling vessel *Chikyu*, and the European Consortium for Ocean Research Drilling (ECOR) will seek approval from its eighteen members for continued funding of its fiscal contribution to program platforms including its own operations of Mission Specific Platforms (MSPs).

Among the significant changes from the current IODP outlined in the Framework document is the replacement of central program management and platform scheduling by IODP Management International with three Facility Governing Boards (FGBs)–one for each platform–to manage and schedule platform operations. NSF will establish an IODP Support Office to coordinate a single scientific advisory structure (SAS) that will advise all three FGBs and support other central program activities. MEXT and *Chikyu* owner JAMSTEC (Japan Agency for Marine-Earth Science and Technology) will establish a Project Partnership Office to develop funding and facilitate large-scale collaborations, and to support SAS processes for the riser drilling projects. A non-executive IODP Forum consisting of community scientists, and to include representatives from funding agencies, platform operators and others, will be the custodian of the Science Plan and a venue for exchanging views on the scientific progress of the Program provided by the three drilling platforms. The Australia-New Zealand IODP consortium (ANZIC), IODP China, IODP India, IODP Korea, and the most recent IODP member, Brazil, are all preparing for their membership in the new IODP under this new program structure.

MEXT MINISTRY OF EDUCATION, CULTURE, SPORTS, SCIENCE AND TECHNOLOGY-JAPAN

ECORD Managing Agency

Australian and New Zealand IODP Consortium

## Lake Ohrid Drilling Postponed

icdp

Coring operations at Lake Ohrid, Macedonia were planned to be conducted from September to November this year. However, during the mobilization from the United States to Europe the container vessel *MSC* Flaminia caught fire and was evacuated almost in the middle of the Atlantic Ocean. The vessel is currently manned by a salvage crew and the fire is extinguished, but the status of the cargo is still unclear. Due to this delay and potential damage or loss of coring equipment, a start of the drilling operation at Lake Ohrid this year is highly unlikely and has been postponed.

## *CHIKYU*+10 International IODP Workshop

*21–23 April 2013, Tokyo, Japan*

For more information: Chikyu10@iodp.org

IODP INTEGRATED OCEAN DRILLING PROGRAM

D/V *Chikyu*, owned and operated by the Japan Agency for Marine-Earth Science and Technology (JAMSTEC), is one of the three scientific drilling platforms that will continue to operate in the next phase of the International Ocean Discovery Program (IODP). Its deep riser drilling capability offers unique opportunities to address a number of key challenges described in the IODP New Science Plan.

The workshop aims to identify science priorities to be addressed using the *Chikyu* over the next decade. The outcome of the workshop will influence JAMSTEC's decisions as it prepares *Chikyu*'s next Five-Year Plan under the auspices of the Ministry of Education, Culture, Sports, Science, and Technology (MEXT), as well as *Chikyu*'s long-range plan for the next ten years.

An international steering committee of approximately fifteen carefully chosen members will design and organize the workshop facilitated by IODP Management International (IODP-MI). Truly deep targets such as the ongoing NanTroSEIZE require long-term commitment, careful planning and project management. In the present financial outlook, there can likely be a few riser projects of the

NanTroSEIZE-scale in the coming ten years, thus the need for international prioritization based on realistic projections.

The workshop will discuss all the active riser drilling proposals as well as new ideas with technical feasibility and scientific excellence and will map onto decadal roadmaps. Details will be announced on the IODP website.

## UK Joins the International Continental Scientific Drilling Program

The UK has become a full member of the (International Continental Scientific Drilling Program (ICDP) in effect from April 2012. A launch meeting attended by seventy scientists was held at British Geological Survey, Nottingham, in July 2012. The UK is an active member of the (IODP) but has never before participated as a formal member of the ICDP, despite involvement of scientists in a number of programs. It is important that UK scientists are able to participate fully in ICDP to initiate programs and to allow them to collaborate with existing international science programs involving ICDP. The British Geological Survey (BGS) agreed to invest in membership of the ICDP on behalf of the UK's Earth Science Sector to build national capability for the geoscience community. The membership allows UK scientists to now fully participate within ICDP projects via access to key geological sections in order to gain information on processes that cause global change, to understand the controls on resource development, to help monitor and model natural hazards (through borehole monitoring), and to investigate fluid-related biological processes in the sub-surface. UK geoscientists will now be fully engaged in both IODP and ICDP and can help develop synergies between these programs. At the same time the BGS will work with European countries through ECORD-IODP (http://www.ecord.org/) and the European members of ICDP to create a European infrastructure for scientific research drilling.

More information can be found on the ICDP-UK web site: http://www.bgs.ac.uk/icdp.

A small group of UK scientists have been nominated to act as ICDP-UK convenors for the first phase of UK membership. UK geoscientists can contact members from this group for further information: http://www.bgs.ac.uk/icdp/ukconvenors.html.

## First Generation of MeBo-CORKs Extend Capabilities

The seafloor drill rig MeBo, developed at MARUM, University of Bremen, Germany, is capable of drilling 75 mbsf in water depths exceeding 2000 meters (http://www.marum.de/en/Sea_floor_drill_rig_MeBo.html). Recently, the first long-term observatories for MeBo boreholes were developed and deployed on active mud volcanoes in the Nankai Trough subduction zone, Japan. After hole completion with MeBo, the drill pipe is left in the ground as casing, and an instrument of the same geometry is added to "CORK" the hole (i.e. obviate communication between hole and overlying seafloor). The self-contained instrument comprises pressure and temperature transducers, a data logger, and an acoustic modem so that data download from ships of opportunity is feasible. In the lower portion of the instrumented rod, a needle-shaped drop weight connected to hydraulic tubes is released acoustically and penetrates the sediment at the bottom of the hole. The tubes primarily serve to monitor pressure but can also be utilized to extract fluid samples if the simple MeBo-CORK instrument is disconnected at its hot stab and gets replaced by a seafloor unit with an osmo-sampler. The hot stab connection ensures that instruments can be replaced easily by remotely operated underwater vehicle (ROV) and allows holes to be used repeatedly. Given that seafloor drills are envisaged as an affordable mission-specific tool in IODP from 2013 onwards, these first generation MeBo-CORKs offer many opportunities for the scientific community.

Left: MeBo seafloor drill rig; Right: MeBo-CORK seafloor unit after ROV visit. Plastic tube right of pressure housing hosts osmo-sampler.

## CDEX Successfully Completes Japan Trench Fast Drilling Project (J-FAST)

IODP Expeditions 343 and 343T were successfully completed by Center for Deep Earth Exploration (CDEX), with the Expedition Science Party overcoming technical and weather challenges to enable D/V *Chikyu* to set a new record drilling depth in scientific ocean drilling with a total of 7768.5 m measured from the rig floor (7740 m below sea level). Most importantly, the science party was able to obtain core samples from the fault zone and install borehole observatories to monitor conditions at the fault.

The J-FAST Expedition demonstrated IODP's ability to respond quickly to learn more about the devastating Great East Japan Earthquake of 11 March 2011. Planning for J-FAST started soon after the earthquake and after repairs to *Chikyu* from damage caused by the tsunami, the Expedition began at the port of Shimizu on 1 April 2012. Expedition 343T, J-FAST II, was an extension of the project approved by IODP SAS and funded by MEXT/JAMSTEC to give the Science Party time to install the borehole observatories after earlier delays prevented their deployment during the initial Expedition.

The J-FAST expedition represents the first time that frictional heat produced by the fault slip of a great subduction zone earthquake could be measured at the fault zone. The installed borehole observatories will enable scientist to monitor the temperature of the fault zone and study the physics of the earthquake at the fault.

## DEBI RCN Meeting at Bremen

On 7–9 June 2012, sixty researchers from five countries gathered at MARUM, University of Bremen, Germany (www.marum.de), for this year's Dark Energy Biosphere Institute (DEBI) Research Coordination Network (RCN) Meeting. Lectures and training workshops including IODP cores focused on ocean crust processes and consequences for subseafloor life. The science conference part of the Bremen meeting provided a forum for presenting and discussing the most recent results in ocean crust processes and identifying the most pressing challenges that lie ahead. Specific focus was on (1) crustal heterogeneity and fluid flow, (2) ocean-crust interactions, and (3) role of microbes in rock alteration.

The 3-day meeting had equal time for a science conference and a training workshop. Both parts were focused on recent developments in understanding ocean crust formation and evolution, ocean-crust exchange, and detection of sub-basement life and microbe-rock interactions. The conference portion provided a forum to present research activities and findings to a broadly trained but scientifically focused audience. The training workshop served as a means to further educate scientists and students. The aim was to present key research techniques and methods commonly employed, to discuss the advantages and drawbacks for specific applications, to produce consensus recommendations, and to make available detailed lab and field protocols. A visit to the IODP Bremen Core Repository (BCR) (http://www.marum.de/en/IODP_Bremen_Core_

## Southwest Pacific Ocean IODP Workshop

*8–11 October 2012, Sydney, Australia*
For more information, write: Neville.Exon@anu.edu.au

This workshop, to be held at Sydney University, will address global geoscience problems in the Southwest Pacific Ocean by building on existing and new geophysical and geological information including earlier scientific drilling. This news item reports on progress since the announcement of the workshop in the April 2012 edition of *Scientific Drilling*. The workshop themes are largely those in the new science plan:

**1) Climate and Ocean Change: Reading the Past, Informing the Future**
Co-chairs: Tim Naish (NZ) and Jim Kennett (USA)

**2) Biosphere Frontiers: Deep Life, Biodiversity, and Environmental Forcing of Ecosystems**
Co-chairs: Steven D'Hondt (USA) and Ken Takai (Japan)

**3) Earth Connections: Deep Processes and Their Impact on Earth's Surface Environment**
Co-chairs: Richard Arculus (Australia) and Mike Gurnis (USA)

**4) Earth in Motion: Processes and Hazards on Human Time Scales**
Co-chairs: Laura Wallace (USA) and Jim Mori (Japan)

**5) Marine Resources: Opportunities and Responsibilities**
Co-Chairs: Alex Malahoff (New Zealand) and Clinton Foster (Australia)

Funding has been made available by IODP-MI, U.S. Science Support Program (USSSP), ANZIC and Japan Drilling Earth Science Consortium (J-DESC). The event was advertised in *EoS* in May, and on the IODP (www.iodp.org) and ANZIC (www.iodp.org.au) web sites, and it has generated considerable interest. As of early August, we expected more than eighty participants, with a large proportion being from outside Australia and New Zealand.

The workshop will provide a forum to review the latest work in the region, briefly outline possible future IODP expeditions, coordinate activities associated with schedul-ed and proposed geoscience research cruises in the area, and set up working groups to develop proposals for post-2013 IODP expeditions. It will also explore co-investment opportunities between IODP partners and other parties, including industry. There is no doubt that the Workshop will succeed in its main aim of starting to build coherent and well-integrated proposals for the International Ocean Discovery Program starting in late 2013.

Repository.html) was also included to show first-hand the rock samples currently available for study. Workshop participants learned details of shipboard core recovery, sampling, and other techniques pertinent to subseafloor biosphere expeditions.

Further information: http://www.darkenergybiosphere.org/RCN/meetings/2012.html

## ECORD Summer School 2012 in Canada

The joint ECORD/ICDP/IODP-Canada Summer School on cryosphere dynamics took place from 5–21 July in Montréal, Québec. Nineteen participants gathered from universities in Belgium, Canada, Denmark, Greece, Sweden, the Netherlands, and the UK.

The first week consisted of classes and workshops presented by invited lecturers at GEOTOP (University of Québec in Montréal), focused on reconstructing the cryosphere and climate change in the Cenozoic from different perspectives and time scales, using modeling, geomorphology, palaeomagnetism, and terrestrial and marine core records. During the second week, the summer school travelled to the north shore of the St. Lawrence River for a 5-day field excursion. At Lac Walker, an ancient fjord now the deepest known lake in Québec, participants carried out surveys using CTD profiles, took underwater gravity cores and explored the nearshore surface sediments with a remotely operated submarine. Students also conducted sub-bottom acoustic profiling and high-resolution multibeam bathymetry; and they were guided through numerous outcrops of Quaternary sediments. Upon returning from the field, students participated in interactive exercises and classes on microfossil analysis, sedimentary analysis, seismic interpretation, time series analysis, radiogenic and stable isotope methods.

The Canadian summer school offered an excellent opportunity for students of diverse nationalities and academic backgrounds to collaborate and learn about the current understanding of cryosphere dynamics and methods used to reconstruct past climate change.

Section of Holocene deposits along the north coast of the St. Lawrence estuary.

## IODP/ICDP at the International Geological Congress (IGC) in Brisbane

The International Geological Congress in Brisbane (6–10 August 2012) attracted more than 6000 participants and over 200 exhibitors, including a joint IODP-ICDP exhibition booth. It was a successful conference, and many international scientists turned out for IODP- and ICDP-related events.

There were thirty-six major themes covered in the symposia and some excellent plenary talks. The IODP Symposium was part of the Marine Geoscience and Paleoceanography Theme, and attracted thirteen oral presentations and several poster presentations. The audience size varied from forty to seventy, and the talks generated favorable comments. Keynote addresses were by Neville Exon on general aspects of IODP and its

Miyuki Otomo, Neville Exon, Catherine Beasley and Carol Cotterill at the IODP booth at IGC.

future, Dorrik Stow on the Mediterranean Outflow Expedition 339, and Mike Mottl on porewater chemistry in ocean drilling. The audience was particularly large when Jody Webster spoke on aspects of the Great Barrier Reef Expedition 325. A number of other IODP-based talks were given in various other symposia, thus getting the word about IODP out to a wider audience. ICDP held an evening ICDP Primer that was well attended.

The IODP/ICDP booth was staffed by Catherine Beasley of ANZIC, Thomas Wiersberg and Uli Harms of ICDP, and Miyuki Otomo and Jamus Collier of IODP-MI. Many people were deeply interested, and many scientists with ocean drilling experience came by and enlivened the discussions. Carol Cotterill of ECORD deserves particular thanks for her generous help. Many visitors asked about the planning for a Mohole in the next phase of IODP. Visitors helped themselves to all of the available information and scientific material, a sign of widespread interest.

25

# IODP Expedition 327 and *Atlantis* Expedition AT 18-07: Observatories and Experiments on the Eastern Flank of the Juan de Fuca Ridge

by Andrew T. Fisher, Takeshi Tsuji, Katerina Petronotis, C. Geoff Wheat, Keir Becker, Jordan F. Clark, James Cowen, Katrina Edwards, Hans Jannasch, and the IODP Expedition 327 and *Atlantis* Expedition AT18-07 Shipboard Parties

## Abstract

Integrated Ocean Drilling Program (IODP) Expedition 327 (summer 2010) was designed to resolve the nature of fluid-rock interactions in young, upper volcanic crust on the eastern flank of the Juan de Fuca Ridge. Expedition 327 drilled, cased and cored two new basement holes, conducted hydrogeologic experiments, and installed subseafloor borehole observatories (*Circulation Obviation Retrofit Kits*, CORKs). These CORKs were intended to allow borehole conditions to recover to a more natural state after the dissipation of disturbances caused by drilling, casing, and other operations; provide a long-term monitoring and sampling presence for determining fluid pressure, temperature, composition, and microbiology; and facilitate the completion of active experiments to resolve crustal hydrogeologic conditions and processes. Expedition 327 was followed (summer 2011) by R/V *Atlantis* Expedition AT18-07, with the remotely-operated vehicle (ROV) *Jason*, to service these CORKs, collect subseafloor pressure data, recover and deploy autonomous fluid and microbial samplers, collect large volumes of borehole fluids, and initiate a cross-hole hydrogeologic experiment using an electromagnetic flow meter. In addition, *Atlantis* Expedition AT18-07 refurbished an old CORK that could not be replaced during IODP Expedition 327, completing a critical part of the three-dimensional observation network that is currently being used to monitor a large-scale, directional formation response to long-term fluid flow from the crust.

## Introduction and Goals

Fluid flow within volcanic ocean crust influences the thermal and chemical evolution of oceanic lithosphere and lithospheric fluids; subseafloor microbial ecosystems; diagenetic, seismic, and magmatic activity along plate-boundary faults; and the creation of ore deposits on and below the seafloor (Coggon et al., 2010; Huber et al., 2006; Parsons and Sclater, 1977). The global hydrothermal fluid mass flux through the upper oceanic crust rivals the global riverine fluid flux to the ocean, passing the volume of the oceans through the crust once every $10^5$–$10^6$ y (Johnson and Pruis, 2003; Mottl, 2003; Wheat et al., 2003). Most of this flow occurs at relatively low temperatures, far from volcanically active seafloor-spreading centers where new ocean floor is created. This "ridge flank" circulation can be influenced by off-axis volcanic or tectonic activity but is driven mainly by the flow of lithospheric heat from below the crust.

Despite the importance of fluid-rock interactions in the crust, little is known about the magnitude and distribution of critical hydrologic properties; the rates and spatial extent of ridge-flank fluid circulation; the extent to which crustal compartments are well connected or isolated (laterally and with depth); or the links between ridge-flank circulation, crustal alteration, and geomicrobial processes. IODP Expedition 327 (Fisher, et al., 2011b) was part of a long-term experimental program that has included multiple survey, drilling, submersible, and ROV expeditions; observatory and laboratory testing, sampling, and monitoring; and modeling of coupled fluid-thermal-chemical-microbial processes. Expedition 327 built from the technical and scientific achievements and lessons learned during Ocean Drilling Program (ODP) Leg 168 (Davis et al., 1997), which focused on hydrothermal processes within uppermost basement rocks and sediments along an age transect, and IODP Expedition 301 (Fisher et al., 2005a), which penetrated deeper into the crust at the eastern end of the Leg 168 transect (Fig. 1). During both expeditions, CORKs were installed in basement holes to allow borehole conditions to recover to a more natural state after the dissipation of disturbances caused by drilling, casing, and other operations; to provide a long-term monitoring and sampling presence for determining fluid pressure, temperature, composition, and microbiology; and to facilitate the completion of active experiments to resolve crustal hydrogeologic conditions and processes (Fisher et al., 2005b, 2011c; Orcutt et al., 2010b; Wheat et al., in review). During subsequent ROV and submersible expeditions, borehole pressure and thermal data were collected, fluid and microbial samples were recovered, batteries, data loggers, and sampling systems were replaced, and long-term, cross-hole experiments were initiated.

We present operational results from IODP Expedition 327, which sailed on the D/V *JOIDES Resolution*, and R/V *Atlantis* and ROV *Jason* expedition AT18-07 (completed in 2011), focusing on the installation and use of CORKs. Data and samples gathered during the second expedition show that newly-installed CORKs are working as intended, and the ambitious monitoring and cross-hole experimental goals that motivated these experiments are being achieved.

## Experimental Setting and Selected Earlier Work

Many studies have summarized geology, geophysics, basement-fluid composition, and hydrogeology within young seafloor on the eastern flank of the Juan de Fuca Ridge (Fig. 1; Davis et al., 1992; Elderfield et al., 1999; Hutnak et al., 2006; Wheat et al., 2000). Topographic relief associated with the Juan de Fuca Ridge axis and abyssal hill bathymetry on the ridge flank has helped trap turbidites flowing west from the continental margin (Underwood et al., 2005), resulting in the rapid burial of young oceanic basement rocks. Sediment cover is regionally thicker and more continuous towards the continental margin, but there are basement outcrops north and south of the Expedition 327 work area, up to 100 km east of the spreading center.

During ODP Leg 168, a transect of eight sites was drilled across 0.9–3.6 Ma seafloor; sediment, rock, and fluid samples were collected; thermal, geochemical, and hydrogeologic conditions in basement were determined; and CORKs were installed in the uppermost crust (Davis et al., 1997). Two of the Leg 168 observatories were placed in 3.5–3.6 Ma seafloor in Holes 1026B and 1027C, near the eastern end of the drilling transect (Fig. 1). IODP Expedition 301 returned to this area and drilled deeper into basement; sampled additional sediment, basalt, and microbiological materials; replaced the borehole observatory in Hole 1026B; and established two additional CORK observatories at Site U1301 (Fisher et al., 2005a).

Before Leg 168 there was a largely two-dimensional view of the dominant fluid circulation pathways in this area, with recharge thought to occur close to the ridge where basement was exposed (near the western end of the Leg 168 transect), and flowing toward the east. Some results from Leg 168 were consistent with this view, including seafloor heat flow and basement temperatures that increase and basement fluids that are warmer and more altered with progression from west to east (Davis et al., 1999; Elderfield et al., 1999; Stein and Fisher, 2003). However, Leg 168 results and related surveys showed that ridge-parallel fluid transport is important in this region. For example, although basement fluids are

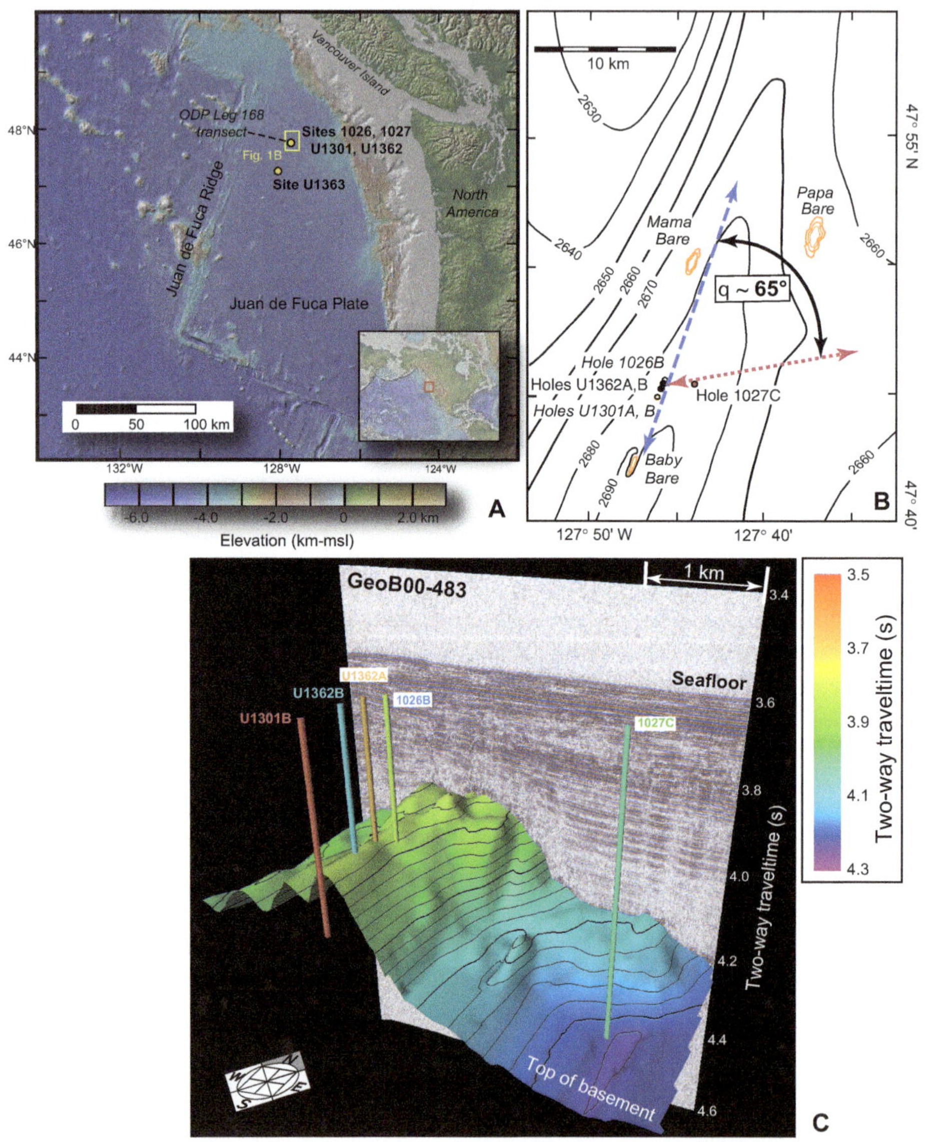


Figure 1. [A] Regional map of field area, eastern flank of the Juan de Fuca Ridge, with inset showing global location. Locations of the ODP Leg 168 transect and sites of IODP Expeditions 301 and 327 and Expedition AT18-07 are shown. [B] Seafloor bathymetric contour map of area around IODP Expedition 301 and 327 sites showing locations of CORK observatories (circles) in Holes 1026B, 1027C, U1301A, U1301B, U1362A, and U1362B, and nearby basement outcrops (gold bathymetric contours) where hydrothermal discharge has been observed (Davis et al., 1992; Wheat et al., 2004b). The blue dashed arrow shows the general structural trend of basement rocks, running sub-parallel to the active spreading center 100 km to the west. The red dotted arrow shows the heading between Holes U1362B and 1027C, which is oriented about 65° clockwise from the primary structural trend in basement. Depth contours in meters. [C] Three-dimensional perspective views of basement relief around CORK sites discussed in this paper. Relief is shown as two-way traveltime.

warmer and older with increasing distance from the ridge along the western end of the drilling transect, fluids are younger with respect to $^{14}C$ at additional Leg 168 sites to the east, although the eastern fluids are also warmer and more altered (Elderfield et al., 1999; Walker et al., 2007). Bathymetric data from near the western end of the Leg 168 transect show exposed basement north and south of the drill sites, which may allow hydrothermal fluids to recharge and discharge (Hutnak et al., 2006). Heat flow data from this area support this interpretation, as do geochemical and thermal data from the eastern end of the Leg 168 transect (Fisher et al., 2003; Wheat et al., 2000). Numerical models of outcrop-to-outcrop hydrothermal circulation between recharging and discharging seamounts show that rapid flow can be sustained (as a "hydrothermal siphon") and basement temperatures matched to observations if basement permeability along the flow path is about $10^{-11}$ m$^2$ (Hutnak et al., 2006).

Hydrologic (packer) experiments conducted in upper basement during Expedition 301 indicate a layered crustal structure with permeability on the order of $10^{-12}$–$10^{-11}$ m$^2$ (Becker and Fisher, 2008). Fisher et al. (2008) conducted additional hydrogeologic analyses using the formation pressure response to the long-term flow of cold bottom seawater into basement at Site U1301 in the thirteen months after drilling, as observed at Site 1027 (2.4 km away). Their data suggest large-scale permeability at the low end of or below values indicated by single-hole packer testing. This result was unexpected because cross-hole tests should be representative of a much larger crustal volume than single-hole tests, yielding higher apparent permeability values. One possible explanation is that basement permeability is azimuthally anisotropic, a hypothesis that is being tested with Expedition 327 CORKs (as described below).

CORKs installed in this area during ODP Leg 168 and IODP Expedition 301 have produced valuable fluid and microbial samples indicating the nature of linked geochemical and microbiological conditions in basement (Cowen et al., 2003; Orcutt et al., 2010a; Smith et al., in review; Wheat et al., 2004a, 2010). Cells collected from the Hole 1026B wellhead included bacteria and archaea whose closest known phylogenetic neighbors comprise nitrate reducers, thermophilic sulfate reducers, and thermophilic fermentative het-erotrophs. Fluid and microbial samples collected at depth in Holes 1026B and Hole U1301A indicate strongly reducing conditions within basement once recovery to a pre-drilling geochemical state was achieved, and with colonization of incubation substrate by the thermophilic anerobe Fermicutes and mesophilic oligotrophs. The highest cell counts within flow-through experiments from Hole U1301A were found on samples of olivine, and culturable organisms from these samples were capable of nitrate reduction and iron oxidation. Expedition 327 CORKs should provide even better samples of endemic microbial communities from the crust, as newer observatories were designed (a) to reduce the extent of contamination (by using inert materials where possible and coating reactive materials with nonreactive coatings) and (b) to simplify and speed the extraction of large quantities of fluid (using larger diameter tubing and custom fittings).

## IODP Expedition 327

The primary goals of Expedition 327 were (1) to drill two new basement holes at Site U1362, core and wireline log one of these holes, run and initiate hydrogeologic tests, and install CORK observatories; (2) to recover an existing CORK from Hole 1027C, deepen the hole and install a new CORK; and (3) to recover and replace an instrument string deployed in the CORK in Hole U1301B (Fisher, et al., 2011b). Secondary objectives included sampling and analyzing sediments and pore fluids near an area of suspected hydrothermal recharge to assess patterns and rates of ridge-flank hydrothermal circulation in basement.

Holes U1362A and U1362B were drilled and cased through the upper ~100 m of basement, without coring, in order to avoid unstable layers and to allow greater depth penetration into the crust (Fig. 2a). Hole U1362A was drilled and cored to 528 meters below seafloor (mbsf), 292 meters sub-basement, 200 m south of Hole 1026B (Fig. 1). Core recovery was ~30%, typical for upper crustal rocks, and eight lithologic units were identified. Numerous samples were collected for post-expedition petrologic, alteration, physical properties, and microbiological analyses. A single wireline logging string helped to determine formation prop-erties and to identify suitable intervals for hydrologic testing and CORK installation. Caliper logs revealed an enlarged borehole over most of the open interval, but there were several near-gauge sections generally corresponding to intervals that produced good core recovery and/or drilled slowly. Short-term packer experiments were completed at depth in Hole U1362A to assess permeability adjacent to the borehole, with hourly injection and recovery periods.

Hole U1362B was drilled to 359 mbsf (117 meters sub-basement), 300 m south of Hole U1362A and 500 m north of Hole U1301B (Fig. 1). A 24-h pumping and tracer injection experiment was initiated prior to CORK installation in this hole (Fisher et al., 2011a). Pressure gauges and fast-sampling reverse osmotic pumps were run in a specially designed "stinger" that extended into the open hole just below the base of casing, which was sealed with a casing running tool. During this experiment, seawater was pumped into the formation at ~7 L s$^{-1}$. Tracers added during pumping included $SF_6$ gas (injected continuously for ~22 h), CsCl and $ErCl_3$ salts (during a brief period 3 h into the experiment) and CsCl and $HoCl_3$ salts (at 19 h), fluorescent microspheres (at 20 h), and stained bacteria (at 21 h) extracted from sea-surface water. Fresh water was pumped rather than seawater during two, hour-long periods. Evaluation of tracer transport in the

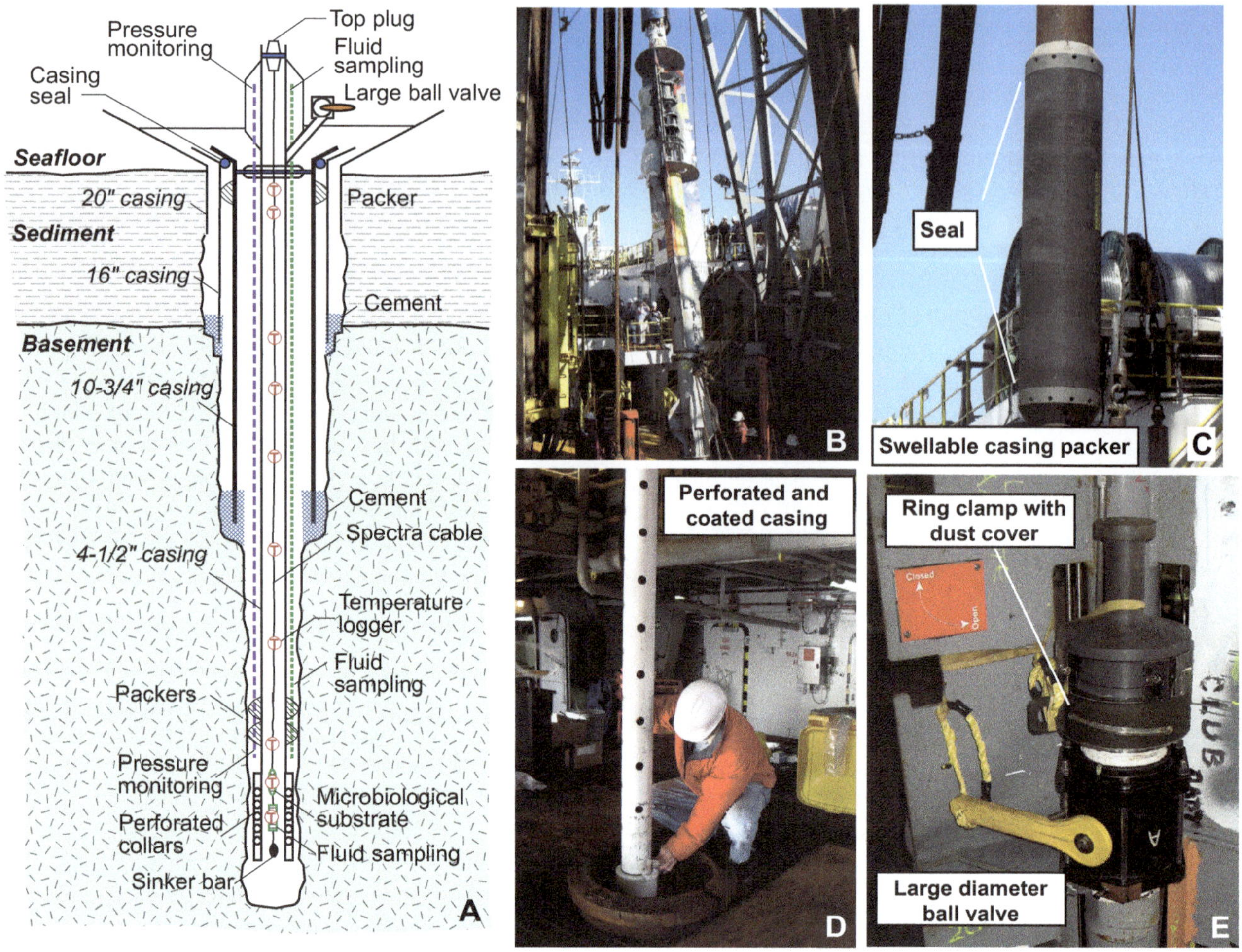


Figure 2. Diagram and photographs showing features of CORK systems deployed during IODP Expedition 327. [A] Schematic of hole completion and design of CORKs deployed in Holes U1362A and U1362B. The upper ~100 m of basement was drilled and cased to improve hole stability. Expedition 327 CORKs are built around two concentric casing strings, with multiple seals at different depths, instruments mounted on the wellhead and downhole, and the ability to monitor and sample from more than one depth interval. They also have perforated and coated drill collars and casing at depth, two kinds of CORK and casing packers (inflatable and swellable), and a ball valve above a lateral pipe that can be used for long-term free-flow experiments. Downhole temperatures are recorded with autonomous sensor and logging instruments incorporated into the fluid and microbiological samplers or hung independently from a cable. [B] CORK wellhead being hoisted on the rig floor during Expedition 327. [C] Swellable packer element, which expands during reaction with fluid in the borehole to seal against casing or the open formation. [D] Perforated and coated casing being cleaned with alcohol prior to deployment. [E] Ball valve in wellhead to be used for free-flow experiment and sampling.

crust will require recovery of borehole samplers from nearby CORKs, as discussed below.

The CORK systems deployed during Expedition 327 to isolate and allow monitoring and sampling of basement fluids and microbes include multiple types of casing and borehole seals (Fig. 2a). The part of the CORK that sits above the seafloor when installed—referred to as the "wellhead"—is constructed from concentric 4½-inch and 10¾-inch casing sections, with parts of the larger casing omitted or cut away between horizontally oriented bulkheads (Fig. 2b). Seafloor sampling and valve manifolds, sensor packages, data loggers, and samplers offset by 120°are arranged within three bays, and separated by vertical gussets. The gussets provide strength to the wellhead, help to guide the ship's camera system and the submersible platform around the bays during CORK installation, and serve as submersible and ROV handholds for stability and leverage. For each of the Expedition 327 CORKs, one bay is dedicated to monitoring and logging pressure data, another is configured for fluid sampling, and a third designed to house a flowmeter and permit microbiological sampling and other experiments.

Expedition 327 CORKs use two kinds of packers to isolate one or more borehole intervals: hydraulic (inflatable) and swellable packers (Fig. 2c). Inflatable packer elements provide an immediate seal, but their long-term reliability is unknown. Swellable elements react with seawater and expand over several months, and should provide a seal that lasts for years. The lower parts of these CORKs (extending down from the base of the lowermost set of packers) are extensively coated with non-reactive materials (Fig. 2d). Perforated and coated casing and drill collars in these CORKs house downhole instruments (fluid samplers, temperature sensors, microbiological growth substrate) that remain in good communication with the surrounding fluids and rocks and are protected from collapse of unstable formation (ensuring future recovery). The Expedition 327 CORKs also have a pipe that comes off the central casing near the wellhead, ending with a ball valve and ring clamp (Fig. 2e). Opening this valve allows overpressured formation fluids to flow from the wellhead without having to remove the top plug. These fluids can be sampled and the rate of flow from the CORK measured, as described below.

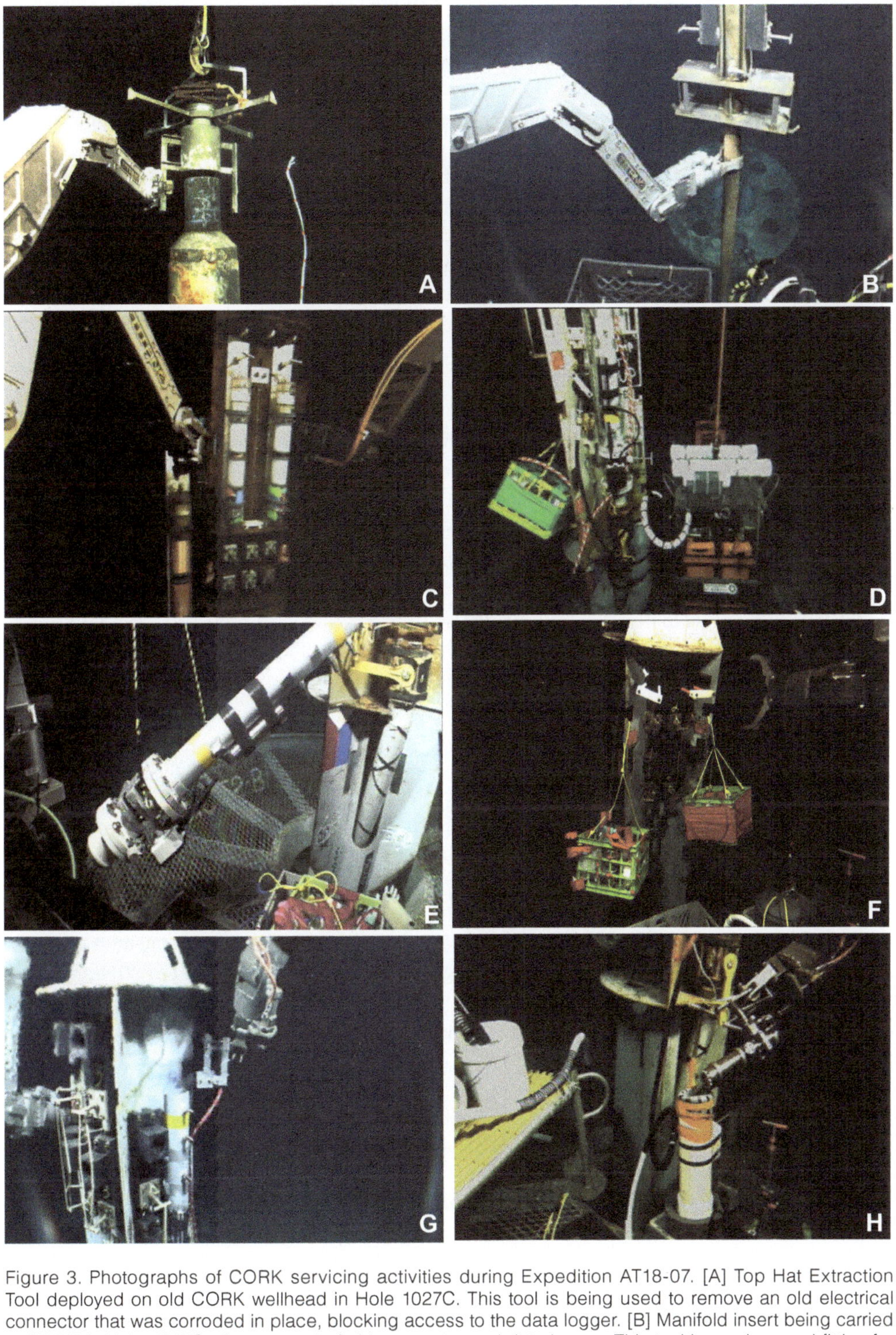


Figure 3. Photographs of CORK servicing activities during Expedition AT18-07. [A] Top Hat Extraction Tool deployed on old CORK wellhead in Hole 1027C. This tool is being used to remove an old electrical connector that was corroded in place, blocking access to the data logger. [B] Manifold insert being carried to CORK in Hole 1027C after removal of old connector and data logger. This tool has valves and fitting for pressure monitoring and a large ball valve and ring clamp (Fig. 2e) that can be used for later instrument deployments. [C] Exchanging OsmoSamplers on CORK in Hole U1301B. Samplers are (left to right): copper (gas tight), geochemistry (Teflon), microbiology. [D] OsmoSamplers in milk crate and GeoMICROBE sled deployed on ROV platform in Hole U1362B. [E] Flowmeter being lifted for deployment on Hole U1362B. [F] Multiple OsmoSampler crates deployed on CORK in Hole U1362A. [G] Fluid sampling inlet being deployed at top of chimney discharging shimmering hydrothermal fluids on Hole U1362B. [H] Gas-tight sampler being used with gas trap to collect fluids from Hole U1362A.

The Hole U1362A CORK was configured to monitor two basement intervals. Pressure in both intervals is monitored through stainless steel tubing connected to mini-screens installed just below inflatable packers at the top of distinct crustal intervals. Stainless steel fluid sampling lines terminate at two depths. A single polytetrafluoroethylene (PTFE) microbiological sampling line ends in a titanium mini-screen. The downhole instrument string includes six OsmoSamplers and microbial growth incubators positioned within coated and perforated casing and collars, and eleven autonomous tem-perature probes.

The Hole U1362B CORK monitors a single basement interval that extends ~75 m below swellable and inflatable packers positioned just inside the base of casing. Pressure is monitored continuously in this basement interval and also in the cased interval above to assess the quality of hydraulic sealing of this observatory. This CORK also contains stainless steel fluid sampling lines and a dedicated PTFE microbiological sampling line. The downhole instrument string comprises six OsmoSamplers, and microbial growth incubators and eight autonomous temperature probes, including two installed in OsmoSamplers suspended inside the perforated and coat-ed casing and collars at depth.

Despite the use of a special unlatching tool and considerable force to pull, push, and rotate, Expedition 327 failed to recover the old CORK deployed in Hole 1027C and to install a new, multi-level CORK in the deepened well. But as described in the next section, we were able to retrofit the Hole 1027C CORK with new pressure monitoring equip-ment in summer 2011, allowing this CORK to be used for crustal-scale, multi-directional, hydrogeologic experiments.

## R/V *Atlantis* and ROV *Jason* Expedition AT18-07

During *Atlantis* Expedition AT18-07, we planned and accomplished the following goals.

- recovery of wellhead OsmoSampling systems deployed on pre-Expedition 327 CORKs, and deployment of new OsmoSampling systems on the new CORKs (for geochemical sampling and microbial growth experiments)
- downloading of long-term pressure data
- assessment of the condition and retrofitting the old CORK in Hole 1027C for pressure monitoring
- recovery and deployment of a long-term GeoMICROBE sled used for autonomous, large-volume fluid sampling and electrochemistry
- collection of large volumes of fluids from depth below the CORKs using a variety of pumping, sampling, and filtering systems
- deployment of an autonomous flowmeter on one of the Site U1362 CORKs, followed by opening of a ball valve to initiate a long-term, cross-hole flow experiment.

Of the planned operations, our ability to retrofit the CORK in Hole 1027C was the least certain when Expedition AT18-07 began. We did not know if we had damaged the CORK with the vigorous recovery attempt during Expedition 327, and we rushed to design and built a series of tools to upgrade measurement and sampling capabilities in time for the subsequent expedition. The first necessary step was to extract a brass connector that was corroded in place on top of the wellhead (Fig. 3a). This eventually required using a customized harness hanging from below the ROV control vehicle *Medea* to pull out the connector. The old data logger was recovered at the same time, and it subsequently produced a complete pressure record from the previous two years (including the pressure response to nearby drilling during Expedition 327). We were then able to install a manifold insert with valves and fittings (Fig. 3b) and connect a new pressure logger for high-resolution monitoring. We were gratified to see almost immediate pressure equilibration in Hole 1027C following these operations, indicating that the Hole 1027C CORK is sealed and providing reliable information on formation fluid pressure.

Pressure data downloaded from the new CORKs at Site U1362 show that they are functioning as intended. We can see different equilibrium pressures at different depths in Hole 1362A (which isolates distinct crustal intervals) and the setting of the swellable packer element at the base of casing in Hole U1362B (where we have continuous monitoring of the cased interval). OsmoSampler systems were recovered and replaced on older CORKs (Fig. 3c), and new systems were successfully deployed on new CORKs in Holes U1362A and U1362B (Fig. 3d, f). The GeoMICROBE sampling and electrochemical analysis sled was recovered from Hole U1301A (where it had been deployed) and redeployed on the CORK in Hole U1362B (Fig. 3d), where it will filter and collect large volumes of fluid and make electrochemical measurements over the next year. A new electromagnetic flow-meter was deployed above the large-diameter ball valve on the Hole U1362B wellhead (Fig. 3e); opening this valve resulted in a jet of warm (~63°C) hydrothermal fluid from the crust below. The flowmeter is recording fluid discharge hourly, and we will return to recover the tool and download the data in summer 2012. Fluid flow from Hole U1362B is expected to decrease with time over the next year, as the excess pressure in the sealed borehole is reduced. This record of time-varying flow, and the associated pressure responses in nearby observatories, will provide the first direct assessment of azimuthal hydrologic properties in the upper oceanic crust.

The inlets to fluid sampling lines were placed in the top of the "chimney" above the flowmeter after the valve was opened, and discharging fluids are being collected with geochemical and microbiological OsmoSamplers (Fig. 3g). Large volumes of high quality formation fluids were recovered from the new CORKs using pump and manifold systems in 2011 (Fig. 3h), and these CORKs will be visited again for fluid sampling in 2012. Samples recovered in 2011 are being analyzed to learn about formation chemistry and microbiology, and to assess the possible arrival of tracers pumped during the 24-hour injection experiment during IODP Expedition 327. The best tracer samples are likely to come from downhole samplers, which will be recovered over the next several years.

## Acknowledgements

We thank the staff of Transocean and the U.S. Implementing Organization for IODP for their dedicated work in preparing for IODP Expedition 327. CORK servicing was skillfully supported by the crew, technicians and officers of the R/V *Atlantis* and ROV *Jason*. These experiments were supported by the U.S. National Science Foundation grants OCE-0727952 and OCE-1031808 (ATF), OCE-0727119 and OCE-1030061 (CGW), OCE-0726887 and OCE-1030350 (KB), OCE-0726563 and OCE-1031352 (JFC), and MCB-0604014 and OCE-0726838 (JPC). Additional support provided by the Gordon and Betty More Foundation (KJE and CGW). This is contribution #111 from the Center for Dark Energy Biosphere Investigations.

## The IODP Expedition 327 Scientists

A. Fisher, T. Tsuji, K. Petronotis, K. Becker, J. Cowen, J.M. Gautier, A. Haddad, J. Kane, S. Keske, M. Harris, S. Hulme, F. Ji, R. Masui, H. Miyamoto, S. Morvan, S. Mrozewski, B. Orcutt, L. Peart, B. Richardson, J. Rutter, B. Thiberge, C.G. Wheat, and D. Winslow

## *Atlantis* Expedition AT18-07 Shipboard Parties

A. Fisher, K. Becker, J. Clark, S. Cooper, J. Cowen, K. Edwards, B. Glazer, S. Hulme, B. Orcutt, C.G. Wheat, A. Ausejo-Robador, R. Brennon, A. Gross, K. Hamner, C.-C. Hseih, S. Jungbluth, J. Kane, H.-T. Lin, G. Ramirez, J. Ringlein, A. Slovacek, and L. Strong

## References


Becker, K., and Fisher, A.T., 2008. Borehole tests at multiple depths in 3.5-Ma seafloor resolve distinct hydrologic intervals. *J. Geophys. Res.*, 113:B07105. doi:10.1029/2007JB005446

Coggon, R.M., Teagle, D.A.H., Smith-Duque, C.E., Alt, J.C., and Cooper, M.J., 2010. Reconstructing past seawater Mg/Ca and Sr/Ca from mid-ocean ridge flank calcium carbonate veins. *Science*, 327(5969):1114–1117. doi:10.1126/science.1182252

Cowen, J.P., Giovannoni, S.J., Kenigm, F., Johnson, H.P., Butterfield, D., Rappé, M.S., Hutnak, M., and Lam, P., 2003. Fluids from aging ocean crust that support microbial life. *Science*, 299:120–123. doi:10.1126/science.1075653

Davis, E.E., Chapman, D.S., Mottl, M.J., Bentkowski, W.J., Dadey, K., Forster, C., Harris, R., Nagihara, S., Rohr, K., Wheat, G., and Whiticar, M., 1992. FlankFlux: An experiment to study the nature of hydrothermal circulation in young oceanic crust. *Can. J. Earth Sci.*, 29(5):925–952.

Davis, E.E., Chapman, D.S., Wang, K., Villinger, H., Fisher, A.T., Robinson, S.W., Grigel, J., Pribnow, D., Stein, J., and Becker, K., 1999. Regional heat-flow variations across the sedimented Juan de Fuca Ridge eastern flank: Constraints on lithospheric cooling and lateral hydrothermal heat transport. *J. Geophys. Res.*, 104(B8):17675–17688.

Davis, E.E., Fisher, A.T., and Firth, J., 1997. *Proc. ODP, Init. Repts.* 168, College Station, TX (Ocean Drilling Program).

Elderfield, H., Wheat, C.G., Mottl, M.J., Monnin, C., and Spiro, B., 1999. Fluid and geochemical transport through oceanic crust: A transect across the eastern flank of the Juan de Fuca Ridge. *Earth Planet. Sci. Lett.*, 172:151–165. doi:10.1016/S0012-821X(99)00191-0.

Fisher, A.T., Cowen, J., Wheat, C.G., and Clark, J.F., 2011a. Preparation and injection of fluid tracers during IODP Expedition 327, eastern flank of Juan de Fuca Ridge. *In* Fisher, A.T., Tsuji, T., and Petronotis, K., *Proc. IODP*, 327: Tokyo (Integrated Ocean Drilling Program Management International, Inc.). doi:10.2204/iodp.proc.2327.2108.2011

Fisher, A.T., Davis, E.E., and Becker, K., 2008. Borehole-to-borehole hydrologic response across 2.4 km in the upper oceanic crust: implications for crustal-scale properties. *J. Geophys. Res.*, 113:B07106. doi:10.1029/2007JB005447

Fisher, A.T., Davis, E.E., Hutnak, M., Spiess, V., Zühlsdorff, L., Cherkaoui, A., Christiansen, L., et al., 2003. Hydrothermal recharge and discharge across 50 km guided by seamounts on a young ridge flank. *Nature*, 421:618–621. doi:10.1038/nature01352

Fisher, A.T., Tsuji, T., and Petronotis, K., 2011b. *Proc. IODP*, 327: Tokyo (Integrated Ocean Drilling Program Management International, Inc.). doi:10.2204/iodp.proc.327.2011

Fisher, A.T., Urabe, T., and Klaus, A., 2005a. *Proc. IODP*, 301: Washington, DC (Integrated Ocean Program Management International, Inc.). doi:10.2204/iodp.proc.2301.2005 pp

Fisher, A.T., Wheat, C.G., Becker, K., Cowen, J., Orcutt, B., Hulme, S., Inderbitzen, K., et al., 2011c. Design, deployment, and status of borehole observatory systems used for single-hole and cross-hole experiments, IODP Expedition 327, eastern flank of the Juan de Fuca Ridge. *In* Fisher, A.T., Tsuji, T., and Petronotis, K., *Proc. IODP*, 327: Tokyo(Integrated Ocean Drilling Program Management International, Inc.). doi:10.2204/iodp.proc.327.107.2011

Fisher, A.T., Wheat, C.G., Becker, K., Davis, E.E., Jannasch, H., Schroeder, D., Dixon, R., et al., 2005b. Scientific and technical design and deployment of long-term, subseafloor observatories for hydrogeologic and related experiments, IODP Expedition 301, eastern flank of Juan de Fuca Ridge. *In* Fisher, A.T., Urabe, T., and Klaus, A., *Proc. IODP*, 301: Washington, DC (Integrated Ocean Drilling Program Management International, Inc.). doi:10.2204/iodp.proc.2301.2103.2005

Huber, J.A., Butterfield, D.A., Johnson, H.P., and Baross, J.A., 2006. Microbial life in ridge flank crustal fluids. *Env. Microbiol.*, 88:88–99. doi:10.1111/j.1462-2920.2005.00872.x

Hutnak, M., Fisher, A.T., Zühlsdorff, L., Spiess, V., Stauffer, P., and Gable, C.W., 2006. Hydrothermal recharge and discharge guided by basement outcrops on 0.7-3.6 Ma seafloor east of the Juan de Fuca Ridge: Observations and numerical models. *Geochem. Geophys. Geosys.*, 7:Q07O02. doi:10.1029/2006GC001242

Johnson, H.P., and Pruis, M.J., 2003. Fluxes of fluid and heat from the oceanic crustal reservoir. *Earth Planet. Sci. Lett.*, 216:565–574. doi:10.1016/S0012-821X(03)00545-4

Mottl, M., 2003. Partitioning of energy and mass fluxes between mid-ocean ridge axes and flanks at high and low temperature. *In* Halbach, P., Tunnicliffe, V., and Hein, J. (Eds.), *Energy and Mass Transfer in Submarine Hydrothermal Systems:* Berlin (Dahlem University Press), 271–286.

Orcutt, B.N., Bach, W., Becker, K., Fisher, A.T., Hulme, S., Toner, B.M., Wheat, C.G., and Edwards, K.J., 2010a. Microbial borehole observatories deployed within the oceanic crust: Design considerations and initial results from long-term colonization experiments. [Invited paper presented at American Geophysical Union 2010 Fall Meeting, San Francisco, CA, 12–17 December 2010]. doi:10.1080/

01490450903456772

Orcutt, B.N., Wheat, C.G. and Edwards, K., 2010b. Subseafloor ocean crust microbial observatories: Development of FLOCS (FLow-through Osmo Colonization System) and evaluation of borehole construction materials. *Geomicrobiol. J.*, 27:143–157. doi:10.1080/01490450903456772

Parsons, B., and Sclater, J.G., 1977. An analysis of the variation of ocean floor bathymetry and heat flow with age. *J. Geophys. Res.*, 82:803–829. doi:10.1029/JB082i005p00803

Smith, A., Popa, R., Fisk, M., Nielsen, M., Wheat, C.G., Jannasch, H., Fisher, A.T., Becker, K., Sievert, S., and Flores, G., in review. A novel method for studying *in situ* microbial colonization of igneous minerals and glasses in ocean crust. *Geochem. Geophys. Geosys.*

Stein, J.S., and Fisher, A.T., 2003. Observations and models of lateral hydrothermal circulation on a young ridge flank: Numerical evaluation of thermal and chemical constraints. *Geochem. Geophys. Geosys.*, 4:1026. doi:10.1029/2002GC000415

Underwood, M., Hoke, K.D., Fisher, A.T., Giambalvo, E.G., Davis, E.E., and Zühlsdorff, L., 2005. Provenance, stratigraphic architecture, and hydrogeologic effects of turbidites in northwestern Cascadia Basin, Pacific Ocean. *J. Sediment. Res.*, 75(1):149–164. doi:10.2110/jsr.2005.012

Walker, B.D., McCarthy, M.D., Fisher, A.T., and Guilderson, T.P., 2007. Dissolved inorganic carbon isotopic composition of low-temperature axial and ridge-flank hydrothermal fluids of the Juan de Fuca Ridge. *Mar. Chem.*, 108(1–2):123–136. doi:10.1016/j.marchem.2007.1011.1002

Wheat, C.G., Elderfield, H., Mottl, M.J., and Monnin, C., 2000. Chemical composition of basement fluids within an oceanic ridge flank: Implications for along-strike and across-strike hydrothermal circulation. *J. Geophys. Res.*, 105(B6):13437–13447. doi:10.1029/2000JB900070

Wheat, C.G., Jannasch, H.W., Fisher, A.T., Becker, K., Sharkey, J., and Hulme, S.M., 2010. Subseafloor seawater-basalt-microbe reactions: Continuous sampling of borehole fluids in a ridge flank environment. *Geochem. Geophys. Geosys.*, 11:Q07011. doi:10.1029/2010GC00305

Wheat, C.G., Jannasch, H.W., Kastner, M., Hulme, S., Cowen, J., Edwards, K., Orcutt, B., and Glazer, B., in review. Fluid sampling from oceanic borehole observatories: Design and methods for CORK activities (1990–2010). *In* Fisher, A.T., Tsuji, T., and Petronotis, K., *Proc. IODP,* 327: Washington, DC (Integrated Ocean Drilling Program Management International, Inc.). doi:10.2204/iodp.proc.327.109.2011

Wheat, C.G., Jannasch, H.W., Kastner, M., Plant, J.N., DeCarlo, E., and Lebon, G.T., 2004a. Venting formation fluids from deep-sea boreholes in a ridge flank setting: ODP Sites 1025 and 1026. *Geochem. Geophys. Geosys.*, 5(8):Q08007. doi:08010.01029/02004GC000710.

Wheat, C.G., McManus, J., Mottl, M., and Giambalvo, E.G., 2003. Oceanic phosphorus imbalance: Magnitude of the mid-ocean ridge flank hydrothermal sink. *Geophys. Res. Lett.*, 30:1895. doi:10.1029/2003GL017318

Wheat, C.G., Mottl, M.J., Fisher, A.T., Kadko, D., Davis, E.E., and Baker, E., 2004b. Heat flow through a basaltic outcrop on a sedimented young ridge flank. *Geochem. Geophys. Geosys.*, 5(12):Q12006. doi:10.1029/2004GC000700

## Authors

**Andrew T. Fisher**, Earth and Planetary Sciences Department and Institute for Geophysics and Planetary Physics, University of California at Santa Cruz, 1156 High Street, Santa Cruz, CA 95064, U.S.A., e-mail: afisher@ucsc.edu.

**Takeshi Tsuji**, Graduate School of Engineering, Kyoto University, C1-1-110 Kyotodaigaku-Katsura, Nishikyo-ku, Kyoto 615-8540, Japan.

**Katerina Petronotis**, Integrated Ocean Drilling Program, Texas A&M University, 1000 Discovery Drive, College Station, TX 77845-9547, U.S.A.

**C. Geoff Wheat**, Global Undersea Research Unit, P.O. Box 475, Moss Landing, CA 95039, U.S.A.

**Keir Becker**, Rosenstiel School of Marine and Atmospheric Science, University of Miami, Division of Marine Geology and Geophysics, 4600 Rickenbacker Causeway, Miami, FL 33149-1098, U.S.A.

**Jordan F. Clark**, Department of Earth Science and Program of Environmental Studies, University of California Santa Barbara, 1006 Webb Hall, Santa Barbara, CA 93106-9630, U.S.A.

**James Cowen**, Department of Oceanography, University of Hawaii at Manoa, 1000 Pope Road, Marine Sciences Building, Honolulu, HI 96822, U.S.A.

**Katrina Edwards**, Center for Earth Sciences, University of Southern California, 3651 Trousdale Parkway, Los Angeles, CA 90089-0740, U.S.A.

**Hans Jannasch**, Monterey Bay Aquarium Research Institute, 7700 Sandholt Road, Moss Landing, CA, 95039, U.S.A.

**and the IODP Expedition 327 and *Atlantis* Expedition AT18-07 Shipboard Parties**

## Photo Credits

Photographs were provided courtesy of Andrew Fisher (UCSC), the AT18-07 Shipboard Party, and the U.S. National Deep Submergence Facility (WHOI)

# 26

# IODP Expedition 328: Early Results of Cascadia Subduction Zone ACORK Observatory

by Earl Davis, Martin Heesemann, and the IODP Expedition 328 Scientists and Engineers

## Abstract

Integrated Ocean Drilling Program (IODP) Expedition 328 was devoted to the installation of an "Advanced CORK" (Circulation Obviation Retrofit Kit) in the Cascadia subduction zone accretionary prism to observe the physical state and properties of the formation as they are influenced by long-term and episodic deformation and by gas hydrate accumulation. Pressures are monitored at four levels on the outside of a standard 10¾-inch casing string, two above and two below the base of the gas-hydrate stability zone at 230 mbsf (m below seafloor). The casing was sealed at the bottom, leaving the inside open down to 302 mbsf for installation of a tilt meter, seismometer, and thermistor cable (scheduled for 2013). The initial data, recovered in July 2011, document an initially smooth recovery from the drilling perturbation followed by what may be a sequence of hole-collapse events. Pressure at the deepest screen is roughly 40 kPa above hydrostatic; higher pressures (80 kPa) are observed at the two screens close to the level of hydrate stability. Tidal variations at the deepest screen are in phase with ocean tides, and define a loading efficiency of 0.6, which is reasonable in light of the consolidation state of the for-mation (porosity ~0.5). Tidal signals near the level of gas hydrate stability display large phase lags, probably as a consequence of hydraulic diffusion stimulated by the large contrast in interstitial fluid compressibility at the gas-hydrate boundary. The degree of isolation among the screens, the anticipated good coupling, and the estimated strain-to-pressure conversion efficiency (~5 kPa $\mu strain^{-1}$) indicate that this installation will serve well to host a variety of hydrologic, seismic, and geodynamic experiments.

## Introduction

Operations carried out during IODP Expedition 328 were devoted to the installation of a borehole hydrologic observatory near ODP Site 889, the location chosen for one of the first CORK observatories during ODP Leg 146 in 1992 (Westbrook et al., 1994; Davis et al., 1995). In that early attempt, the relatively unstable sediments of the formation intruded rapidly through the perforations and bottom of an open-ended casing liner and prevented proper sealing of the hole. The new installation utilized the ACORK design, developed initially for installations at the Nankai subduction zone during ODP Leg 196 to permit pressure monitoring at multiple formation levels on the outside of a 10¾-inch casing string. The casing is sealed at the bottom, leaving the inside fully isolated from the formation but available for future installations of additional monitoring instruments.

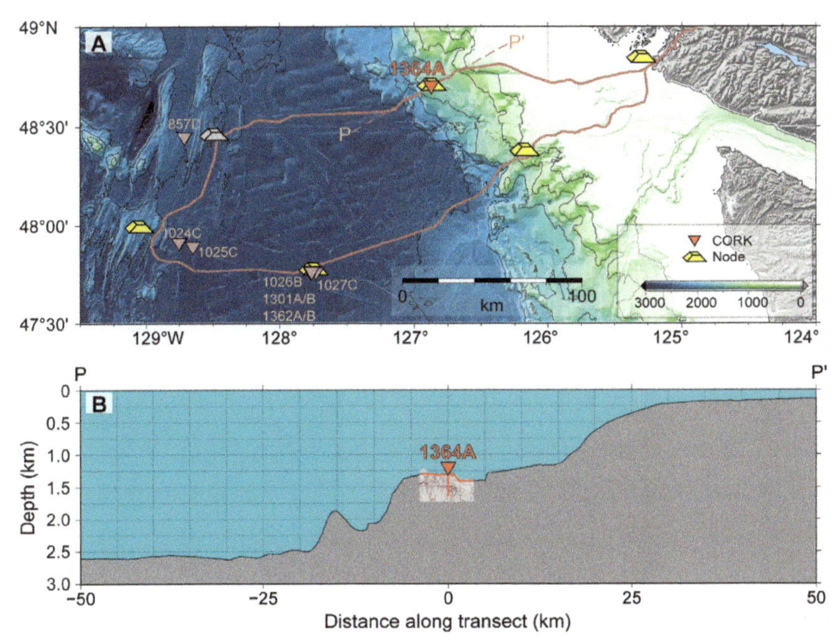


Figure 1. [A] Regional map of East Pacific near Vancouver Island and [B] topographic/bathymetric profile showing context of IODP Hole U1364A roughly 18 km landward of the toe of the accretionary prism, and other "CORKed" boreholes where pressure monitoring is underway. Hole 1026B is currently connected to the NEPTUNE Canada fiber-optic cable (route and nodes shown), and a connection at Hole U1364A is planned for 2013. Location of bathymetric profile P-P' is shown in [A], and location of seismic reflection profile shown in Fig. 2 is indicated on the bathymetric profile as a thumbnail image.

A broad range of objectives will be addressed with monitoring over the decades to come, some the same as those to have been addressed by the original CORK installation in Hole 889C including documenting the average state of pressure and constraining the vertical component of fluid flow from the consolidating sediments in the frontal part of the Cascadia accretionary prism, and investigating the mode of formation of gas hydrates. Other objectives have been added

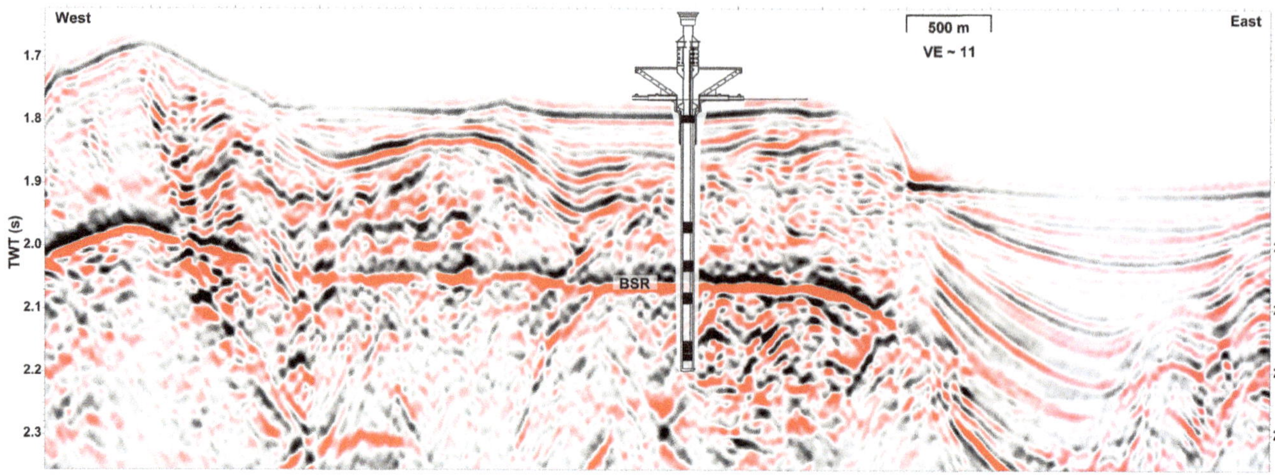


Figure 2. Seismic reflection profile crossing Hole U1364A, showing gently deformed slope basin sediments, strongly deformed accretionary prism sediments, and a bottom-simulating reflector (BSR) marking the boundary between hydrate-bearing sediments above and gas-bearing pore water below. Position of the ACORK with its monitoring screens is indicated. Profile location is shown in Fig. 1b.

subsequently as a result of the multiple discrete monitoring levels provided by the ACORK configuration, advances in measurement resolution and sampling frequency, and knowledge gained from other monitoring experiments. These include determining the influence of gas hydrates and free gas on the mechanical properties of their host lithology, the response of the formation to seismic ground motion, and the magnitude of strain at the site caused by any episodic seismic or aseismic slip in this subduction setting. Instrumentation deployed at the time of drilling includes autonomously recording seafloor and formation pressure and seafloor temperature sensors. Sensors planned for deployment inside the sealed casing at a later date will measure temperature, tilt, and seismic ground motion. In 2013, all instruments will be connected to the NEPTUNE Canada fiber-optic cable for power and real-time communications. This will provide a greater sampling rate, time accuracy, and monitoring lifetime than what is possible with autonomous operations using battery power and local data storage.

## Geologic Setting

As shown in Fig. 1, IODP Site U1364 lies 18 km landward of the toe of the Cascadia subduction zone accretionary prism, where much of the thick section of turbidite and hemipelagic sediments deposited on the eastern flank of the Juan de Fuca Ridge are scraped off the underthrusting oceanic crust (Davis and Hyndman, 1989; Hyndman et al., 1990; Westbrook et al., 1994). Convergence of the Juan de Fuca oceanic plate relative to the North American continental plate occurs in a direction roughly normal to the continental margin and at a rate of roughly 42 mm $y^{1}$ (DeMets et al., 1990). A topographic trench at this subduction zone is absent as a consequence of the extremely high rate of glacial sediment supply from the continent during the Pleistocene. The majority of the supply has been impounded by the elevated igneous crustal topography of the Juan de Fuca Ridge. At the accretionary prism toe (also referred to as the deformation front, where the seawardmost thrust fault of the accretionary complex intersects the seafloor), the sediments that bury the eastern Juan de Fuca Ridge flank are approximately 2.7 km thick. At Site U1364 the accreted sedimentary section is nearly doubled to a thickness of approximately 5 km (Yuan et al., 1994). With tectonic thickening and compaction, pore fluids are expelled, and gas—primarily biogenic methane—is transported upward to contribute to the formation of gas hydrates in the upper few hundred meters of the sediment section (Davis et al., 1990; Hyndman and Davis, 1992; Haacke et al., 2007; Riedel et al., 2010).

The location for Site U1364 was chosen for the same reasons as Site 889. It lies at a position landward of the prism toe where the fluid expulsion rate, estimated on the basis of the rates of compaction and vertical growth of the prism, reaches a cross-margin maximum, and where a clearly developed bottom simulating reflector marks the base of the gas hydrate stability field (Davis et al., 1990). Other holes drilled earlier during ODP Leg 146 and IODP Expedition 311 (Westbrook et al., 1994; Riedel et al., 2010) have documented the nature of the incoming undeformed sediments, the compaction history during accretion, the details of the lithology, and the distribution and composition of gas hydrates across the area. This information, along with detailed site survey studies (Scherwath et al., 2006; Riedel et al., 2010), provided an excellent basis for planning the depth, location, and other details of the ACORK installation at Site U1364 and established a valuable context for ACORK observations.

## Instrumentation

Hole U1364A was drilled to a total depth of 336 mbsf through roughly 90 m of gently deformed slope-basin deposits and underlying sediments of the accretionary prism that

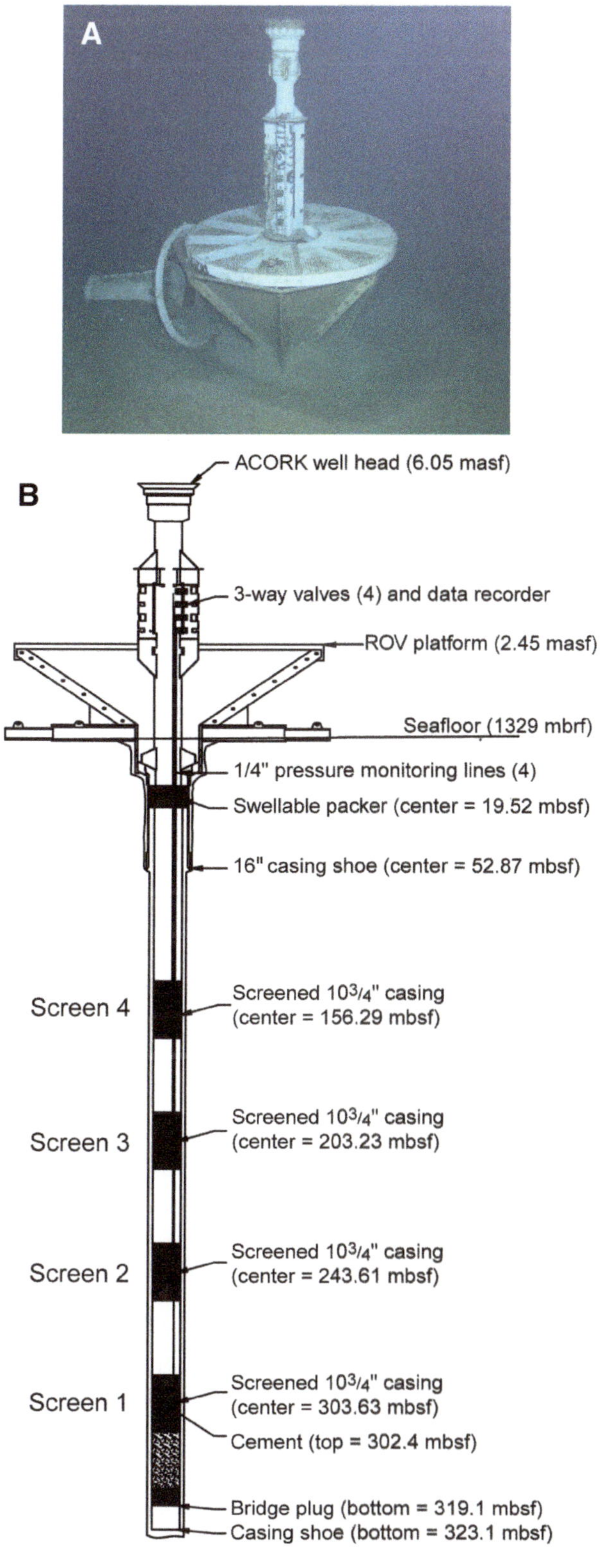


Figure 3. [A] The photograph of the reentry cone and ACORK wellhead (inset) was taken during the July NEPTUNE Canada servicing trip by the ROV ROPOS. The secondary cone seen lying on the seafloor to the left of the wellhead was intended to guide reentry into the casing for the bridge plug and cement installation, but it failed to land properly. Reentry was accomplished through the 24-inch diameter of the uppermost part of the ACORK structure. [B] Schematic illustration of the ACORK system installed in Hole U1364A, with component depths given in meters above and below seafloor (masf, mbsf).

are folded and faulted on a scale too small to be resolved in seismic reflection profiles (Fig. 2). The configuration of the ACORK, constructed around solid 10¾-inch casing and installed immediately after drilling, is shown schematically in Fig. 3b. Formation fluid pressures are transmitted to sensors at the wellhead (inset Fig. 3a) via 2.03-m-long circumferential sand-packed filter screens and 3-mm-diameter (1/8-inch i.d.) stainless steel hydraulic tubing mounted on the outside of the casing. The solid casing is open at the seafloor but sealed at the bottom with a bridge plug backed with cement, leaving the interior open to a depth of 302 mbsf for instruments requiring thermal or mechanical—but not direct—contact with the formation. Screens (numbered S1 to S4 from bottom to top) are positioned at depths of 304 mbsf, 244 mbsf, 203 mbsf, and 156 mbsf. The middle two pressure monitoring screens of the ACORK system (S2, S3) were positioned 14 m below and 27 m above the limit of gas hydrate stability at 230 mbsf to observe the effects of free gas and gas hydrate in the sediment matrix, and diffusive signals originating at the hydrate-gas boundary. The lowermost and uppermost screens (S1, S4) were placed 74 m below and above the boundary, a distance that was anticipated to be sufficient to avoid hydrologic complications originating at the boundary.

Instrumentation installed with the ACORK included five pressure sensors, four plumbed to be switchable between the formation screens and the ocean and the fifth plumbed permanently to the ocean, and a platinum thermometer sensing ocean temperature at the wellhead. Frequency output of the pressure sensors is measured with a low-power, high-precision (1 ppb) period counter. Data are stored in flash memory, and passed through to a communications port (i.e., to fiber-optic cable when connected). Under battery power, data are logged at one-minute intervals (programmable); under cable power, data are recorded at one sample per second. An on-board voltage detector automatically switches between the two modes. Calibration among sensors was done at the seafloor during deployment before the screens entered the hole, and will be done periodically to check for sensor drift though the use of three-way valves that can close the formation lines and connect the formation sensors to the open ocean. Temperature at the wellhead is determined with a highly stable platinum thermometer inside the instrument pressure case, and with one of the pressure sensors that contains a temperature-sensitive quartz oscillator. Further details about the instrumentation, the ACORK hardware, and the geologic setting are provided in Davis et al. (2010).

## Early Data

Connection of the instrumentation at Hole U1364A to the NEPTUNE Canada fiber-optic cable system is planned for 2013; in the meantime, data are being logged autonomously using on-board battery power. The site was first visited with the remotely operated vehicle *ROPOS* in 2011 July (dive

R1430) as part of a NEPTUNE Canada service cruise. The early portion of the downloaded record (with the full length of the ACORK assembly in the water column) is shown in Fig. 4. The 1-minute-interval data record of the deployment shows a highly aliased view of the effects of heave resulting from both depth variations and acceleration of the water in the hydraulic tubing. The latter produced sys-tematically increasing amplitudes from the uppermost to lowermost screens. Average values of the pressure difference between the various screens and the seafloor (SF) allow correction for inaccuracy of the sensor calibrations (see Fig. 4).

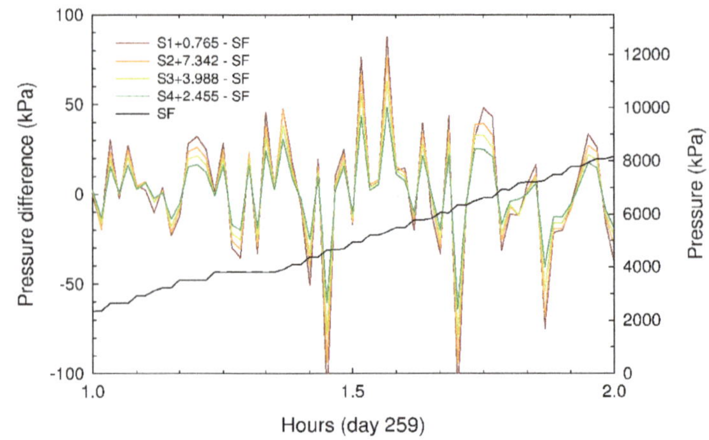


Figure 4. Pressure variations measured during the decent of the ACORK system. Screen pressures are shown relative to the seafloor sensor pressure (scale on left). Offsets given in legend are applied in this and other figures. Absolute pressure is shown for the seafloor sensor (scale on right). Units used for differential and absolute pressure in this and other plots are kPa, equivalent to a seawater head of 0.1 m..

Pressures recorded during the ACORK entry and installation (Fig. 5) show growing offsets and modest overpressures as the screens enter the hole. Peak pressures exceeded those that would be produced by the drilling mud used to prevent premature hole collapse ($\rho$=10.4 lb gal$^{-1}$=1200 kg m$^{-3}$) by 50%–80%. This may be a result of charging of the formation by water and mud pumped during drilling and casing installation. Pressure variations present during the early installation phase abated when the drill string was disconnected, and briefly returned after reentry for the bridge plug and cement installations. The drop in pressure at the time of cementing at screen S2, and to a lesser extent at screen S3, may have been caused by the temporary cooling effects of circulating drill mud during that operation, possibly with free gas being driven into solution.

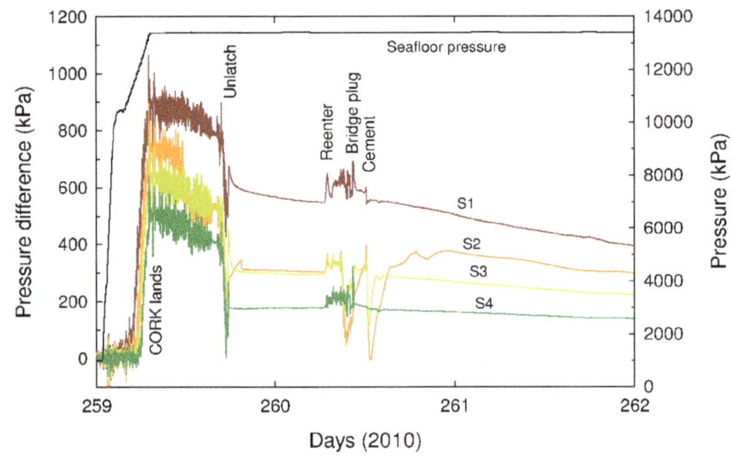


Figure 5. Record of pressure variations (screen pressures relative to seafloor, seafloor pressure absolute) measured during the Site U1364 ACORK installation, annotated with operational events.

After completion of the installation, pressures began a relatively smooth decay for roughly one week, then showed multiple, generally sympathetic perturbations (tentatively inferred to be the consequence of hole collapse) before entering a period of relative stability (Fig. 6). Hydrologic isolation between adjacent screens and between screens and the seafloor is apparent from the attenuation of the tidal load at the seafloor. Ratios of the formation to seafloor signal amplitudes, referred to as the tidal loading efficiency, were roughly 0.7, 0.5, 0.6, and 0.9 at screens S1–S4, respectively, early in the recording period (Fig. 7), and declined to roughly 0.6, 0.3, 0.4, and 0.7 by the end of the recording period. The value for screen S1 at 304 mbsf is comparable to values observed elsewhere in sediments of equivalent porosity (~50%; Davis and Villinger, 2006; Davis et al., 2009)

The greater degree of attenuation and phase lag of the signal at screen S2 relative to the loading at the seafloor is indicative of free gas in the sediments below the level of gas hydrate stability. Attenuated amplitude would result from the high compressibility of gas-bearing water in the sediment pore volume, and a change of phase would result from a diffusive signal generated by the large contrast in fluid and possibly matrix compressibilities at the gas/gas-hydrate boundary (Wang and Davis, 1996; Wang et al., 1998). Lack of a phase difference at screens S1 and S4 relative to the seafloor loading suggests that they are sufficiently far from the gas/gas-hydrate boundary for diffusive effects to be absent. The scale over which this is true depends on the period of the loading signal and the hydraulic diffusivity (permeability and storage compressibility) of the formation. A full analysis of the nature of loading will be done once high-sampling-rate data are available via the connection to the NEPTUNE Canada cable system, and an examination over a broad frequency band (e.g., ranging from microseismic to seasonal) is possible.

The full history (Fig. 8a) reveals complexities that are currently not understood. During the first eighty days of the recording period, pressures varied over days to weeks. Variations are largest at the uppermost screen (S4) and smallest at the lowermost (S1), and variations at the lower three screens mirror those at the uppermost. Thermal perturbations that might be caused by vertical flow in the annulus outside the casing (warm water transported up, cool solids down) could produce this difference (i.e., with sediment settling into the lower part of the hole causing cooling and a temporary decline in pressure arising from thermal contraction of the fluid in the screens and hydraulic tubing, and expelled water warming the uppermost interval), although any heating above the uppermost screen would affect all hydraulic lines equally. Fortunately, towards the end of the initial 160-day recording period, pressures settled to levels similar to those prior to the first major perturbations, although an elevated level of "noise" persists, most notably at the upper two screens. Highest average pressures are present at the two screens immediately above and below the base of the gas-hydrate stability field, and lowest at the uppermost screen.

On 21 February (day 417 on plot), the seafloor pressure sensor suffered a failure possibly caused by crevice corrosion penetrating beneath the O-ring seal of its electrical connector. Excess current to the seafloor sensor was detected by the instrument, current was limited, and the problem was flagged in the stored data, but after roughly thirty minutes the leakage produced a short circuit, leaving all channels dysfunctional. On 24 July (*ROPOS* dive R1452) the cable to the faulty sensor was cut, which restored the system functionality. A brief interval of on-site logging showed the remaining sensors to be fully opera-tional and the formation pressure state to be similar to that seen in February.

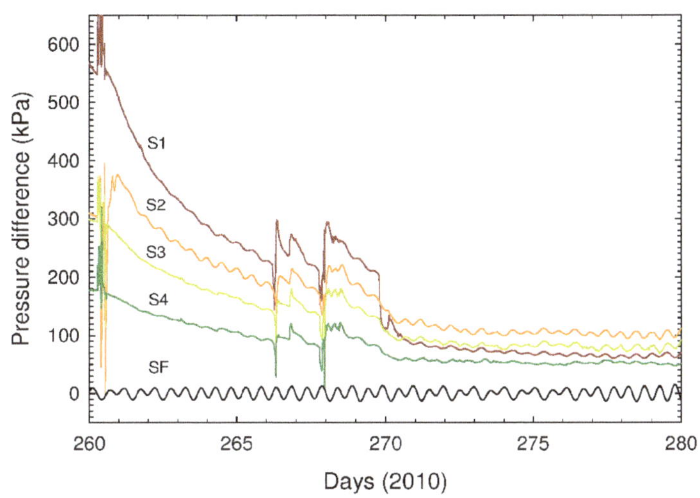


Figure 6. Three-week record of the decay of screen pressures (relative to seafloor) following the ACORK installation at Hole U1364A. Absolute seafloor pressure is shown with the same scaling as the relative pressures at screens but with the mean seafloor pressure removed.

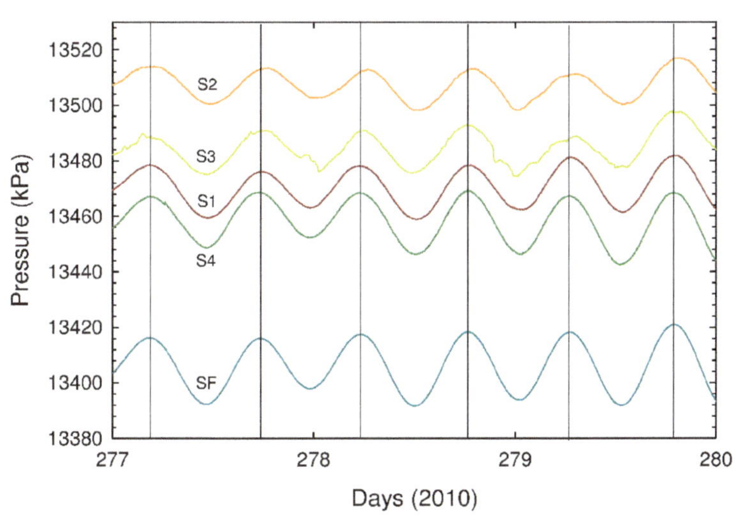


Figure 7. Three-day record of absolute seafloor and screen pressures, showing the dominant tidal component of variation. Differences in amplitudes and phases of pressure variations in the formation relative to pressure at the seafloor are discussed in the text. Lines at times of seafloor tidal peaks highlight the large phase lag of the signal at screen S2.

Temperatures were measured throughout the history of the recording period with the temperature-dependent oscillator in the seafloor pressure sensor and with the platinum thermometer (Fig. 8b). Absolute accuracy is limited to a few tenths of a degree Celsius; calibration will be improved by comparing values to an accurately calibrated conductivity-temperature-depth (CTD) sensor during the next *ROPOS* visit. Precision of both sensors is much better than the accuracy; variations in the two records follow one another within a few milliKelvin. Small differences are primarily the consequence of the thermal time constant of the pressure sensor being longer than that of the platinum thermometer. Temporal variations are seen from diurnal to seasonal periods.

## Summary and Discussion

Data from the first five months of seafloor and formation pressure recording at Hole U1364A at the toe of the Cascadia subduction zone accretionary prism have demonstrated that the IODP ACORK multi-level hydrologic monitoring system will provide well isolated hydrologic observations in the sedimentary formation via hole closure around external casing screens at relatively shallow levels (~300 mbsf). However, persistent noise, particularly at the uppermost screen (156 mbsf), suggests variable leakage in the annulus outside the casing. Borehole temperature monitoring, planned with other instrument installations inside the casing in 2013, will provide a sensitive test for any long-term or episodic vertical flow.

Formation pressures measured at the monitoring screens range from 40 kPa to 80 kPa above hydrostatic. These are significant but not high, i.e., a small fraction of the lithostatic potential at the deepest screen of roughly 2 MPa. High pressures are not expected at this relatively shallow level in the prism, given the relatively high formation permeability and the long time that has elapsed (~300 years) since the last large subduction earthquake when large contractional strain may have been produced in the outermost part of the prism. Highest pressures are observed at the screens immediately above and below the level of gas hydrate stability (marked both at this location and regionally by a clear bottom simulating seismic reflector). This may be a consequence of ongoing gas production.

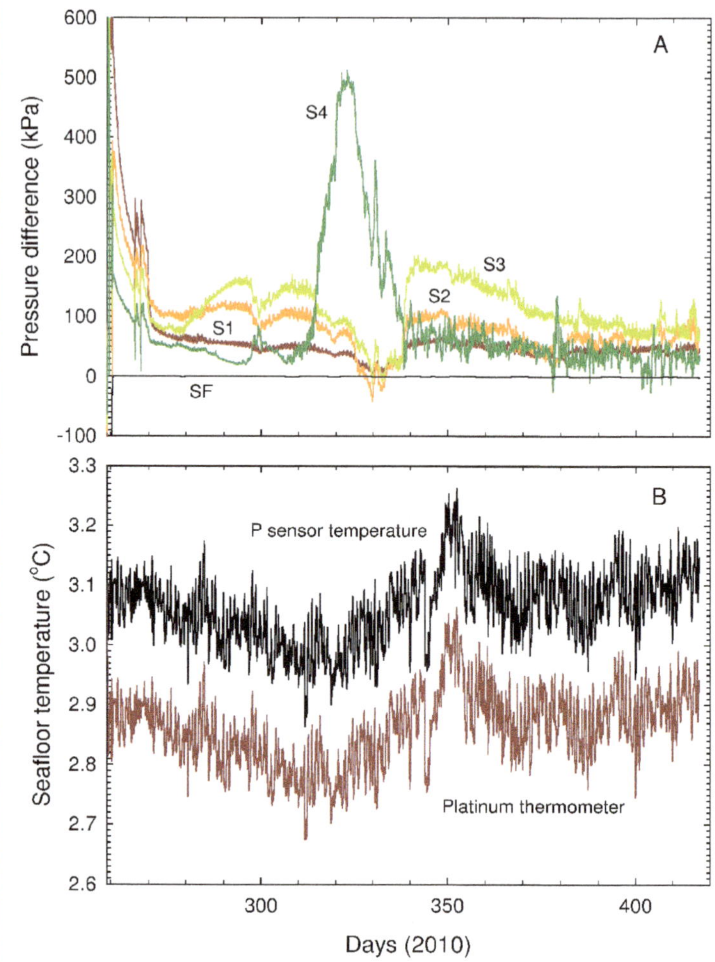


Figure 8. Full history of [A] seafloor (tides filtered) and relative formation pressures and [B] seafloor temperature observed prior to failure of the seafloor pressure sensor on 21 February 2011 (day 417). Scaling convention for pressures is the same as in Figs. 4–6. Temperatures (offset for plotting clarity; absolute values to be calibrated during the next *ROPOS* visit) are measured inside the main instrument pressure case with a platinum thermometer, and inside the seafloor pressure sensor with a quartz oscillator.

While information about dynamic response to seafloor loading over a broad range of frequencies will not be available until the instrumentation is connected to the NEPTUNE Canada power and communications cable, variations in the amplitude and phase of loading at tidal frequencies provide some preliminary insights. There is a general decline in the tidal amplitude at all screens with time, suggesting a progressive increase in the degree of closure of the formation around the casing with time. At the deepest screen, formation pressure variations are in phase with the seafloor load throughout the record, demonstrating that there is no significant hydrologic diffusion or leakage at this level. The tidal loading efficiency (formation/seafloor = 0.6 late in the record) is consistent with observations in other sedimentary sections with similar porosity, and suggests that the formation is sufficiently rigid for volumetric strain to be detected with pressure (i.e., with a conversion efficiency of roughly 5 kPa $\mu$strain$^{-1}$; Davis et al., 2009).

The tidal signal at the screen just below the hydrate stability boundary (S2) is anomalously low in amplitude (formation/seafloor = ~0.3 late in the record) and lags the seafloor load by roughly two hours. The low amplitude is probably the consequence of the high compressibility of interstitial free gas, and the large phase lag is likely the consequence of a diffusive signal propagating from the gas/gas-hydrate boundary, where there may be a sharp contrast in both matrix and interstitial fluid compressibilities.

## Acknowledgments

Installation of the CORK observatory at Hole U1364A seemed like a routine operation as a result of the skilled help of IODP shipboard technical, operations, and engineering personnel and the crew on D/V *JOIDES Resolution*, led by Mitch Malone, Bill Rhinehart, and Bob Aduddell. Michael Riedel provided invaluable information about local seismic and lithologic structure. Zest was added to the short cruise by the energetic leaders and participants of the IODP School of Rock. Data downloading operations and remediation of the problem created by the seafloor pressure sensor failure were carried out by the crew of the remotely operated vehicle *ROPOS*, led by Keith Shepherd and assisted by Martin Scherwath and Reece Hasanen of NEPTUNE Canada. Support for the design and construction of the wellhead instrumentation prior to the expedition was provided by Robert Macdonald, Robert Meldrum, James Guenther, John Bennest, Jonathan Schmidt, and David Miles. Pressure sensors, sensor frequency counters, and the NEPTUNE-compatible data logger were provided by Paroscientific Inc., Bennest Enterprises Ltd., and Minerva Technologies Ltd., respectively.

## The IODP Expedition 328 Scientists and School of Rock Educators

E. Davis, M. Malone, M. Heesemann, M. Riedel, D. Divins, J. Collins, K. Ludwig, S. Hovan, M. Reagan, and S. Slough


## References


Davis, E., Becker, K., Wang, K., and Kinoshita, M., 2009. Co-seismic and post-seismic pore-fluid pressure changes in the Philippine Sea plate and Nankai decollement in response to a seismogenic strain event off Kii Peninsula, Japan. *Earth Planets Space*, 61:649–657.

Davis, E.E., and Hyndman, R.D., 1989. Accretion and recent deformation of sediments along the Cascadia subduction zone. *Geol. Soc. Am. Bull.*, 101:1465–1480. doi:10.1130/0016-7606(1989)101%3C1465:AARDOS%3E2.3.CO;2

Davis, E.E., and Villinger, H., 2006. Transient formation fluid pressures and temperatures in the Costa Rica forearc prism and subducting oceanic basement: CORK monitoring at ODP sites 1253 and 1255. *Earth Planet. Sci. Lett.*, 245:232–244.

Davis, E.E., Becker, K., Wang, K., and Carson, B., 1995. Long-term observations of pressure and temperature in Hole 892B, Cascadia accretionary prism. *In* Carson, B., Westbrook, G.K., Musgrave, R.J., and Suess, E. (Eds.), *Proc. ODP, Sci. Results*, 146 (Pt. 1): College Station, TX (Ocean Drilling Program), 299–311. doi:10.2973/odp.proc.sr.146-1.219.1995

Davis, E.E., Hyndman, R.D., and Villinger, H., 1990. Rates of fluid expulsion across the northern Cascadia accretionary prism: Constraints from new heat flow and multichannel seismic reflection data. *J. Geophys. Res.*, 95:8869–8889. doi:10.1029/JB095iB06p08869

Davis, E.E., Malone, M.J., and the Expedition 328 Scientists and Engineers, 2010. Integrated Ocean Drilling Program Expedition 328 Preliminary Report: Cascadia subduction zone ACORK observatory. *IODP Prel. Rept.*, 328. doi:10.2204/iodp.pr.328.2010

DeMets, C., Gordon, R.G., Argus, D.F., and Stein, S., 1990. Current plate motions. *Geophys. J. Int.*, 101:425–478. doi:10.1111/j.1365-246X.1990.tb06579.x

Haacke, R.R., Westbrook, G.K., and Hyndman, R.D., 2007. Gas hydrate, fluid flow and free gas: Formation of the bottom-simulating reflector. *Earth Planet. Sci. Lett.*, 261:407–420. doi:10.1016/j.epsl.2007.07.008

Hyndman, R.D., and Davis, E.E., 1992. A mechanism for the formation of methane hydrate and seafloor bottom-simulating reflectors by vertical fluid expulsion. *J. Geophys. Res.*, 97:7025–7041. doi:10.1029/91JB03061

Hyndman, R.D., Yorath, C.J., Clowes, R.M., and Davis, E.E., 1990. The northern Cascadia subduction zone at Vancouver Island: Seismic structure and tectonic history. *Can. J. Earth Sci.*, 27:313–319. doi:10.1139/e90-030

Riedel, M., Collett, T.S., and Malone, M., 2010. Expedition 311 synthesis: Scientific findings. *In* Riedel, M., Collett, T.S., Malone, M.J., and the Expedition 311 Scientists, *Proc. IODP*, 311: Washington, DC (Integrated Ocean Drilling Program Management International, Inc.). doi:10.2204/iodp.proc.311.213.2010

Scherwath, M., Riedel, M., Spence, G.D., and Hyndman, R.D., 2006. Data report: Seismic structure beneath the north Cascadia drilling transect of IODP Expedition 311. *In* Riedel, M., Collett, T.S., Malone, M.J., and the Expedition 311 Scientists, *Proc. IODP*, 311: Washington, DC (Integrated Ocean Drilling Program Management International, Inc.). doi:10.2204/iodp.proc.311.110.2006

Wang, K., and Davis, E.E., 1996. Theory for the propagation of tidally induced pore pressure variations in layered subseafloor formations. *J. Geophys. Res.*, 101:11483–11495. doi:10.1029/96JB00641

Wang, K., Davis, E.E., and van der Kamp, G., 1998. Theory for the effects of free gas in subsea formations on tidal pore pressure variations and seafloor displacements. *J. Geophys. Res.*, 103:12339–12354. doi:10.1029/98JB00952

Westbrook, G.K., Carson, B., Musgrave, R.J., et al., 1994. *Proc. ODP, Init. Repts.*, 146 (Pt. 1): College Station, TX (Ocean Drilling Program). doi:10.2973/odp.proc.ir.146-1.1994

Yuan, T., Spence, G.D., and Hyndman, R.D., 1994. Seismic velocities and inferred porosities in the accretionary wedge sediments at the Cascadia margin. *J. Geophys. Res.*, 99:4413–4427. doi:10.1029/93JB03203


## Authors


**Earl Davis**, Pacific Geoscience Centre, Geological Survey of Canada, 9860 West Saanich Road, Sidney, BC V8L 4B2, Canada, e-mail: edavis@nrcan.gc.ca.

**Martin Heesemann**, NEPTUNE-Canada, University of Victoria, P.O. Box 1700 Stn CSC, Victoria, BC V8W 2Y2, Canada, e-mail: mheesema@uvic.ca.

**and the IODP Expedition 328 Scientisits and Engineers**

# 27

# IODP Expedition 331: Strong and Expansive Subseafloor Hydrothermal Activities in the Okinawa Trough

by Ken Takai, Michael J. Mottl, Simon H.H. Nielsen and
the IODP Expedition 331 Scientists

## Abstract

Integrated Ocean Drilling Program (IODP) Expedition 331 drilled into the Iheya North hydrothermal system in the middle Okinawa Trough in order to investigate active subseafloor microbial ecosystems and their physical and chemical settings. We drilled five sites during Expedition 331 using special guide bases at three holes for reentry, casing, and capping, including installation of a steel mesh platform with valve controls for postcruise sampling of fluids. At Site C0016, drilling at the base of the North Big Chimney (NBC) mound yielded low recovery, but core included the first Kuroko-type black ore ever recovered from the modern subseafloor. The other four sites yielded interbedded hemipelagic and strongly pumiceous volcaniclastic sediment, along with volcanogenic breccias that are variably hydrothermally altered and mineralized. At most sites, analyses of interstitial water and headspace gas yielded complex patterns with depth and lateral distance of only a few meters. Documented processes included formation of brines and vapor-rich fluids by phase separation and segregation, uptake of Mg and Na by alteration minerals in exchange for Ca, leaching of K at high temperature and uptake at low temperature, anhydrite precipitation, potential microbial oxidation of organic matter and anaerobic oxidation of methane utilizing sulfate, and methanogenesis. Shipboard analyses have found evidence for microbial activity in sediments within the upper 10–30 m below seafloor (mbsf) where temperatures were relatively low, but little evidence in the deeper hydrothermally altered zones and hydrothermal fluid regime.

## Introduction and Goals

Active seafloor hydrothermal systems at mid-ocean ridges, volcanic arcs, backarc basins, and hotspots are environments with extraordinarily high fluxes of energy and matter. The "subvent biosphere" is the subseafloor biosphere that is predicted to exist just beneath active hydrothermal vents and fluid discharge zones and is sustained from the hydrothermal energy and matter inputs (Deming and Baross, 1993; Takai et al., 2001). The existence of a subvent biosphere has been inferred from many microbiological and geochemical investigations of vent chimney structures and diffuse hydrothermal fluids (Nunoura and Takai, 2009; Nunoura et al., 2010; Takai et al., 2008; 2009; and references in Takai et al., 2006, and Huber and Holden, 2008). In the Iheya North field, a typical deep-sea hydrothermal system in the Okinawa Trough, it has been suggested that a variety of microbial communities based on different chemolithoautotrophic primary producers is present in subseafloor habitats (Nakagawa et al.,

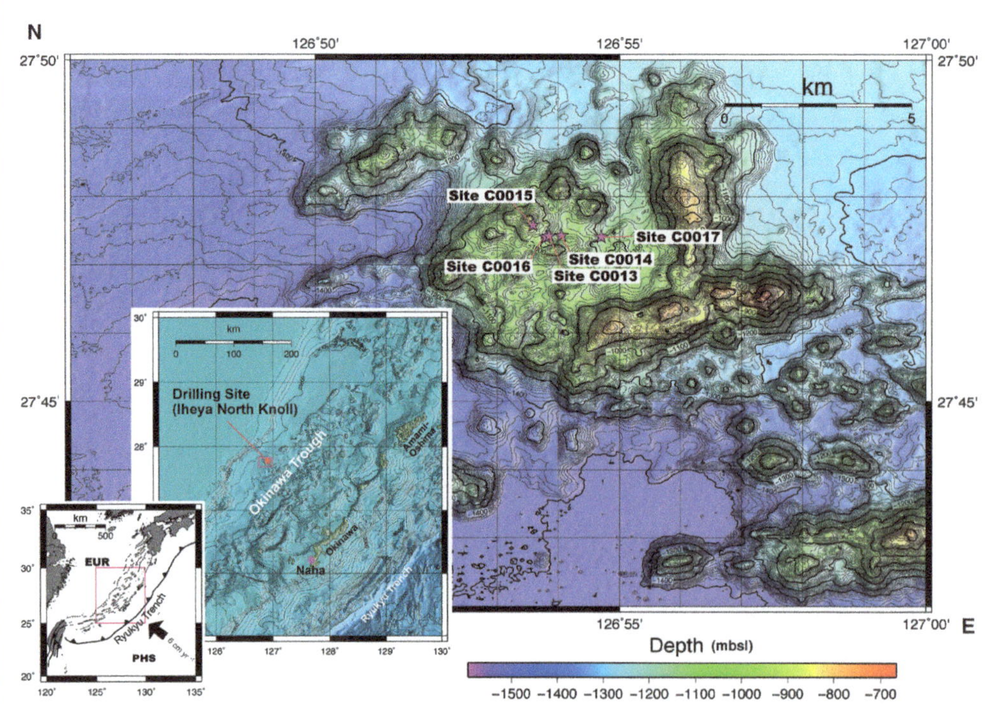


Figure 1. Area map of Iheya North Knoll showing Sites C0013–C0017 drilled during Expedition 331. Insets show the Iheya North Knoll in relation to Okinawa and Okinawa in relation to major tectonic components. EUR= Eurasian Plate, PHS = Philippine Sea Plate. Modified from Takai et al. (2011).

2005). Variability in potential subseafloor microbial communities is likely associated with physical and chemical variation of hydrothermal fluids, controlled by phase-separation and phase-partition of hydrothermal fluid beneath the seafloor. In addition, the overall hydrothermal environments associated with organic-rich sediments provide unusual amounts of C1 compounds ($CO_2$ and $CH_4$) in hydrothermal fluids as carbon sources, as well as unique microbial habitats affected by liquid $CO_2$ and gas hydrates (Nakagawa et al., 2005; Kawagucci et al., 2011). Thus, the abundant supply of energy and carbon and the richness of the habitats supported by physical and chemical variations in the Iheya North field provide an ideal setting for the formation of functionally and metabolically diverse subseafloor microbial communities associated with hydrothermal activity.

There were three major scientific objectives of Expedition 331 drilling were.

1. to test for the existence of a functionally active, metabolically diverse subvent biosphere associated with subseafloor hydrothermal activity;
2. to clarify the architecture, function, and impact of subseafloor microbial eco-systems and their relationship to physical, geochemical, and hydrogeologic variations within the hydrothermal mixing zones around the discharge area; and
3. to establish artificial hydrothermal vents in cased holes from potential subseafloor hydrothermal flows, and to prepare a research platform at each cased hole for later study of fluids tapped from various parts of the hydrothermal system and their associated microbial and macrofaunal communities.

## Geological Setting and Earlier Work

The Okinawa Trough is a backarc basin extending for ~1200 km, between the Ryukyu arc-trench system and the Asian continent (Fig. 1; Lee et al., 1980; Letouzey and Kimura, 1986). Seismic reflection data suggest a typical backarc structure (Letouzey and Kimura, 1986) with a high-velocity mantle below ~6000 mbsf overlain by potentially young basalt with an average velocity of 5.8 km $s^{-1}$ between ~3000 mbsf and 6000 mbsf, an igneous rock layer (4.9 km $s^{-1}$) between ~1000 mbsf and 3000 mbsf, and ~1000 m of sediment immediately beneath the seafloor. Since the discovery of submarine hydrothermal activity at Iheya Ridge and Izena Hole in the middle Okinawa Trough in 1988 (Halbach et al., 1989; Sakai et al., 1990), six active hydrothermal fields have been discovered—Minani-Ensei Knoll, Iheya North, Iheya Ridge, Izena Hole, Hatoma Knoll, and Yonaguni Knoll IV.

Characteristics of the tectonic setting in this active hydrothermal system are reflected in the chemical composition of sulfide deposits. Sulfide samples collected from the Iheya North field are distinctly more enriched with Pb than mid-oceanic-ridge sulfides (Fig. 2). The polymetallic Zn-Pb-Cu chemical signature of Iheya North sulfides is similar to that of Kuroko-type hydrothermal deposits formed during the Tertiary in northeast Japan.

The chemistry of hydrothermal fluids collected from active sulfide chimneys in the Okinawa Trough is characterized by higher concentrations of $CO_2$, $CH_4$, $NH_4$, I, and K and higher alkalinity than those in typical sediment-free mid-ocean-ridge hydrothermal fluids (Sakai et al., 1990; Gamo et al., 1991; Konno et al., 2006; Takai and Nakamura, 2010; Kawagucci et al., 2011). The distinctive hydrothermal fluid chemistry is strongly linked with the geologic setting and the thick terrigenous sediments of the Okinawa Trough. Philippine Plate subduction along the Ryukyu arc-trench system supplies dacitic-rhyolitic magma rich in K and volatile components to the Okinawa Trough (Sakai et al., 1990; Gamo et al., 2006). Organic-rich terrigenous sediment filling the Okinawa Trough (Narita et al., 1990) supplies not only the sedimentary chemical inputs ($NH_4$, I, etc.; Gamo et al., 1991; You et al., 1994), but also promotes the widespread occurrence of functionally active microbial communities that impact hydrothermal fluid chemistry and circulation (Nakagawa et al., 2005; Inagaki et al., 2006; Nunoura and Takai, 2009; Nunoura et al., 2010; Takai and Nakamura, 2010; Kawagucci et al., 2011). In addition to the chemical aspects, the relatively shallow water depth of many Okinawa Trough hydrothermal systems serves to induce subcritical phase separation (Suzuki et al., 2008) and subsequent phase segregation, as the boiling temperature of seawater decreases steeply with decreasing pressure at ~100 bar. Phase separation and segregation sometimes

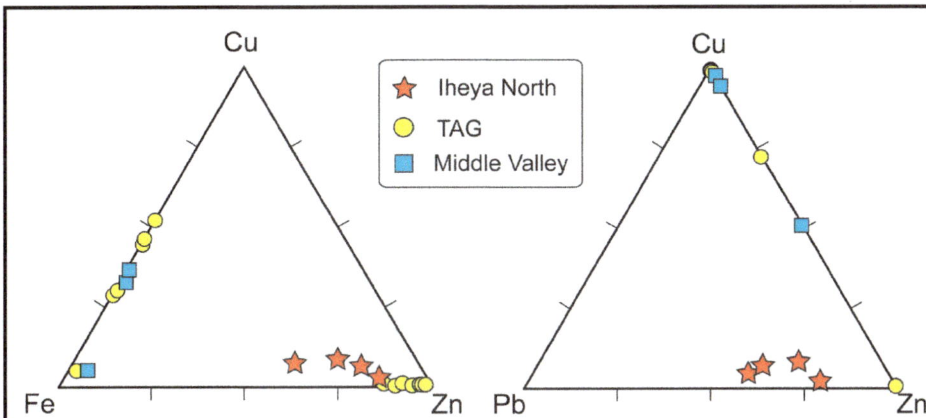


Figure 2. Ternary diagrams of the composition of sulfide samples from active hydrothermal fields. Iheya North data from Ueno et al. (2003). TAG = TAG mound, Mid-Atlantic Ridge. Middle Valley = Middle Valley, Juan de Fuca Ridge. Modified from Takai et al. (2011).

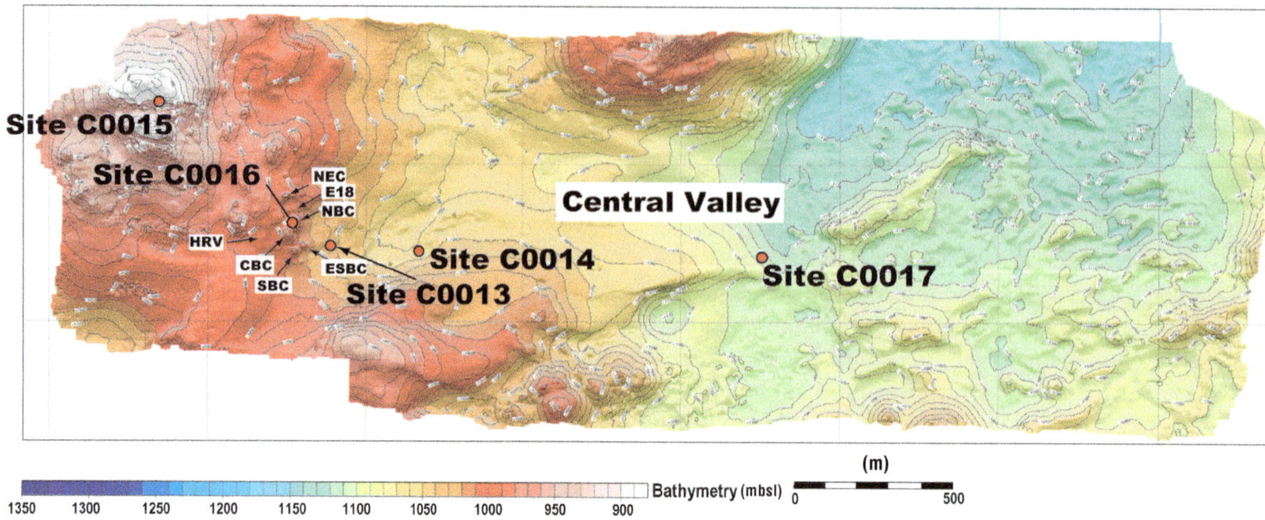


Figure 3. Detailed bathymetry of the Iheya North hydrothermal field and the central valley at Iheya North Knoll, with the location of Sites C0013–C0017 drilled during Expedition 331. HRV=High Radioactive Vent mound, CBC=Central Big Chimney mound, SBC=South Big Chimney mound, NEC=North Edge Chimney mound, E18=event Marker 18 mound, NBC = North Big Chimney mound, ESBC="Ese" South Big Chimney mound. Modified from Takai et al. (2011).

produce hydrothermal fluids of quite different chemical composition at different vent sites in the same hydrothermal field, even though they are derived from the same source fluid (Kawagucci et al., 2011).

## Drilling, Temperature Measurement, Sampling, and Installation of Vents

We drilled five sites during Expedition 331: the active hydrothermal vent site and sulfide-sulfate mound at North Big Chimney (NBC) (Site C0016, Fig. 3), three sites east of NBC at distances of ~100 m (C0013), 450 m (C0014), and 1550 m (C0017) from the active vents, and one site (C0015) on a hill ~600 m northwest of the active vents that represents a potential migration path for hydrothermal fluid.

The NBC hydrothermal mound at Site C0016 is 20 m high and 6 m in diameter. When we attempted to core this active high-temperature (311°C) vent at its summit, the pipe broke, and core recovery failed (Hole C0016A). We drilled a second hole 20 m away, immediately at the base of the mound on its western side (Hole C0016B). We used conventional hard rock drilling equipment supplied by Baker-Hughes Inteq (BHI) specifically for Expedition 331 due to high temperatures and expectation of hard rock. The BHI system collects 4-inch diameter core in aluminum liners in lengths of 9 m, 18 m, or 27 m, but requires a time-consuming pipe trip for each core. It penetrated to 45 mbsf in three runs of 9 m, 18 m, and 18 m. Each run recovered several large pieces of core, but the total recovery was only 2.095 meters, nearly all of it hard rock. The only evidence for the actual depths of the recovered rock thus comes from the drilling records. Hole C0016B was not cased, but it was fitted with a corrosion cap with 3 m of 5.5-inch pipe hanging beneath and extending 0.4 m into the seafloor (Fig. 4). ROV video images showed vigorous black smoker discharge from the corrosion cap outlet immediately after its deployment. This hydrothermal emission began only after the third coring run, which penetrated 27–45 mbsf, and was probably derived from a depth below 38 mbsf.

At Sites C0013, C0014, and C0017, we drilled the relatively high, moderate, and low heat flow areas, respectively, to the east of the Iheya North hydrothermal field to investigate subseafloor microbial habitats and communities within broad gradients of physical and chemical variation, both laterally and vertically, that could be affected by mixing between discharging hydrothermal solutions and recharging ambient bottom seawater. These sites were drilled using the hydraulic piston coring system (HPCS) to first refusal and then again in any softer intervals encountered deeper in the hole, alternating as necessary with the extended punch (EPCS) and extended shoe (ESCS) coring systems, and, for one run at Site C0013, the BHI system, to penetrate the harder layers. Both the EPCS and ESCS systems were able to penetrate the harder layers, with a slight advantage to the ESCS for the hardest layers, though the EPCS was generally much better at core recovery. We penetrated the margin of the local discharge-recharge zone to depths of 55 mbsf (C0013), 137 mbsf (C0014), and 151 mbsf (C0017), coring variably hydrothermally altered sediment and pumiceous deposits (Figs. 5, 6). We were able to measure *in situ* temperature at two of these sites using the advanced piston corer temperature tool (APCT3) shoe (upper calibration limit = 55°C) as part of the HPCS, combined with commercial thermoseal strips (Nichiyu Giken Co., Ltd.) taped to the outer surface of the core liner. We were not able to measure temperature at proximal flank Site C0013, but the gradient was likely higher than at the other two flank sites.

Eight holes (C0013A–C0013H) were drilled at Site C0013, and core was recovered from all but Hole C0013A. Hole C0013E was the deepest (54.5 mbsf) and was cased down to 40.2 mbsf and fixed with a corrosion cap (open outlet pipe) mounted on the guide base (Fig. 4). During drilling and coring operations at Site C0013, we encountered many operational and sample handling problems. These problems were due to the unexpectedly high temperature gradient at the site and the presence of repeated hard layers that appear to behave as cap rocks alternating with soft and sticky clay-rich layers. Porosity measurements on the core clearly document the repeated occurrence of low-porosity harder layers (e.g., 0–2 mbsf, 7–10 mbsf, and 20–30 mbsf). We observed at several depths that when a hard cap rock was drilled through into softer underlying layers, subseafloor hydrothermal fluid began to outflow from the hole, where it was imaged by the ROV video camera. To tap this fluid, we used slotted, perforated casing pipe over the depth interval 21–39.8 mbsf in Hole C0013E (Fig. 4). Immediately after casing and capping this hole, we observed in the ROV video image strong hydrothermal fluid discharge from the casing pipe that was hung in the guide base. Thermoseal temperature-sensitive strips on the corrosion cap outlet pipe showed in the ROV video imagery that the discharging water temperature was >250°C.

Seven holes were drilled at Site C0014 (Holes C0014A–C0014G). Hole C0014G was the deepest (136.7 mbsf) and was cased down to 117.8 mbsf and fixed with a corrosion cap (open outlet pipe) mounted on the guide base. As at Site C0013, we encountered repeated hard layers that behaved as cap rock, and discharge from the holes, beyond what may be only expelled drilling fluid, after penetrating these layers (35–44.5 mbsf in Hole C0014B, 25.5–35 mbsf in Hole C0014E, and 37.7–47.2 mbsf and 89.2–93.7 mbsf in Hole C0014G). We again cored multiple low-porosity layers (e.g., in Holes C0014B, C0014E, and C0014G). Based on pore water chemistry, density, and porosity (and indicated by low recovery during drilling), we inferred lateral hydrothermal flow at 31–42.5 mbsf and 90–95 mbsf at Site C0014. We therefore installed slotted, perforated casing pipe at 29.8–49.2 mbsf, 78.3–97.8 mbsf, and 107.5–117.2 mbsf in Hole C0014G (Fig. 4). After casing and capping, we saw in the ROV-mounted video diffuse hydrothermal fluid discharge, not from the corrosion cap outlet but from the seafloor, through the annulus, the space between the wall of the hole and the casing pipe. The temperature of the diffusing fluids was found to be >240°C based on exposure of thermoseal strips mounted on the corrosion cap outlet pipe. We measured temperature at Site C0014 using the APCT3 temperature shoe on the HPCS core barrel for lower temperatures (0°C–55°C) and thermoseal temperature-sensitive strips for higher temperatures (75°C–250°C). The temperature-depth profile at Site C0014 is shown in Fig. 7a. Temperature increases nearly linearly at $3^{\circ}C\ m^{-1}$ to 145°C ± 5°C at 47 mbsf and then increases abruptly to >210°C at 50 mbsf, below a hard layer near that depth. The drilled sequence at Site C0014 comprises interbedded, variably altered, and consolidated volcaniclastic gravels and breccias, as well as hemipelagic mud (Fig. 5). Given the location of the site on the upper flank of an active volcanic complex, mass wasting and debris flows are likely to be important sedimentary processes, potentially leading to high rates of redeposition of hemipelagic and volcaniclastic material. The deeper portion of the rock volume cored has been hydrothermally altered. Differing degrees and styles of hydrothermal alteration form the basis for the division of Site C0014 sediments and lithologies into lithostratigraphic units.

Hole C0017D is the deepest (150.7 mbsf) of four holes drilled at Site C0017 and is located 1550 m east of the high-temperature vents. Based on its low heat flow, it was inferred to be a location of probable recharge of the hydrothermal system. We observed no evidence for discharge of water from any of these holes, but the concave-upward tempera-

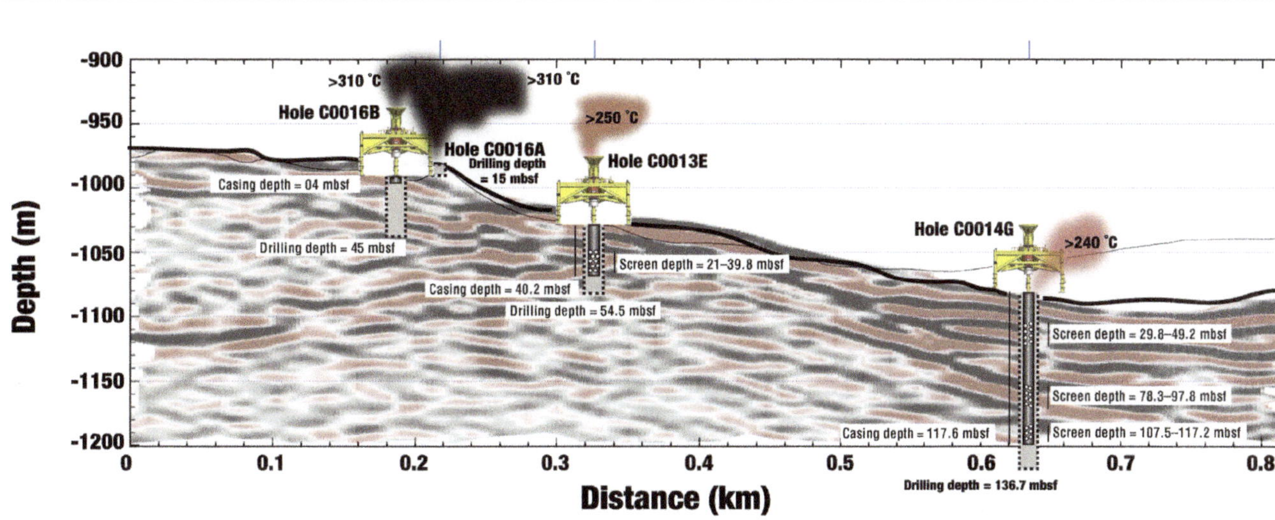


Figure 4. Schematic illustration of post-drilling natural and artificial hydrothermal vents created during Expedition 331. Outflow temperatures measured by ROV (Hole C0013E) and thermal stickers (Holes C0014G and C0016B). Modified from Takai et al. (2011).

ture profile we measured (Fig. 7b) can be fit reasonably well with an exponential function and is consistent with overall downwelling, with distinct perturbations in the temperature gradient that suggest localized lateral flow that may be influenced by the variable lithologies we cored (Fig. 6). We measured temperature successfully at seven depths in Holes C0017B–C0017D over the interval 18.3–150.7 mbsf; combined with the ocean bottom water temperature at the site of 4.9°C ± 0.5°C, our profile is defined by eight points. Six of the downhole measurements were made using the APCT3 temperature shoe on the HPCS. The seventh and deepest measurement, at 150.7 mbsf, was made in triplicate using three identical thermoseal temperature-sensitive strips with chemically impregnated beads for 75°C, 80°C, 85°C, 90°C, and 95°C taped to the bottom outer surface of the plastic core liner. These beads were darkened and thus exposed at maximum temperatures of 85°C, 90°C, and 90°C on the three strips, which we reported as 90°C ± 5°C. The APCT3 shoe recorded a temperature in excess of its maximum range of 55°C for this core. Down to ~50 mbsf, where we encountered a hard layer that was probably pumice (no core was recovered, Fig. 6), temperature remained low, <15°C. Beneath this hard layer, temperature jumped to 25°C–39°C at 69–85 mbsf and then appeared to level off for a short interval, reaching only 44°C at 112 mbsf before increasing nearly exponentially to 90°C at the bottom of Hole C0017D at 150.7 mbsf.

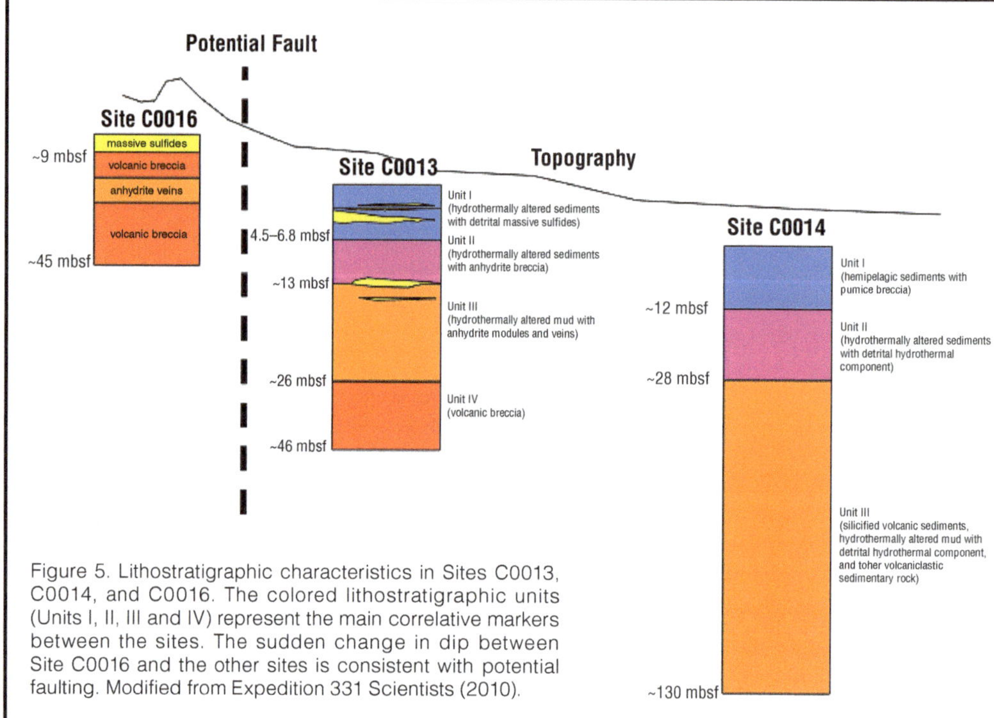

Figure 5. Lithostratigraphic characteristics in Sites C0013, C0014, and C0016. The colored lithostratigraphic units (Units I, II, III and IV) represent the main correlative markers between the sites. The sudden change in dip between Site C0016 and the other sites is consistent with potential faulting. Modified from Expedition 331 Scientists (2010).

Three holes were drilled at Site C0015 (Holes C0015A–C0015C). Although a relatively short interval was cored (0–9.5 mbsf), a broad diversity of sediment types was drilled, with coarse pumiceous gravel and grit, siliciclastic sand, hemipelagic mud, bioclastic gravel, and foraminiferal sediment all recovered. In contrast to the other sites, the samples of Site C0015 do not have any significant hydrothermal input, nor do they appear at present to support a robust microbial community associated with hydrothermal activity. Thus, Site C0015 is excluded from further description and discussion.

## Preliminary Scientific Results

Drilling and coring operations during Expedition 331 provided insight into the hydrothermal flow regime at Iheya North Knoll. While the thermal gradient was known to be high at Site C0016 at NBC mound, Sites C0013 and C0014 have steeper thermal gradients than we expected. Site C0013 was located 100 m east of the vigorous high-temperature vents and mounds (Fig. 3). Drilling to our maximum depth of 54.5 mbsf, we penetrated several hard, low-porosity layers that could function as a cap rock and found thick, porous sediment hydrothermally altered at high temperature between the harder, less permeable layers. The interstitial water showed large changes in composition both laterally and vertically over short distances, suggesting chaotic lateral flow in permeable horizons separated by impermeable barriers. The lithostratigraphy, physical properties of the sediment and rock, and interstitial water chemistry thus all provided insight into the hydrothermal flow regime at Site C0013.

Site C0014 was located ~450 m east of the high-temperature vents and mounds (Fig. 3). The temperature gradient was roughly linear from 0 mbsf to 47 mbsf, increasing from the bottom water temperature of 4.5°C to 145°C over that depth range, but it deviated greatly from this line at 0–9 mbsf and 47–50 mbsf, where it was clearly affected by high-temperature fluid pooling or lateral flow. Interstitial water chemistry demonstrated vertical stratification, from water with seawater chloride values at 0–25 mbsf, to a vapor at 29–38 mbsf, and to a brine from 48 mbsf to the deepest sample at 114 mbsf. Within the upper sediments, which consisted of pelagic sediments and pumiceous gravel, there was considerable lateral variability in the intensity of microbial sulfate reduction between the four holes that were only a few meters apart; clearly there was a functionally robust and metabolically diverse subseafloor biosphere here. At 20 mbsf, where the temperature was ~70°C, and to the bottom of the deepest Hole C0014G at 137 mbsf, where a linear fit would have the temperature exceed 400°C, the sediment and rock we recovered was intensely hydrothermally altered. As at Site C0013, chaotic flow should be occurring through permeable formations and along fault structures, and was likely separated vertically by impermeable beds that behave as cap rocks.

Site C0017 was located 1550 m east of the high-temperature vents of the Iheya North hydrothermal field, in an area of low heat flow (Fig. 3). The overall temperature profile was exponential and concave upward, consistent with downwelling of cold water, implying that this was an area of recharge to the hydrothermal system. We reached a maximum temperature of 90°C ± 5°C at the bottom of the deepest Hole C0017D at 151 mbsf. Deviations from a smooth temperature profile indicated the presence of a discrete zone of cold water recharge, consistent with interstitial water chemistry and the presence of a highly oxidized layer at 26–35 mbsf that supported a microbial community. We found only a small amount of hydrothermally-altered sediment deep in Hole C0017D.

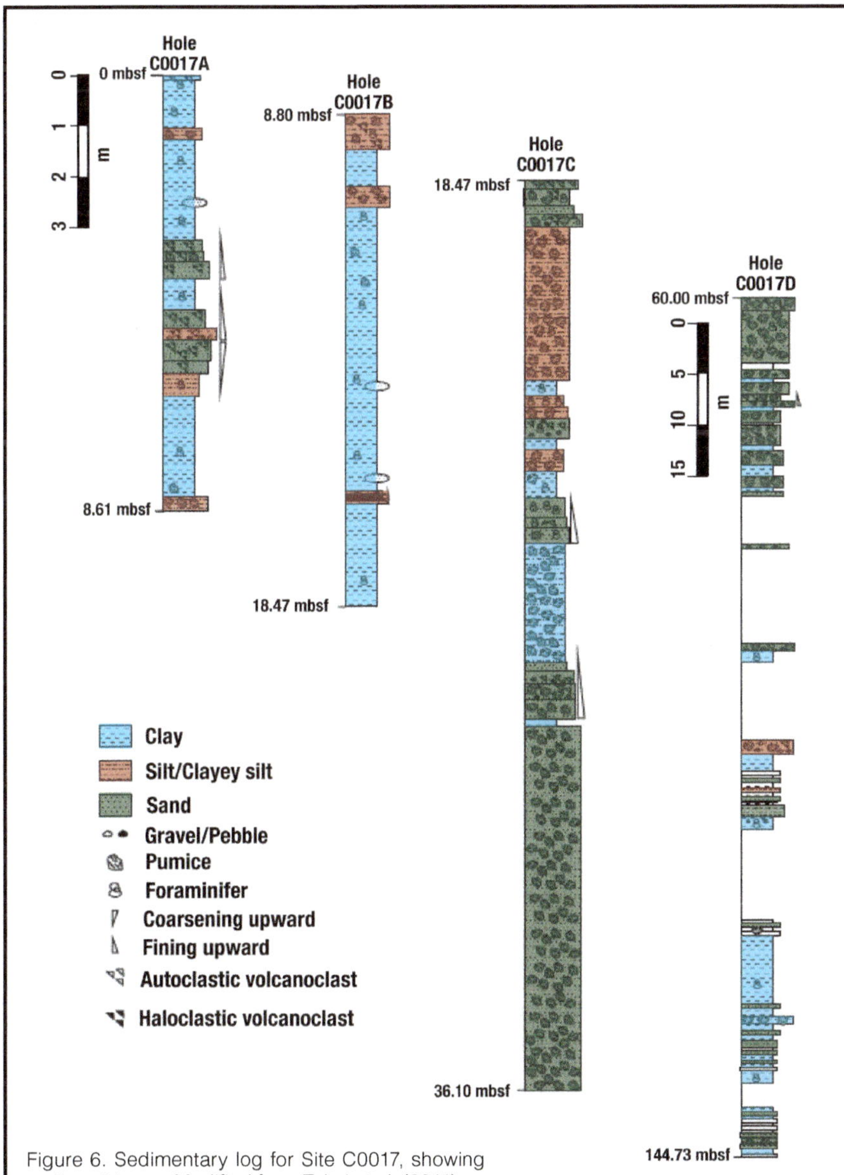


Figure 6. Sedimentary log for Site C0017, showing sediment types. Modified from Takai et al. (2011).

These results, taken together, provided a more detailed look at the large-scale hydrogeology of the Iheya North Knoll hydrothermal system than previously observed at subseafloor hydrothermal systems elsewhere (Takai et al., 2011). The hydrological regime at Iheya North Knoll was characterized by large-scale hydrothermal alteration, deposition, and fluid migration within permeable rocks and sediments hosted by the Iheya North Knoll volcanic complex. Prior to Expedition 331, we knew little about the spatial and temporal scales and patterns of hydrothermal circulation in subseafloor hydrothermal systems, particularly those in subduction zone settings such as volcanic arcs and backarc spreading centers.

Much of the sediment and rock we cored at Sites C0013, C0014, and C0016 was intensely hydrothermally altered, and mineralized (Fig. 5). Exceptional among the disseminated and vein, stockwork-type sulfide we recovered was the massive sphalerite-rich ore we cored in Hole C0016B. This marked the first time this type of massive sulfide, which closely resembles the Kuroko black ore, has been recovered from the subseafloor environment of an active deep-sea hydrothermal system. We recovered only 2.1 meters of core from 45 m of penetration in this hole, which made reconstruction of the lithostratigraphy impossible, but the recovered core included a wide diversity of lithologies that were typically associated with volcanic-hosted massive sulfide (VHMS) mineralization. In addition to the sphalerite-rich black ore, recovered core included a boulder-sized, coarsely crystalline piece of anhydrite veined by sphalerite-pyrite, as well as pyrite-veined and pyrite-altered volcanic rock.

The shipboard microbiological analyses and experiments provided little evidence for the existence of a hot subvent biosphere beneath the Iheya North hydrothermal field, though direct microscopic cell count and cultivation from a colder, diffusely venting site and a site of lateral recharge provided evidence for subseafloor psychrophilic to mesophilic microbial communities. Prior to this expedition, it was hypothesized that subseafloor mixing between hydrothermal fluids and recharging seawater was sustained by a fine-scale network of narrow hydrothermal fluid flow paths within the subseafloor along the eastern flank of the hydrothermal system. The hydrothermal regimes we intersected by drilling, including the permeable reservoirs and flow paths, appeared to be larger in scale than we had envisioned. Most of the sediment and rock we cored at Sites C0013, C0014, and C0016 were exposed to much higher temperatures than the microbiologically habitable temperature range (<122°C). However, smaller and more limited zones that were habitable by microbes were sampled, as well as a larger one at the hydrothermal recharge zone at Site C0017. Recharge zones were, of course, an essential part of all hydrothermal systems.

## Plans for Future Investigations

A total of twenty-four holes were drilled during Expedition 331, of which twenty-one resulted in recovered cores. We drilled 708 m and recovered 312 meters of core from 560 m attempted, yielding an overall recovery of 56%. This compares highly favorably with the only other attempt to drill a submarine felsic-hosted hydrothermal system during ODP Leg 193, which recorded an overall core recovery of 10.7% at the PACMANUS site in the Eastern Manus Basin, Papua New Guinea (Shipboard Scientific Party, 2002). The improved core recovery during Expedition 331 will provide excellent opportunities to assess the existence of a functionally active, metabolically diverse subvent biosphere and to clarify the architecture and function of subseafloor microbial ecosystems and the physical, geochemical, and hydrogeologic variations.

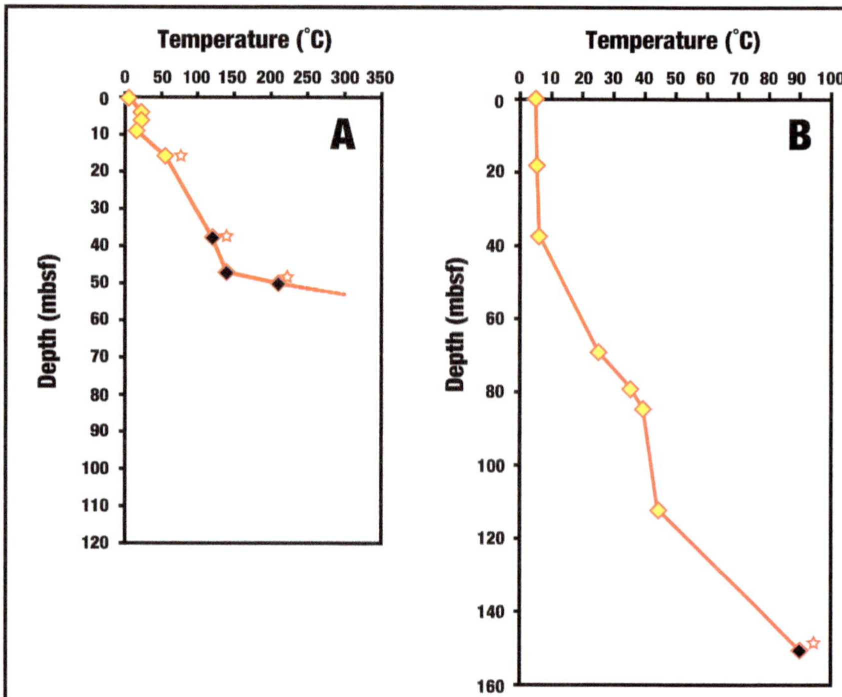


Figure 7. Depth profile of temperature in the subseafloor at Sites [A] C0014 and [B] C0017 as measured by the APCT3 shoe and thermoseal chemically impregnated temperature-sensitive strips. Yellow=APCT3, black=thermoseal strip. Stars=minimum values. Modified from Takai et al. (2011).

Holes C0013E and C0014G were cased with stainless steel pipe and capped with stainless steel corrosion caps with open outlets, and Hole C0016B was capped but not cased above a 3-m insertion pipe. Four post-drilling natural and artificial hydrothermal vents were created in Holes C0013E, C0014G, C0016A, and C0016B, in which hydrothermal fluid formerly trapped in the subseafloor ascended up the hole and exited into the ocean (Fig. 4). Temperature at all hydrothermal vent emissions was confirmed to be >240°C. These newly created hydrothermal vents will serve as windows into the subseafloor and will facilitate post-drilling, long-term studies of any associated microbial communities, fluid composition and flow, and *in situ* microbial and macrofaunal colonizations.

Interstitial water collected during Expedition 331 suggested that the subseafloor hydrothermal fluid regime of Iheya North field is chemically stratified with respect to chloride concentration. Pore water near the seafloor had the chloride concentration of seawater. It was underlain by a thin and possibly discontinuous vapor-rich layer at ≥5 mbsf at Site C0013 and at 29–38 mbsf at Site C0014. Beneath the vapor-rich layer lays a chloride-enriched brine, at least to the maximum depth of the holes drilled at these two sites. Previous surveys of the Iheya North hydrothermal vents always found that the discharging fluids were low in chloride but rich in vapor (Kawagucci et al., 2011). It was a puzzle where the complementary brines resided. Several theoretical calculations predicted that the greater densities of brines caused them to sink to greater depths within subseafloor hydrothermal reservoirs (Fontaine and Wilcock, 2006). Expedition 331 provided tentative evidence of subseafloor stratification of hydrothermal fluids that had separated phases. The post-drilling investigation of the hydrothermal fluids discharging from the natural and artificial hydrothermal vents may provide further evidence of the chemically stratified subseafloor hydrothermal fluid regime of Iheya North field.

At Site C0016, drilling at the summit of the active hydrothermal mound failed to recover core, and drilling at the base of the mound yielded only 2.1 meters of core from 45 m of penetration, but the core included the first Kuroko-type, sphalerite-rich black ore ever recovered from the modern subseafloor. The other four sites yielded interbedded hemipelagic and strongly pumiceous volcaniclastic sediment, along with volcanogenic breccias that are variably hydrothermally altered and mineralized, in the zeolite to greenschist facies. Detailed studies of these lithologies, in the context of all that is known about the active hydrothermal system at Iheya North Knoll, will provide direct evidence of how VHMS mineral deposits in general, and the Kuroko ores in particular, are formed.

Shipboard analyses did not confirm the presence of an active deep hot biosphere. Cell abundances were much lower than those found in previous Ocean Drilling Program/IODP sites on continental margins ($8.7 \times 10^5$–$2.4 \times 10^7$ cells $mL^{-1}$ sediment at Site C0013, $1.3 \times 10^6$–$4.1 \times 10^8$ cells $mL^{-1}$ sediment at Site C0014 and $5.8 \times 10^5$–$2.4 \times 10^7$ cells $mL^{-1}$ sediment at Site C0017; Takai et al., 2011), and attempts at culturing (hyper)thermophiles were generally unsuccessful. In fact, we found evidence for microbial activity in sediments within the upper 10–30 mbsf where temperatures were relatively low but little evidence in the deeper hydrothermally altered zones and hydrothermal fluid regime. Future shore-based, more detailed investigations may provide multiple lines of evidence for functionally active, metabolically diverse subseafloor microbial communities within the environments of the Iheya North hydrothermal system. In addition, the future investigations will seek to clarify the depth limits for the existence of microbial cells, metabolic activities, and living and fossil organic biomarkers in the subseafloor environment. These limits may represent a realistic boundary between the habitable and the uninhabitable zones in the subseafloor environments, namely a thermal barrier of the subseafloor biosphere.

## Acknowledgements

We thank the captains, offshore installation managers, operation superintendents, and all the crew, expedition staff and technicians on *Chikyu* from Mantle Quest Japan and CDEX, JAMSTEC, who assisted us in drilling, sampling, installation, and measurements during Expedition 331. We also thank the proponents and all involved in preparation of IODP Proposal 601. Without their cooperation, this expedition could have never been fulfilled. The curatorial team provided us with their support during the expedition and upon the post-expedition sample requests.

## The IODP Expedition 331 Scientists

K. Takai (Co-Chief Scientist), M.J. Mottl (Co-Chief Scientist), S.H. Nielsen (Expedition Project Manager), J.L. Birrien, S. Bowden, L. Brandt, A. Breuker, J.C. Corona, S. Eckert, H. Hartnett, S.P. Hollis, C.H. House, A. Ijiri, J. Ishibashi, Y. Masaki, S. McAllister, J. McManus, C. Moyer, M. Nishizawa, T. Noguchi, T. Nunoura, G. Southam, K. Yanagawa, S, Yang, and C. Yeats.

## References

Deming, J.W., and Baross, J.A., 1993. Deep-sea smokers: Windows to a subsurface biosphere? *Geochim. Cosmochim. Acta*, 57(14):3219–3230. doi:10.1016/0016-7037(93)90535-5

Expedition 331 Scientists, 2010. Deep hot biosphere. *IODP Prel. Rept.*, 331. doi:10.2204/iodp.pr.331.2010

Fontaine, F.J., and Wilcock, W.S.D., 2006. Dynamics and storage of brine in mid-ocean ridge hydrothermal systems. *J. Geophys. Res.*, 111(B6):B060102. doi:10.1029/2005JB003866

Gamo, T., Ishibashi, J., Tsunogai, U., Okamura, K., and Chiba, H., 2006. Unique geochemistry of submarine hydrothermal fluids from arc-back-arc settings of the Western Pacific. *In* Christie, D.M., Fisher, C.R., Lee, S.-M., and Givens, S. (Eds.), *Back-arc Spreading Systems: Geological, Biological, Chemical, and Physical interactions*. Geophys. Monogr., 166:147–161. doi:10.1029/166GM08

Gamo, T., Sakai, H., Kim, E.-S., Shitashima, K., and Ishibashi, J., 1991. High alkalinity due to sulfate reduction in the CLAM hydrothermal field, Okinawa Trough. *Earth Planet. Sci. Lett.*, 107(2):328–338. doi:10.1016/0012-821X(91)90080-2

Halbach, P., Nakamura, K., Wahsner, M., Lange, J., Sakai, H., Käselitz, L., Hansen, R.-D., et al., 1989. Probable modern analogue of Kuroko-type massive sulfide deposits in the Okinawa Trough back-arc basin. *Nature*, 338(6215):496–499. doi:10.1038/338496a0

Huber, J.A., and Holden, J.F., 2008. Modeling the impact of diffuse vent microorganisms along mid-ocean ridges and flanks. *In* Lowell, R.P., Seewald, J.S., Metaxas, A., and Perfit, M.R. (Eds.), *Magma to Microbe: Modeling Hydrothermal Processes at Oceanic Spreading Centers*. Geophys. Monogr., 178:215–231. doi:10.1029/178GM11

Inagaki, F., Kuypers, M.M.M., Tsunogai, U., Ishibashi, J.-I., Nakamura, K.-I., Treude, T., Ohkubo, S., et al., 2006. Microbial community in a sediment-hosted $CO_2$ lake of the southern Okinawa Trough hydrothermal system. *Proc. Natl. Acad. Sci. U.S.A.*, 103(38):14164–14169. doi:10.1073/pnas.0606083103

Kawagucci, S., Chiba, H., Ishibash, J., Yamanaka, T., Toki, T., Muramatsu, Y., Ueno, Y., et al., 2011. Hydrothermal fluid geochemistry at the Iheya North field in the mid-Okinawa Trough: Implication for origin of methane in subseafloor fluid circulation systems. *Geochemical. J.*, 45(2):109–124.

Konno, U., Tsunogai, U., Nakagawa, F., Nakaseama, M., Ishibashi, J., Nunoura, T., and Nakamura, K., 2006. Liquid $CO_2$ venting on the seafloor: Yonaguni Knoll IV hydrothermal system, Okinawa Trough. *Geophys. Res. Lett.*, 33(16):L16607. doi:10.1029/2006GL026115

Lee, C.-S., Shor, G.G., Jr., Bibee, L.D., Lu, R.S., and Hilde, T.W.C., 1980. Okinawa Trough: Origin of a back-arc basin. *Mar. Geol.*, 35(1–3):219–241. doi:10.1016/0025-3227(80)90032-8

Letouzey, J., and Kimura, M., 1986. The Okinawa Trough: Genesis of a backarc basin developing along a continental margin. *Tectonophysics*, 125(1–3):209–230. doi:10.1016/0040-1951(86)90015-6

Nakagawa, S., Takai, K., Inagaki, F., Chiba, H., Ishibashi, J., Kataoka, S., Hirayama, H., Nunoura, T., Horikoshi, K., and Sako, Y., 2005. Variability in microbial community and venting chemistry in a sediment-hosted backarc hydrothermal system: Impacts of subseafloor phase-separation. *FEMS Microbiol. Ecol.*, 54(1):141–155. doi:10.1016/j.femsec.2005.03.007

Narita, H., Harada, K., and Tsunogai, S., 1990. Lateral transport of sediment particles in the Okinawa Trough determined by natural radionuclides. *Geochem. J.*, 24:207–216. doi:10.2343/geochemj.24.207

Nunoura, T., and Takai, K., 2009. Comparison of microbial communities associated with phase-separation-induced hydrothermal fluids at the Yonaguni Knoll IV hydrothermal field, the Southern Okinawa Trough. *FEMS Microbiol. Ecol.*, 67:351–370. doi:10.1111/j.1574-6941.2008.00636.x

Nunoura, T., Oida, H., Nakaseama, M., Kosaka, A., Ohkubo, S.B., Kikuchi, T., Kazama, H., et al., 2010. Archaeal diversity and distribution along thermal and geochemical gradients in hydrothermal sediments at the Yonaguni Knoll IV hydrothermal field in the southern Okinawa Trough. *Appl. Environ. Microbiol.*, 76(4):1198–1211. doi:10.1128/AEM.00924-09

Sakai, H., Gamo, T., Kim, E.-S., Shitashima, K., Yanagisawa, F., Tsutsumi, M., Ishibashi, J., et al., 1990. Unique chemistry of the hydrothermal solution in the mid-Okinawa Trough backarc basin. *Geophys. Res. Lett.*, 17(12):2133–2136. doi:10.1029/GL017i012p02133

Shipboard Scientific Party, 2002. Leg 193 summary. *In* Binns, R.A., Barriga, F.J.A.S., Miller, D.J., et al., *Proc. ODP, Init. Repts.*, 193: College Station, TX (Ocean Drilling Program). doi:10.2973/odp.proc.ir.193.101.2002

Suzuki, R., Ishibashi, J., Nakaseama, M., Konno, U., Tsunogai, U., Gena, K., and Chiba, H., 2008. Diverse range of mineralization induced by phase separation of hydrothermal fluid: Case study of the Yonaguni Knoll IV hydrothermal field in the Okinawa Trough back-arc basin. *Res. Geol.*, 58(3):267–288. doi:10.1111/j.1751-3928.2008.00061.x

Takai, K., and Nakamura, K., 2010. Compositional, physiological and metabolic variability in microbial communities associated with geochemically diverse, deep-sea hydrothermal vent fluids. *In* Barton, L., Mendl, M., and Loy, A. (Eds.), *Geomicrobiology: Molecular and Environmental Perspective:* New York (Springer), 251–283. doi:10.1007/978-90-481-9204-5_12

Takai, K., Komatsu, T., Inagaki, F., and Horikoshi, K., 2001. Distribution of archaea in a black smoker chimney structure. *Appl. Environ. Microbiol.*, 67(8):3618–3629. doi:10.1128/AEM.67.8.3618-3629.2001

Takai, K., Mottl, M.J., Nielsen, S.H., and the Expedition 331 Scientists, 2011. *Proc. IODP, 331*: Washington, DC (Integrated Ocean Drilling Program Management International, Inc.). doi:10.2204/iodp.proc.331.2011

Takai, K., Nakagawa, S., Reysenbach, A.L., and Hoek, J., 2006. Microbial ecology of mid-ocean ridges and backarc basins. *In* Christie, D.M., Fisher, C.R., Lee, S.M., and Givens, S. (Eds.), *Back-Arc Spreading Systems: Geological, Biological, Chemical and Physical Interactions*. Geophys. Monogr., 166:185–213. doi:10.1029/144GM23

Takai, K., Nunoura, T., Horikoshi, K., Shibuya, T., Nakamura, K., Suzuki, Y., Stott, M., et al., 2009. Variability in microbial communities in black smoker chimneys at the NW caldera vent field, Brothers Volcano, Kermadec arc. *Geomicrobiol. J.*, 26(8):552–569. doi:10.1080/01490450903304949

Takai, K., Nunoura, T., Ishibashi, J., Lupton, J., Suzuki, R., Hamasaki, H., Ueno, Y., et al., 2008. Variability in the microbial communities and hydrothermal fluid chemistry at the newly discovered Mariner hydrothermal field, southern Lau Basin. *J. Geophys. Res.*, 113:G02031. doi:10.1029/2007JG000636

Ueno, H., Hamasaki, H., Murakawa, Y., Kitazono, S., and Takeda, T., 2003. Ore and gangue minerals of sulfide chimneys from the North Knoll, Iheya Ridge, Okinawa Trough, Japan. *JAMSTEC J. Deep Sea Res.*, 22:49–62, http://docsrv.godac.jp/MSV2_DATA/23/22_06.pdf.

You, C.-F., Butterfield, D.A., Spivack, A.J., Gieskes, J.M., Gamo, T., and Campbell, A.J., 1994. Boron and halide systematics in submarine hydrothermal systems: Effects of phase separation and sedimentary contributions. *Earth Planet. Sci. Lett.*, 123(1–3):227–238. doi:10.1016/0012-821X(94)90270-4

## Authors

**Ken Takai**, Subsurface Geobiology Advanced Research (SUGAR) Project, Japan Agency for Marine-Earth Science & Technology (JAMSTEC), 2-15 Natsushima-cho, Yokosuka 237-0061, Japan, e-mail: kent@jamstec.go.jp.

**Michael J. Mottl,** Department of Oceanography, School of Ocean and Earth Science and Technology (SOEST), University of Hawaii at Manoa, 1000 Pope Road, Honolulu, HI 96822, e-mail: mmottl@soest.hawaii.edu.

**Simon H.H. Nielsen,** Center for Deep Earth Exploration (CDEX), Japan Agency for Marine-Earth Science & Technology (JAMSTEC), 3173-25 Showa-machi Kanazawa-ku, Yokohama 236-0001, Japan, e-mail: micropora@gmail.com.

**and the IODP Expedition 331 Scientists**

## Related Web Link

http://www.jamstec.go.jp/chikyu/eng/Expedition/okinawa/exp331.html

28

# IODP Expedition 335: Deep Sampling in ODP Hole 1256D

by Damon A.H. Teagle, Benoit Ildefonse, Peter Blum and the IODP Expedition 335 Scientists

## Abstract

Observations of the gabbroic layers of untectonized ocean crust are essential to test theoretical models of the accretion of new crust at mid-ocean ridges. Integrated Ocean Drilling Program (IODP) Expedition 335 ("Superfast Spreading Rate Crust 4") returned to Ocean Drilling Program (ODP) Hole 1256D with the intention of deepening this reference penetration of intact ocean crust a significant distance (~350 m) into cumulate gabbros. Three earlier cruises to Hole 1256D (ODP 206, IODP 309/312) have drilled through the sediments, lavas, and dikes and 100 m into a complex dike-gabbro transition zone.

Operations on IODP Expedition 335 proved challenging throughout, with almost three weeks spent re-opening and securing unstable sections of the hole. When coring commenced, the comprehensive destruction of the coring bit required further remedial operations to remove junk and huge volumes of accumulated drill cuttings. Hole-cleaning operations using junk baskets were successful, and they recovered large irregular samples that document a hitherto unseen sequence of evolving geological conditions and the intimate coupling between temporally and spatially intercalated intrusive, hydrothermal, contact-metamorphic, partial melting, and retrogressive processes.

Hole 1256D is now clean of junk, and it has been thoroughly cleared of the drill cuttings that hampered operations during this and previous expeditions. At the end of Expedition 335, we briefly resumed coring before undertaking cementing operations to secure problematic intervals. To ensure the greatest scientific return from the huge efforts to stabilize this primary ocean lithosphere reference site, it would be prudent to resume the deepening of Hole 1256D in the nearest possible future while it is open to full depth.

## Why Study Ocean Crust Forming at Fast Spreading Rates?

The vast majority (~70%) of magma derived from the mantle is brought into the Earth's crust at the mid-ocean ridges, and approximately two-thirds of that magma cools

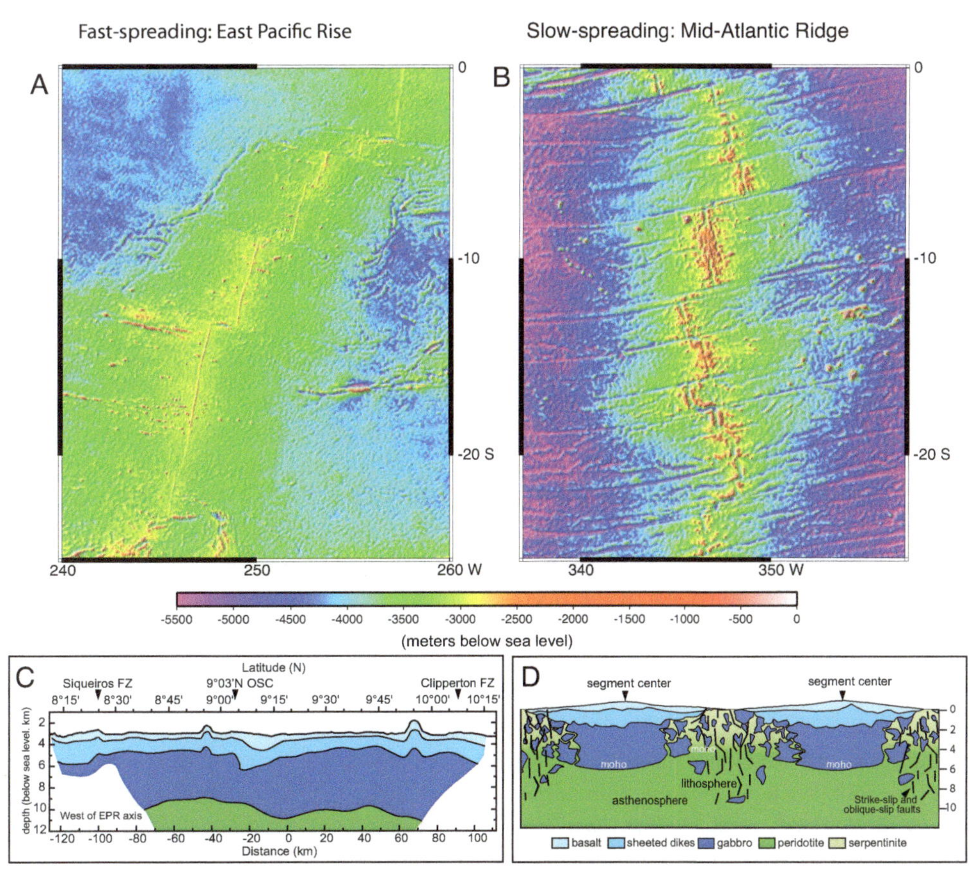


Figure 1. Bathymetric maps of the [A] fast spreading southern East Pacific Rise (EPR) compared to the [B] slow spreading Mid-Atlantic Ridge (MAR) from the equator to 25°S. Note the more pronounced bathymetric relief and axial graben along the MAR as well as the more closely spaced transform faults. Interpreted along-axis sections for the [C] EPR from ~8° to 10° N (after Canales et al., 2003) and [D] MAR (after Cannat et al., 1995).

and crystallizes in the lower portion of the ocean crust. Seismic, bathymetric, and marine geological observations indicate that ocean crust formed at fast spreading rates (full rate >80 mm yr$^{-1}$) is much less variable than crust formed at slow spreading rates (<40 mm yr$^{-1}$; Fig. 1) and is closer to the ideal "Penrose" pseudo-stratigraphy developed from ophiolites. Hence, understanding fast-spreading accretion processes at a few sites might reasonably be extrapolated to describe a significant portion of the Earth's surface. Although <20% of modern ridges are moving apart at fast spreading rates (Fig. 2), nearly 50% of present-day ocean crust and ~30% of the Earth's surface were produced by this pace of spreading. The great majority of crust that subducted back into the mantle over the past ~200 m.y. formed at fast-spreading ridges (Müller et al., 2008), making characterizing this style of crust most relevant for understanding the recycling of crustal and ocean-derived components back into the mantle.

The rate of spreading of the oceanic lithosphere has profound effects on the style of accretion at mid-ocean ridges due to changing balances between plate motion, magma production, conductive and hydrothermal cooling, detachment tectonics, and serpentinization of the upper mantle (Fig. 1c vs. 1d; Dick, 1989; Cannat et al., 1995, 2004, 2009; Chen and Phipps Morgan, 1996; Dick et al., 2003; Escartin et al., 2008). Although insights on formation of intrusive crust at detachment-dominated, slow-spread lithosphere have been obtained (e.g., Holes 735B, U1309D, and other sites; ODP Legs 118, 153, 176, 209, and IODP Expeditions 304–305; Dick et al., 2000; Ildefonse et al., 2007; Kelemen et al., 2007), the thermal regime and the melt supply and delivery in these settings differ significantly from those of the axial zone in fast-spreading lithosphere. Detailed understanding of the relatively uniform mechanisms operating at fast spreading ridges would provide a vital benchmark against which heterogeneous accretion on slow-spreading ridges could be compared.

## Geological Setting of Site 1256 and Science Objectives of IODP Expedition 335

IODP Expedition 335 (13 April to 3 June 2011) was the fourth scientific drilling cruise of the Superfast campaign, and it returned to ODP Hole 1256D (6°44.163'N, 91°56.061'W) to deepen this ocean crust reference penetration a significant distance into cumulate gabbros. ODP Hole 1256D is located on 15-Ma-crust in the eastern equatorial Pacific Ocean in oceanic basement that formed during a sustained episode of superfast ocean ridge spreading (>200 mm yr$^{-1}$; Wilson, 1996). Ocean crust formed at a superfast spreading rate was deliberately targeted because there is strong evidence from mid-ocean ridge seismic experiments that gabbros occur at shallower depths in intact ocean crust with higher spreading rates. Therefore, the often difficult-to-drill upper ocean crust should be relatively thin. Expedition 335 follows on from ODP Leg 206 in 2002 and IODP Expeditions

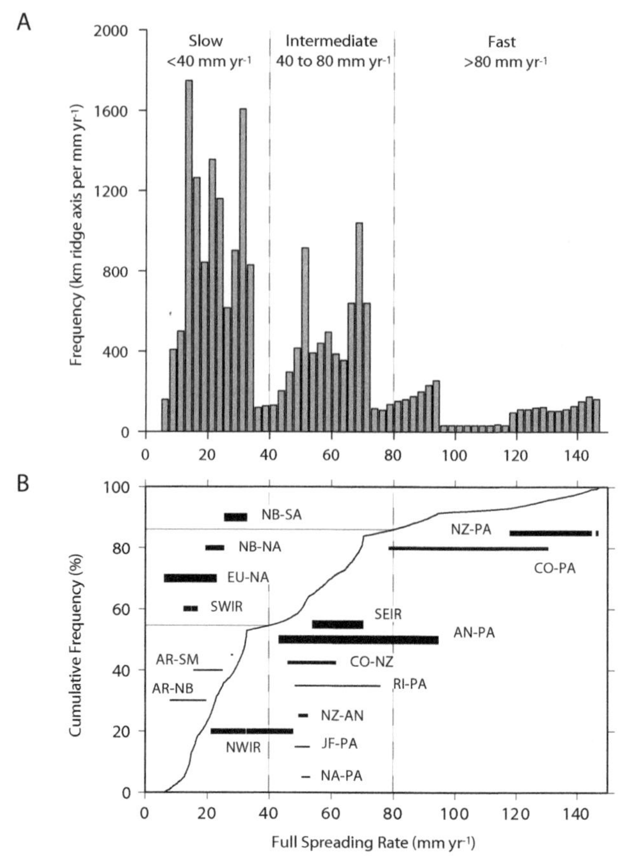


Figure 2. Distribution of spreading rates for major active plate boundaries, presented in [A] histogram form and as [B] cumulative distribution. The rates are the best-fit rates of the MORVEL model (DeMets et al., 2010), and ridge length is measured as the component perpendicular to spreading direction. Horizontal lines on the cumulative plot show the range of spreading rate for each plate pair, with line width scaled approximately to plate boundary length. Plate-pair labels follow the MORVEL convention, except in the Indian Ocean where the southeast, southwest, and northwest branches of the ridge system are grouped for 2–3 plate pairs to simplify labeling

309/312 in 2005 (Wilson et al., 2003; Teagle et al., 2006; Wilson et al., 2006) that prepared the first scientific borehole for deep drilling by installing a large reentry cone secured with almost 270 meters of 16-inch casing through the 250-m-thick sedimentary overburden and cemented into the uppermost basement. Hole 1256D was then deepened through a ~810-m-thick sequence of lavas and a thin (~346 m) sheeted dike complex, the lower 60 meters of which is strongly contact metamorphosed to granoblastic textures (Fig. 4). The first gabbroic rocks were encountered at 1407 meters below seafloor (mbsf) where the hole entered a complex dike-gabbro transition zone that includes two 20- to 50-m thick gabbro lenses intruded into granoblastic dikes. As of the end of Expedition 312 in December 2005, Hole 1256D had a total depth of 1507.1 mbsf, and it was open to its full depth.

It was expected that even with a shortened cruise (~6 weeks), coring on IODP Expedition 335 would be able to deepen Hole 1256D ~350 m, completely through the dike-gabbro transition zone and varitextured gabbros and a

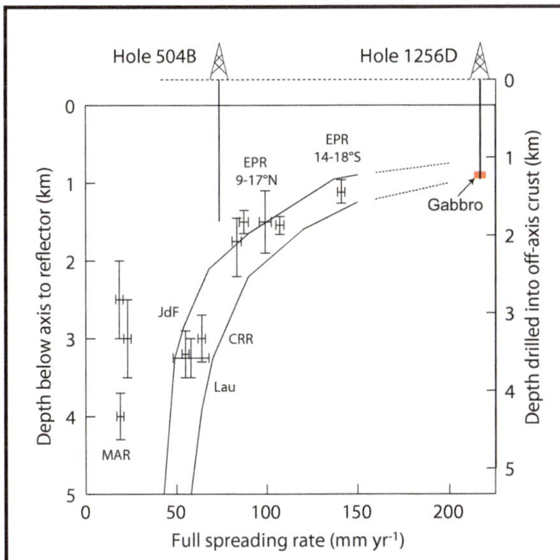


Figure 3. Depth to axial low-velocity zone plotted against spreading rate (Purdy et al., 1992; Carbotte et al., 1997). Depth versus spreading rate predictions from two models of Phipps Morgan and Chen (1993) are shown, extrapolated subjectively to 200 mm y$^{-1}$. Penetration to date in Holes 504B and 1256D is shown, with the depth at which gabbros were intersected indicated by the red box. Following core descriptions, a thickness of ~300 m of off-axis lava is shown for Hole 1256D and assumed for Hole 504B. EPR=East Pacific Rise, JdF=Juan de Fuca Ridge, Lau=Valu Fa Ridge in Lau Basin, CRR=Costa Rica Rift, MAR=Mid-Atlantic Ridge.

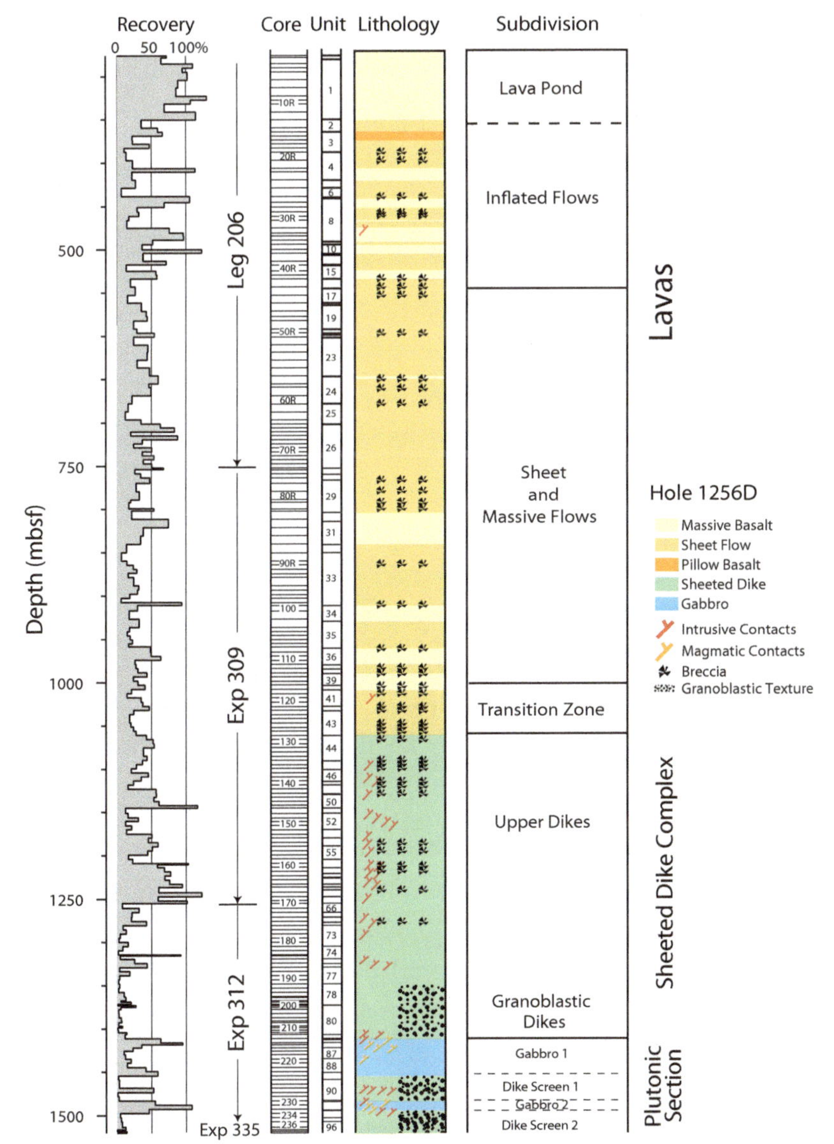


Figure 4. Stratigraphic column for Hole 1256D at the end of Expedition 335, showing the major and minor lithologic divisions of the upper ocean crust.

significant distance into true cumulate gabbros. Listed below are the specific scientific objectives of IODP Expedition 335 that were to be addressed by deepening Hole 1256D a significant distance into cumulate gabbros.

What is the major mechanism of magmatic accretion in crust formed at fast spreading rates? Is the lower crust formed by gabbro glaciers or sheeted sills (Korenaga and Kelemen, 1998; Fig. 5) or by some mixed or unknown mechanism?

How is heat extracted from the lower ocean crust? Specifically, what is the role of hydrothermal circulation in extracting latent heat from the lower ocean crust?

What is the geological significance of the seismic layer 2/3 boundary at Site 1256?

What is the magnetic contribution of the gabbro layer? Can the magnetic polarity structure of the lower crust be used to constrain cooling rates?

## Operations

Expedition 335 reentered Hole 1256D more than five years after the last expedition to this site. The expedition encountered and overcame a series of significant engineering challenges, including a major obstruction at a depth of ~922 mbsf that had initially prevented reentry into the hole to its full depth of 1507 meters. At the bottom of the hole, coring resumed in the very hard granular rocks recovered during the previous Superfast expedition. Although there may only be a few tens of meters of these particularly tenacious granoblastic basalts formed by the contact metamorphism of sheeted dikes, their extreme toughness once more proved challenging to sample, resulting in the grinding down of one of the hardest formation coring bits (C9) into a smooth stump (Fig. 6a,b).

Action was then undertaken to clear the bottom of the hole of metal debris from the failed coring bit and drilling cuttings. These efforts returned hundreds of kilograms of rocks and drill cuttings (Fig. 6d), including large blocks (up to 3.5 kg) of the

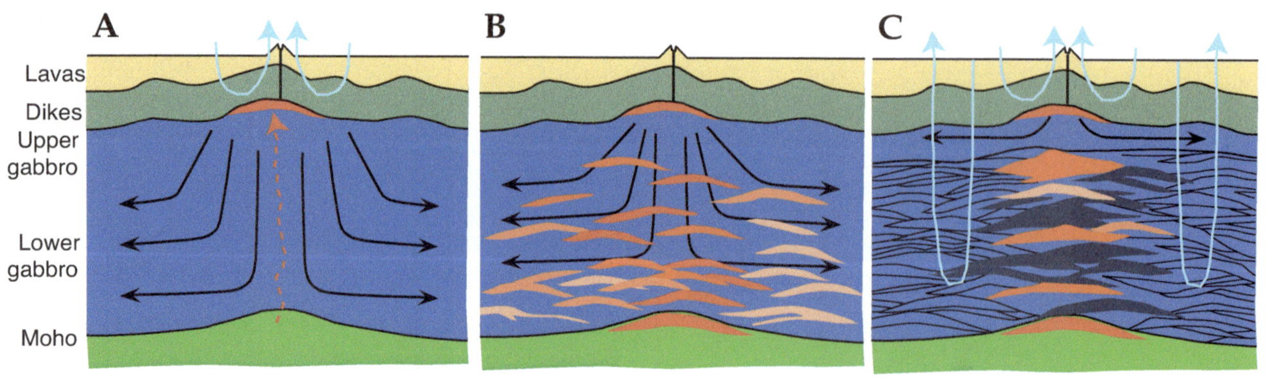


Figure 5. Diagrams of crustal accretion models (modified from Korenaga and Kelemen, 1998). Black arrows show the movement of the solid lower crust; blue arrows show the dominant zones where hydrothermal circulation will remove latent and sensible heat; red arrows show the movement of magma (unknown in all models). [A] Gabbro glacier ductile flow model (Henstock et al., 1993; Phipps Morgan and Chen, 1993; Quick and Denlinger, 1993). Ductile flow down and outward from a high level axial magma chamber constructs the lower crust. [B] A hybrid model of ductile flow with sill intrusions (Boudier et al., 1996). [C] "Sheeted" or "stacked" sill model of *in situ* formation of the lower crust by on-axis sill intrusions (Bédard et al., 1988; Kelemen et al., 1997; Kelemen and Aharonov, 1998; MacLeod and Yoauancq, 2000). (Original figure by Rosalind Coggon).

culprit granoblastic basalts that hitherto had only been very poorly recovered through coring . A limited number of gabbro boulders were also recovered, indicating again that we are tantalizingly close to breaking through into the gabbroic layer.

## Scientific Results

Hole opening and remediation operations resulted in a major loss of time from the coring and wireline activities planned for Expedition 335. Coring on this expedition deepened Hole 1256D only modestly, from 1507.1 mbsf to 1521.6 mbsf (Fig. 4) at low rates of penetration (0.9 m $hr^{-1}$) and recovery (11%).

Hole cleaning operations at the bottom of Hole 1256D, particularly those runs that deployed the reverse-circulation junk basket, brought back a unique collection of large cobbles, angular rubble, and fine cuttings of principally strongly to completely recrystallized granoblastic basalt with minor gabbroic rocks and evolved plutonic rocks. The granoblastic basalts record high-temperature (granulite facies) metamorphism of previously hydrothermally altered sheeted dikes, with unequivocal evidence for hydrothermal alteration before and after the formation of granoblastic textures (Teagle et al., 2006; Koepke et al., 2008; France et al., 2009; Alt et al., 2010). The large blocks exhibit intrusive, structural, and textural relationships, and overprinting and cross-cutting hydrothermal alteration and metamorphic paragenetic sequences. Some of the rubble and ~30% of the fine cuttings recovered by fishing and cleaning operations most likely came from the lava sequences at the top of the hole on the basis of igneous textures and low-temperature alteration minerals (Mg-saponite, amorphous silica). In contrast, the high extent of metamorphic recrystallization exhibited by the granoblastic basalts, and operational factors (e.g., pipe movements), provide strong evidence that the granoblastic

Figure 6. Photographs of some drilling, fishing, and milling tools illustrating the multiple operations achieved during IODP Expedition 335. [A] & [B] Remains of the Ulterra C9 RCB coring bit used during Run 9. The bit was probably used for ~10 h after it disintegrated, which resulted in this spectacularly abraded and sculptured bit ("Stumpy"), something never seen before by the drillers. [C] Bottom of the Bowen Reverse Circulation Junk Basket (RCJB), showing its hard-facing structure and the junk catcher spring fingers inside. This tool recovered a series of large-scale cobbles (up to 5 kg) of granoblastic basalt. [D] A series of RCJB runs returned a bottom hole assembly packed by fine-grained cuttings. This contributed efficiently to cleaning Hole 1256D. [E] Heavily worn and under-gauge 9-inch flat-bottomed milling tool. This tool worked at the bottom of the hole for 12 h; its final state indicates the very abrasive nature of metal debris and/or rocks at the bottom of the hole and an under-gauge lowermost portion (tens of centimeters) of the hole. Note for comparison the hard-facing structure of the next milling tool on the right side of the picture. [F] 9-inch Smith FH3VPS tricone bit used for Run 16. This armored bit was efficient to ream and clean the under-gauge bottom of the hole.

basalts, minor gabbros, and evolved plutonic rocks were sourced from the lowermost reaches of Hole 1256D (1494.9–1521.6 mbsf), most probably from below the lower-most recovered gabbroic interval (Fig. 4). Hence, the rocks recovered during Expedition 335 represent a ~15-m interval of the upper crust/lower crust transition, occurring below the ~90-m section, recovered during Expedition 312, of two gabbroic bodies (Gabbro 1 and Gabbro 2) separated by Dike Screen 1 (Fig. 4).

When the textural and contact relationships exhibited by the rocks recovered from the junk baskets are placed in the geological context of the Hole 1256D stratigraphy, a vision emerges of a complex, dynamic thermal boundary layer zone. This region of the crust between the principally hydrothermal domain of the upper crust and the intrusive magmatic domain of the lower crust is one of evolving geological conditions. There is intimate coupling between temporally and spatially intercalated magmatic, hydrothermal, partial melting, intrusive, metamorphic, and retrograde processes.

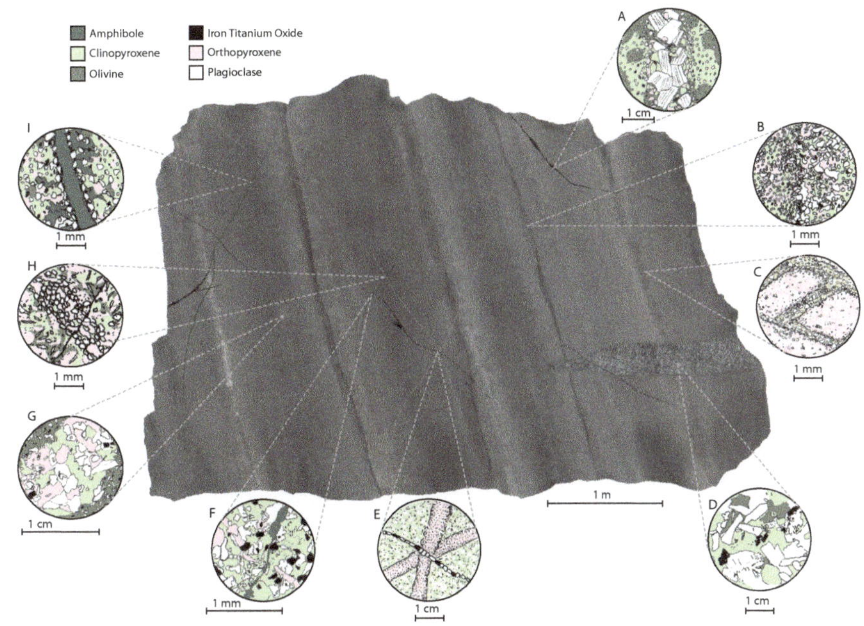


Figure 7. Sketch of a hypothetical outcrop in a future Site 1256 ophiolite of the dike-gabbro transition zone showing the contact relationships observed in the cores and junk basket rocks recovered during Expedition 335. Small sketches are drawn from sample and thin-section descriptions done during Expedition 335. [A] Diorite dikelet, comprised predominantly of primary igneous amphibole and plagioclase, cross-cutting granoblastic basalt. [B] Fine-grained dike chilled against coarser grained dike. The entire sample is recrystallized to granoblastic assemblages of plagioclase, orthopyroxene, clinopyroxene, magnetite, and ilmenite. Later fine post-contact-metamorphic hydrothermal amphibole veins (not shown) cut across the contact and both dikes. [C] Chilled and brecciated dike margin recrystallized to granoblastic assemblages. Angular clasts consist of granoblastic plagioclase with minor orthopyroxene, clinopyroxene, ilmenite, and magnetite. They are recrystallized from chilled dike margin breccia protolith. Clast matrix is orthopyroxene rich with minor clinopyroxene, plagioclase, and oxides; it is recrystallized from hydrothermal minerals (amphibole and chlorite) that veined and cemented the breccia protolith. [D] Subophitic texture in gabbro. [E] Diorite vein cross-cutting a conjugate set of metamorphic orthopyroxene veins. [F] Post-contact-metamorphic hydrothermal hornblende vein cutting granoblastic basalt. [G] Granoblastic basalt with a diorite patch. [H] Granoblastic orthopyroxene vein, recrystallized from precursor hydrothermal vein, in granoblastic dike. Orthopyroxene vein is cut by small post-contact-metamorphic hydrothermal amphibole vein. [I] Post-contact-metamorphic hydrothermal amphibole vein cutting granoblastic basalt, with 1-mm-wide amphibole-rich alteration halo where pyroxenes are replaced by amphibole.

With only a minor depth advance in Hole 1256D, we are yet to recover samples of cumulate gabbros required to test models of ocean ridge magmatic accretion and the intensity of hydrothermal cooling at depth. The total vertical thickness of granoblastic basalts is over 114 m, and Dike Screen 2 is now about the same thickness (so far) as Dike Screen 1. Highly perched, isolated gabbro intrusions are uncommon in ophiolites. The energy requirements for the granoblastic recrys-tallization at granulite facies condition of a >114-m-thick zone of sheeted dikes massively exceeds the thermal capacity of Gabbros 1 and 2 (Coggon et al., 2008; Koepke et al., 2008) if a simple subhorizontal arrangement of these layers is assumed. The enormous heat requirements for such extensive granulite facies recrystallization, the evidence for partial melting, together with the tantalizing presence of minor but not uncommon gabbroic rocks and felsic intrusives, dikelets, and veins, strongly indicates that the layer of purely plutonic rocks should be at most only a few tens of meters deeper in the hole.

## Conclusions

Although the extensive remedial operations on Expedition 335 precluded significant deepening of Hole 1256D, significant progress was made in improving the stability of the borehole. The most problematic out-of-gauge zone at ~920–960 mbsf that caused reentry problems on Expeditions 312 and 335 has been stabilized with cement. The bottom of the hole has been cleared of rubble and junk, and there appears to be only a short (<1 m) under-gauge zone. What is important is that the regular, large sweeps of high viscosity mud have finally overcome and expunged the vast amount of fine cuttings recirculating in the hole, some most likely resident since ODP Leg 206. This is shown by the absence of soft fill in the final ~5 reentries, compared to upwards of 50 meters of soft fill at the end of Expedition 312. The engineering efforts on Expedition 335 have repaired and prepared Hole 1256D for further deep drilling, following five years of neglect. Hole 1256D is 1500 m of hard rock coring closer to cumulate gabbros than any other options in intact ocean crust and once more poised to answer fundamental questions about the formation of new crust at fast spreading mid-ocean ridges. This would be best achieved by a timely return to the site.

## Acknowledgments

IODP Expedition 335 was not an easy cruise. Engineering difficulties throughout the expedition presented a continuous series of challenges. The Shipboard Science Party is extremely grateful for the immense efforts of the transocean crew aboard *JOIDES Resolution* for their perseverance, innovation, open discourse, and "can-do" attitude. This team was aboard when Hole 1256D was first spudded in and cased into basement on ODP Leg 206; they were not going to allow the hole to fail. The Texas A&M University (TAMU) Operations Superintendent Ron Grout has sailed on three of the four cruises to Hole 1256D, and his advice in testing situations proved invaluable. The proponents of the Superfast campaign thank the IODP Science Advisory Structure (SAS) for their extension of Expedition 335; without the extra two weeks there could have been the catastrophic loss of Hole 1256D, but this extra time allowed the clearing of junk and cuttings and preservation of the hole for future deepening into the cumulate gabbro layers.

## IODP Expedition 335 Scientists

Damon A.H. Teagle (Co-Chief Scientist), Benoit Ildefonse (Co-Chief Scientist), Peter Blum (Staff Scientist), Natsue Abe, Bénédicte Abily, Yoshiko Adashi, Jeffrey C. Alt, Ryo Anma, Graham Baines, Jeremy Deans, Henry J.B. Dick, Daisuke Endo, Eric C. Ferré, Lydéric France, Marguerite Godard, Gilles Guérin, Michelle Harris, Yoon-Mi Kim, Juergen H. Koepke, Mark D. Kurz, C. Johan Lissenberg, Sumio Miyashita, Antony Morris, Ryo Oizumi, Betchaida D. Payot, Marie Python, Parijat Roy, Jessica L. Till, Masako Tominaga, Douglas S. Wilson, and Natalia Zakharova.

## References

Alt, J.C., Laverne, C., Coggon, R.M., Teagle, D.A.H., Banerjee, N.R., Morgan, S., Smith-Duque, C.E., Harris, M., and Galli, L., 2010. Subsurface structure of a submarine hydrothermal system in ocean crust formed at the East Pacific Rise, ODP/IODP Site 1256. *Geochem., Geophys., Geosyst.* 11:Q10010–Q10038. doi:10.1029/2010GC003144

Bédard, J.H., Sparks, R.S.J., Renner, R., Cheadle, M.J., and Hallworth, M.A., 1988. Peridotite sills and metasomatic gabbros in the Eastern layered series of the Rhum complex. *J. Geol. Soc. London*, 145(2):207–224. doi:10.1144/gsjgs.145.2.0207

Boudier, F., Nicolas, A., and Ildefonse, B., 1996. Magma chambers in the Oman ophiolite: Fed from the top or from the bottom? *Earth Planet. Sci. Lett.*, 144:239–250. doi:10.1016/0012-821X(96)00167-7

Canales, J.P., Detrick, R.S., Toomey, D.R., and Wilcock, W.S.D., 2003. Segment-scale variations in the crustal structure of 150-300 kyr old fast spreading oceanic crust (East Pacific Rise, 8 degrees 15' N-10 degrees 5' N) from wide-angle seismic refraction profiles. *Geophys. J. Int.*, 152:766–794.

Cande, S.C., and Kent, D.V., 1995. Revised calibration of the geomagnetic polarity timescale for the Late Cretaceous and Cenozoic. *J. Geophys. Res.*, 100: 6093–6095. doi:10.1029/94JB03098

Cannat, M., Cann, J.R., and Maclennan, J., 2004. Some hard rock constraints on the heat supply to mid-ocean ridges. *In* German, C., Lin, J., and Parson, L. (Eds.), *Mid-Ocean Ridges: Hydrothermal Interactions Between the Lithosphere and Oceans*. Geophys. Monogr. Ser., 148:111–150. Washington, D. C. (American Geophysical Union). doi:10.1029/148GM05

Cannat, M., Manatschal, G., Sauter, D., and Peron-Pinvidic, G., 2009. Assessing the conditions of continental breakup at magma-poor rifted margins: What can we learn from slow spreading mid-ocean ridges? *Compt. Rend. Geosci.* 341:406–427. doi:10.1016/j.crte.2009.01.005

Cannat, M., Mevel, C., Maia, M., Deplus, C., Durand, C., Gente, P., Agrinier, P., et al., 1995. Thin crust, ultramafic exposures, and rugged faulting patterns at Mid-Atlantic Ridge (22°–24°N). *Geology*, 23:49–52. doi:10.1130/0091-7613(1995)023%3C0049:TCUEAR%3E2.3.CO;2

Carbotte, S., Mutter, C., Mutter, J., and Ponce-Correa, G., 1997. Influence of magma supply and spreading rate on crustal magma bodies and emplacement of the extrusive layer: Insights from the East Pacific Rise at lat 16°N. *Geology*, 26: 455–458. doi:10.1130/0091-7613(1998)026<0455:IOMSAS>2.3.CO;2

Chen, Y.J., and Phipps Morgan, J., 1996. The effects of spreading rate, the magma budget, and the geometry of magma emplacement on the axial heat flux at mid-ocean ridges. *J Geophys. Res.*, 101:11475–11482. doi:10.1029/96JB00330

Coggon, R.M., Alt, J.C., and Teagle, D.A., 2008. Thermal history of ODP Hole 1256D lower sheeted dikes: Petrology, chemistry and geothermometry of the granoblastic dikes. *Eos, Trans. Am. Geophys. Union*, 89 (53, Suppl.): V44B-08 (Abstract) http://www.agu.org/meetings/fm08/waisfm08.html

DeMets, C., Gordon, R.G., and Argus, D.F., 2010. Geologically current plate motions, *Geophys. J. Int.*, 181:1–80. doi:10.1111/j.1365-246X.2009.04491.x

Dick, H.J.B., 1989. Abyssal peridotites, very slow spreading ridges and ocean ridge magmatism. *In* Saunders, A.D., Norry, M.J. (Eds.), *Magmatism in the ocean basins*. Geol. Soc. Spec. Pub., London, 42:71–105. Oxford (Blackwell Scientific Publications).

Dick, H.J.B., Lin, J., Schouten, H., 2003. An ultraslow-spreading class of ocean ridge. *Nature*, 426:405–412. doi:10.1038/nature02128

Dick, H.J.B., Natland, J.H., Alt, J.C., Bach, W., Bideau, D., Gee, J.S., Haggas, S., et al., 2000. A long *in situ* section of the lower ocean crust: Results of ODP Leg 176 drilling at the Southwest Indian Ridge. *Earth Planet. Sci. Lett.*, 179:31–51. doi:10.1016/S0012-821X(00)00102-3

Escartin, J., Smith, D.K., Cann, J., Schouten, H., Langmuir, C.H., and Escrig, S., 2008. Central role of detachment faults in accretion of slow-spreading oceanic lithosphere. *Nature*, 455:790–795. doi:10.1038/nature07333

France, L., Ildefonse, B., and Koepke, J., 2009. Interactions between magma and hydrothermal system in Oman ophiolite and in IODP Hole 1256D: Fossilization of a dynamic melt lens at fast spreading ridges. *Geochem. Geophys. Geosys.* 10:Q10O19. doi:10.1029/2009GC002652

Henstock, T.J., Woods, A.W., and White, R.S., 1993. The accretion of oceanic crust by episodic sill intrusion. *J. Geophys. Res.*, 98(B3): 4143–4161. doi:10.1029/92JB02661

Ildefonse, B., Blackman, D.K., john, B.E., Ohara, Y., Miller, D.J., MacLeod, C.J., and the IODP Expeditions 304–305 Party, 2007. Oceanic core complexes and crustal accretion at slow-spreading ridges. *Geology*, 35:623–626. doi:10.1130/G23531A.1

Kelemen, P.B., and Aharonov, E., 1998. Periodic formation of magma fractures and generations of layered gabbros in the lower crust beneath oceanic spreading ridges. *In* Buck, W.R., Delaney, P.T., Karson, J.A., and Lagabrielle, Y. (Eds.), *Faulting and Magmatism at Mid-Ocean Ridges.* Geophysical Monograph, 106:267–289. Washington, DC (American Geophysical Union). doi:10.1029/GM106p0267

Kelemen, P.B., Koga, K. and Shimizu, N., 1997. Geochemistry of gabbro sills in the crust-mantle transition zone of the Oman ophiolite: implications for the origin of the oceanic lower crust. *Earth Planet. Sci. Lett.*, 146:475–488.

Kelemen, P.B., Kikawa, E., Miller, D.J., and Shipboard Scientific Party, 2007. Leg 209 summary: Processes in a 20-km-thick conductive boundary layer beneath the Mid-Atlantic Ridge, 14°–16°N. *In* Kelemen, P.B., Kikawa, E., and Miller, D.J. (Eds.), *Proc. ODP, Sci. Results*, 209:1–33. College Station, TX (Ocean Drilling Program). doi:10.2973/odp.proc.sr.209.001.2007

Koepke, J., Christie, D.M., Dziony, W., Holtz, F., Lattard, D., Maclennan, J., Park, S., Scheibner, B., Yamasaki, T., and Yamazaki, S., 2008. Petrography of the dike–gabbro transition at IODP Site 1256 (equatorial Pacific): the evolution of the granoblastic dikes. *Geochem., Geophys., Geosyst.*, 9(7):Q07O09–Q07O37. doi:10.1029/2008GC001939

Korenaga, J., and Kelemen, P.B., 1998. Melt migration through the oceanic lower crust: A constraint from melt percolation modeling with finite solid diffusion. *Earth Planet. Sci. Lett.*, 156:1–11. doi:10.1016/S0012-821X(98)00004-1

MacLeod, C.J., and Yaouancq, G., 2000. A fossil melt lens in the Oman ophiolite: Implications for magma chamber processes at fast spreading ridges. *Earth Planet. Sci. Lett.*, 176:357–373. doi:10.1016/S0012-821X(00)00020-0

Müller, R.D., Sdrolias, M., Gaina, C., and Roest, W.R., 2008. Age, spreading rates, and spreading asymmetry of the world's ocean crust. *Geochem. Geophys. Geosys.*, 9:Q04006. doi:10.1029/2007GC001743

Phipps Morgan, J., and Chen, Y.J., 1993. The genesis of oceanic crust: Magma injection, hydrothermal circulation, and crustal flow. *J. Geophys. Res.*, 98:6283–6297. doi:10.1029/92JB02650

Purdy, G.M., Kong, L.S.L., Christeson, G.L., and Solomon, S.C., 1992. Relationship between spreading rate and the seismic structure of mid-ocean ridges. *Nature*, 355:815–817. doi:10.1038/355815a0

Quick, J.E., and Denlinger, R.P., 1993. Ductile deformation and the origin of layered gabbro in ophiolites. *J. Geophys. Res.*, 98:14015–14027. doi:10.1029/93JB00698

Teagle, D.A.H., Alt, J.C., Umino, S., Miyashita, S., Banerjee, N.R., Wilson, D.S., and Expedition 309/312 Scientists, 2006. *Proc. IODP*, 309/312: Washington, DC (Integrated Ocean Drilling Program Management International, Inc.). doi:10.2204/iodp.proc.309312.2006

Wilson, D.S., 1996. Fastest known spreading on the Miocene Cocos-Pacific Plate Boundary. *Geophys. Res. Lett.*, 23:3003–3006. doi:10.1029/96GL02893

Wilson, D.S., Teagle, D.A.H., Acton, G.D., et al., 2003. *Proc. ODP, Init. Repts.*, 206: College Station, TX (Ocean Drilling Program). doi:10.2973/odp.proc.ir.206.2003.

Wilson, D.S., Teagle, D.A.H., Alt, J.C., Banerjee, N.R., Umino, S., Miyashita, S., Acton, G.D., et al., 2006. Drilling to gabbro in intact ocean crust. *Science*, 312(5776):1016–1020. doi:10.1126/science.1126090

## Authors

**Damon A.H. Teagle**, National Oceanography Centre, Southampton, University of Southampton Waterfront Campus, European Way, Southampton SO14 3ZH, U.K., e-mail: Damon.Teagle@southampton.ac.uk.

**Benoit Ildefonse**, Géosciences Montpellier, CNRS, Université Montpellier 2, CC 60, 34095 montpellier cédex 05, France.

**Peter Blum**, Integrated Ocean Drilling Program, Texas A&M University, 1000 Discovery Drive, College Station, TX 77845-9547, U.S.A.

**and the IODP Expedition 335 Scientists**

## Photo Credit

Fig. 6: Photos [A] to [D], and [F]: Benoit Ildefonse, CNRS; Photo [E]: Johan Lissenberg, Cardiff University.

29

# Operational Review of the First Wireline *In Situ* Stress Test in Scientific Ocean Drilling

**by Moe Kyaw Thu, Takatoshi Ito, Weiren Lin, Mai-Linh Doan, David Boutt, Yoshihisa Kawamura, Chee Kin Khong, Lisa McNeill, Timothy Byrne, Demian Saffer, Eiichiro Araki, Nobu Eguchi, Ikuo Sawada, Peter Flemings, Yasuyuki Kano, Casey Moore, Masataka Kinoshita, and Harold Tobin**

## Introduction

Scientific ocean drilling's first *in situ* stress measurement was made at Site C0009A during Integrated Ocean Drilling Program (IODP) Expedition 319 as part of Nankai Trough Seismogenic Zone Experiment (NanTroSEIZE) Stage 2. The Modular Formation Dynamics Tester (MDT, Schlumberger) wireline logging tool was deployed in riser Hole C0009A to measure *in situ* formation pore pressure, formation permeability (often reported as mobility=permeability/viscosity), and the least principal stress (S3) at several isolated depths (Saffer et al., 2009; Expedition 319 Scientists, 2010).

The importance of *in situ* stress measurements is not only for scientific interests in active tectonic drilling, but also for geomechanical and well bore stability analyses. Certain *in situ* tools were not previously available for scientific ocean drilling due to the borehole diameter and open hole limits of riserless drilling. The riser-capable drillship, D/V *Chikyu*, now in service for IODP expeditions, allows all of the techniques available to estimate the magnitudes and orientations of 3-D stresses to be used. These techniques include downhole density logging for vertical stress, breakout and caliper log analyses for maximum horizontal stress, core-based anelastic strain recovery (ASR, used in the NanTroSEIZE expeditions in 2007–2008), and leak-off test (Lin et al., 2008) and minifrac/hydraulic fracturing (NanTroSEIZE Expedition 319 in 2009). In this report, the whole operational planning process related to *in situ* measurements is reviewed, and lessons learned from Expedition 319 are summarized for efficient planning and testing in the future.

## Background

Borehole studies can provide information about vertical stress by downhole density logging, about minimum horizontal stress by extended leak-off test and hydraulic fracturing, and about maximum horizontal stress by breakout analyses (Zoback et al., 2003). During the expeditions of NanTroSEIZE Stage 1, a transect of eight sites was drilled in the regions of the frontal thrust, the midslope megasplay fault, and the Kumano forearc basin using a full suite of Measurement While Drilling - Logging While Drilling (MWD-LWD), coring, and downhole measurements (Kinoshita et al., 2009; Moe et al., 2009). The overarching objective of the LWD program was to provide borehole data that will be used in conjunction with cores to document the geology, physical properties, mechanical state, fluid content, and stress conditions at the drill sites (Kinoshita et al., 2009). Preliminary experiments to determine the orientations and magnitudes of principal stresses in the Nankai Trough were undertaken during the NanTroSEIZE Stage 1 expeditions using borehole image analysis (stress-induced breakouts and tensile fractures; Kinoshita et al., 2008) and indirect, core-based methods such as anelastic strain recovery about three-dimensional stress (Lin et al., 2006; Byrne et al., 2009). An extensive logging program at Site C0009A as part of NanTroSEIZE Stage 2 (Fig. 1) included conventional wireline logging in riser mode with formation stress measurements and a wide-angle Walkaway Vertical Seismic Profile (VSP) with the longest offset in scientific ocean drill-

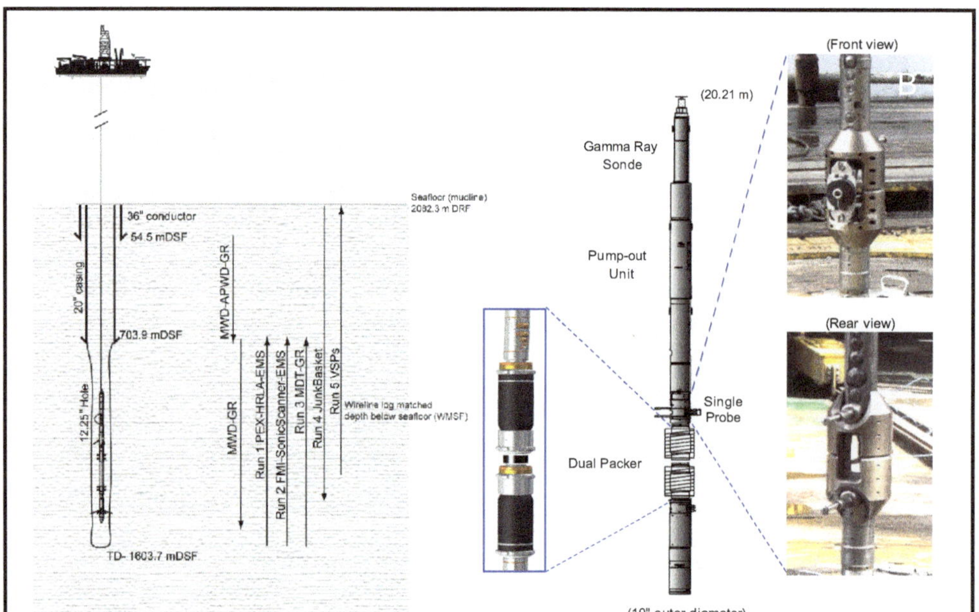


Figure 1. Modular Formation Dynamics Tester (MDT) in operation and illustration of all logging runs from Expedition 319. Measurement While Drilling to total-depth was followed by five runs of wireline toolstrings.

ing (30 km) to improve velocity information of the deep plate boundary and mega-splay faults—the ultimate targets of the project (Moe et al., 2009; Saffer et al., 2009; McNeill et al., 2010; Expedition 319 Scientists, 2010). In a vertical borehole, the orientation of breakouts is a well-established indicator of the orientation of the horizontal maximum principal stress in the present-day stress field (Zoback et al., 2003). Borehole breakouts have been used for stress estimation in NanTroSEIZE Stage 1 and previous expeditions on the Nankai margin (McNeill et al., 2004; Ienaga et al., 2006; Chang et al., 2010). This is the case for stress orientation but not for stress magnitude. Thus, for further understanding the state of stress, the hydraulic fracturing (HF) method was used to measure stress magnitude in Expedition 319 (Ito et al., 2009; Saffer et al., 2009; McNeill et al., 2010; Expedition 319 Scientists, 2010).

Data taken from HF measurements are not only scientifically important but also necessary for operational planning of deep riser holes. In this regard, *in situ* stress is useful for geomechanical and wellbore stability studies. The Center for Deep Earth Exploration (CDEX) has collaborated with GeoMechanics International, Inc. on such studies since 2007 in pore pressure estimation using seismic traces extracted from the high-resolution 3-D seismic data cube (Moore et al., 2009) and calibrated using LWD data as well as geomechanical studies using logging and core analyses data taken from the NanTroSEIZE Stage 1 expeditions (Kinoshita et al., 2009). The absence of leak-off and hydraulic fracturing information in this preliminary study increases the uncertainty in estimating the fracture gradient, and hence can be inconsistent with wellbore failure observations (Castillo et al., 2008).

## *In Situ* Stress Measurement Methods

Stress magnitudes can be inferred from continuous logs (e.g., from sonic-based measurements), but these must be calibrated to obtain direct measurements of stress. At present, the only reliable and accurate measurements of *in situ* stress come from micro-fracturing (Thiercelin et al., 1996). There are several techniques to measure horizontal stress magnitudes and/or stress orientations through data obtained from a well bore. Core experiments include performing a variety of strain experiments. Checking with other results, such as from the micro-fracturing technique, provides a more reliable stress determination. Logging techniques rely on semi-empirical relationships between rock properties and stresses both existing at or near the well bore face, or the interference of stress via a mechanistic model which uses measurements of the borehole breakout (e.g., breakout width). Sonic-based measurements are perhaps the best known logging technique for obtaining stresses. Downhole techniques which fracture the formation are the Leak-Off Test (LOT) and the micro-fracturing technique. A wireline micro-fracturing tool can be used on a wireline (e.g., MDT) or on the drill pipe, where high performance gauges measure the packer internal pressure and the pressure of the fluid between the packers (Carnegie et al., 2002) (Fig. 1).

The importance of strict criteria on borehole conditions and selecting target depths should be noted. Limitations for the MDT wireline tool include the following.

- a hole diameter within 29.21 cm (11.5 inches) and 37.47 cm (14.75 inches) for easy packer insertion and efficient inflation
- hole ovality (Dmax/Dmin) <130% for efficient packer sealing
- expected permeability of <1 mD for the HF test by seawater injection using standard pump and >0.1 mD for the drawdown test (for formation pore pressure and permeability measurement) to prevent no pressure buildup and poroelastic disturbance within the formation caused by too large permeability and too long drawdown test from too small permeability (Note: Permeability derived in the Expedition 319 was ~ 0.1 mD (Boutt et al., in press).)
- zone thickness of >3 m for the HF and >2 m for the drawdown test
- avoidance of pre-existing fractures for both tests.
- good mud cake formation to avoid super-charging and enabling reliable pore pressure measurement and single probe permeability testing
- good borehole stability since the tool is significantly larger than usual logging tools

A classic stress test procedure consists of a series of injection stages or cycles as described below.

Packer inflation is meant to isolate the test interval and results in the pressure in the interval starting to rise after the tool has been properly positioned. A subsequent pressure decline is then observed to check the quality of the packer seal, with further pressurization if the seal is not satisfactory.

Leak-off cycles include injections of fluid into the interval at a constant flow rate. This pressurizes the wellbore below the breakdown of the formation to check that the downhole pump can deliver enough flow rate to overcome fluid diffusion through the mudcake and into the formation.

In the hydraulic fracturing cycles, fluid is again injected into the interval at a constant flow rate and up to the initiation of a tensile fracture, which is recognized either by a breakdown or a pressure plateau. The fracture is then extended for 1–5 minutes before the interval is isolated and the pump stopped (shut-in), the pressure decline then observed, and then followed by a series of injection/fall-off cycles (two to five) to reopen, further propagate, and then close the fracture to both check the test is repeatable and possibly change the injection parameters (flow rate and injected volume).

Once the operator is satisfied that good quality data have been acquired, the packers are deflated, and the tool is moved to the next interval. A typical test can take from 20 minutes to 1.5 hours, depending on the number of injection cycles that are performed (Desroches and Kurkjian, 1998).

## Logging Planning and Operations for Site C0009, Nankai Subduction Zone

A CDEX expedition team was formed in July 2008 following the scheduling of Expedition 319, and logging planning began with assessment of several new tools for scientific ocean drilling, such as MDT, Formation MicroImager (FMI), SonicScanner and High-Resolution Laterolog Array (HRLA), and walkaway VSP (utilizing the Vertical Seismic Imager tool). The call for participation left relatively little time between initiating operations and logging planning and the start of the expedition. An MDT test was agreed upon with the condition that the hole stability was first confirmed by the drilling and the logging results of two wireline runs.

In the process of preparing the MDT measurement plan, four major steps (before and during the expedition) were carried out. The first step was a series of seminars about the new tools and measurements, where expedition scientists attended and led the preparation for MDT draft measurement plan together with the CDEX Logging Staff Scientist (LSS) before expedition. The second step was working on details of measurement plan by logging scientists and Schlumberger engineers onboard with the LSS, Co-Chiefs (CCs), Expedition Project Manager (EPM), and operations geologists from CDEX. Daily updates were shared with the onshore LSSs, Co-chief Scientists, EPM and a Schlumberger MDT specialist. The third step was interpretation of wireline logs taken from logging runs prior to the MDT experiment. Caliper, FMI resistivity image, resistivity, gamma-ray, and sonic velocity logs were thoroughly examined to finalize the plan and select the target zones and depths for each test (Fig. 2). The fourth step was making final decisions on operational details in regard to safety among the onboard Executive Committee (i.e., Ship Captain, Operations Superintendent (OSI), EPM, CCs, and Offshore Installation Manager).

Major adjustments were made during steps onboard. From the cost factors, both dual packer HF tests and drawdown tests were reduced. For safety, the single probe tool was added to run together with the dual packer to reduce the risk of differential pressure sticking, for quick pore pressure and drawdown tests in the high or medium permeability formations, and for selection of dual packer test depths (ultimately providing further scientific data), This imposed a time limit for each MDT measurement.

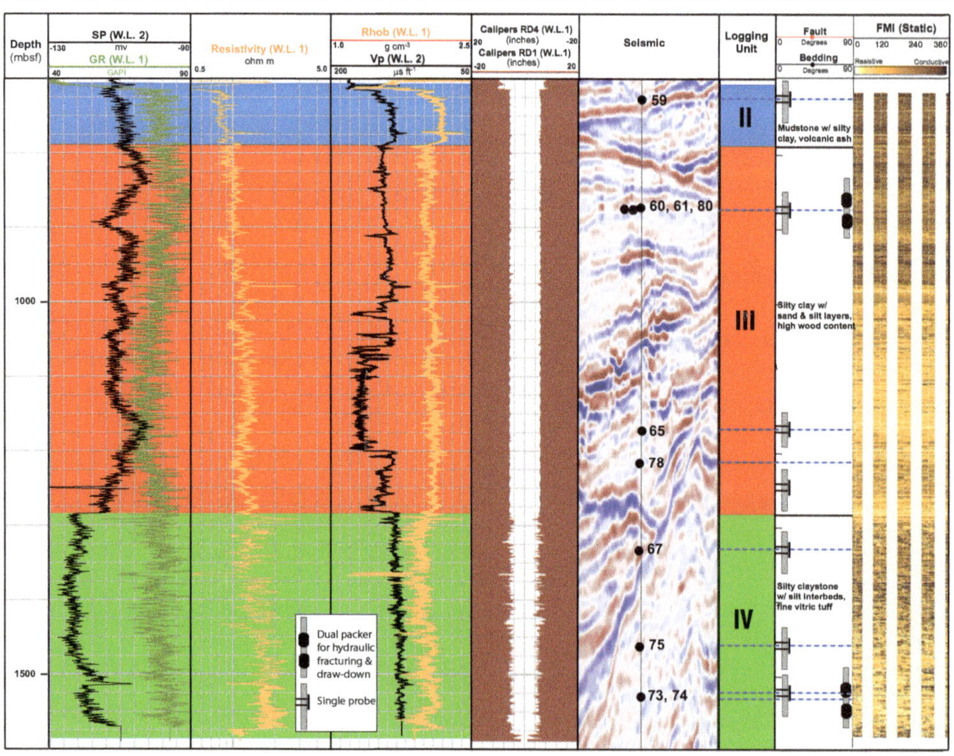


Figure 2. MDT test points on the seismic cross-section together with some important logging data and their interpretation. (From left to right) Depth (meter below seafloor, mbsf), wireline logging runs (W.L.) [spontaneous potential (SP), gamma-ray (GR)], resistivity, density (Rhob), p-wave velocity (Vp), EMS calipers, MDT test points on seismic, logging units from cuttings/core-log-seismic integration, dip picks from FMI data. and static-normalized FMI image.

After two runs of wireline logging, logging scientists and chief scientists identified test depths from logs during a short time of wiper trip in the hole. As a primary focus was making tests in the cored section at the bottom of the hole (where sample analyses would complement the *in situ* measurements), one HF and one drawdown test were made within the cored section at ~1500 mbsf with the dual packer. In addition, one HF test was conducted in the shallower part of the borehole, traveling to the bottom of the hole and four tests on return, and two tests were conducted at the depths of the dual packer tests as a pre-test check on the site selection (Fig. 2). These sites are distributed throughout the borehole to give a range of data in different environments and formations. The single probe provided critical data for selecting the final MDT sites (both for HF and permeability) and in evaluating risks, as well as providing additional data on formation pore pressure and permeability.

## Summary

Deployment of the MDT tool deeper within the forearc and in the vicinity of major fault zones in future riser holes will constitute a major breakthrough in understanding subduction zone fault mechanics, and it is a critical part of the NanTroSEIZE program (Araki et al., 2009; Saffer et al., 2009; McNeill et al., 2010). Operation of this tool was complex both in the planning and execution phase, but valuable experience has been gained by both CDEX and NanTroSEIZE scientists. Unlike other logging measurements, MDT requires considerable involvement of the scientists during planning stage, in decision making during the experiment through real-time monitoring, and constant communication with the MDT specialist, engineer, and drilling operations team. The following factors summarized from the experience and lessons learned from Expedition 319 should inform future MDT deployment on riser expeditions for efficient and scientifically reliable operations.

- A minimum of two LSSs are necessary for the logging operations of riser expeditions.
- Sufficient time is necessary for the dual packer test to generate useful results. A high-performance packer with auto retract mechanism may be an option for future experiments.
- Additional use of single probe at various depths was useful for range of scientific objectives: depth profile for hydrologic interpretation, evaluating the role of erosional and structural events on major transitions/unconformities, lithologic comparison, understanding how the deformation influences the permeability of accretionary prisms, and pore pressure estimation, which is a crucial parameter to understand the mechanics of the prism and décollement.
- Borehole imaging prior to and following the HF test is necessary to determine orientation of the fracture induced or activated by hydraulic fracturing.

## Acknowledgements

Thanks are due to the marine and drilling crew of the *Chikyu* and many others for their hard work for the success of IODP Expedition 319. Funding from Ministry of Education, Culture, Sports, Science and Technology (MEXT), Japan, National Science Foundation (NSF), United States, and twenty-two other countries support this project as part of the Integrated Ocean Drilling Program (IODP) for successful drilling in the Nankai Trough.

## References

Araki, E., Byrne, T., McNeill, L., Saffer, D., Eguchi, N., Takahashi, K., and Toczko, S., 2009. NanTroSEIZE Stage 2: NanTroSEIZE riser/riserless observatory. *IODP Sci. Prosp.,* 319. doi:10.2204/iodp.sp.319.2009.

Byrne, T.B., Lin, W., Tsutsumi, A., Yamamoto, Y., Lewis, J., Kanagawa, K., Kitamura, Y., Yamaguchi, A. and Kimura, G., 2009. Anelastic strain recovery reveals extension across SW Japan subduction zone. *Geophy. Res. Lett.*, 36:L23310. doi:10.1029/2009/GL040749

Boutt, D., Saffer, D., Doan, M-L., Lin, W., Ito, T., Kano, Y., Flemings, P., McNeil, L., Byrne, T., and Moe, K.T., in press. Scale dependence of *in situ* permeability measurements in the Nankai accretionary prism: The role of fractures. *Geophy. Res. Lett.*

Carnegie, A., Thomas, M., Efnik, M.S., Hamawi, M., Akbar, M., and Burton, M., 2002. An advanced method of determining *in situ* reservoir stresses: Wireline conveyed micro-fracturing. *Soc. Petrol. Eng. Conference Paper,* 78486-MS.

Castillo, D., Magee, M., and Dhakur, S., 2008. NanTroSEIZE Geomechanical analysis of the Stage 1 drilling and Data Acquisition Program: Phase 1. GeoMechanics International, Inc. (CDEX Internal Report).

Chang, C., McNeill, L.C., Moore, J.C., Lin, W., Conin, M., and Yamada, Y., 2010. *In situ* stress state in the Nankai accretionary wedge estimated from borehole wall failures. *Geochem. Geophys. Geosys.*, 11:Q0AD04. doi:10.1029/2010GC003261

Desroches, J., and Kurkjian, A.L., 1998. Applications of wireline stress measurements. *SPE Res. Eval. & Eng.*, 2(5):451–461.

Expedition 319 Scientists, 2010. Expedition 319 summary. *In* Saffer, D., McNeill, L., Byrne, T., Araki, E., Toczko, S., Eguchi, N., Takahashi, K., and the Expedition 319 Scientists, *Proc. IODP,* 319: Tokyo (Integrated Ocean Drilling Program Management International, Inc.). doi:10.2204/iodp.proc.319.101.2010

Ienaga, M., McNeill, L.C., Mikada, H., Saito, S., Goldberg, D., and Moore, J.C., 2006. Borehole image analysis of the Nankai accretionary wedge, ODP Leg 196: Structural and stress studies. *Tectonophysics,* 426:207–220. doi:10.1016/j.tecto.2006.02.018

Ito, T., Lin, W., Doan, M-L., Boutt, D., Kano, Y., Flemings, P., Ito, H., and IODP Expedition 319 Scientists, 2009. Outline of the in-situ stress tests using MDT tool at 2.7–3.6 km below sea level in the first riser hole on the NanTroSEIZE operation. [The 15th Formation Evaluation Symposium of Japan, 1–2 October 2009]. (extended abstract)

Kinoshita, M., Tobin, H., Moe, K.T., and the Expedition 314 Scientists, 2008. NanTroSEIZE Stage 1A: NanTroSEIZE LWD transect. *IODP Prel. Rept.,* 314. doi:10.2204/iodp.pr.314.2008

Kinoshita, M., Tobin, H., Ashi, J., Kimura, G., Lallemant, S., Screaton, E.J., Curewitz, D., Masago, H., Moe, K.T., and the Expedition 314/315/316 Scientists, 2009. NanTroSEIZE Stage 1 expeditions: Introduction and synthesis of key results. *Proc. IODP,* 314/315/316: Washington, DC (Integrated Ocean

Drilling Program Management International, Inc.).

Lin, W., Kwasniewski, M., Imamura, T., and Matsuki, K., 2006. Determination of three-dimensional *in situ* stresses from anelastic strain recovery measurement of cores at great depth. *Tectonophysics*, 426:221–238. doi:10.1016/j.tecto.2006.02.019

Lin, W., Yamamoto, K., Ito, H., Masago, H., and Kawamura, Y., 2008. Estimation of minimum principal stress from an extended leak-off test onboard the *Chikyu* drilling vessel and suggestions for future test procedures. *Sci. Drill.*, 6:43–47.

McNeill, L.C., Ienaga, M., Tobin, H., Saito, S., Goldberg, D., Moore, J.C., and Mikada, H., 2004. Deformation and *in situ* stress in the Nankai Accretionary Prism from resistivity-at-bit images, ODP Leg 196. *Geophys. Res. Lett.*, 31:L02602. doi:10.1029/2003GL018799

McNeill, L.C., Saffer, D., Byrne, T., Araki, E., Toczko, S., Eguchi, N., Takahashi, K., and Expedition 319 Scientists, 2010. IODP Expedition 319, NanTroSEIZE Stage 2: First IODP riser drilling operations and observatory installation towards understanding subduction zone seismogenesis. *Sci. Drill.*, 10:4–13.

Moe, K.T., Sanada, Y., Kido, Y., Kawamura, Y., Cukur, D., Doan, M.-L., Kano, Y., et al., 2009. Overview of the NanTroSEIZE riser logging: Operational planning and reality. [The 15th Formation Evaluation Symposium of Japan, 1–2 October 2009]. (extended abstract)

Moore, G.F., Park, J.-O., Bangs, N.L., Gulick, S.P., Tobin, H.J., Nakamura, Y., Sato, S., et al., 2009. Structural and seismic stratigraphic framework of the NanTroSEIZE Stage 1 transect. *In* Kinoshita, M., Tobin, H., Ashi, J., Kimura, G., Lallement, S., Screaton, E.J., Curewitz, D., Masago, H., Moe, K.T., and the Expedition 314/315/316 Scientists, *Proc. IODP,* 314/315/316. doi:10.2204/iodp.proc.314315316.102.2009

Saffer, D., McNeill, L., Araki, E., Byrne, T., Eguchi, N., Toczko, S., Takahashi, K., and the Expedition 319 Scientists, 2009. NanTroSEIZE Stage 2: NanTroSEIZE riser/riserless observatory. *IODP Prel. Rept.,* 319. doi:10.2204/iodp.pr.319.2009

Thiercelin, M.J., Plumb, R.A., and Desroches, J., 1996. A new wireline tool for *in situ* stress measurements. *SPE Format. Eval.*, 11(1):19–25.

Zoback, M.D., Barton, C.A., Brudy, M., Castillo, D.A., Finkbeiner, T. Grollimund, B.R., Moos, D.B., Peska, P., Ward, C.D., and Wiprut, D.J., 2003. Determination of stress orientation and magnitude in deep wells. *Int. J. Rock Mech. Min. Sci.*, 40(7–8):1049–1076. doi:10.1016/j.ijrmms.2003.07.001

## Authors

**Moe Kyaw Thu, Nobu Eguchi, and Ikuo Sawada,** Center for Deep Earth Exploration (CDEX)—Japan Agency for Marine-Earth Science and Technology (JAMSTEC), 3173-25 Showa-machi, Kanazawa-ku, Yokohama, Kanagawa 236-0001, Japan, e-mail: moe@jamstec.go.jp.

**Yoshihisa Kawamura**, *Ippan Shadan Hojin* IODP-MI, Tokyo University of Marine Science and Technology, Office of Liaison and Cooperative Research, 3rd Floor, 2-1-6 Etchujima, Koto-ku, Tokyo, 135-8533 Japan.

**Chee Kin Khong,** ECA Services Techniques Schlumberger, Le Palatin 1, 1 cour du Triangle, 92936 La Defense –France.

**Takatoshi Ito**, Institute of Fluid Science, Tohoku University, 2-1-1 Katahira, Aoba-ku, Sendai, Miyagi 980-8577, Japan.

**Weiren Lin**, Kochi Institute for Core Sample Research, Japan Agency for Marine-Earth Science and Technology, 200 Mononobe-Otsu, Nankoku, Kochi 783-8502, Japan.

**Mai-Linh Doan**, ISTerre Université de Grenoble 1, CNRS, F-38041 Grenoble, France.

**David Boutt**, Department of Geosciences, University of Massachusetts Amherst, 611 North Pleasant Street, 233 Morrill Center, Amherst, MA 01002, U.S.A.

**Lisa McNeill**, School of Ocean and Earth Science, National Oceanography Centre, Southampton, University of Southampton, Southampton SO14 3ZH, U.K.

**Demian Saffer**, The Pennsylvania State University, 0310 Deike Building, University Park, PA 16802, U.S.A.

**Eiichiro Araki**, Earthquake and Tsunami Research Project for Disaster Prevention, Japan Agency for Marine-Earth and Technology, 2-15 Natsushima-cho, Yokosuka, Kanagawa 237-0061, Japan.

**Timothy Byrne**, Center for Integrative Geosciences, University of Connecticut, U-2045, 354 Mansfield Road, Storrs, CT 06269, U.S.A.

**Yasuyuki Kano**, Disaster Prevention Research Institute, Kyoto University, Gokasho, Uji 611-0011, Japan.

**Peter Flemings**, Institute for Geophysics, Jackson School of Geosciences, University of Texas at Austin, 10100 Burnet Road (R2200), Austin, TX 78758-4445, U.S.A.

**Casey Moore**, University of California Santa Cruz, 1156 High Street, Santa Cruz, CA 95064, U.S.A.

**Masataka Kinoshita**, Institute for Research on Earth Evolution, Japan Agency for Marine-Earth Science and Technology, 2-15 Natsushima-cho, Yokosuka, Kanagawa 237-0061, Japan.

**Harold Tobin**, Department of Geology and Geophysics, University of Wisconsin-Madison, 1215 West Dayton Street, Madison, WI 53706, U.S.A.

## Related Web Links

http://www.jamstec.go.jp/chikyu/eng/Expedition/NantroSEIZE/exp319.html

http://publications.iodp.org/preliminary_report/319/index.html

http://publications.iodp.org/proceedings/319/103/103_.htm

## Photo Credit

Fig. 1: Photo courtesy of T. Ito, Tohoku University

# 30

# Kochi Core Center Provides New Information Services

by Lallan Gupta, Toshio Hisamitsu, and Noriaki Masui

## Introduction

During the last four decades of scientific drilling of seafloor all over the world, an enormous amount (~360,000 m) of marine cores has been collected. These cores are being curated in the three Integrated Ocean Drilling Program (IODP) core repositories located in Germany, Japan, and the U.S.A.. The IODP core repository in Japan is housed in the Kochi Core Center (KCC), Kochi on the island of Shikoku. The total amount of IODP core currently being curated in the KCC is about 93,000 m. These cores were collected from the Indian and west Pacific Oceans, and they consist of a diverse range of geological material. Detailed description of the core material is available in preliminary reports of each drilling expedition. This report details three new services being offered through the KCC website (www.kochi-core.jp): Core Summary, Routine Microbiological Sample (RMS), and Virtual Core Library (VCL).

## Core Summary

Cores contain a wide variety of material for the earth science community. Although much of the research is carried out by taking bulk samples from cores, some researchers have specific interests for subsamples, such as ash, calcite nodules, or plankton fossils. However, such detail was so far only provided in individual volumes of preliminary reports or proceedings of expeditions of the IODP core material. For an electronic access to these data, the KCC curatorial staff compiled major features of sample material recovered through Deep Sea Drilling Project (DSDP), Ocean Drilling Program (ODP) and IODP, and made the "Core Summary" information publically available through the KCC web page (http://www.kochi-core.jp/cs/).

The core summary web page lists basic parameters like expedition or leg number, site number, hole, geographical location of the drilled site, drilled interval depth, total length of recovered core, approximate geological age range covered by the core sediments, an overview of lithology, age model, and magnetostratigraphy, together with the link for the preliminary report or proceedings of each drilling expedition. The list can be filtered using various combinations of keywords and parameters. This action will facilitate short-listing of the information related to the core material, which is dispersed in the individual volumes of the preliminary report or proceedings of each leg/expedition. For quickly locating the drill site, a click on the site name takes the user to geographical map services with the site clearly marked on it. Lithological summary of core material recovered from each hole can be quickly viewed in a pop-up window or a new window. Age model, core recovery, and magnetostratigraphy information extracted from preliminary reports or proceedings of each leg/expedition are available in PDF format, which can easily be downloaded.

## Routine Microbiological Samples (RMS)

Subsequent to the discovery of life and biosphere in subseafloor environments, the study of biomass, activity, and function of subseafloor life and its ecological roles in biogeochemical cycles has become one of the major scientific objectives in the IODP. In addition, biochemical analytical techniques based on DNA are rapidly developing. Therefore, IODP's Science Advisory Structure (SAS) recommended routine sampling for geomicrobiology, and core repositories were asked to develop the capabilities for storage of core samples at very low temperature. The Japanese Center for Deep Earth Exploration (CDEX) initiated routine collection of microbiological sub-samples of cores in 2009, and these samples are being curated in the KCC.

Storage of the RMS under deep-freezing conditions is critical for preventing degradation—caused by abiotic hydrolysis, enzymatic reactions, and possible contamination—of fragile bio-molecules such as DNA, RNA, enzymes, sugar chains, and intact polar lipids. Moreover, aseptic sub-sampling and distribution of the RMS without thawing is desired for quality assurance purposes. KCC is equipped with deep freezers (-80°C) and liquid nitrogen-cooled tanks (-160°C) for long-term storage and with an aseptic band saw system for "clean" processing of the frozen core sample (Masui et al., 2009). Using these geobiological sampling devices, aliquots of the RMS can be distributed to the science community worldwide according to IODP policy. To spread information about IODP and RMS, KCC prepared a leaflet and a handbill for distribution at academic meetings. In addition, an RMS promotion site has been added to the KCC website. Overview of RMS curation and a list of the RMS have been made available through this website. The list can be filtered for sample depth, expedition, and site name in order to find suitable RMS.

## Virtual Core Library (VCL)

The drilling vessel *Chikyu* is equipped with an X-ray Computed Tomography (XCT) scanner, which provides the capability to visualize the internal structure of cores to aid in discerning important geological features prior to splitting and discrete sampling of the cored materials. Standard core flow on board *Chikyu* involves XCT scanning of cores before they are split and archiving of the digital core images (Fig. 1). These data are used by shipboard scientists to identify key features within a core section and to decide if whole round core (WRC) sampling and preservation of part of an unsplit core is necessary for future investigations. The XCT scanning of one core section produces a file that contains data for previewing a 2-D image and for reconstructing a 3-D image of the core section. The files are saved into a format known as DICOM (Digital Imaging and Communication in Medicine). The J-CORES data server stores all files related to the XCT-scanned images, and provides access to the files via the Internet. Files are freely available after the expedition moratorium period.

KCC designed a new web page named "Virtual Core Library" (VCL) that introduces and promotes use of the XCT data for researchers and educators, and displays digital 3-D images of typical core sections on the gallery page. Users can also access preliminary reports of *Chikyu* expeditions and understand scientific details related to a particular cruise, site, hole, etc. The VCL provides a link to the coronal images (2-D images) of core sections that can be quickly viewed. Researchers can then request a copy of DICOM files for core sections of interest. After examining details of the request, curatorial staff will provide the data either by copying them on a DVD or through a file transfer protocol (FTP) setup.

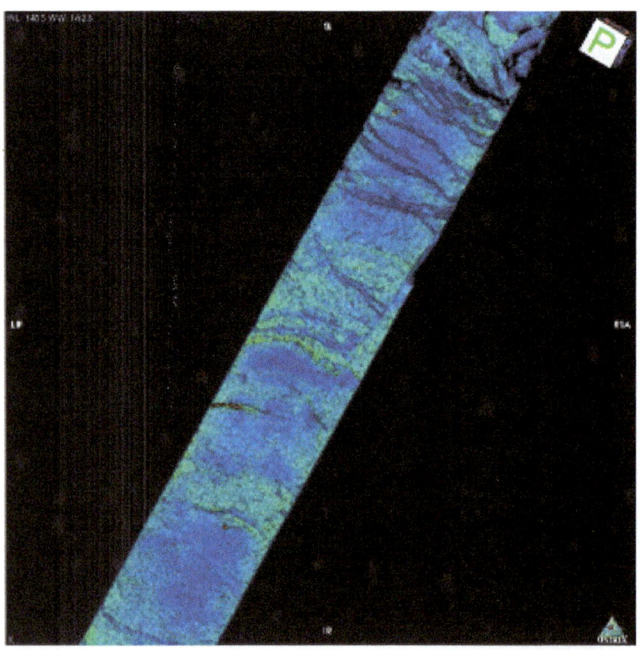

Figure 1. The OsiriX-rendered false colored image of a core section based on XCT scanning data. Fractures and gaps are easily visualized in this image. Change in color from light blue to red can be interpreted to be a reflection of change in density of core material, although calibration is a must to conclude about density.

Recently, DICOM viewing software has become freely available on the Internet (e.g., OsiriX, ApolloView Lite, Onis Viewer). Using such software, researchers can render 3-D images of core sections from DICOM files and can visualize the internal structure of cores by changing transparency and color and by rotating or slicing the image. General DICOM viewers have functions like rendering, cutting, or slicing 3-D images, and they provide an opportunity for users to "virtually sample". The VCL will be improved to accelerate use of the XCT data and promote new approaches in geological studies.

The KCC curatorial staff will continue to improve these Internet-based services to allow for the best possible utilization of IODP samples by researchers around the world, and thus contribute in enhancing our current understanding of the Earth and its environments.

## References

Masui, N., Morono, Y., and Inagaki, F., 2009. Bio-Archive core storage and subsampling procedure for subseafloor molecular biological research. *Sci. Drill.*, 8:35–37. doi:10.2204/iodp.sd.8.05.2009

## Authors

**Lallan Gupta, Toshio Hisamitsu, and Noriaki Masui**, Kochi Core Center, 200 Monobe Otsu, Nankoku, Kochi 783-8502, Japan, e-mail: gupta@jamstec.go.jp.

## Related Web Links

http://www.kochi-core.jp/cs/
http://www.kochi-core.jp/rms/
http://www.kochi-core.jp/VCL/

# Umbria-Marche Basin, Central Italy: A Reference Section for the Aptian-Albian Interval at Low Latitudes

by Rodolfo Coccioni, Luigi Jovane, Giuseppe Bancalà, Carla Bucci, Gerson Fauth, Fabrizio Frontalini, Liliane Janikian, Jairo Savian, Renato Paes de Almeida, Grasiane Luz Mathias, and Ricardo Ivan Ferreira da Trindade

## Introduction

Within the Cretaceous Period, the Aptian-Albian interval (125–99.6 Ma, Ogg et al., 2008) was a critical time on a global scale. This is evident from 1) changes in the nature of the ocean-climate system brought about by increased ocean crust production coupled with active midplate and plate margin volcanism in a shifting paleogeography (Skelton et al., 2003); 2) cyclic deposition and preservation of common "black shales", some of them termed Oceanic Anoxic Events (OAE1a to OAE1d) (Schlanger and Jenkyns, 1976; Arthur et al., 1990); 3) periodic changes in redox conditions at the ocean bottom (Oceanic Red Beds, ORBs) (Wang et al., 2009); and 4) rapid biotic radiations and turnovers (Leckie et al., 2002). The Aptian-Albian time is also of interest for one of the most noteworthy geomagnetic events, namely the post-M0r "Cretaceous Quiet Zone". This long and constant normal polarity superchron without any convincing true reversal to date (Satolli et al., 2008) precludes usage of reversals magnetostratigraphy from the Aptian through the Santonian.

The Poggio le Guaine core was designed to provide a high-resolution age model and a high-resolution relative magnetic paleointensity reference curve for the Aptian-Albian interval of the long normal Cretaceous superchron; it was also designed to understand the causal linkages among geological, biogeochemical, oceanographic and climatic events as well as their consequences. The core was drilled at Poggio le Guaine, where the most continuous, complete, and best preserved Aptian-Albian succession is exposed throughout the Umbria-Marche Basin (UMB) of the northern Apennines of central Italy (Fig. 1). It represents a continuous record of fossiliferous pelagic rocks extending from the Albian-Cenomanian boundary down to the uppermost Barremian (99.6–126 Ma). In this progress report we present the first preliminary findings of this ongoing project.

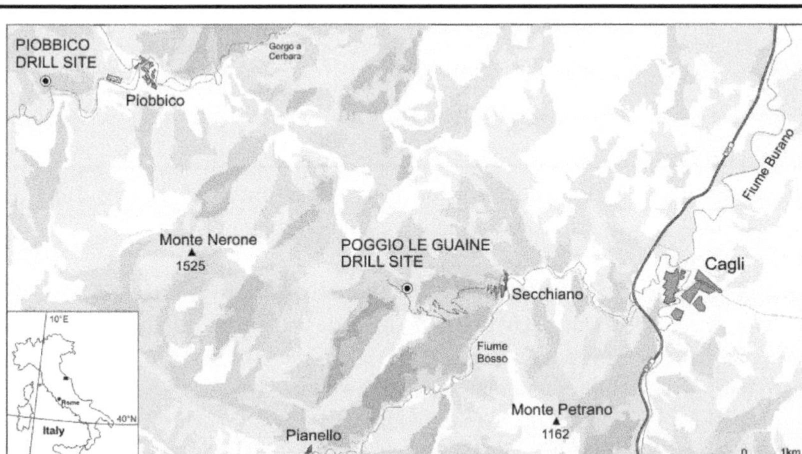


Figure 1. Location of the Poggio le Guaine drill site in the Umbria-Marche Basin (northern Apennines, central Italy). The location of the Piobbico drill site—where an 84-m-thick core extending from the upper Albian down to the uppermost Barremian was drilled in 1982—is also shown.

## Geological and Stratigraphic Setting

After three decades of dedicated research, the Aptian-Albian pelagic succession of the UMB has become a classical reference section for studies at a regional to global scale. It deposited well above the calcite compensation depth at middle to lower bathyal depths (1000–1500 m) and at ~20°N paleolatitude over the southern margin of the western Tethys Ocean (Arthur and Premoli Silva, 1982; Coccioni et al., 1987, 1989, 1990, 1992; Erba, 1988, 1992; Cresta et al., 1989; Tornaghi et al., 1989; Coccioni, 1996; Coccioni and Galeotti, 1993; Satolli et al., 2008; Tiraboschi et al., 2009; Turchyn et al., 2009).

This succession extends from the uppermost part of the Maiolica Formation (Tithonian to lower Aptian) to the lower part of the Scaglia Bianca Formation (uppermost Albian to lowermost Turonian) and includes the entire Marne a Fucoidi Formation (Fig. 2). The uppermost part of the Maiolica Formation is represented by thin-bedded white to gray limestones interbedded with black shales. The lower part of the Scaglia Bianca Formation is characterized by thin-bedded yellowish-gray lime-stones with subordinate reddish limestones and some discrete thin black shales, which are the regional sedimentary expression of the latest Albian OAE1d (Fig. 2). Within the Cretaceous succession of the UMB, the Aptian-Albian Marne a Fucoidi Formation represents a distinctive varicolored interlude with more shale. This formation consists of thinly interbedded pale reddish to dark reddish, pale olive to dark reddish brown and pale olive to grayish olive marlstones and calcareous marlstones together with dark gray to black organic carbon-rich shales, usually with a low carbonate content, and yellowish-gray to light gray marly limestones and lime-stones (Fig. 2). Several distinctive organic-rich black shale and marl marker beds occur within the Aptian-Albian interval, some of which have been identified as the regional sedimentary expression of OAE1a to OAE1d (Coccioni et al., 1987, 1989, 1990; Coccioni, 1996, 2001).

Following the timescale of Ogg et al. (2008), the sediment accumulation rate is estimated to be ~0.24 cm $kyr^{1}$ and ~0.5 cm $kyr^{1}$ in the Aptian and Albian, respectively.

## Drilling Site Location and Drilling Operations

The Poggio le Guaine succession was selected as the reference section for the Aptian-Albian interval following previous detailed stratigraphic investigations (Coccioni et al., 1987, 1989, 1990; Coccioni, 1996).

The Poggio le Guaine drill site (lat. 43°32'42.72"N; long. 12°32'40.92"E) is located on the Monte Nerone ridge at 888 m above sea level, 6 km west of the town of Cagli (Regione Marche, Italy) (Fig. 1). The outcropping beds show a strike of 310° and a dip of 30° to the northeast.

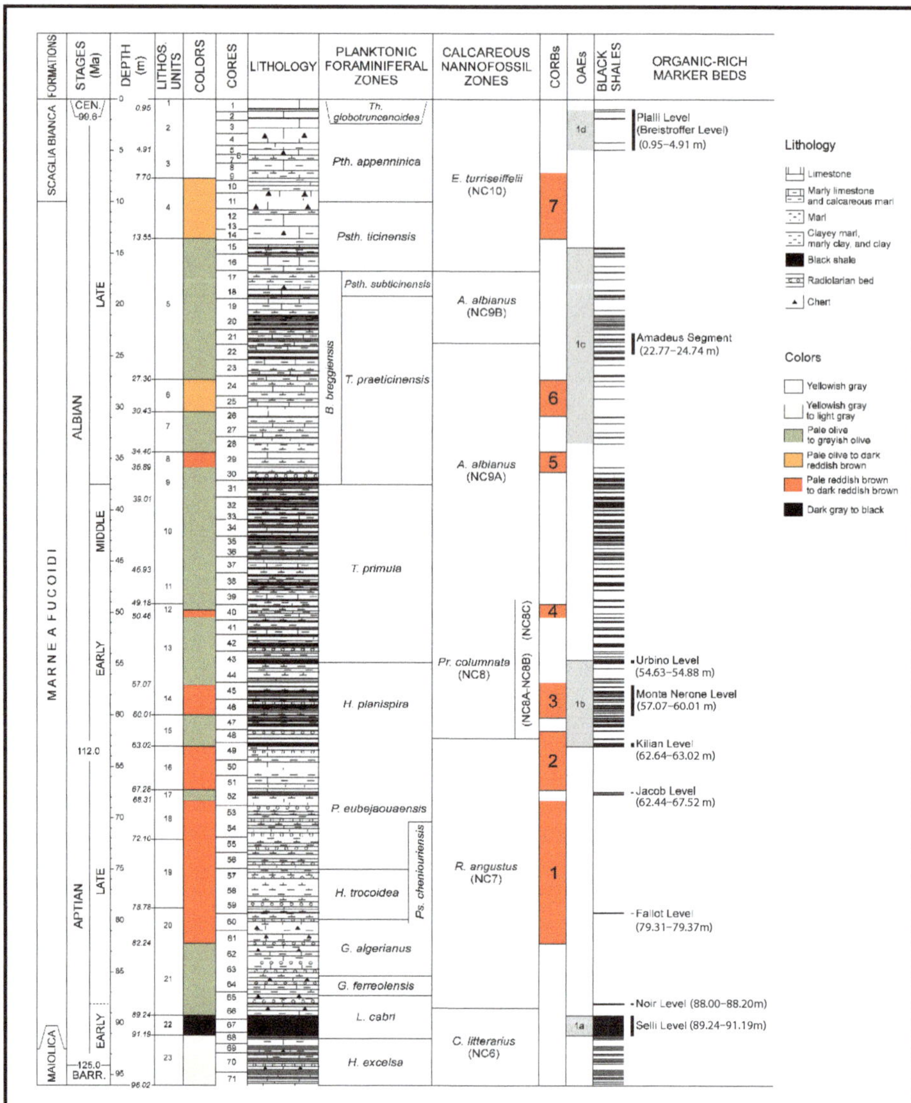


Figure 2. Stratigraphic framework of the Poggio le Guaine core with real stratigraphic depths according to bed dip measurements. Also shown are the occurrence and distribution of the organic-rich black shales including the marker beds resulting from Oceanic Anoxic Events (OAEs) and the discrete intervals where reddish colored beds become dominant (Cretaceous Oceanic Red Beds, CORBs). Integrated plankton biostratigraphy is adapted after Coccioni et al. (1987, 1989, 1990, 1992, 2007), Erba (1988, 1992), Tornaghi et al. (1989), Tiraboschi et al. (2009) and unpublished data. Abbreviations: Lithos.=Lithostratigraphic, Barr.=Barremian, Cen.=Cenomanian, H.=*Hedbergella*, L.=*Leupoldina*, G.=*Globigerinelloides*, P.=*Paraticinella*, Ps.=*Pseudoplanomalina*, T.=*Ticinella*, B.=*Biticinella*, Psth.=*Pseudothalmanninella*, Pth.=*Parathalmanninella*, Th.=*Thalmanninella*, C.=*Chiastozygus*, Pa.=*Parhabdolithus*, N.=*Nannoconus*, Pr.=*Predicosphaera*, A.=*Axopodorhabdus*, R.=*Rhagodiscus*, E.=*Eiffellithus*. Timescale is after Ogg et al. (2008). Following Huber and Leckie (2011), the Aptian-Albian boundary is placed at the highest occurrence of *Paraticinella eubejaouaensis*.

The drilling campaign took place on 1–14 September 2010. Drilling operations were performed with the Atlas Copco Mustang 5-F4 surface core drilling rig (Fig. 3). Coring was accomplished with a T2 double corer, using narrow-kerf, sawtoothed drill bits that cut a 10.1-cm-diameter hole and approximately 8-cm-diameter cores. The target interval was successfully encountered below the weathering zone, which is only a couple of decimeters thick. Lithologies were logged and digital color photographs taken, as each core section was recovered (Fig. 4). Later, the seventy-one cores drilled were packed, labeled, and put in PVC plastic boxes to prevent contamination and loss of moisture. During core packing, care was taken in collecting the fragments from the few fractured portions in plastic bags. The entire core is stored at the core repository of the Department of Earth, Life and Environmental Sciences, University of Urbino.

The Poggio le Guaine cored interval extends from the Albian-Cenomanian boundary down to the uppermost Barremian including the upper transition to the Scaglia Bianca Formation and the lower transition to the Maiolica Formation. The total length of the core is 98.72 m with essentially 100% recovery of excellent quality material throughout the drilling. For each core a set of dip measurements was taken directly at the drill site with four measurements on average for each meter of core. Marked changes of dip angle were observed in the uppermost and lowermost portions of the core, the latter particularly affected by folding and minor faulting. Taking into account the dip measurements, the corrected thickness of the drilled section is 96.02 m, with 82.53 m corresponding to the Marne a Fucoidi Formation, 3.51 m corresponding to the underlying Maiolica Formation, and 9.98 m corresponding to the overlying Scaglia Bianca Formation (Fig. 2).

## Litho-, Bio-, and Chronostratigraphy of the Poggio le Guaine Core

On the basis of dominant colors determined with the Munsell Rock Color Chart, the pattern of alternating colors, the calcium carbonate content, the occurrence of black shales, chert and radiolarian-rich beds, and by direct comparison with previous results from an adjacent outcrop (Coccioni et al., 1987, 1989, 1990), the Poggio le Guaine core was subdivided into twenty-three lithological units, which are recognizable on a regional scale (Coccioni, 1996; Fig. 2). Unit 1 to part of Unit 4 corresponds to the Scaglia Bianca Formation, and part of Unit 4 to the upper part of Unit 23 corresponds to the Marne a Fucoidi Formation. The Maiolica Formation is equated with most of Unit 23.

A total of two hundred forty organic-rich black shales and marls, millimeters to decimeters thick, is recorded in the cored sequence. These layers are not evenly distributed throughout the core. They are frequent to common in Units 2, 5 to 7, 9 to 15, 22 and 23 that are dominated by lithotypes that are pale olive to grayish olive and yellowish gray to light gray, and occasionally by pale olive and pale reddish brown to dark reddish brown (Fig. 2). Some of these organic-rich beds record OAE1a to OAE1d, namely the lower Aptian Selli Level (OAE1a, ~124 Ma), the uppermost Aptian Jacob Level, the lowermost Albian Kilian Level, the lower Albian Monte Nerone Level and the Urbino Level (all added to OAE1b, ~110–113 Ma), the upper Albian Amadeus Segment (equivalent to part of OAE1c, ~105 Ma) and the uppermost Albian Pialli Level (OAE1d, ~100.5 Ma). Some others are interpreted as regional equivalents of widely distributed OAEs, namely the lower Aptian Noir Level and the middle to upper Aptian Fallot Level (Herrle al., 2004; Figs. 2, 4).

Figure 3. Drilling operations at Poggio le Guaine.

Seven discrete intervals dominantly reddish are also recognized in the cored sequence, which correlate well with those identified as ORBs in the Aptian-Albian succession of the UMB (Hu et al., 2005; Fig. 2).

In agreement with previous detailed stratigraphic investigations from Aptian-Albian land-based sections throughout the UMB (Coccioni et al., 1987, 1989, 1990, 1992; Cresta et al., 1989; Coccioni and Galeotti, 1993; Coccioni 1996, 2001, 2002) and from an 84-m-thick core drilled at Piobbico (Fig. 1) in 1982 and extending from the upper Albian down to the uppermost Barremian (Erba, 1988, 1992; Tornaghi et al., 1989), the cored succession biostratigraphically represents the interval from the *Hedbergella excelsa* to the base of the *Thalmanninella globotruncanoides* planktonic foraminiferal zones and from the *Chiastozygus litterarius* (NC6) to the *Eiffellithus turriseiffelii* (NC10) calcareous nannofossil zones. Accordingly, from a chronostratigraphic point of view the Poggio le Guaine core encompasses the latest Barremian to earliest Cenomanian interval (Fig. 2).

## Objectives

The Poggio le Guaine core, which can be considered as a reference section for the Aptian-Albian interval at low latitudes, is designated by a consortium of Italian and Brazilian researchers to provide high-ranking informative records for that critical time interval through high-resolution multiproxy studies. The studies already in progress include integrated stratigraphic and sedimentological analyses, multidisciplinary magneto-, bio-, and chemostratigraphy, as well as clay and organic geochemistry investigations.

The data set collected will supply the framework for the following three activities. 1) Computing a high-resolution magnetostratigraphy, in particular a relative paleointensity reference curve for the Aptian–Albian interval of the long normal Cretaceous superchron, will allow correlation of relative paleointensity curves from other places in the world. 2) A unique and original age model can be built up, at high-resolution and using absolute dating and astronomical tuning. Into this the sequence of the geological, biogeochemical, oceanographic and climatic events that are recorded throughout the Aptian–Albian critical time can be placed. 3) The causal linkages, as well as their consequences, among these events can be defined.

## Acknowledgments

The Poggio le Guaine core is an integral part of the Magnetic Anomaly of Relative Intensity for the Aptian (MARIA) project. Coring was funded by Petróleo Brasileiro S.A. - Petrobras. The UniversitàAgraria di Secchiano (Cagli, Regione Marche, Italy) generously granted access and facilitated drilling. We warmly thank Glen Hill, Thomas Wiersberg, Mika Saido and the Editor Kevin Johnson for their helpful comments.

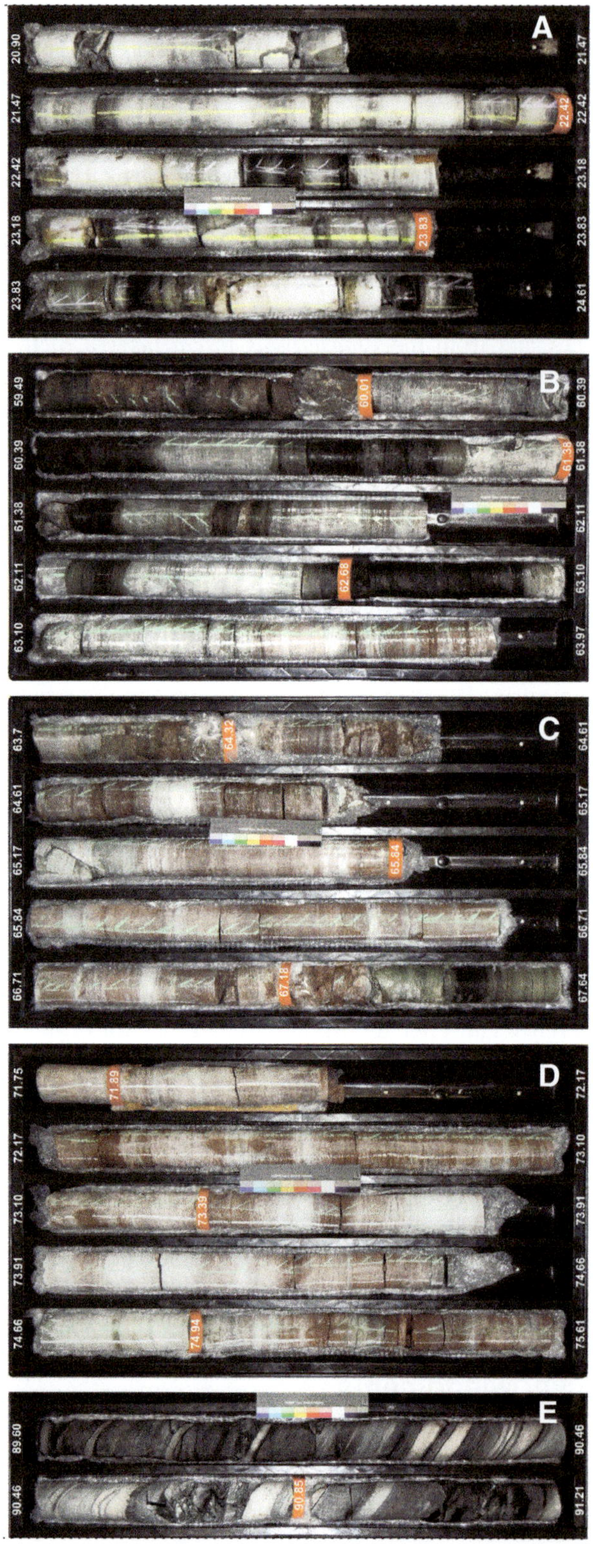

Figure 4. Photographs of selected cores recovered at Poggio le Guaine. [A] Cores 20 (20.90–22.45 m) to 22 (23.83–25.39 m) penetrated the marly limestones, marls, and black shales of the upper Albian Amadeus Segment (OAE1c); [B] Cores 46 (58.42–60.01 m) to 49 (62.68–64.33 m) recovered the marly clays, clays, marls, and black shales of the lower Albian Monte Nerone and Kilian Levels (OAE1b); [C] Cores 49 (62.68–64.33 m) to 52 (67.18–68.77 m) penetrated latest Aptian marls and calcareous marls together with the black shales of the uppermost Aptian Jacob Level (OAE1b); [D] Cores 54 (70.32–71.89 m) to 57 (74.94–76.45 m) recovered late Aptian marls, marly clays, and calcareous marls; [E] Cores 66 (88.00–89.60 m) to 68 (90.85–91.99 m) penetrated the lower Aptian Selli Level (OAE1a).

## References


Arthur, M.A., and Premoli Silva, I., 1982. Development of widespread organic carbon-rich strata in the Mediterranean Tethys. *In* Schlanger, S.O., and Cita, M.B. (Eds.), *Nature of Cretaceous Carbon-Rich Facies*: San Diego (Academic Press), 7–54.

Arthur, M.A., Jenkyns, H.C., Brumsack, H.-J., and Schlanger, S.O., 1990. Stratigraphy, geochemistry, and paleoceanography of organic-carbon-rich Cretaceous sequences. *In* Ginsburg, R.N., and Beaudoin, B. (Eds.), *Cretaceous Resources, Events and Rhythms. NATO ASI Ser. 304*: Dordrecht, Netherlands (Kluwer Acad.), 75–119.

Coccioni, R., 2002. Stop 5 – The proposed global boundary stratotype section and point (GSSP) for the Barremian–Aptian boundary at Gorgo a Cerbara. In Santantonio, M. (Ed.), *General Field Trip Guidebook*. 6th International Symposium on the Jurassic System, 12–22 September 2002, Palermo, 239–241.

Coccioni, R., 1996. The Cretaceous of the Umbria-Marche Apennines (Central Italy). *Jost Wiedmann Symposium on Cretaceous Stratigraphy, Paleobiology and Paleobiogeography*, Tübingen, 7–10 March 1996, 129–136.

Coccioni, R., 2001. The "Pialli Level" from the latest Albian of the Umbria-Marche Apennines (Italy). *Federazione Italiana di Scienze della Terra, Geoitalia 2001*, 192–193.

Coccioni, R., and Galeotti, S., 1993. Orbitally induced cycles in benthonic foraminiferal morphogroups and trophic structures distribution patterns from the Late Albian "Amadeus Segment" (Central Italy). *J. Micropaleontol.*, 12:227–239. doi:10.1144/jm.12.2.227

Coccioni, R., Erba, E., and Premoli Silva, I., 1992. Barremian-Aptian calcareous plankton biostratigraphy from the Gorgo a Cerbara section (Marche, Central Italy) and implications for plankton evolution. *Cret. Res.*, 13:517–537. doi:10.1016/0195-6671(92)90015-I

Coccioni, R., Franchi, R., Nesci, O., Wezel, F.C., Battistini, F., and Pallecchi, P., 1989. Stratigraphy and mineralogy of the Selli Level (Early Aptian) at the base of the Marne a Fucoidi in the Umbro-Marchean Apennines, Italy. In Wiedmann, J. (Ed.), *Cretaceous of the Western Tethys. Proceedings 3rd International Cretaceous Symposium*, Stuttgart (E. Schweizerbart'sche Verlagsbuchhandlung), 563–584.

Coccioni, R., Franchi, R., Nesci, O., Perilli, N., Wezel, F.C., and Battistini, F., 1990. Stratigrafia, micropaleontologia e mineralogia delle Marne a Fucoidi delle sezioni di Poggio le Guaine e del Fiume Bosso (Appennino umbro-marchigiano). Atti 2° Convegno Internazionale "Fossili, Evoluzione, Ambiente", Pergola, 25–30 ottobre 1987, Tecnostampa, 163–201.

Coccioni, R., Nesci, O., Tramontana, M., Wezel, F.C., and Moretti, E., 1987. Descrizione di un livello-guida "radiolaritico-bituminoso-ittiolitico" alla base delle Marne a Fucoidi nell'Appennino umbro-marchigiano. *Boll. Soc. Geol. Italiana*, 106:183–192.

Coccioni, R., Premoli Silva, I., Marsili, A., and Verga, D., 2007. First radiation of Cretaceous planktonic foraminifera with radially elongate chambers at Angles (Southeastern France) and biostratigraphic implications. Rev. micropaléontologie, 50:215–224.

Cresta, S., Monechi, S., and Parisi, G., 1989. Stratigrafia del Mesozoico al Cenozoico nell'area Umbro-Marchigiana. *Mem. Descr. Carta Geol. Italia*, 34:185.

Erba, E., 1988. Aptian-Albian calcareous nannofossil biostratigraphy of the Scisti a Fucoidi cored at Piobbico (central Italy). *Riv. Ital. Paleont. Strat.*, 94:249–284.

Erba, E., 1992. Calcareous nannofossil distribution in pelagic rhythmic sediments (Aptian-Albian Piobbico core, central Italy). *Riv. Ital. Paleont. Strat.*, 97:455–484.

Herrle, J.O., Kössler, P., Friedrich, O., Erlenkeuser, H., and Hemleben, C., 2004. High-resolution carbon isotope records of the Aptian to Lower Albian from SE France and the Mazagan Plateau (DSDP Site 545): A stratigraphic tool for paleoceanographic and paleobiologic reconstruction. *Earth Planet. Sci. Lett.*, 218:149–161. doi:10.1016/S0012-821X(03)00646-0

Hu, X., Jansa, L., and Sarti, M., 2005. Mid-Cretaceous oceanic red beds in the Umbria-Marche Basin, central Italy: Constraints on paleoceanography and paleoclimate. *Palaeogeogr. Palaeoclimatol. Palaeoecol.*, 233:163–186. doi:10.1016/j.palaeo.2005.10.003

Huber, B.T., and Leckie, R.M., 2011. Planktic foraminiferal species turnover across deep-sea Aptian/Albian boundary sections. *J. Foram. Res.*, 41:53–95. doi:10.2113/gsjfr.41.1.53

Leckie, R.M., Bralower, T.J., and Cashman, R., 2002. Oceanic anoxic events and plankton evolution: Biotic response to tectonic forcing during the mid–Cretaceous. *Paleoceanogr.*, 17:13-1–13-29. doi: 10.1029/2001PA000623

Ogg, J.G., Ogg, G., and Gradstein, F.M., 2008. *The Concise Geologic Time Scale*: Cambridge (Cambridge University Press).

Satolli, S., Besse, J., and Calamita, F., 2008. Paleomagnetism of Aptian-Albian sections from the Northern Apennines (Italy): Implications for the 150–100 Ma apparent polar wander of Adria and Africa. *Earth Planet. Sci. Lett.*, 276:115–128. doi:10.1016/j.epsl.2008.09.013

Schlanger, S.O., and Jenkyns, H.C., 1976. Cretaceous oceanic anoxic events: Causes and consequences. *Geol. Mijnbouw*, 55:179–184.

Skelton, P.W., Spicer, R.A., Kelley, S.P., and Gilmour, I., 2003. *The Cretaceous World*: Cambridge (Cambridge University Press).

Tiraboschi, D., Erba, E., and Jenkyns, H.C., 2009. Origin of rhythmic Albian black shales (Piobbico core, central Italy): Calcareous nannofossil quantitative and statistical analyses and paleoceanographic reconstructions. *Paleoceanogr.*, 24:PA2222. doi:10.1029/2008PA001670

Tornaghi, M.E., Premoli Silva, I., and Ripepe, M., 1989. Lithostratigraphy and planktonic foraminiferal biostratigraphy of the Aptian-Albian "Scisti a Fucoidi" in the Piobbico core, Marche, Italy: Background for cyclostratigraphy. *Riv. Ital. Paleont. Strat.*, 95:223–264.

Turchyn, A.V., Schrag, D.P., Coccioni, R., and Montanari, A., 2009. Stable isotope analysis of the Cretaceous sulfur cycle. *Earth Planet. Sci. Lett.*, 285:115–123.

Wang, C., Hu, X., Huang, Y., Scott, R.W., and Wagreich, M., 2009. Overview of Cretaceous Oceanic Red Beds (CORBs): A window on global oceanic and climate change. *In* Hu, X., Wang, C., Scott, W., Wagreich, M., and Jansa, L. (Eds.), *Oceanic Red Beds: Stratigraphy, Composition, Origins, and Paleoceanographic and Paleoclimatic Significance. Society for Sedimentary Geology, Special Publication*, 91:13–33.

## Authors

**Rodolfo Coccioni,** Dipartimento di Scienze della Terra, della Vita e dell'Ambiente dell'Università degli Studi "Carlo Bo," Campus Scientifico "E. Mattei", Località Crocicchia, 61029 Urbino, Italy, e-mail: rodolfo.coccioni@uniurb.it.

**Luigi Jovane,** Departamento de Oceanografia Física, Instituto Oceanográfico, Universidade de São Paulo, 05508-900 São Paulo, Brazil.

**Giuseppe Bancalà,** Dipartimento di Scienze della Terra, della Vita e dell'Ambiente dell'Università degli Studi "Carlo Bo," Campus Scientifico "E. Mattei", Località Crocicchia, 61029 Urbino, Italy.

**Carla Bucci,** Dipartimento di Scienze della Terra, della Vita e dell'Ambiente dell'Università degli Studi "Carlo Bo," Campus Scientifico "E. Mattei", Località Crocicchia, 61029 Urbino, Italy.

**Gerson Fauth,** Centro de Ciências Exatas e Tecnológicas, Área de Conhecimento e Aplicação de Geociências, Universidade do Vale do Rio dos Sinos, São Leopoldo, Brazil.

**Fabrizio Frontalini,** Dipartimento di Scienze della Terra, della Vita e dell'Ambiente dell'Università degli Studi "Carlo Bo," Campus Scientifico "E. Mattei", Località Crocicchia, 61029 Urbino, Italy.

**Liliane Janikian,** Departamento de Geofísica, Instituto de Astronomia, Geofísica e Ciências Atmosféricas, Universidade de São Paulo, 05508-090 São Paulo, Brazil.

**Jairo Savian,** Departamento de Geofísica, Instituto de Astronomia, Geofísica e Ciências Atmosféricas, Universidade de São Paulo, 05508-090 São Paulo, Brazil.

**Renato Paes de Almeida,** Departamento de Geologia Sedimentar e Ambiental, Instituto de Geociências, Universidade de São Paulo, 05508-090 São Paulo, Brazil.

**Grasiane Luz Mathias,** Departamento de Geofísica, Instituto de Astronomia, Geofísica e Ciências Atmosféricas, Universidade de São Paulo, 05508-090 São Paulo, Brazil

**Ricardo Ivan Ferreira da Trindade,** Departamento de Geofísica, Instituto de Astronomia, Geofísica e Ciências Atmosféricas, Universidade de São Paulo, 05508-090 São Paulo, Brazil.

## Photo Credits

Figs. 3 and 4: Rodolfo Coccioni, Urbino University

# Comparison of Rhizon Sampling and Whole Round Squeezing for Marine Sediment Porewater

by Heather N. Schrum, Richard W. Murray, and Britta Gribsholt

## Introduction

The collection and chemical analysis of sedimentary porewater is central to many marine studies. Porewater alkalinity, dissolved inorganic carbon (DIC), sulfate, nitrate, and other dissolved ions are used to identify and determine rates of geochemical reactions and microbial respiration pathways, such as sulfate reduction and denitrification (Froelich et al., 1979; Berner, 1980; Gieskes et al., 1986; D'Hondt et al., 2004; Schulz, 2006; Martin and Sayles, 2007). Ammonium is critical for understanding microbial respiration and the nitrogen cycle (Blackburn, 1988). Chloride is used to reconstruct ocean salinity variations, constrain flow rates, and estimate gas hydrate concentrations (Paull et al., 1996; Adkins et al., 2002; Spivack et al., 2002). Each of these studies requires the recovery of porewater that is not compromised by sampling artifacts.

A number of methods are used to extract fluids from sediments. Here we report our experimental comparison of two methods used by the Integrated Ocean Drilling Program (IODP): extraction with Rhizon samplers and sediment squeezing (Seeberg-Elverfeldt et al., 2005; Manheim and Sayles, 1974).

Rhizon samplers are inserted directly into an intact sediment core; they collect porewater by vacuum filtration through a thin porous tube attached to a syringe (Seeberg-Elverfeldt et al., 2005; Dickens et al., 2007). Rhizons allow for high-resolution sampling because they may be spaced as closely as 2–3 cm apart. They also only minimally mechanically disturb the sediment and are typically time efficient. Porewater and soil moisture studies report that Rhizon sampling produces accurate results for dissolved ammonium, sulfate, chloride, and iron (Gribsholt and Kristensen, 2002; Song et al., 2003; Seeberg-Elverfeldt et al., 2005; Backman et al., 2006; Dickens et al., 2007). A disadvantage of Rhizon sampling is that it is often restricted to the upper portion of the sediment column (tens of meters, depending on lithology), where sediment is not too compacted. Also, the partial vacuum may lead to significant sampling artifacts for certain chemical species.

Whole round squeezing mechanically pressurizes a sediment sample, forcing the porewater through a sampling port. Squeezing has been shown to effectively collect porewater to depths of ~1600 m below seafloor (Saffer et al., 2010). Whole round squeezing also minimizes the contact of porewater with air. The disadvantages of squeezing include pressure-related additions through the destruction of particulate detritus or microbial cells (Bollinger et al., 1992), oxygen contamination (Robbins and Gustinis, 1976), and temperature artifacts (Bischoff et al., 1970; Fanning and Pilson, 1971). Many of these have been studied, however, and were deemed minimal (Gieskes, 1974). Dickens et al. (2007) compared Rhizon and squeezing methods for a suite of dissolved chemicals and concluded that the chemical concentrations did not vary significantly between methods.

We evaluated the use of Rhizon sampling and whole round squeezing for analyzing alkalinity, DIC, ammonium, sulfate, chloride, and nitrate in porewater from equatorial and South Pacific sites. In these experiments we directly compared the dissolved species collected by both methods in the same core and/or location and used these data to interpret potential sampling artifacts.

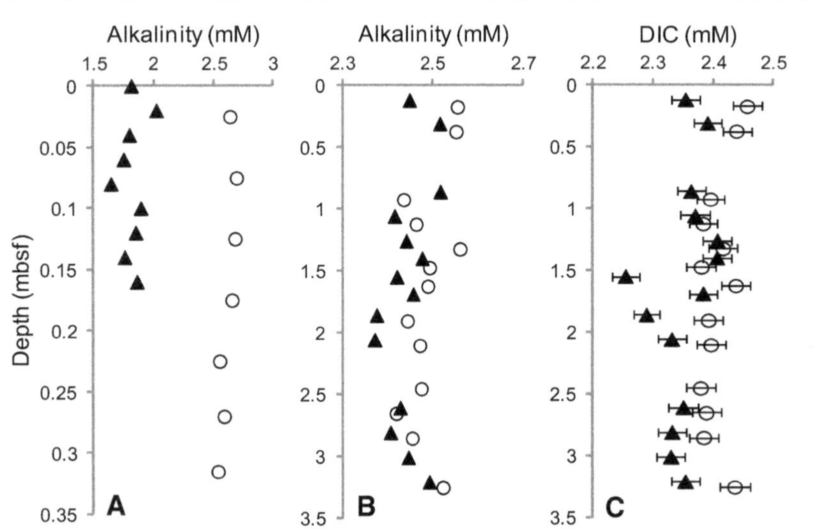


Figure 1. Alkalinity profiles at [A] SPG-9 and [B] EQP-9, and [C] DIC profile from EQP-9. Data from squeezing (open circles) and Rhizons (black triangles). Note differences in scales between sites. 0.5% error bars on alkalinity profiles are smaller than symbols. 1% error bars on DIC.

## Materials and Procedures

We collected porewater from five sites in the Pacific Ocean with variable amounts of clay and biogenic ooze. At Site SPG-9 in the South Pacific Gyre we compared the Rhizon and squeezing method for alkalinity from sediment retrieved by a multi-core system, and we compared the porewater extraction methods for alkalinity and DIC from a gravity core at the Pacific Site EQP-9 (Fig. 1).

DIC, alkalinity, ammonium, sulfate, and chloride were measured from gravity cores at EQP-1, EQP-6, and EQP-6A (Fig. 2). Ammonium concentrations were compared for both extraction methods from samples at EQP-1. Two separate gravity cores at EQP-6 and -6A were retrieved at nearly the same location, and they have nearly identical lithologies. We compared both porewater extraction methods for sulfate and chloride at these two sites.

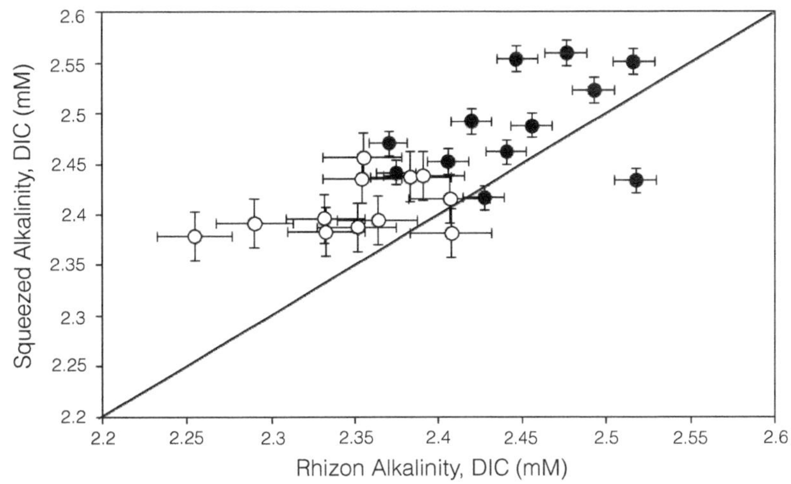


Figure 2. EQP-9 alkalinity (filled circles) and DIC (open circles) concentrations from Rhizon and squeezing methods. Line is 1:1. 0.5% and 1% error bars on alkalinity and DIC, respectively.

For SPG and EQP sites, Rhizon CSS-F 5-cm core solution samplers with 0.1µm microporous filters were used to collect porewater. The squeezing apparatus and method on both cruises were identical to those used by the Ocean Drilling Program (ODP) and Integrated Ocean Drilling Program (IODP) (see Methods in Backman et al., 2006).

For SPG and EQP samples, we measured alkalinity by Gran-titration following Gieskes et al. (1991). DIC was measured by acid extraction and infrared measurement of purged $CO_2$. Sulfate, chloride, and nitrate were quantified by ion chromatography. Dissolved ammonium was determined spectrophotometrically based on Solórzano (1969).

## Comparative Assessment of Rhizon and Squeezing Protocols

At SPG-9 and EQP-9, alkalinities from porewater collected by Rhizon sampling are consistently lower than those measured from squeezed samples by approximately 0.8 mM and 0.06 mM, respectively. At EQP-9, DIC concentrations from Rhizons are approximately 0.06 mM on average less than squeezed concentrations (Fig. 1). These differences in alkalinity and DIC are larger than can be explained by analytical uncertainty.

Alkanity and DIC concentrations collected by Rhizon sampling have more scatter at SPG-9 and EQP-9. Rhizon alkalinities ranges from 1.64–2.02 mM and 2.37–2.52 mM at SPG-9 and EQP-9, respectively, while squeezed alkalinities range from 2.53–2.69 mM and 2.42–2.56 mM at these respective sites. Rhizon DIC values ranges from 2.25–2.41 mM versus 2.38–2.46 mM for squeezed samples at EQ-9.

In contrast to these findings, measured concentrations of ammonium, sulfate, and chloride are all within the error of the analytical methods, for both squeezing and Rhizon extraction at EQP-1, and EQP-6 and -6A (Fig. 3). This is consistent with results from Dickens et al. (2007) and Backman et al. (2006).

While neither the squeezing nor the Rhizon method for porewater extraction showed measurable sampling artifacts for ammonium, sulfate, and chloride, during the equatorial Pacific expedition we found repeated evidence that the squeezing method introduced nitrate contamination on the scale of tens of micromoles. We suspect that the Whatman #1 paper filters used inside the squeezer contained trace amounts of nitrate that were not removed even after soaking the filters in 18 MOhm water for ~24 hours and allowing them to dry. We filtered 18 MOhm water through the squeezer and found that a nitrate blank was present. Pressure-related destruction of cells may also have contributed to nitrate contamination during squeezing. Risgaard-Petersen et al. (2006) showed that there are considerable amounts of nitrate stored in foraminifera. These concentrations (~60 nmol $cm^{-3}$ sediment) may not fully explain the contamination but may contribute to the erroneous nitrate values.

## Discussion

Vacuum filtration is commonly used for fluid degassing. During Rhizon sampling into a low pressure headspace, subatmospheric pressure creates a pressure differential that forces the porewater through the filter. The filter creates a flow disturbance, causing minute bubbles of entrained gas to coalesce or to break the fluid into small droplets to provide a large surface-to-volume ratio for enhancing the effect of the vacuum degassing the fluid (Chambers et al., 1998).

$CO_2$ degassing during Rhizon sample extraction leads to alkalinity and DIC concentrations that are consistently lower than squeezed sample concentrations (Fig. 1). $CO_2$ degassing drives an increase in carbonate ion concentration, which increases the tendency for calcium car-bonate to precipitate, thereby decreasing alkalinity. In contrast, whole round squeezing applies pressure to the sediment sample, pushing the porewater out of the sediment into the filter-capped syringe, and it should have minimal impact on the carbonate system during sampling. It is possible, however, that pressure-related destruction of cells may cause some carbonate contamination during squeezing. When squeezing is used for sample extraction, care should be taken to store whole rounds at ***in situ*** temperature prior to squeezing, as carbonate saturation has retrograde solubility, and cation exchange between clay minerals and interstitial water is temperature dependent (Bischoff et al., 1970; Masuzawa et al., 1980).

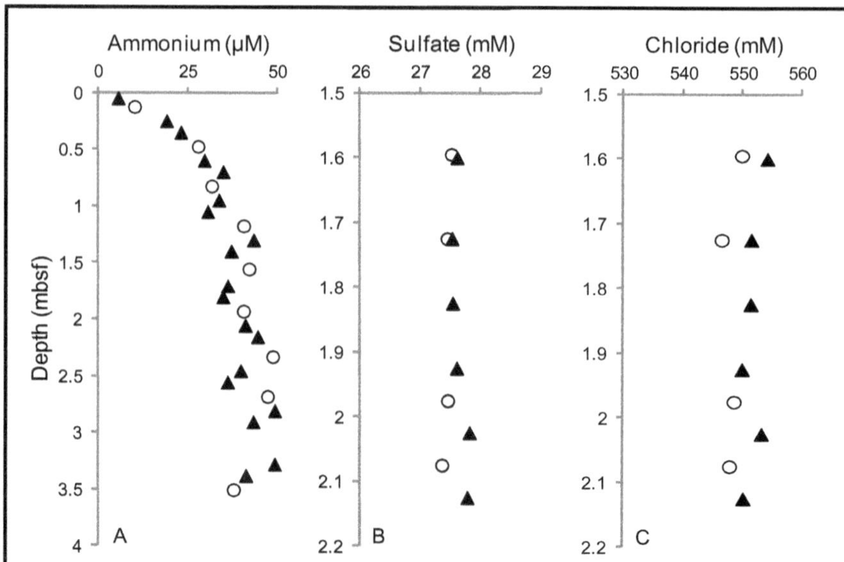


Figure 3. [A] Ammonium profiles at EQP-1 from squeezing (open circles) and Rhizons (black triangles). [B] Sulfate profiles at EQP-6 from squeezing (open circles) and at EQP-6A from Rhizons (black triangles). [C] Chloride profiles at EQP-6 from squeezing (open circles) and at EQP-6A from Rhizons (black triangles). EQP-6 and EQP-6A are nearly at the same location. Note differences in scales between sites. 3%, 0.15%, and 0.2% error bars on ammonium, sulfate and chloride profiles, respectively, are smaller than symbols.

The increased scatter in the alkalinity and DIC concentrations collected by Rhizon sampling also indicates compromised data. Based on visual inspection, if these concentrations were taken as true values, the data would lead to unrealistic concentration gradients in the sediment with improbable fluxes and rapidly changing sources and sinks (Fig. 1).

Our alkalinity results are not consistent with Backman et al. (2006), who convincingly reported no significant difference in alkalinity between samples collected by Rhizon and whole round squeezing. The differences in our findings are possibly due to sediment composition and porewater carbonate concentrations. Dissolved calcium carbonate concentrations are higher at most of the Pacific sites than in Arctic sediment collected by Backman et al. Porewater degassing during Rhizon sampling likely caused relatively greater calcium carbonate precipitation, which led to a greater decrease in alkalinity compared to squeezed samples. This is also the likely explanation for the greater difference in alkalinity concentrations between the Rhizon and squeezed samples at SPG-9 compared to EQP-9 (Fig. 1).

## Recommendations

Our study shows that a combined sampling protocol of Rhizon sampling for ammonium (this study and Backman et al., 2006), nitrate, and metals (e.g., Mn, Backman et al., 2006) and squeezing for alkalinity, DIC, and pH will lead to a complete characterization of the dissolved chemical species in multiple environments. Future scientific coring and drilling expeditions that rely on high precision chemical profiles in marine sediment would greatly benefit from this combined sampling protocol.

While Seeberg-Elverfeldt et al. (2005) suggest that pH may be a viable measurement from Rhizon samplers, we argue that the degassing of $CO_2$ by Rhizons will produce higher than realistic pH and should be avoided. In this context, critical differences between soil systems and marine sediments appear to compromise comparison between soil-based and sediment-based studies. It is important that Rhizon samplers are employed to collect porewater that will be used to measure nitrate, as squeezing potentially introduces nitrate contamination on the scale of tens of micromoles.

## Acknowledgements

We thank the captains and crew of the R/V *Roger Revelle* and R/V *Knorr*, and the Knox-02RR (South Pacific Gyre) and KNR 195-3 (equatorial Pacific) shipboard science party members. Special thanks to Steven D'Hondt and Arthur Spivack for their support and helpful comments and to Timothy Ferdelman, Benjamin Brunner, and Dennis Graham for fruitful conversations, sample collection, and chemical analyses. This project was funded by the NSF Biological Oceanography Program and the Integrated Ocean Drilling Program.

## References

Adkins, J.F., McIntyre, K., and Schrag, D.P., 2002. The salinity, temperature, and $\delta^{18}O$ of the glacial deep ocean. *Science*, 298:1769–1773, doi:10.1126/science.1076252.

Backman, J., Moran, K., McInroy, D.B., Mayer, L.A., and the Expedition 302 Scientists., 2006. *Proc. IODP*, 302: Washington, DC (Integrated Ocean Drilling Program Management International, Inc.).

Berner, R.A., 1980. *Early Diagenesis: A Theoretical Approach:* Princeton, NJ (Princeton University Press).

Bischoff, J.L., Greer, R.E., and Luistro, A.O., 1970. Composition of interstitial waters of marine sediments: Temperature of squeezing effect. *Science*, 167(3922):1245–1246, doi:10.1126/science.167.3922.1245.

Blackburn, T.H., 1988. Chapter 8: Benthic mineralization and bacterial production, *In* Blackburn, T.H., and Sørensen, J. (Eds.), *Nitrogen Cycling in Coastal Marine Environments:* Chichester (John Wiley and Sons), 175–190.

Bollinger, R., Brandl, H., Höhener, P., Hanselmann, K.W., and Bachofen, R., 1992. Squeeze-water analysis for the determination of microbial metabolites in lake sediment - Comparison of methods. *Limnol. Oceanogr.*, 37:448–455, doi:10.4319/lo.1992.37.2.0448.

Chambers, A., Fitch, R.K., and Halliday, B.S., 1998. *Basic Vacuum Technology, second edition:* London (Institute of Physics Publ.), doi:10.1201/NOE0750304955.

D'Hondt, S., Jørgensen, B.B., Miller, D.J., Batzke, A., Blake, R., Cragg, B.A., Cypionka, H., et al., 2004. Distributions of microbial activities in deep subseafloor sediments. *Science*, 30:2216–2221, doi:10.1126/science.1101155.

Dickens, G.R., Koelling, M., Smith, D.C., Schneiders, L., and the IODP Expedition 302 Scientists, 2007. Rhizon sampling of pore waters on scientific drilling expeditions: An example from the IODP Expedition 302, Arctic Coring Expedition (ACEX). *Sci. Drill.*, 4:22–25.

Fanning, K.A., and Pilson, M.E.Q., 1971. Interstitial silica and pH in marine sediments: Some effects of sampling procedures. *Science*, 173:1228–1231, doi:10.1126/science.173.4003.1228.

Froelich, P.N., Klinkhammer, G.P., Bender, M.L., Luedtke, N.A., Heath, G.R., Cullen, D., Dauphin, P., Hammond, D., Hartman, B., and Maynard, V., 1979. Early oxidation of organic matter in pelagic sediments of the eastern equatorial Atlantic: Suboxic diagenesis. *Geochim. Cosmochim. Acta*, 43:1075–1090.

Gieskes, J.M., 1974. Interstitial water studies, Leg 25. *In* Simpson, E.S.W., Schlich, R., et al., *Init. Repts. DSDP*, 25: Washington, DC (U.S. Govt. Printing Office), 361–394.

Gieskes, J.M., Kastner, M., Erzinger, J., Boulegue, J., and Hart, S.R., 1986. Geochemical studies Hole 504B, Leg 92. *In* Leinen, M., Rea, D.K., et al., *Init. Repts. DSDP*, 92: Washington, DC (U.S. Govt. Printing Office), 547.

Gieskes, J.M., Gamo, T., and Brumsack, H., 1991. Chemical methods for interstitial water analysis aboard *JOIDES Resolution. ODP Tech. Note*,15.

Gribsholt, B., and Kristensen, E., 2002. Impact of sampling methods on sulfate reduction rates and dissolved organic carbon (DOC) concentrations in salt marsh sediment. *Wetlands Ecol. Manage.*, 10:371–379, doi:10.1023/A:1020940314010.

Manheim, F.T., and Sayles, F.L., 1974. Composition and origin of interstitial waters of marine sediments, based on deep sea drill cores. *In* Goldberg, E.D. (Ed.), *The Sea (Vol. 5): Marine Chemistry: The Sedimentary Cycle.* New York (Wiley-Interscience), 527–568.

Martin, W.R., and Sayles, F.L., 2007. The recycling of biogenic material at the seafloor. *In* Mackenzie, F.T. (Ed.), *Treatise on Geochemistry. Vol. 7: Sediments, Diagenesis, and Sedimentary Rocks:* Oxford (Elsevier Science), 37–65.

Masuzawa, T., Kanamori, S., and Kitano, Y., 1980. The reversible effect of temperature on the chemical composition of interstitial water of marine sediment. *J. Oceanogr. Soc. Jpn.*, 36:68–72, doi:10.1007/BF02209357.

Paull, C.K., Matsumoto, R., Wallace, P.J., and Expedition 164 Scientists, 1996. *Proc. ODP, Init. Repts.*, 164: College Station, TX (Ocean Drilling Program).

Risgaard-Petersen, N., Langezaal, A.M., Ingvardsen, S., Schmid, M.C., Jetten, M.S.M., Op den Camp, H.J.M., Derksen, J.W.M., et al., 2006. Evidence for complete denitrification in a benthic foraminifer. *Nature*, 443:93–96, doi:10.1038/nature05070.

Robbins, J.A., and Gustinis, J., 1976. A squeezer for efficient extraction of pore water from small volumes of anoxic sediment. *Limnol. Oceanogr.*, 21:905–909, doi:10.4319/lo.1976.21.6.0905.

Saffer, D., McNeill, L., Byrne, T., Araki, E., Toczko, S., Eguchi, N., Takahashi, K., and the Expedition 319 Scientists, 2010. *Proc. IODP*, 319: Tokyo (Integrated Ocean Drilling Program Management International, Inc.).

Schulz, H.D., 2006. Quantification of early diagenesis: Dissolved constituents in pore water and signals in the solid phase. *In* Schulz, H.D., and Zabel, M. (Eds.), *Marine Geochemistry, second edition:* Berlin (Springer-Verlag), 73–124.

Seeberg-Elverfeldt, J., Schlüter, M., Feseker, T., and Kölling, M., 2005. Rhizon sampling of porewater near the sediment-water interface of aquatic systems. *Limnol. Oceanogr.: Methods*, 3:361–371, doi:10.4319/lom.2005.3.361.

Solórzano, L., 1969. Determination of ammonia in natural waters by the phenolhypochlorite method. *Limnol. Oceanogr.*, 14:799–801, doi:10.4319/lo.1969.14.5.0799.

Song, J., Luo, Y.M., Zhao, Q.G., and Christie, P., 2003. Novel use of soil moisture samplers for studies on anaerobic ammonium fluxes across lake sediment-water interfaces. *Chemosphere*, 50(6):711–715, doi:10.1016/S0045-6535(02)00210-2.

Spivack, A.J., Kastner, M., and Ransom, B., 2002. Elemental and isotopic chloride geochemistry and fluid flow in the Nankai Trough. *Geophys. Res. Lett.*, 29:1661–1665, doi:10.1029/2001GL014122.

## Authors

**Heather N. Schrum**, Massachusetts Maritime Academy, 101 Academy Drive, Buzzards Bay, MA 02532, U.S.A., e-mail: hschrum@maritime.edu

**Richard W. Murray**, Department of Earth Sciences, Boston University, 1 Silber Way, Boston, MA 02215, U.S.A.

**Britta Gribsholt**, Department of Biological Sciences, Center for Geomicrobiology, University of Aarhus, Ny Munkegade 114, DK-8000 Aarhus C, Denmark.

# 33

# Climate Evolution in Central Asia during the Past Few Million Years: A Case Study from Issyk Kul

by Hedi Oberhänsli and Peter Molnar

## Introduction

The lake Issyk Kul occupies a deep basin within the Earth's most active intracontinental mountain belt, the Tien Shan, far from any oceanic influence. It offers a record of continental climate spanning millions of years that is likely unmatched by any other source.

A three-day workshop, with the same title as this report and sponsored by the International Continental Drilling Project and German Science Foundation, was held on 12–17 June 2011 on the shore of Issyk Kul to discuss the scientific justification for and the logistical aspects of scientific drilling of the lake. A two-day geological field trip followed the workshop. Forty-five scientists from twelve countries discussed three obvious targets for paleoclimatic study, a related study of erosion, and a study of how microbial life has evolved within the basin. The conclusion was that these research topics justify further consideration of deep continental drilling at Issyk Kul.

The premise underlying the paleoclimate questions is that climates within continents need not follow that recorded by, and studied well with, marine sediment. A particularly clear example is that glaciation within alpine settings is not synchronized with that of continental ice sheets; maximum advances of alpine glaciers in many regions precede the last glacial maximum (LGM) at ~20 kyr BP.

The major outcome of the meeting was the identification of two drilling targets: one within the lake that would reach as deep into Quaternary time as possible and certainly as far as 100 kyr BP, and a second on shore that could penetrate sediment as old as 6–10 Ma.

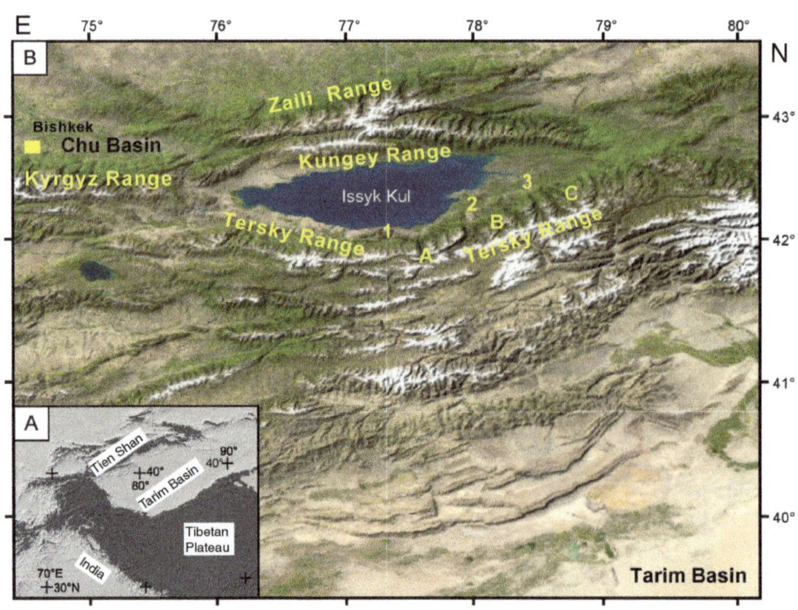


Figure 1. Topographic map of the Tien Shan showing the location of Issyk Kul and range-and-basin nature of the belt.

## Geologic Background

Issyk Kul (literally "hot lake" in Kyrgyz) occupies a deep topographic depression within the Tien Shan (Fig. 1). Observatories using GPS measurements (Fig. 2), modern seismicity, and active faulting indicate relatively rapid deformation and suggest that the Tien Shan is Earth's most active intracontinental mountain belt.

Seismicity in the Tien Shan is high, with four earthquakes since 1889 assigned magnitudes as large as 8 (Abdrakhmatov et al., 2002; Kondorskaya and Shebalin, 1977). Fault plane solutions of earthquakes consistently show reverse or thrust faulting. Evidence of active faulting is widespread (Abdrakhmatov et al., 2007; Chedia, 1986; Laverov and Makarov, 2005), and rates of slip on several faults exceed 1 mm $yr^{-1}$ (Chen et al., 2007; Thompson et al., 2002).

At present the Tarim Basin, which lies to the south of the Tien Shan, moves northward at 20 ± 2 mm $yr^{-1}$ with respect to the Eurasian Plate to the north (Reigber et al., 2001; Zubovich et al., 2010). Deformation is not localized on the margins of the belt, but occurs throughout it, consistent with seismicity and studies of active faulting (Fig. 2). The present-day convergence rate between the Terskey Alatau south of Issyk Kul and the Kungey Alatau to its north is ~5 mm $yr^{-1}$.

Mineral cooling ages (Bullen et al., 2003) suggest accelerated exhumation of parts of the Tien Shan since ~10 Ma, and magnetostratigraphic dating of sediment in adjacent basins (Abdrakhmatov et al., 2001; Charreau et al., 2009; Ji et al., 2008;

Sun and Zhang, 2009) suggest accelerated deposition of sediment since that time. Both findings are consistent with enhanced erosion due to growing relief beginning near or shortly before 10 Ma. How deformation has developed since that time, either as an average across the belt or in individual mountain ranges within the Tien Shan, remains poorly constrained.

The Cenozoic stratigraphy of Kyrgyzstan is commonly divided into four units that overlie a surface that had little relief prior to deposition onto it (Abdrakhmatov et al., 2001; Burgette, 2008). At the bottom is the Kokturpak group, commonly tens of meters thick, fine-grained, deeply weathered, and typically dark red in color, perhaps deposited from Cretaceous to Eocene and possibly as young as Miocene. Disconformably overlying it is the Shamsi Group with thickness varying from hundreds of meters to ~2000 m. It is relatively well-sorted, reddish sediment, rich in quartz and feldspar with abundant conglomerate. Magnetostratigraphy suggests that deposition began 12–13 Ma ago. The Chu Group overlies the Shamsi group, with assigned dates ranging from 8 Ma to 4 Ma; in the Issyk Kul area, reported thicknesses range up to ~2800 m. It includes tabular beds with sandstone and conglomerate lenses interpreted as river channels of moderate size as well as lacustrine sediment and evaporites. Overlying the Chu group in many regions is coarse conglomerate of the Sharpyldak Formation. Thicknesses vary from tens to hundreds of meters to more than 1000 m in places. At the east end of the Issyk Kul basin, however, lacustrine sediment of Quaternary age is exposed.

At present, Issyk Kul sediment consists of a mixture of riverine sediment and endogenic carbonates (De Batist et al., 2001; Giralt et al., 2002). The steep northern and southern margins present evidence of slope failure, and mass-flow

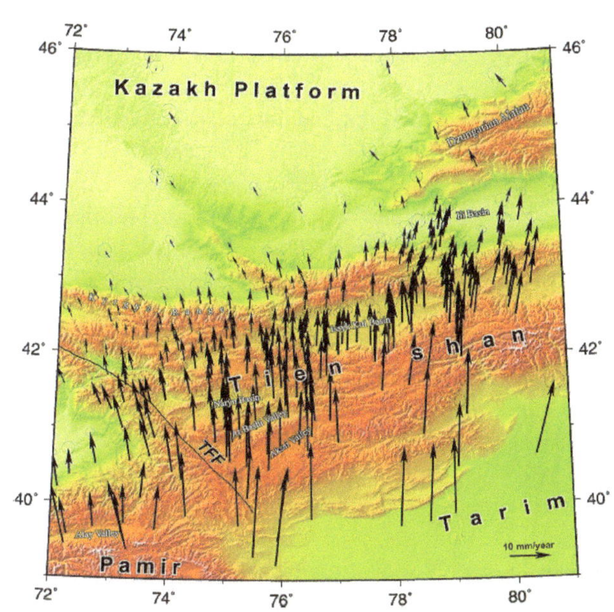


Figure 2. Map of GPS velocities across the Tien Shan. Error ellipses show 95% confidence. (from Zubovich et al., 2010).

deposits are found at the foot of these steep slopes. Near the steep borders, only centimeter-thick fining-upwards sequences of small gravel and sand are present, whereas the central deep basin is filled with a mixture of carbonate hemipelagic sediment and fine terrigenous sediment that includes distal turbidite facies. The southern part of this deep central basin contains a topographic high, and available high-resolution seismic data and short gravity cores indicate that sedimentation there is dominated by bottom currents migrating up-slope towards the crest of the high. By contrast, deltaic sedimentation processes have dominated deposition on the eastern and western margins. Near the eastern and western shores, proximal deltaic sediment is found, whereas distal deltaic sediment (laminated silty-clay sediment) has been deposited closer to the central deep plain. Slope failure deposits dominate the fronts of both deltas.

## Modern Climate

Issyk Kul and the Tien Shan lie within the mid-latitudes, far from the tropics where monsoons dominate climate, and where precipitation interacts profoundly with atmospheric circulation. At the same time, the Tien Shan is far from high-latitude conditions where extremes in the seasonal cycle play a crucial role, and where warming over the past century and a half has been most pronounced. Considerably more attention has been given to the Arctic than the mid-latitudes, and many view the study of mid-latitude climate as a gap in our understanding of climate dynamics.

The mid-latitudes, and the Issyk Kul region in particular, offer often under-appreciated features for climate study. For instance, changes in vegetation, either with continued global warming or on paleoclimate time scales, are much larger than in the tropics or high latitudes. Changes from forests to grassland, or vice versa, occur largely in mid-latitudes, and such changes affect both albedo and moisture availability. In the deep tropics, where changes in precipitation and the variability in temperature are small, large changes in vegetation are unlikely except in extreme climates. Similarly, at high latitudes, where trees are sparse, an expansion of woody vegetation is unlikely except in an extreme change in climate. Moreover, interannual variability of modern climate in and around the Tien Shan also is among the greatest on Earth (Aizen et al., 1997, 2001).

The Issyk Kul region offers some simple aspects that make its study attractive. For example, the most reliable part of General Circulation Models (GCMs) solves the Navier-Stokes equation to obtain the dynamics of flow, and although this solution depends on numerous parameterizations for local processes, such as latent and sensible heat transport from the surface, precipitation and latent heating of the remaining atmosphere, etc., many of these interactions can be described reliably. GCMs resolve well the physics of mid-latitude weather patterns and their associated mechanisms of precipitation. Precipitation in mid-latitudes results from advection of moisture by the large-scale flow and by the subsequent lifting of moisture in mid-latitude cyclones and by the flow incident on large topographical obstacles, all of which are governed by processes that are explicitly and well represented in climate models.

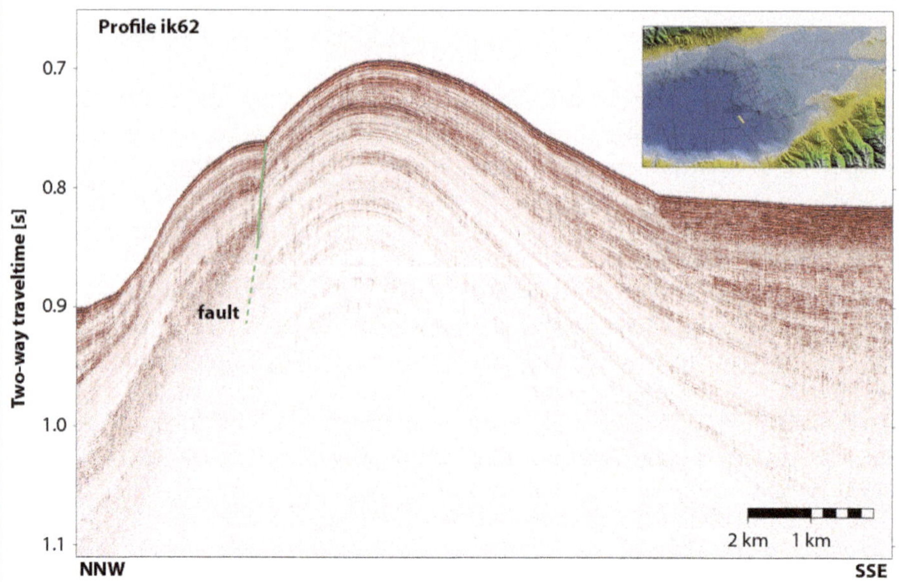

Figure 3. Seismic section that shows a subtle rise in bottom topography and with undisturbed sediment. (De Batist, M., unpublished; compiled by C. Gebhardt)

The Issyk Kul sedimentary archive represents a potential source of record from an intra-continental region whose climate neither responds directly to monsoons nor is forced by adjacent marine conditions, making it an attractive target for isolating mid-latitude atmospheric circulation components of past climate changes on a range of timescales. The climate of the Issyk Kul region is relatively simple but highly sensitive to vari-ations in the moisture supply, which affect both lake levels and the extent of Tien Shan alpine glaciation. It is possible to recover a long continuous sediment record from the Issyk Kul basin to test series of hypotheses at the junction of geomorphology, tectonics, paleoclimate, and biotic evolution.

## Scientific Questions That Merit Drilling of the Issyk Kul Basin

We first discuss five questions that motivate scientific drilling of the Issyk Kul Basin: three bear directly on climate, a fourth concerns the role that climate change plays in erosion of high terrain, and a fifth tackles microbiological properties in an old, possibly endemic lake, which, due to changes in its water chemistry, promoted halotolerant organisms and eventually endemism. These questions overlap to varying degrees, as each in part concerns differences between climates of high central Asia and global climate. We then pose other additional questions that deep drillholes will help resolve and thus will broaden the scope of a major scientific drilling effort at Issyk Kul.

Hypothesis 1. A semi-arid, continental climate characterizes the Issyk Kul watershed deep within a continental interior and far from large marine sources of water vapor. Today, its main moisture sources are the Mediterranean and the North Atlantic, with moisture arriving mostly during the spring. Issyk Kul is ideally suited to examine differing responses to advection of moisture from these sources and to direct temporal variations in insolation on an otherwise nearly pure continental climate.

Hypothesis 2. Seismic data (Fig. 3) from Issyk Kul reveal a bathymetric high that holds a stratigraphically continuous section, 150–200 m thick, of fine and at least partly laminated sediment. Based on prior stratigraphic studies in the basin (Giralt et al., 2002; Ricketts et al., 2001), it is likely that this sediment is calcareous and proxy-rich, with a sedimentation rate that will enable a high-resolution (roughly decadal to centennial) climate reconstruction back perhaps to 300–400 kyr BP. Turbidites, mass flow, and slumps seem not to have disturbed this section (Fig. 4). Millenial-scale abrupt climate climate changes, like Dansgaard-Oeschger (D-O) stadials whose origin seems to be in the North Atlantic, have been recoded well by speleothems in China, but recent work suggests that the connection is through the South Asian Monsoon (Pausata et al., 2011). The Issyk Kul region offers an ideal, intracontinental region to examine the direct response to such changes in the North Atlantic, without the strong modulation by the South Asian Monsoon.

Hypothesis 3. As noted earlier, alpine glaciers and continental ice sheets seem to respond differently to orbital scale forcing (Abramowski et al., 2006; Gillespie and Molnar, 1995; Owen et al., 2005), including the Tien Shan (Koppes et al., 2008; Xu et al., 2010), but such evidence, both global and specific to the Tien Shan, spans only the last glacial period, since ~100 kyr BP. In particular, due to the present-day aridity of the Issyk Kul region, glacial mass balance is most sensitive to amount of precipitation. A high-resolution, accurately dated history of glacial sediment input into the Issyk Kul basin will allow a reconstruction of the glacial history of its surroundings in the Tien Shan for several glacial cycles and hence an examination of how Alpine glaciers and continental ice sheets advance asynchronously over many cycles of orbital variability, if indeed they do. A detailed study of such asynchroneity could allow insights into how orbital variations affect not just ice sheets but continental climate in general.

Hypothesis 4. The geomorphological, tectonic, and paleoclimate communities differ, both among each other and within each group, on the relative roles played by tectonics (whether active or dead) and climate (and climate change) in affecting erosion rates (Molnar and England, 1990; Willenbring and von Blanckenburg, 2010; Zhang et al., 2001). Tectonic movement is needed for terrain to lie above sea level, but present-day elevations need not be results of active tectonics. Erosive processes differ in different climates, and changing climates (especially in the statistics of precipitation variability) might alter erosion rates so that current landscapes are not in equilibrium with the present-day climate. Testing the relative importance of climate and tectonics as they manifest themselves in erosion requires finding a region where changes in both climate and tectonics can be dated well. Issyk Kul might provide an excellent archive not only to date changes in erosion rates, but also to evaluate the roles of tectonics and climate change in affecting such erosion rates.

Hypothesis 5. A variety kinds of evidence, such as the periodic alternation of endogenic Ca- and Mg-bearing carbonates and changes in the concentration of noble gases in sediment pore water (Brennwald et al., 2004), suggests that the Issyk Kul has alternated between being fresh and saline at many times in the past. Moreover, seismic surveys clearly show evidence of several dramatic lowstands of Issyk Kul, and historic records indicate repeated episodes of outflow from what is today a closed-basin lake. As various organisms thrive better in fresh than saline water, and vice versa, salinity changes may have profoundly affected taxonomic diversity in the basin and the development of endemism. At present, Issyk Kul is species-poor compared to most freshwater lakes and the oceans. Thus, Issyk Kul offers a field laboratory for studying how organisms respond to larger temporal variations in salinity, and perhaps evolve in response to such changes.

## Other Questions Potentially Solved with Cores from the Issyk Kul Basin

The tectonics community at the workshop posed a number of questions that do not by themselves merit the high cost of drilling, but whose questions might be answered, or partly answered, with results from cores taken from the Issyk Kul basin. Moreover, some questions bear on local geology and local hazards, and all participants agreed that if a drilling program pokes holes into Kyrgyz territory, the Kyrgyz people ought to reap benefits from such a program.

*Dating of sedimentary rock.* As noted above, the Cenozoic stratigraphy of Kyrgyzstan is commonly divided into four units. Resolving when and how the Issyk Kul basin formed and when sedimentary facies changed is important for interpreting the contemporaneous climatic record archived within the basin.

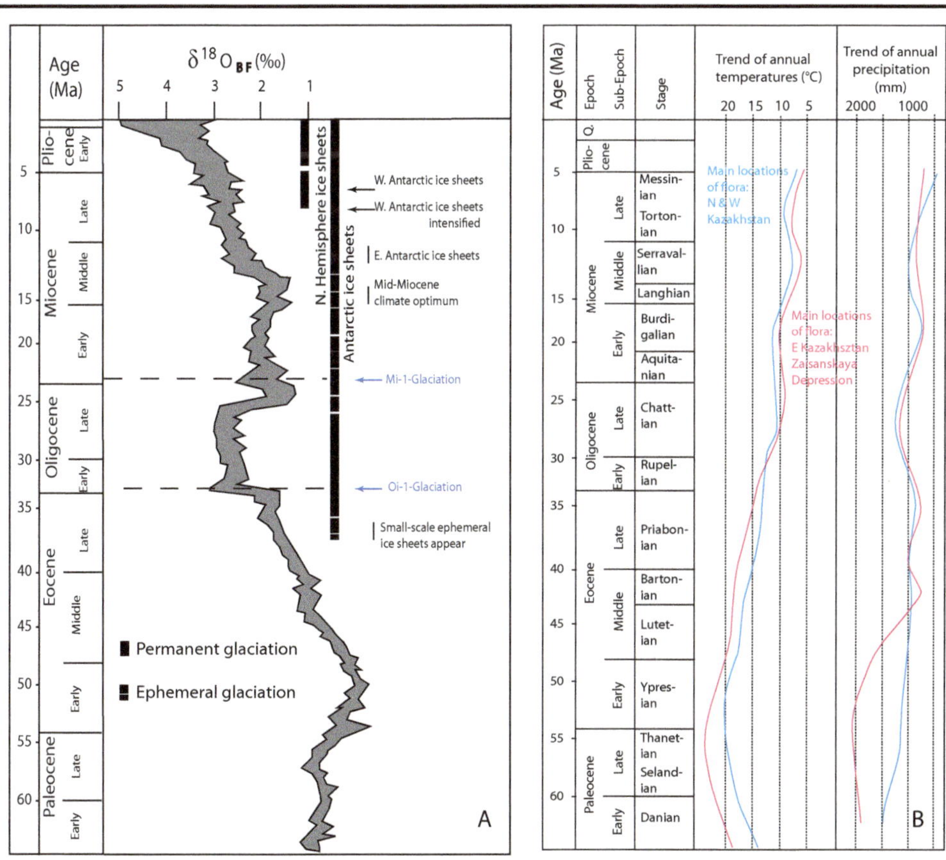

Figure 4. [A] The marine $\delta^{18}O$ synthesis of Zachos et al. [2001] compared with [B] the Central Asian climate history from Akhmetyev et al. (2005).

*Seismites and paleoearthquakes.* The Kyrgyz people live with a serious threat of earthquake disaster. For instance, in the northern Tien Shan of Kyrgyzstan and Kazakhstan, several major earthquakes occurred in historical times: in 1885 (Ms 6.9), 1887 (Ms 7.3), 1889 (Ms 8.3), and 1911 (Ms 8.1) (Abdrakhmatov et al., 2002). The latter two events are among the largest known intracontinental earthquakes worldwide, but the recurrence intervals of such earthquakes, which could be thousands to tens of thousands of years, are still unconstrained. If seismites can be found in lake sediment, they might provide a long record of earthquake activity far beyond instrumental records and recorded history. If found, seismites from the Issyk Kul basin could significantly enhance our knowledge of the earthquake history of the Tien Shan, as well as serving as a case study for other low-strain regions worldwide.

*Accurate measurements of heat flux.* The state of the lower crust—brittle or ductile, and in other respects, too—depends largely on the temperature (but also on the degree of hydration) at relevant depths. Accurate measurements of temperatures in deep holes could address two quite different problems: one relevant to geodynamics and the other to climate over the past several thousand to one hundred thousand years.

*Changes in the locus and rate of deformation.* Finally, a topic that was discussed at some length and that interested many participants concerns the following questions. (1) How have changes in the locus and rate of deformation over the past 5–10 Ma both enabled sediment production and generated sediment accommodation space in the Issyk Kul basin, which thereby conditioned the landscape for climate-induced change? (2) Can we find evidence that Issyk-Kul surface processes have played an active role in shaping tectonic deformation, rather than just passively responding to it (Garcia-Castellanos, 2007; Whipple, 2009)?

## Key Questions for Drill Site Selection

Several questions motivated the workshop discussion. What drilling targets allow the climate and tectonic-related questions to be answered most effectively? What are the geometry, depth, and timing of basin formation? What is the sedimentary architecture? Where in the stratigraphy are regressions and transgressions located? Preliminary data allow partial answers to these questions, but further seismic profiling planned for 2012 should answer at least some of these questions.

Available background data include bathymetric surveys and a detailed seismic survey carried out in 1982 and earlier in Soviet times. In addition, three seismic surveys were carried out in 1997, 2001, and 2002, using digital recording, with accompanying retrieval of several short cores (Giralt et al., 2002; Ricketts et al., 2001). In addition, 100 (or more) wells of up to 5-km depth have been drilled on land (Duchkov et al., 2001). Most of these cores are at least in part accessible. In addition to geologic maps made in Soviet times, a new atlas to be published soon and geologic maps of selected portions of the north slope of the range south of the lake (Burgette, 2008) are available. Furthermore, elevation data with 70-m horizontal resolution are digitally available for the region.

Given these data, the following additional scientific investigations have been deemed necessary.

- In selected areas, further bathymetric surveys are needed, particularly around the possible drill sites to acquire accurate water depths and bottom topographies. These surveys would help to delineate features of interest in the deepest basin.
- In regions with depths shallower than 120 m, which is the minimum drop in lake level during lowstands and where river incision may have removed sediment, denser seismic coverage than currently exists is needed.
- Seismic data from selected deeper parts and/or land is needed to extend existing coverage.
- Seismic refraction is necessary to determine seismic velocities and to define the geometry where layers are not flat in intervals between reflectors.
- A map of recent lake sediment based on cores that have been examined would be helpful.
- Finally, much data exists from drilling around the lake. Acquiring those data, digitizing them, and correlating them with seismic and outcrop data would help decide where a hole would be most likely to yield good results. A plan is underway to digitize logs from some such wells.

These pre-site surveys will help to define specific drilling locations and depths, as well as informing some the following questions.

- How deep is the basin and/or the bedrock?
- What are the geometry of the basin floor and timing of its formation?
- To what extent is the basin the result of recent or ancient tectonics?
- What is the nature of the infill; how much is terrestrial and how much lacustrine? What has been its stratigraphic development?
- Where has sediment that was transported by rivers been deposited? How much has been deposited by plumes and as turbidites?
- To what extent is the bathymetry the result of progradation from the east and the west? How have regressions, with possible lowering of the level down to 300 m, affected the bathymetry and sediment distribution?
- How large and when in the past were lake level fluctuations? Did any major incision/erosion remove sediment? Have there been catastrophic floods? Why are there high lake terraces? What are the ages and distribution of these lake high-stand terraces?

## Possible Drill Sites

Four possibilities were discussed. They are listed below not in any order but along with the most important pros and cons.

**Site A:** In the deep abyssal plain. If the Drilling, Observation and Sampling of the Earth's Continental Crust (DOSSEC) rig were used, then with a water depth of 650 m, a penetration of a 850-m deep well, and a sedimentation rate of 0.3 mm $yr^{-1}$ in this region, it could reach sediment with a maximum age of 2.8 Ma.
**Virtues:** Such a core could provide a complete record of undisturbed sediment since continental ice sheets first developed in North America.
**Weaknesses:** Such drilling will challenge existing drilling techniques, as the season for drilling is short. The dynamic positioning system would likely be required for the drilling platform to hold position. This would increase costs dramatically. Moreover, the core might contain thick layers of sand, deposited as turbidites, which would be hard to penetrate.

**Site B:** A well at a water depth of 300 m. Again, if sediment had been deposited at (only) 0.2 mm $yr^{-1}$ (Vermeesch et al., 2004), a penetration of 1200 m, as with a DOSSEC rig, would reach a sediment with a maximum age of 6 Ma.
**Virtues:** Such a core might contain a long, complete but perhaps condensed record.
**Weaknesses:** Again, a short drilling season would challenge existing drilling techniques, and again anchoring of the drill rig would be required. Also again, there is the risk of drilling sand. Furthermore, the sedimentation rate might be much higher.

**Site C:** A well drilled on land on the plain just east of the lake. Penetration to 1500 m depth with sedimentation at a high average rate of 1 mm $yr^{-1}$ would reach a maximum of 1.5 Ma, but at 0.2 mm $yr^{-1}$, it would reach 7.5 Ma. Such a well could penetrate farther, if a commercial rig rather than the DOSSEC rig were to be used.
**Virtues:** Logistics should be relatively easy, and costs should be relatively low. Other nearby wells drilled in Soviet times demonstrate feasibility. Moreover, downhole logging should be easier than on the lake.
**Weaknesses:** Potentially large stratigraphic gaps might exist, and sediment might be coarse. Before such an undertaking, seismic reflection profiling on land is needed.

**Site D:** A well sited on an anticline at 500 m water depth at the eastern end of the deep central basin. Assuming a DOSSEC rig is used, a penetration of 1000 m into sediment deposited at 0.3 mm $yr^{-1}$ would reach a maximum age of 3.3 my.
**Virtues:** A long, complete but perhaps slightly condensed record would be obtained. The likelihood of drilling through coarse-grained material/turbidites is lowest among the sites considered.
**Weaknesses**: Again, a short drilling season would challenge existing drilling techniques, and again dynamic positioning system would likely be required for the drilling platform to hold position. This would increase costs dramatically. In addition, from the existing seismic profiling, a well might intersect a fault at depth.

The workshop participants did not discuss at length the virtues and weaknesses of all sites in detail, but the general consensus was that the best option for reaching sediment several million years old was Site C, and the best for paleoclimatic studies on the 100-ky timescales was Site D.

## References


Abrakhmatov, K.E., Djanuzakov, K.E., and Delvaux, D., 2002. Active tectonics and seismic hazard of the Issyk-Kul basin in the Kyrgyz Tian-Shan, *In* Klerkx, J., and Imanackunov, B. (Eds.) *Lake Issyk-Kul: Its Natural Environment*: Dordrecht (NATO Science Series, Kluwer), 147–160.

Abdrakhmatov, K.E., Thompson, S., and Weldon, R., 2007. *Active Tectonics of the Tien Shan*: Bishkek, Kyrgyzstan (Ilim), in Russian.

Abdrakhmatov, K.E., Weldon, R., Thompson, S., Burbank, D., Rubin, Ch., Miller, M., and Molnar, P., 2001. Origin, direction, and rate of modern compression in the central Tien Shan, Kyrgyzstan. *Geologiya i Geofizika (Russian Geology & Geophysics)*, 42:1585–1609.

Abramowski, U., Bergau, A., Seebach, D., Zech, R., Glaser, B., Sosin, P., Kubik, P.W., and Zech, W., 2006. Pleistocene glaciations of Central Asia: Results from $^{10}$Be surface exposure ages of erratic boulders from the Pamir (Tajikistan), and the Alay-Turkestan range (Kyrgyzstan). *Quat. Sci. Rev.*, 25:1080–1096. doi:10.1016/j.quascirev.2005.10.003

Aizen, E.M., Aizen, V.B., Melack, J.M., Nakamura, T., and Ohta, T., 2001. Precipitation and atmospheric circulation patterns at mid-latitudes of Asia. *Int. J. Climatology*, 21:535–556. doi:10.1002/joc.626

Aizen, V.B., Aizen, E.M., Melack, J., and Dozier, J., 1997. Climatic and hydrologic changes in the Tien Shan, Central Asia. *J. Climate*, 10:1393–1404. doi:10.1175/1520-0442(1997)010<1393:CAHCIT>2.0.CO;2

Akhmetyev, M.A., Dodoniv, A.E., Somikova, M.V., Spasskaya, I.I., Kremenetsky, K.V., and Klimanov, V.A., 2005. Kazakhstan and Central Asia (plains and foothills). In Velichko, A.A., and Nechaev, V.P. (Eds.), *Cenozoic Climatic and Environmental Changes in Russia, Geol. Soc. Am. Special Paper*, 382:139–161.

Brennwald, M.S., Peeters, F., Imboden, D.M., Giralt, S., Hofer, M., Livingstone, D.M., Klump, S., Strassmann, K., and Kipfer, R., 2004. Atmospheric noble gases in lake sediment pore water as proxies for environmental change. *Geophys. Res. Lett.*, 31:L04202. doi:10.1029/2003GL019153

Bullen, M.E., Burbank, D.W., and Garver, J.I., 2003. Building the northern Tien Shan: Integrated thermal, structural, and topographic constraints. *J. Geol.*, 111:149–165. doi:10.1086/345840

Burgette, R.J., 2008. Uplift in response to tectonic convergence: The Kyrgyz Tien Shan and Cascadia subduction zone [Ph.D. dissert.]. University of Oregon, Eugene, OR.

Charreau, J., Chen, Y., Gilder, S., Barrier, L., Dominguez, S., Augier, R., Sen, S., et al., 2009. Neogene uplift of the Tian Shan Mountains observed in the magnetic record of the Jingou River section (northwest China). *Tectonics*, 28:TC2008. doi:10.1029/2007TC002137

Chedia, O.K., 1986. Morphostructures and Neotectonism of the Tien Shan: Frunze (Ilim), in Russian.

Chen, J., Heermance, R., Burbank, D.W., Scharer, K.M., Miao, J., and Wang, C., 2007. Quantification of growth and lateral propagation of the Kashi anticline, southwest Chinese Tian Shan. *J. Geophys. Res.*, 112:B03S16. doi:10.1029/2006JB004345

De Batist, M., Imbo, Y., Vermeesch, P., Klerkx, J., Giralt, S., Delvaux, D., Lignier, V., Beck, C., Kalugin, I., and Abdrakhmatov, K.E., 2001. Bathymetry and sedimentary environments of Lake Issyk-Kul, Kyrgyz Republic (Central Asia): A large, high-altitude, tectonic lake. *In* Klerkx, J., and Imanackunov, B. (Eds.), *Lake Issyk-Kul: Its Natural Environment*: Dordrecht (NATO Science Series, Kluwer), 101–123.

Duchkov, A.D., Shvartsman, Yu.G., and Sokolova, L.S., 2001. Deep heat flow in the Tien Shan: Advances and drawbacks. *Geologiya i Geofizika (Russian Geology & Geophysics)*, 42:1436–1452.

Garcia-Castellanos, D., 2007. The role of climate during high plateau formation. Insights from numerical experiments. *Earth Planet. Sci. Lett.*, 257:372–390. doi:10.1016/j.epsl.2007.02.039

Gillespie, A., and Molnar, P., 1995. Asynchronous maximum advances of mountain and continental glaciers. *Rev. Geophys.*, 33:311–364. doi:10.1029/95RG00995

Giralt, S., Klerkx, J., Riera, S., Julia, R., Lignier, V., Beck, C., De Batist, M., and Kalugin, I., 2002. Recent paleoenvironmental evolution of Lake Issyk-Kul. *In* Klerkx, J., and Imanackunov, B. (Eds.), *Lake Issyk-Kul: Its Natural Environment*: Dordrecht (NATO Science Series, Kluwer), 125–145.

Ji, J., Luo, P., White, P., Jiang, H., Gao, L., and Ding, Z., 2008. Episodic uplift of the Tianshan Mountains since the late Oligocene constrained by magnetostratigraphy of the Jingou River section, in the southern margin of the Junggar Basin, China. *J. Geophys. Res.*, 113:B05102. doi:10.1029/2007JB005064

Kondorskaya, N.V., and Shebalin, N.V., 1977. *New Catalog of Strong Earthquakes in the Territory of the USSR*: Moscow (Nauka), in Russian.

Koppes, M., Gillespie, A.R., Burke, R.M., Thompson, S.C., and Stone, J., 2008. Late Quaternary glaciation in the Kyrgyz Tien Shan. *Quat. Sci. Rev.* 27:846–866.

Laverov, N.P., and Makarov, V.I., 2005. *Recent Geodynamics of Intracontinental Areas of Collision Mountain Building (Central Asia)*: Moscow (Sci. World).

Molnar, P., and England, P., 1990. Late Cenozoic uplift of mountain ranges and global climate change: Chicken or egg? Nature, 346:29–34. doi:10.1038/346029a0

Owen, L.A., Finkel, R.C., Barnard, P.L., Haizhou, M., Asahi, K., Caffee, M.W., and Derbyshire, E., 2005. Climatic and topographic controls on the style and timing of Late Quaternary glaciation throughout Tibet and the Himalaya defined by $^{10}$Be cosmogenic radionuclide surface exposure dating. *Quat. Sci. Rev.*, 24:1391–1411. doi:10.1016/j.quascirev.2004.10.014

Pausata, F.S.R., Battisti, D.S., Nisancioglu, K.H., and Bitz, C.M., 2011. Chinese stalagmite $\delta^{18}$O controlled by changes in the Indian monsoon during a simulated Heinrich event. *Nature* Geoscience, 4:474–480.

Reigber, Ch., Michel, G.W., Galas, R., Angermann, D., Klotz, J., Chen, J.Y., Papschev, A., Arslanov, R., Tzurkov, V.E., and Ishanov, M.C., 2001. New space geodetic constraints on the distribution of deformation in Central Asia. *Earth Planet. Sci. Lett.*, 191:157–165. doi:10.1016/S0012-821X(01)00414-9

Ricketts, R.D., Johnson, T.C., Brown, E.T., Rasmussen, K.A., and Romanovsky, V.V., 2001. The Holocene paleolimnology of Lake Issyk-Kul, Kyrgyzstan: Trace element and stable isotope composition of ostracodes. *Palaeogeogr. Palaeoclimatol. Palaeoecol.*, 176:207–227. doi:10.1016/S0031-0182(01)00339-X

Sun, J., and Zhang, Z., 2009. Syntectonic growth strata and implications for late Cenozoic tectonic uplift in the northern Tian Shan, China. *Tectonophysics*, 463:60–68. doi:10.1016/j.tecto.2008.09.008

Thompson, S.C., Weldon, R.J., Rubin, C.M., Abdrakhmatov, K., Molnar, P., and Berger, G.W., 2002. Late Quaternary slip rates across the central Tien Shan, Kyrgyzstan, central Asia. *J. Geophys. Res.*, 107(B9):2203. doi:10.1029/2001JB000596

Vermeesch, P., Poort, J., Duchkov, A.D., Klerkx, J. and De Batist, M., 2004. Lake Issyk-Kul (Tien Shan): Unusually low heat flow in an active intermontane basin. *Geologiya I Geofizika*, 45(5): 616–625.

Whipple, K.X., 2009. The influence of climate on the tectonic evolution of mountain belts. *Nature Geosci.*, 2:97–104. doi:10.1038/ngeo413

Willenbring, J.K., and von Blanckenburg, F., 2010. Long-term stability of global erosion rates and weathering during late-Cenozoic cooling. *Nature*, 465:211–214. doi:10.1038/nature09044

Xu, X., Kleidon, A., Miller, L., Wang, S., Wang, L., and Dong, G., 2010. Late Quaternary glaciation in the Tianshan and implications for palaeoclimatic change: A review. *Boreas*, 39:215–232. doi:10.1111/j.1502-3885.2009.00118.x

Zachos, J., Pagani, M., Sloan, L., Thomas, E., and Billups, K., 2001. Trends, thythms, and aberrations in global climate 65 Ma to present. *Science*; Apr 27, 2001; 292, 551. doi:10.1126/science.1059412

Zhang, P., Molnar, P., and Downs, W.R., 2001. Increased sedimentation rates and grain sizes 2–4 Myr ago due to the influence of climate change on erosion rates. *Nature*, 410:891–897. doi:10.1038/35073504

Zubovich, A.V., Wang, X., Scherba, Y.G., Schelochkov, G.G., Reilinger, R., Reigber, C., Mosienko, O.I., et al., 2010. GPS velocity field for the Tien Shan and surrounding regions. *Tectonics*, 29:TC6014. doi:10.1029/2010TC002772

## Authors

**Hedi Oberhänsli**, Museum of Natural History Berlin, Invalidenstrasse 43, 10115 Berlin, Germany, e-mail: hedi.oberhaensli@mfn-berlin.de.

**Peter Molnar**, Department of Geological Sciences and Cooperative Institute for Research in Environmental Sciences (CIRES), University of Colorado at Boulder, Boulder, CO 80309-0216, U.S.A., e-mail: molnar@colorado.edu.

# Workshop on Drilling of Lake Junin, Peru: Potential for Development of a Continuous Tropical Climate Record

by Donald T. Rodbell, Mark B. Abbott, and the 2011 ICDP Lake Junin Working Group

Lake Junin (11.0°S, 76.2°W; Fig. 1) is one of the few lakes in the tropical Andes that predates the last glacial maximum. Located at 4000 masl, it covers approximately 300 km$^2$ between the eastern and western cordillera of the central Peruvian Andes. Moraines from alpine glaciers that emanated from the eastern cordillera reach the eastern margin of the lake, but at no time in at least the last several hundred thousand years has the lake been overridden by ice.

Existing sediment cores (20–25 m long) from Lake Junin span ~50 ka and reveal intervals of high glacigenic siliclastic sediment input that alternate with interstadial intervals dominated by the rapid accumulation (~1 mm yr$^{-1}$) of authigenic marl (Fig. 2; Hansen et al., 1984; Seltzer et al., 2000). The last deglaciation is marked by a ~6‰ increase in the $\delta^{18}O$ of bulk marl that is nearly identical to the isotopic shift reported from the Huascaran ice core (Fig. 3), ~300 km to the north. Records of $\delta^{18}O$ from Junin and from other carbonate lakes and speleothems in the region are strongly affected by the hydrological balance of the central Andes, which, in turn, is controlled by the strength of the South American Summer Monsoon (SASM). Controls on the vigor of the SASM include long-term insolation receipts, the mean position of the intertropical convergence zone over the tropical Atlantic Ocean, and the mean state of El Niño Southern Oscillation (ENSO).

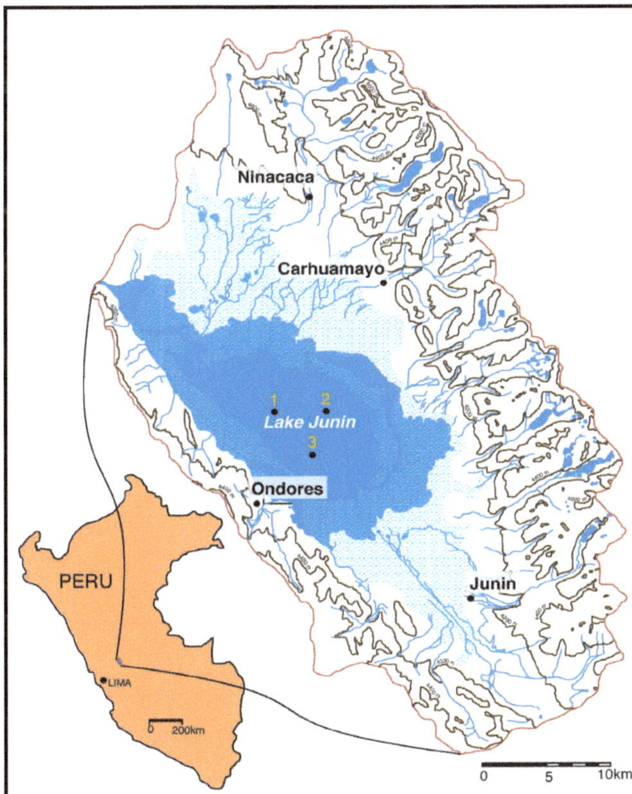


Figure 1. The Lake Junin basin with proposed coring localities 1–3 (Table 1). The eastern cordillera is extensively glaciated, and paleoglaciers reached the eastern edge of the lake on multiple occasions over the past several hundred thousand years. Relatively high rates of glacigenic sedimentation are expected during intervals of expanded ice cover at proposed core sites 1 and 2, whereas site 3 should contain the highest concentration of marl.

Lake Junin contains a sediment record that spans multiple glacial cycles—perhaps covering many hundreds of thousands of years—and the sediments contain high concentrations of authigenic calcite that can be used for both age control through U-Th measurements and production of a subdecadal record of hydrologic variability through stable isotope analyses. However, Lake Junin has never been cored to a depth greater than 25 m. The clear potential of Lake Junin to provide a well-dated high-resolution record of tropical climate evolution spanning several hundred millennia justifies the attention of the scientific community through the International Continental Scientific Drilling Program (ICDP) to ensure the successful acquisition of multiple cores at several locations in the Junin basin (Fig. 1; Table 1).

With this goal in mind, twenty-seven scientists from six countries participated in a workshop in Tarma, Peru on 15–17 June 2011 to discuss scientific priorities and logistical requirements of deep drilling of Lake Junin. Speakers presented overviews of the following topics:

1) the geologic setting of Lake Junin, and the four decades of Quaternary geologic and paleolimnologic research conducted there,
2) the global relevance of a long climate record from Lake Junin,
3) modern climatic controls on stable isotopes in the tropics, and recently-acquired high-resolution $\delta^{18}O$ records from carbonate lakes and speleothems in the Junin region,
4) recently acquired air gun seismic stratigraphy that reveals a total sediment thickness of at least 125 m, no evidence of faulting, and few if any unconformities,
5) the paleoecological potential of Lake Junin sediments,
6) promising new U/Th dates for Junin carbonates, and

Table 1. Proposed drill sites for Lake Junin.

| Coring Site Number (Fig. 1) | Latitude (°S) | Longitude (°W) | Water Depth (m) | Distance to input streams of eastern siliclastic sediment* | Distance to input streams of northern siliclastic sediment* | Distance to input streams of western siliclastic sediment* |
|---|---|---|---|---|---|---|
| 1 | 11.01 | 76.13 | 11 | distal | intermediate | proximal |
| 2 | 11.02 | 76.07 | 8 | proximal | distal | distal |
| 3 | 11.05 | 76.09 | 8 | distal | distal | proximal |

*Proximal indicates less than 5 km; intermediate indicates feature is between 5 km and 10 km, and distal indicates >10 km.

7) paleomagnetism as a potential dating tool for Junin sediments and as a means of fingerprinting glacigenic sediment in lakes.

One full day was spent circumnavigating Lake Junin by bus with stops to review the regional glacial chronology and to consider possible lake access points and base camps for future drilling and science teams.

Several discussion groups were organized to focus on the themes noted above and were charged with writing short reports to identify the major questions that could be addressed by studying Lake Junin sediments. These groups included 1) stratigraphy and geochronology, 2) organic geochemistry, 3) paleoecology, and 4) regional records of climate.

Lake Junin sits within a fault-bounded subsiding intermontane basin between the eastern and western cordillera of the central Peruvian Andes. Bedrock is dominated by late Paleozoic to Mesozoic carbonates, although glacial erosion of the eastern cordillera has exposed pre-Cambrian crystalline rocks. The lake basin, with a maximum water depth of ~15 m, is dammed at its northern and southern ends by coalescing alluvial fans that emanate from glacial valleys in both cordillera. These fans can be traced to moraines that are more than 250 ka, and thus the lake is at least this old. Detailed moraine mapping coupled with nearly 200 cosmogenic radionuclide (CRN; $^{10}$Be, $^{26}$Al) exposure ages from boulders on the crests of moraines clearly indicates that during the maximum extent of late Cenozoic glaciation on the Junin Plain, paleoglaciers reached the lake edge, but at no time in perhaps as much as one million years or more has Lake Junin been overridden by ice (Smith et al., 2005a, 2005b).

Reconnaissance seismic-reflection surveys suggest that the Lake Junin basin contains a thick sedimentary sequence amenable to sampling by drill core. High-resolution chirp (4–24 kHz) data collected in 2008, combined with existing core data, reveal tens of meters of interbedded marl and lacustrine mud, the latter dominated by glacial flour. In shallow water (<6 m), these sediments are incised by chan-

Figure 2. Marl exposed during the dry season along the western edge of Lake Junin; view is looking east across the lake to the eastern cordillera. The marl exposed in this photograph has precipitated on the leaves of rooted macrophytes and has accumulated to form a bioherm several meters thick. The glacier-clad summits in the distance fed valley glaciers that reached the eastern edge of Lake Junin on several occasions in the last million years (Smith et al., 2005a, 2005b).

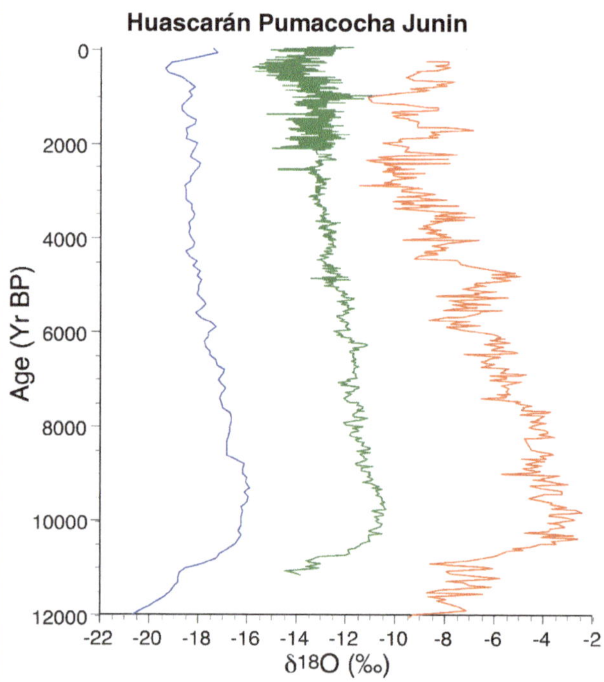


Figure 3. Oxygen isotope records from the central Peruvian Andes: Huascarán Ice Core (Thompson et al., 1995), Laguna Pumacocha (Bird et al., 2011), and Lake Junin (Seltzer et al., 2000 and unpublished data). The age model for the Junin record is based on a combination of radiocarbon and U/Th ages. The steeper Holocene isotope gradient of Junin reflects the hydrologically-closed nature of Junin, which makes Junin especially sensitive to changes in regional hydrologic balance.

nels related to a Holocene lowstand of the lake. An air gun seismic survey in 2011 successfully imaged as much as 150 m of nearly horizontal sediment with no evidence of deformation. Details of the stratigraphy were obscured by ringing in the shallow water column, but several prominent, continuous reflections were observed that could be traced across the lake basin. One of these, reaching a maximum depth of about 125 m below the lake floor, was especially prominent, but several additional deeper reflections were observed below it. The prominent deep reflection could be mapped throughout the survey area, and it is flatter than the modern lake floor. These relationships suggest broad-scale basin subsidence and a lack of local deformation. They also suggest that at the time that the deep reflector was formed, the lake was somewhat larger than at present. No evidence has been observed in the central part of the lake for large-scale lake-level fluctuations, such as wave-cut shorelines, fluvial channels, or low-stand deltas. No local tectonic deformation was observed, and no widespread gas, either volcanic or biogenic, was observed in the sedimentary sequence.

U/Th dating will be an essential element in developing age models for longer cores from Lake Junin. Preliminary analyses of radiocarbon-dated Holocene sediments from the lake indicate that these interglacial sediments are nearly ideal for U/Th measurements, with high carbonate percentages (~90%), very high $^{234}U/^{238}U$ ratios, and low contamination by detrital $^{230}Th$. Though errors on dates are small compared to other lacustrine sediments, errors on Holocene dates are still on the order of 1000 yr and are dominated by uncertainty in the detrital $^{230}Th/^{232}Th$ ratio used to correct for detrital $^{230}Th$. Uncertainty in this ratio is currently estimated at ±100%. Future work will seek to reduce these errors by 1) better characterizing $^{230}Th/^{232}Th$ ratios in the detrital material through analysis of zero-age samples and use of isochrones, and 2) careful selection of small (a few milligrams), extremely clean samples. In previous interglacials, relative errors will be substantially smaller due to decay of the initial detrital $^{230}Th$; even assuming that uncertainty in the detrital $^{230}Th/^{232}Th$ is not reduced, errors in samples from 125 ka should be <1 kyr, assuming the sediments are similar in composition to Holocene sediments.

Lake Junin offers an excellent venue to test current hypotheses of paleoecological responses to climatic change. Abrupt and non-linear responses to change are among the most important issues to be investigated. The transitions of biota ranging from plants to ostracods into and out of previous interglacials, and responses to millennial-scale events will allow us to test a wide range of theories relating to community composition, ecosystem resilience, migratory response, and even extinction.

Workshop participants agreed that the regional-global scale rationale for a concerted effort through the ICDP to acquire multiple long cores from Lake Junin includes the following points. 1) Lake Junin is in a critical location with respect to long-term ENSO-Southern Hemisphere monsoon dynamics. 2) The potential of coupling a marl isotope record from Junin with the emerging nearby $\delta^{18}O$ record from speleothems to produce a paleohumidity record would be a unique contribution to the paleoclimatology and paleoecology of South America. 3) Lake Junin may provide a unique isotopic and paleoecological record of Dansgaard-Oeschger (D-O) and Heinrich events linked to conditions in the core of the SASM region of the Amazon Basin. 4) The glacigenic sediment record from Lake Junin will provide the first continuous record of glaciation and glacial erosion spanning several hundred millennia in tropical South America.

## References

Bird, B, W., Abbott, M. B., Rodbell, D. T., and Vuille, M., 2011, Holocene tropical South American hydroclimate revealed from a decadally resolved lake sediment $\delta^{18}O$ record: *Earth and Planetary Science Letters* 310, 192–202.

Hansen, B.C.S., Wright, H.E., Jr., and Bradbury, J.P., 1984. Pollen studies in the Junín area, central Peruvian Andes. *GA Bull.*, 95:1454–1465, doi:10.1130/0016-7606(1984)95%3C1454:PSITJA%3E2.0.CO;2.

Seltzer, G., Rodbell, D., and Burns, S., 2000. Isotopic evidence for late Quaternary climatic change in tropical South America. *Geology*, 28:35–38, doi:10.1130/0091-7613(2000)28%3C35:IEFLQC%3E2.0.CO;2.

Smith, J.A., Finkel, R.C., Farber, D.L., Rodbell, D.T., and Seltzer, G.O., 2005a. Moraine preservation and boulder erosion in the tropical Andes: Interpreting old surface exposure ages in glaciated valleys. *J. Quat. Sci.*, 20:735–758, doi:10.1002/jqs.981.

Smith, J.A., Seltzer, G.O., Farber, D.L., Rodbell, D.T., and Finkel, R.C., 2005b. Early local last glacial maximum in the tropical Andes. Science, 308:678–677, doi:10.1126/science.1107075.

Thompson, L.G., Mosley-Thompson, E., Davis, M. E., Lin, P. -N, Henderson, K. A., Cole-Dai, J., Bolzan, J. F., and Liu, K. -b., 1995. Late glacial stage and Holocene tropical ice core records from Huascarán, Peru. Science 269, 46–50.

## Authors

**Donald T. Rodbell**, Union College, Schenectady, NY 12308, U.S.A., e-mail: rodbelld@union.edu.

**Mark B. Abbott**, Department of Geology and Planetary Science, University of Pittsburgh, 4107 O'Hara Street, Room 200, SRCC Building, Pittsburgh, PA 15260-3332, U.S.A., e-mail: mabbott1@pitt.edu.

**and the 2011 ICDP Lake Junin Working Group**

## Photo Credit

Fig. 2: Donald T. Rodbell, Union College

# 35

# Continental Transform Boundaries: Tectonic Evolution and Geohazards

by Cecilia M.G. McHugh, Leonardo Seeber, M. Namik Çağatay, Pierre Henry, Christopher Sorlien, Michael Steckler, and Gülsen Uçarkuş

## Introduction

Continental transform boundaries cross heavily populated regions, and they are associated with destructive earthquakes, for example, the North Anatolian Fault (NAF) across Turkey, the Enriquillo-Plantain Garden fault in Haiti, the San Andreas Fault in California, and the El Pilar fault in Venezuela. Transform basins are important because they are typically associated with 3-D fault geometries controlling segmentation—thus, the size and timing of damaging earthquakes—and because sediments record both deformation and earthquakes. Even though transform basins have been extensively studied, their evolution remains controversial because we don't understand the specifics about coupling of vertical and horizontal motions and about the basins' long-term kinematics. Seismic and tsunami hazard assessments require knowing architecture and kinematics of faults as well as how the faults are segmented.

A four-day workshop, 13–16 June 2011 at Istanbul Technical University, Istanbul, Turkey, focused on the seismo-tectonic evolution of the Marmara Sea, which is 150 km long and includes three ~1200-m-deep subsiding basins (Figs. 1, 2). The submerged Marmara segment of the NAF is considered a seismic gap presenting high risk to Istanbul and its surroundings. The workshop was sponsored by the Integrated Ocean Drilling Program (IODP) to lay the groundwork for a proposal in the Marmara Sea.

The NAF extends east-west for over 1600 km across Turkey (Fig. 1). The northern branch of the NAF crosses the Marmara Sea and now accommodates most of the 21–25 mm $y^{-1}$ motion between the Anatolian block and the Eurasian plate (Reilinger et al., 2010). Since 1939, a sequence of seven $M>7$ earthquakes ruptured the NAF progressively westward across Turkey but to the east of Marmara Sea. The most recent two in 1999 were just east of the sea and particularly destructive (~17,000 deaths). West of the Marmara Sea an $M_w$7.4 earthquake ruptured the NAF's Ganos segment in 1912. The only unruptured segments of the NAF in the last century are within the Marmara Sea. They have been identified as a seismic gap with as much accumulated elastic strain as was released in the 1999 sequence (Parsons, 2004).

Details regarding the fault system beneath the Marmara Sea are challenging traditional models for transform tectonics. Some authors account for structural complexities and basins as relics and propose throughgoing transcurrent faults in the current regime (Le Pichon et al., 2001). Alternatively, basins may grow in a steady-state regime. Crustal thinning and subsidence may stem from pull-apart fault geometries and strain partitioning (Armijo et al., 2005) or from oblique slip on non-vertical throughgoing master faults (Okay et al., 2000; Seeber et al., 2006). Particulars of these models determine the segmenta-

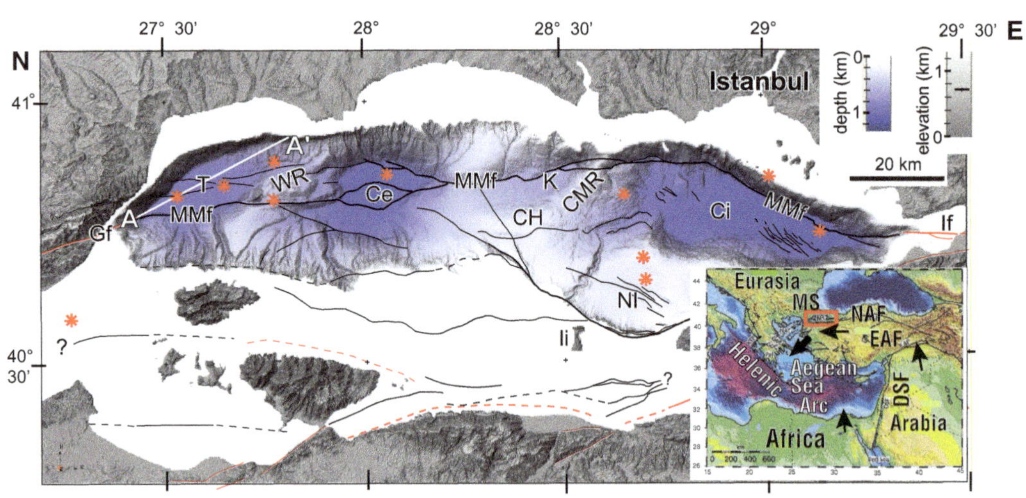


Figure 1. Bathymetry and topography of the Marmara Sea region (bathymetric data from Rangin et al., 2001) showing its basins from east to west: Cinarcik (Ci), North Imrali (NI), Kumburgaz (K), Central (Ce), Tekirdag (T). These basins are separated by the Central Marmara Ridge (CMR) and Western Ridge (WR). The main faults are shown as black lines (Sorlien et al., 2012); those in red are from several published sources as explained in Sorlien et al. (2012): Izmit fault (If), Main Marmara fault (MMf), and Ganos fault (Gf). The red stars mark the locations of the proposed drill sites. Inset shows the topography of the eastern Mediterranean with plate tectonic configuration and GPS-derived vectors (Reilinger et al., 2010). Red box locates Marmara Sea. Also shown are the North Anatolia Fault (NAF), East Anatolia Fault (EAF) and Dead Sea Fault (DSF).

tion of the gap and the size and number of future earthquakes and the hazard facing Istanbul.

Other large metropolitan regions are threatened by continental transforms. This workshop was held to discuss the research opportunities presented by drilling in the Marmara Sea.

## Scientific Presentations

Forty scientists and fifteen graduate and undergraduate students from Europe, Turkey, and the USA attended the IODP Marmara-Trans workshop. The presentations covered the evolution of the plate boundary from subduction to strike slip. Thermal models of the Marmara Sea basins focused on whether the growth was steady, or whether the basins grew initially and passively rode along the transform later. Structural evidence suggests asymmetric fault-bend basins subsiding steadily into the present (Seeber et al., 2004). Unconformities were identified, tentatively dated, and correlated in all the Marmara Sea basins from seismic profiles (Sorlien et al., 2012). Constraining the age model would be critical for evaluating the basins' structural evolution. Large earthquakes and tsunamis have been known for 2000 years from the historical record, and into the upper Holocene from the geological record. Historical and geological data overlap and correlate. Earthquake-triggered deep-water sedimentation in Haiti and Marmara Sea are remarkably similar. They suggest "turbidite-homogenite" units with characteristic structures and chemistry (McHugh et al., 2006; Beck et al., 2007) that could be recognized in the stratigraphy and provide an earthquake record through the growth of the basin. Better understanding of how seismogenesis relates to other geologic observables would improve earthquake forecasting and hazard assessment along the North Anatolian Fault and other continental transforms.

Reconnections between the Black, Marmara, and Aegean Seas in response to eustasy and climate have been much debated in recent years. These freshwater and saltwater exchanges were accompanied by dramatic transitions in flora, fauna, and geochemistry that have been recorded in the sediments of Marmara Sea. But, available piston and gravity cores extend back for only ~130 ka (Çağatay et al., 2009). A longer sedimentary record would recover repeated eustatic and climate cycles and thus distinguish the role of tectonics from sea-level fluctuations.

Seafloor gas emissions and fluid expulsions, observed with ROV and manned submersibles, appear to be focused along active faults in the Marmara Sea (Zitter et al., 2008; Géli et al., 2008). Surveys based on multichannel seismic reflection profiles, chirp, and multibeam data found a widespread distribution of gas in the sediment and the strongest gas discharge in the water column above the topographic highs between basins (Shillington et al., in press). Conduits along the main transform on the Western High expel thermogenic gas and associated brines. 3-D high-resolution seismic data in the same area document the interplay of sediment deformation and formation of fluid conduits (Bourry et al., 2009; Tryon et al., 2010). Authigenic carbonates and associated biogeochemical communities were recently discovered in Marmara Sea. These findings reveal that hydrological and biogeochemical activities along continental transforms are as intense as in other plate boundaries.

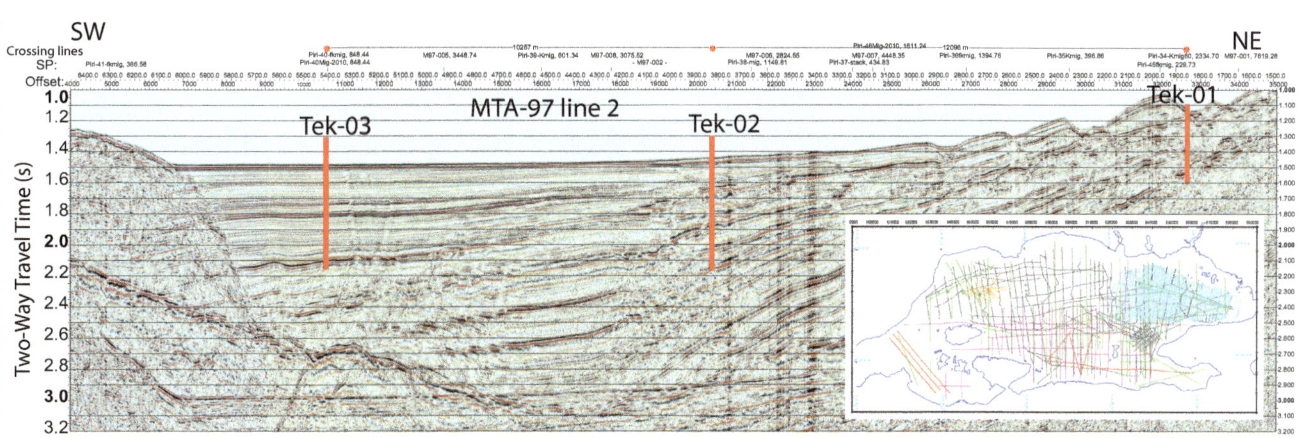


Figure 2. Multichannel seismic reflection (MCS) line MTA97 line 2 across Tekirdag Basin (white A–A' line on Fig. 1; data from Parke et al., 2003). The red rectangles show the locations of potential drilling sites. Tek-03: this SW location was targeted to reach Pleistocene to Holocene strata deposited at high sedimentation rates at the deepest part of the basin (1200 m of water depth). Tek-02 (projected 2.7 km from the SE): targeted to reach Pliocene to Pleistocene strata. Tek-01 (projected 2.6 km from the SE): targeted at the condensed margin of the basin to reach Miocene to Holocene strata.
Inset showing available high-resolution MCS data. Gray lines: TAMAM-2008 and PirMarmara Leg a4 (Sorlien et al., 2012). Orange lines are PirMarmara 2010 Leg a2 (P. Henry, G. Çifçi, and others, unpublished). Red lines are PirMarmara 2010 Legs a1, a3 (G. Çifçi and others, unpublished). The dense cyan grids are from Carton et al., 2007. Magneta are Marathon industry lines (C&K Petroleum, 1974). Green lines are from MTA97 (Parke et al., 2003).

## Links to Other Programs

Several seafloor observatories, which monitor seismic hazards long-term, and continental drilling programs are being carried out in Marmara Sea and surroundings by Turkey and international groups. Observatories have been deployed in Marmara Sea for understanding the long-term relationship between fluids and seismicity as part of the Marmara Demonstration Mission project funded by the European Seafloor Observatory Network and the European Multidisciplinary Seafloor Observatory. The International Continental Scientific Drilling Program (ICDP) project GONAF (a deep Geophysical Observatory at the North Anatolian Fault) plans to drill two boreholes in the Princes Islands, Turkey. Representatives of this program presented their preliminary results on seismotectonics. Future programs include the development and implementation of submarine geodetics as part of seismic risk assessment in collaboration with Istanbul Technical University and the Turkish Navy. Deployment of seafloor drilling rig MeBo is also being considered to investigate the interactions between sediment deformation and fluids.

## Outcome of the Workshop

The working group participants selected locations for drilling sites based on existing data. These were recorded from geophysical investigations (multichannel seismic reflection profiles, multibeam bathymetry, high-resolution sub-bottom profiles, side scan sonar), sediment sampling (piston cores up to 30 m long and many gravity cores), and explorations by manned and unmanned submersibles. The data were acquired during the following cruises and research vessels: Seismarmara, Marmarascarps, Marnaut, Marmesonet, R/V *Marion Dufresene*, R/V *Urania* (2002, 2009), and R/V *Piri Reis* (project TAMAM: 2008, 2010) (Rangin et al., 2001; Le Pichon et al., 2001; Polonia et al., 2004; Armijo et al., 2005; Cormier et al., 2006; Steckler et al., 2008; Géli et al., 2008; Görür and Çağatay, 2010).

These proposed drilling sites target the main Marmara Sea basins with the following goals.

1. Their deepest parts are intended to recover an expanded Holocene to Pleistocene strata with high sedimentation rates.

2. The condensed margins of the basins could reach older, possibly Miocene, strata.

3. They are located at and near fluid emission sites within the main fault zone to characterize the nature of the conduits and the timing of past activity, and to install borehole monitoring systems for fluid pressure and chemistry.

4. Additional plans for borehole instrumentation include monitoring of strain rates and seismic activity. These objectives would address major issues, such as the partitioning between fluid from deep sources possibly relevant to earthquakes and fluid from the compaction and dewatering of sediments.

These scientific and drilling objectives target the new International Ocean Discovery Program focusing on the "Earth's driving tectonics which reveal its inner workings and cause major geological hazards".

## Acknowledgments

We thank IODP-MI, the Eastern Mediterranean Centre for Oceanography and Limnology, and Queens College, City University of New York for their support.

## References

Armijo, R., Pondard, N., Meyer, B., Uçarkus, G., Mercier de Lepinay, B., Malavieille, J., Dominguez, S., et al., 2005. Submarine fault scarps in the Sea of Marmara pull-apart (North Anatolian Fault): Implications for seismic hazard in Istanbul. *Geochem. Geophys. Geosys.*, 6:Q06009. doi:10.1029/2004GC000896

Beck, C., de Lepinay, M., Schneider, J.-L., Cremer, M., Çağatay, N., Wendenbaum, E., Boutareaud, S., Mynot, G., Schmidt, S., and Weber, O., 2007. Late Quaternary co-seismic sedimentation in the Sea of Marmara's deep basins. *Sed. Geol.*, 199:65–89. doi:10.1016/j.sedgeo.2005.12.031

Bourry, C., Chazallon, B., Charlou, J.-L., Donval, J.P., Ruffine, L., Henry, P., Géli, L., Çağatay, M.N., Sedat, İ., and Moreau, M., 2009. Free gas and gas hydrates from the Sea of Marmara, Turkey: Chemical and structural characterization. *Chem. Geol.*, 264:197–206. doi:10.1016/j.chemgeo.2009.03.007

C&K Petroleum, 1974. Report covering geophysics–geology Marmara offshore, Turkey, 71 pp.

Çağatay, M.N., Eriş, K., Ryan, W.B.F., Sancar, Ü., Polonia, A., Akçer, S., Biltekin, D., et al., 2009. Late Pleistocene-Holocene evolution of the northern shelf of the Sea of Marmara. *Mar. Geol.*, 265:87–100. doi:10.1016/j.margeo.2009.06.011

Carton, H., Singh, S.C., Hirn, A., Bazin, S., de Voogd, B., Vigner, A., Ricolleau, A., Cetin, S., Ocakoglu, N., Karakoc, F., and Sevilgen, V., 2007. "Seismic imaging of the three-dimensional architecture of the Cinarcik Basin along the North Anatolian Fault. *J. Geophys. Res.*, 112:B06101, doi:10.1029/2006JB004548.

Cormier, M.-H., Seeber, L., McHugh, C.M.G., Polonia, A., Çağatay, M.N., Emre, Ö., Gasperini, L., et al., 2006. North Anatolian fault in the Gulf of Izmit (Turkey): Rapid vertical motion in response to minor bends of a nonvertical continental transform. *J. Geophys. Res.*, 111:B04102. doi:1029/2005JB003633

Géli, L., Henry, P., Andre, C., Zitter, T., Çağatay, M.N., Mercier de Lepinay, B., Le Pichon, X., et al., 2008. Gas emissions and active tectonics within the submerged section of the North Anatolia Fault zone in the Sea of Marmara. *Earth Planet.*

*Sci. Lett.*, 274(1-2):34–39.

Görür, N., and Çağatay, M.N., 2010. Geohazards rooted from the northern margin of the Sea of Marmara since the late Pleistocene: A review of recent results. *Nat. Hazards*, 54(2):583–603. doi:10.1007/s11069-009-9469-x

Le Pichon, X., Sengör, A.M.C., Demirbag, E., Rangin, C., Imren, C., Armijo, R., Görür, N., et al., 2001. The active main Marmara fault. *Earth Planet. Sci. Lett.*, 192:595–616.

McHugh, C.M.G., Seeber, L., Cormier, M.-H., Dutton, J., Çağatay, M.N., Polonia, A., Ryan, W.B.F., and Görür, N., 2006. Submarine earthquake geology along the North Anatolian Fault in the Marmara Sea, Turkey: A model for transform basin sedimentation. *Earth Planet. Sci. Lett.*, 248:661–684. doi:10.1016/j.epsl.2006.05.038

Okay, A.I., Kaslilar-Ozcan, A., Imren, C., Boztepe-Guney, A., Demirbag, E., and Kuscu, I., 2000. Active faults and evolving strike-slip fault basins in the Marmara Sea, northwest Turkey: A multichannel reflection study. *Tectonophysics*, 321:189–218. doi:10.1016/S0040-1951(00)00046-9

Parke, J.R., White, R.S., McKenzie, D., Minshull, T.A., Bull, J., Kuscu, I., Görür, N., and Sengör, A.M.C., 2003. The Sea of Marmara: A two-dimensional seismic reflection data archive. *Geochem. Geophys. Geosys.*, 4:1084. doi:10.1029/2002GC000493

Parsons, T., 2004. Recalculated probability of M>7 earthquakes beneath the Sea of Marmara, Turkey. *J. Geophys. Res.*, 109:B05304. doi:10.1029/2003JB002667

Polonia, A., Gasperini, L., Amorosi, A., Bonatti, E., Bortoluzzi, G., Çağatay, M.N., Capotondi, L., et al., 2004. Holocene slip rate of the North Anatolian Fault beneath the Sea of Marmara. *Earth Planet. Sci. Lett.*, 227:411–426.

Rangin, C., Demirbag, E., Imren, C., Crusson, A., Normand, A., Le Drezen, E., and Le Bot, A., 2001. *Marine Atlas of the Sea of Marmara (Turkey)*, Ifremer.

Reilinger, R., McClusky, S., Paradissis, D., Ergintav, S., and Vernant, P., 2010. Geodetic constraints on the tectonic evolution of the Aegean region and strain accumulation along the Hellenic subduction zone. *Tectonophysics*, 488:22–30. doi:10.1016/j.tecto.2009.05.027

Seeber, L., Cormier, M.-H., McHugh, C.M.G., Emre, Ö., Polonia, A., and Sorlien, C.C., 2006. Subsidence and sedimentation at a transform bend: The Çinarcik Basin and the North Anatolian Fault in the Marmara Sea, Turkey. *Geology*, 34(11):933–936.

Seeber, L., Emre, Ö., Cormier, M.-H., Sorlien, C.C., McHugh, C.M.G., Polonia, A., and Scientific Team of Marmara, 2004. Uplift and subsidence from oblique slip: The Ganos-Marmara Bend of the North Anatolian Transform, Western Turkey. *Tectonophysics*, 391:239–258. doi:10.1016/j.tecto.2004.07.015

Shillington, D.J., Seeber, L., Sorlien, C.C., Steckler, M.S., Kurt, H., Dondurur, D., Çifç, G., et al., in press. Slow-motion collapse on the flanks of the Sea of Marmara transform basins. *Geology*.

Sorlien, C.C., Akhun, S.D., Seeber, L., Steckler, M., Shillington, D., Kurt, H., Çifçi, G., et al., 2012. Uniform basin growth over the last 500 ka, North Anatolian Fault, Marmara Sea, Turkey. *Tectonophysics*, 518–521:1–16. doi: 10.1016/j.tecto.2011.10.006

Steckler, M.S., Cifci, G., Demirbag, E., Akhun, S.D., Buyukasik, E., Cevatoglu, M., Coskun, S., et al., 2008. High resolution multichannel imaging of basin growth along a continental transform: The Marmara Sea along the North Anatolian Fault in NW Turkey. *Eos Trans. AGU*, 89(53):T21A-1921.

Tryon, M.D., Henry, P., Çağatay, M.N., Zitter, T.A.C., Géli, L., Gasperini, L., Burnard, P., Bourlange, S., and Grall, C., 2010. Pore fluid chemistry of the North Anatolian Fault Zone in the Sea of Marmara: A diversity of sources and processes. *Geochem. Geophys. Geosys.*, 11:Q0AD03. doi:10.1029/2010GC003177

Zitter, T.A.C., Henry, P., Aloisi, G., Delaygue, G., Çağatay, M.N., Mercier de Lepinay, B., Al-Samir, M., et al., 2008. Cold seeps along the main Marmara fault in the Sea of Marmara (Turkey). *Deep Sea Research Part I*, 55:552–570.

## Authors

**Cecilia M.G. McHugh**, School of Earth and Environmental Sciences, Queens College, City University of New York, 65-30 Kissena Boulevard, Flushing, NY 11367, U.S.A., e-mail: cmchugh@qc.cuny.edu.

**Leonardo Seeber**, Lamont-Doherty Earth Observatory of Columbia University, Route 9W, Palisades, NY 10964, U.S.A., e-mail: nano@ldeo.columbia.edu.

**M. Namik Çağatay**, Geology Department, Istanbul Technical University, Ayazaga 80626, Istanbul, Turkey, e-mail: cagatay@itu.edu.tr.

**Pierre Henry**, Centre National de la Recherche Scientifique (CNRS), College de France, Bat Trocadero, Europole de l'Arbois, BP80, 13545 Aix en Provence Cedex 04, France, e-mail: henry@cerege.fr.

**Christopher Sorlien**, Earth Research Institute, 6832 Ellison Hall, University of California, Santa Barbara, CA 93106-3060, U.S.A., e-mail: sorlien@eri.ucsb.edu.

**Michael Steckler,** Lamont-Doherty Earth Observatory of Columbia University, Route 9W, Palisades, NY 10964, U.S.A., e-mail: steckler@ldeo.columbia.edu.

**Gülsen Uçarkuş**, Geology Department, Istanbul Technical University, Ayazaga 80626, Istanbul, Turkey, e-mail: gulsen.ucarkus@gmail.com.

## Related Web Links

http://www.emcol.itu.edu.tr/Icerik.aspx?sid=9300
http://www.ldeo.columbia.edu/TAMAM/
http://cdf.u-3mrs.fr/~henry/marmara

## ECORD Summer School on "Submarine Landslides, Earthquakes and Tsunamis"

*3–14 September 2012,*
*Bremen, Germany*

The 6th ECORD Summer School to be held at the Center for Marine Environmental Sciences (MARUM) at the University of Bremen, Germany, aims to bring Ph.D. students and young postdocs in touch with IODP at an early stage of their careers, inform them about research within this international scientific program, and prepare them for future participation on IODP expeditions. Such training will be achieved by taking the summer school participants on a "virtual ship" by exploiting the unique facilities linked to the IODP Bremen Core Repository (photo below). They will be introduced to a wide spectrum of state-of-the-art analytical technologies and core description methods, including core logging/scanning according to IODP expedition standards. In addition, the topic "Submarine Landslides, Earthquakes and Tsunamis" will be covered by lectures and discussions with leading geoscientists in the field. The latter will include specialists in sedimentology, seismics, tectonics, and sediment transport modeling. This comprehensive approach—combining scientific lectures with practicals on IODP-style "shipboard" measurements—is the blueprint for the Bremen ECORD summer school series, which now rounds off its second three-year cycle of ECORD summer schools covering the three major topics of the IODP Initial Science Plan. For detailed information about the summer school, the application procedure and the scholarship options, visit http://www.marum.de/en/ECORD_Summer_Schools.html.

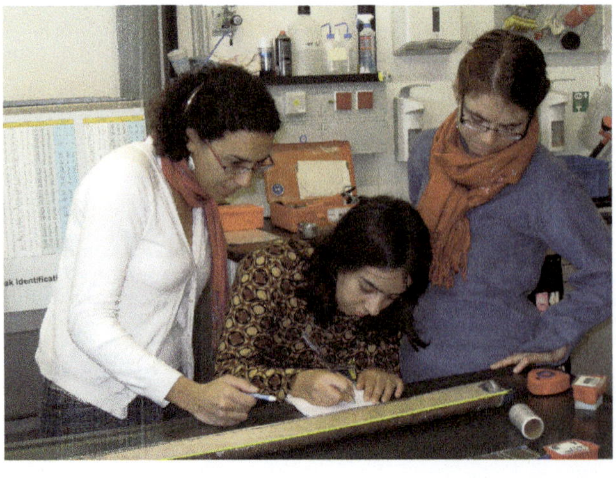

## 2012 IODP-Canada Activities

2012 is shaping up to be a busy year for IODP-Canada! In January, we awarded IODP research grants to three Ph.D. students: John Evangelatos (Dalhousie University)–The origin and evolution of the Canada Basin, Arctic Ocean; Olivia Gibb (Université du Québec à Montréal)–The paleoceanographic conditions of Baffin Bay during the interglacials of MIS 5e and 11; and Matthew Izawa (University of Western Ontario)–Integrated mineral, chemical and isotopic study of microbial ichnofossil preservation in basaltic glass. Student travel grants and summer school scholarships will also be offered this year.

In February, the workshop "Coordinated Scientific Drilling in the Beaufort Sea: Addressing Past, Present and Future Changes in Arctic Terrestrial and Marine Systems" took place in Kananaskis, Alberta. Participants defined and integrated the scientific questions and drilling strategies required to assess environmental change and geohazards in the Beaufort Sea. The workshop was funded by IODP-MI, IODP-Canada, ICDP-Canada and Natural Resources Canada.

IODP-Canada will again share an exhibition booth with ICDP-Canada at the joint annual meeting of the Geological Association of Canada and the Mineralogical Association of Canada in St. John's, Newfoundland and Labrador on 27–29 May 2012. IODP-Canada will also participate in the IODP-ECORD booth at the Goldschmidt conference in Montréal, Quebec on 24–29 June, 2012. In conjunction with Goldschmidt, we plan to host a "town hall" reception to stimulate discussion about Canada's continued participation in IODP post-2013. Planning is also underway to help organize activities for the public and scientific community during the port call of the *JOIDES Resolution* in St. John's in August.

## ECORD/ICDP/IODP-Canada 2012 Summer School

*5–21 July 2012*
*Registration deadline: 1 June 2012*

An ECORD/ICDP/IODP-Canada summer school "Impacts of the cryosphere dynamics from land to ocean" will be held in Québec on 5–21 July 2012 in partnership with ECORD, IODP-Canada and ICDP. The main goal of the summer school is to train graduate students and postdoctoral fellows in the fields of paleoceanography, paleoclimatology, marine geology and their associated methodological approaches and techniques. The program will be centered on four days of fieldwork on the north shore of the St. Lawrence River, which will be complemented with classes given by international invited lecturers, workshops involving the students, and four days of hands-on exercises in Montréal. A special attention will be given to polar and subpolar environments and to ice-ocean-atmosphere interactions and their role in climate dynamics. The school is open to graduate students and postdoctoral fellows from all over the world.

Find more information, visit at IODP-Canada website (http://www.iodpcanada.ca) or contact Diane Hanano (coordinator@mail.iodpcanada.ca).

## Poland Joins ECORD and IODP

Poland became the 18th member of ECORD (European Consortium for Ocean Research Drilling) and the 25th member of IODP on 14 December 2011, when the Director of the Polish Geological Institute – National Research Institute (PGI – NRI) signed the ECORD MoU during a visit to Warsaw by Catherine Mével, ECORD Managing Agency Director.

The PGI – NRI has a long tradition of working in the Baltic Sea, and the prospect of the implementation of the Baltic Sea proposal by the ECORD Science Operator in 2013 has certainly contributed in this excellent decision.

However, there is a move in Poland to expand the scope of marine sciences and a plan to build a Polish research vessel. We therefore expect the Polish science community to be strongly involved in all ocean drilling activities.

As a member of ICDP and now of IODP through ECORD, Poland will definitively play an important role in scientific drilling (More information available at http://www.ecord.org/p/new_members.html)

## International Symposium on Submarine Mass Movements and Their Consequences

In October 2011, 137 researchers from sixteen countires gathered in Kyoto for the IODP-MI co-sponsored 5th International Symposium on Submarine Mass Movements and Their Consequences. The meeting represents the continuation of a series of conferences initiated by UNESCO-IGCP Project 511 and its successor and ongoing project IGCP Project 585 (Earth's continental margins: Assessing the geohazard from sub-marine landslides–www.igcp585.org). The main objective of the conference was to bring a worldwide per-

## Southwest Pacific Ocean IODP Workshop

*Sydney, Australia, 9–12 October 2012*

For more information, write: Neville.Exon@anu.edu.au

This workshop, to be held at Sydney University, will address global geoscience problems in the southwest Pacific Ocean by building on existing and new geophysical and geological information including earlier scientific drilling. Its aim is to commence building coherent and well-integrated IODP proposals. The convenors are Neville Exon, Maria Seton and Stephen Gallagher (Australia), Minoru Ikehara (Japan) and Walter Roest (France), with many others in key roles. At this stage IODP-MI and ANZIC funding is assured. More information will be sent to likely participants soon.

This region has had a complex tectonic history, with plate boundary interactions resulting in an assemblage of deep oceanic basins, volcanic arcs, back-arc and fore-arc basins, continental ribbons, and emerged and submerged carbonate platforms. The workshop will identify the global scientific hypotheses and questions in this region that require ocean drilling to resolve them. The region's extensive plateaus and basins can provide crucial sedimentary records to help understand the developing interactions between the tropics and Antarctica, and between the Pacific and Indian Oceans. Furthermore, Australia has been one of the two major land masses (the other being India) undergoing major northward migration during the Cretaceous and Cenozoic with resulting fundamental changes in the tectonic and climate development of Earth and its biota. The workshop themes are:

1) **Climate and Ocean Change: Reading the Past, Informing the Future**
   This will cover questions related to climate and paleoceanographic change in this complex region, on all timescales.
2) **Biosphere Frontiers: Deep Life, Biodiversity, and Environmental Forcing of Ecosystems**
   Almost nothing is known about the deep biosphere in the region, so pioneering studies of both sediments and basalts should lead to exciting results.
3) **Earth Connections: Deep Processes and Their Impact on Earth's Surface Environment**
   The links between surface lithosphere and deep earth processes are of great interest in this tectonically complex region.
4) **Earth in Motion: Processes and Hazards on Human Timescales**
   This region has its share of earthquakes, tsunamis and submarine slides that have impacted on populations and will continue to do so. Targeted scientific drilling will help address some of these hazards.
5) **Marine Resources: Opportunities and Responsibilities: Opportunities and Responsibilities**

What contribution can IODP make to the exploration, characterization and responsible exploitation of marine resources in the southwest Pacific region and under what arrangements? These resources may include offshore oil and gas, gas hydrates and offshore minerals.

spective of submarine mass movements and their consequences by assembling state-of-the-art contributions from international researchers of academic institutions and the offshore industry. The symposium provided full coverage of the scientific and engineering aspects of this type of marine and coastal geo-hazards. It also highlighted the role of scientific ocean drilling to be well positioned to elucidate submarine slope failure and mass movement processes through direct sampling and *in situ* measurements within active systems, as well as sampling and dating of event deposits to determine timing and size of past geohazards.

The elevated awareness of the need for better understanding of submarine landslides is coupled with great advances in submarine mapping, sampling and monitoring technologies. Laboratory analogue and numerical modeling capabilities have also developed significantly of late. Multibeam sonar, 3-D seismic reflection, and remote and autonomous underwater vehicle technologies provide hitherto unparalleled imagery of the geology beneath the oceans, permitting investigation of submarine landslide deposits in great detail. Increased and new access to drilling, coring, *in situ* measurements and monitoring devices allows for ground truth of geophysical data and provides access to samples for geotechnical laboratory experiments and information on *in situ* strength and effective stress conditions of underwater slopes susceptible to failure. Great advances in numerical simulation techniques of submarine landslide kinematics and tsunami propagation have also led to increased understanding and predictability of submarine landslide consequences.

The symposium provided a refreshing view on these state of the art in submarine landslide research. A total of 103 contributions presented the latest scientific research results by international experts in geological, geophysical, engineering and environmental aspects of submarine mass failures, focused on understanding the full spectrum of challenges presented by submarine mass movements and their consequences. Seven presentations were directly related to recent IODP expeditions, and many more provided links to data and studies resulting from scientific offshore drilling efforts.

As a result of the conference the fourth edition of the 'Submarine Mass Movements and Their Consequences' book containing peer-reviewed manuscripts by symposium presenters was published (Yamada, Y., Kawamura, K., Ikehara, K., Ogawa, Y., Urgeles, R., Mosher, D., Chaytor, J., and Strasser M., (Eds.) 2012. ***Submarine Mass Movement and Their Consequences 5th International Symposium.*** Advances in

## International Geothermal Center in Bochum (Germany) Announces Start of Drilling Services with "BO-Rex"

icdp

Since October 2011 the International Geothermal Center (GZB) in Bochum, Germany started to operate its brand new track-mounted, mobile drill rig for depths up to 2000 m using rod diameters of up to 350 mm. The "BO-Rex" (HRB 207 GT) rig has been designed together by GZB and the manufacturer. It is capable of running any drilling process like mud rotary, DTH hammer drilling with water, mud or air, GeoJetting, geotechnical work, coring and sampling, augering, etc. It is equipped with a double head drilling setup to run and turn two rod strings independently of each other. For all kinds of hydraulic high-pressure drilling applications an external powerful triplex plunger pump is available for "BO-Rex" covering pressure regimes of 100 bar (max. 1000 L min$^{-1}$), 200 bar (max. 600 L min$^{-1}$) and 1500 bar (max. 80 L min$^{-1}$).

Technical Data:

| |
|---|
| **Total gross weight:** 32 t |
| **Rotary heads torque:** 50.000 and 20.000 Nm |
| **Diesel engine:** 250 kW engine |
| **Pull force:** 33 t + 20 t in the clamp |
| **Feed force:** 17 t |
| **Max. drill rod length:** 4.5 m |
| **Mast length overall:** 10 m |
| **Winch system for rod handling:** 2 t |
| **Pumps:** 20 bar robust mud pump, small Triplex water pump |
| **Remote control:** Track unit for full hydraulic operation |
| **Manufacturer:** Hütte GmbH, subsidiary of Casagrande Group, Italy |

Both drill rig and pump are available to be used in scientific drilling, exploration, coring or other kind of project. Terms and conditions will be discussed per case. For more information please contact: Volker Wittig (e-mail:volker.wittig@hs-bochum.de) at the GZB website (www.geothermie-zentrum.de).

Natural and Technological Hazards Research, 31: Dordrecht, The Netherlands (Springer). doi:10.1007/978-94-007-2162-3

Contact details: Yasuhiro Yamada, yamada@earth.kumst.kyoto-u.ac.jp.

## Small, Portable Drilling Rig Available for Research

DOSECC has acquired a small man-portable drilling rig suitable for drilling small coreholes and collecting core samples (1.06-inch/27 mm diameter) in remote areas or in places inaccessible by conventional drilling rigs. The rig weighs just 185 lbs (84 kg) and can be hand-carried to the drilling location. It can be used to collect rock cores to about 200 ft (60 m) deep. The system comes complete with all tools and materials included in a parts trailer. Perfect for small budget projects, it is rented for a low daily fee. Purchased with funding from the U.S. National Science Foundation, it is available to the scientific community for sampling projects and field lab experiments, including field camps. More information and photos can be found at http://www.dosecc.org/index.php/equipment/winkie-drill-system

## New U.S. Continental Scientific Drilling Initiative Report Available

*Follow-up Workshop: 21–22 May 2012, Arlington, Virginia, U.S.A.*

As societies around the world seek an economic balance between environmental and energy needs while planning effectively for a wide range of natural hazards, the need for subsurface information about our planet has never been greater. DOSECC annual workshops in 2009 and 2010 produced the background narrative justifying a new interdisciplinary consortium of research scientists who could lever-age funds more effectively across national and international interests in the geosciences. The DOSECC workshop held on 23–24 May 2011 in Arlington, Virginia focused on the next phase: producing a vision of how the United States, under the auspices of the U.S. National Science Foundation and other federal agencies, would promote scientific drilling as a critical tool in the earth sciences to grapple with major societal issues. To that end, the workshop laid out the initial architecture for a U.S. Continental Scientific Drilling (USCSD) Initiative, to be guided by a revitalized Interagency Coordinating Group linking federal and academic stakeholders.

The workshop produced themes and workshop needs for planning future U.S. continental drilling ranging from pre-site surveys to facilities and data management, emphasizing the synergies to be developed and coordinated with a larger, stronger US CSD program. While large challenges exist in coordinating across mission-based agencies and NSF-program goals, it is clear that the transformative scientific payoffs are far greater and cost effective than the current structure for leveraging both science and mission. The full workshop report will be available at http://www.dosecc.org/index.php/publications/reports-and-brochures.

The follow-up workshop will be held on 21–22 May 2012 in Arlington, Virginia. This workshop will further define the USCSD initiative. More information on this workshop can be found at www.dosecc.org.

## New Drilling and Coring Tools

The Engineering Geology Department of Lund University, Sweden, was recently awarded a Science Council grant for buying, implementing and operating a drilling infrastructure. Currently the implementation and testing phase of a high-capacity, modern diamond wireline coring rig is starting. The rig is lightweight and has a small footprint and a high capacity. It is highly atomized and fully diesel-hydraulic operated under strict safety regulations and complying to the latest environmental noise and exhaust rules. The depth coring capacities are as follows:

| |
|---|
| P-size (hole size 123 mm, core size 85 mm): 1050 m |
| H-size (hole size 96 mm, core size 63 mm): 1600 m |
| N-size (hole size 76 mm, core size 48 mm): 2500 m |

The core rig has a main hoist with 178 kN, an hydraulic feed cylinder with 200 kN and 3.5-m feed stroke. It is equipped with drilling equipment for all coring sizes, at the moment including 11-km long drill rods and casing, fifty diamond core bits, and core barrels as double and single tubes in doublets. Additional drilling equipment such as wellhead, preventer, mud and cementing systems, deviation and fishing tools, logging and MWD-recording capability make the system versatile for all kinds of scientific drilling projects in various environments.

It will be utilized in the Swedish Deep Drilling Program but is also available for the international scientific community.

## History of Controlled-Source Seismology Onshore and Offshore

In March 2012 the Geological Society of America has published its new Memoir 208 entitled "Exploring the Earth's Crust. History and Results of Controlled-Source Seismology", written by Claus Prodehl (University of Karlsruhe, Germany) and Walter D. Mooney (U.S. Geological Survey, Menlo Park, California, U.S.A.). This volume presents a comprehensive, worldwide history of seismological studies of the Earth's crust using controlled sources from 1850 to 2005. Essentially, all major seismic projects on land and the most important oceanic projects are presented. The time period of 1850 to 1939 is presented as a general synthesis, and from 1940 onward the history and results are subdivided into a separate chapter for each decade, with the material ordered by geographical region. Each chapter highlights the major advances achieved during that decade in terms of data acquisition, processing technology, and interpretation methods. For all major seismic projects, we provide specific details regarding the field observations, interpreted crustal cross-section, and key references. The Memoir concludes with global- and continental-scale maps of all field measurements and interpreted Moho contours. An accompanying DVD contains important out-of-print publications and an extensive collection of controlled-source data, location maps, and crustal cross- sections.

Contact: Claus Prodehl, Geophysical Institute, University of Karlsruhe, Karlsruhe Institute of Technology, Hertzstr. 16, 76187 Karlsruhe, Germany, e-mail: claus.prodehl@gmx.net

## Rob McKay Wins the Prime Minister's MacDiarmid Emerging Scientist Prize

Dr. Rob McKay, a postdoctoral researcher at Victoria University's Antarctic Research Centre in Wellington, New Zealand, has been awarded the New Zealand Prime Minister's MacDiarmid Emerging Scientist Prize (worth NZ$200,000) for his work on Antarctic ice sheet response to past climate fluctuations. His research uses glacial deposits and marine sedimentary records to investigate changes in ice volume and thermal regime of the Antarctic ice sheets since their inception approximately 34 million years ago and the resulting influence on eustatic sea level. This has particular relevance today as climate scientists struggle with how the

## The Rodderberg Quaternary Climate Archive

A suite of three shallow boreholes (164 m, 74 m, and 102 m) have explored the filling of the main crater of the Late Quaternary Rodderberg volcanic system near Bonn, Germany. The age of the volcano is dated to 300 ka (Paulick et al., 2008). The drilling activities finished in February, 2012. Two parallel cores were recovered from closely neighboring holes, each of them more than 70 m long, representing the sedimentary and volcano-clastic crater fill. The core material offers the unique opportunity to study a long complete climate archive representative for the Eifel region. Mainly lacustrine and aeolian (loess-like) sediments accumulated in the bowl-shaped crater since its last eruption (Zöller et al., 2010). Intercalated tephra layers of other dated volcanoes of the Eifel give a dense stratigraphic framework. A high-resolution reconstruction of past climate conditions during the last three glacial cycles may be derived from the core samples and the open hole borehole logging data. This is the central objective of the DUST TRAP project. It will study the geometry, volcanic evolution, and tectonic setting of the Rodderberg crater as well as the sedimentology, geophysics, and geochronology of its loess and loess-derived sediments. DUST TRAP is coordinated by the Steinmann Institute of Bonn University, the Leibniz Institute for Applied Geophysics, Hanover, and the Geological Survey NRW, Krefeld, and comprises groups from the Universities of Bayreuth, Braunschweig, Bremen, and Cologne.

### References

Paulick, H., Ewen, C., Blanchard, H., and Zöller, L., 2008. The Middle Pleistocene Rodderberg (~300 ka) maar-scoria cone volcanic complex (Bonn, Germany): Eruptive history, geochemistry, and thermoluminescence dating. – *Int. J. Earth Sci. (Geol. Rundsch.)*, 98:1879–1899. doi:10.1007/s 00531-008-0341-0

Zöller, L., Hambach, U., Blanchard, H., Fischer, S., Köhne, S., and Stritzke, R., 2010. Der Rodderberg-Krater bei Bonn. *E&G – Quat. Sci. J.*, 59:1–2

### Authors

**Franz Binot**, Leibniz Institute for Applied Geophysics, Stilleweg 2, D-30655 Hannover, Germany, e-mail: franz.binot@liag-hannover.de

**Nikolaus Froitzheim**, Steinmann Institute of Bonn University, Poppelsdorfer Schloss, D-53115 Bonn, Germany, e-mail: niko.froitzheim@uni-bonn.de

**and the DUST TRAP science team**

### Related Web Link

www.rodderberg.org

Antarctic ice sheets will respond to future climate change.

Much of his recent work has been a result of his participation on two scientific drilling expeditions to Antarctica. The first was the ANDRILL McMurdo Ice Shelf Project (2006–2007), which collected a 1280-m-long core representing the last 13 million years to study ice shelf response to past climate forcing. Rob's research focuses on investigating the glacial-interglacial cycles preserved in this record and the concomitant change in thermal regime of the Antarctic ice sheet. He is also working to improve the chronology of West Antarctic Ice Sheet changes in the Ross embayment over the last 18,000 years to better assess Antarctica's role in global sea-level change.

Rob returned to Antarctica in 2010 when he sailed as a sedimentologist on Integrated Ocean Drilling Program (IODP) Expedition 318 to the Wilkes Land margin of Antarctica. The primary goal of the expedition is to collect long-term sedimentary records in a nearshore to offshore transect off of the Wilkes Land margin to examine the influence of Cenozoic Antarctic glaciation on global climate. As part of the Wilkes Land science team, he is currently developing a cyclostratigraphy and geochemical record of the past ice sheet, sea ice and oceanographic variability at the East Antarctic margin of the Southern Ocean for the past 5 million years.

Another avenue of Rob's research utilizes Ocean Drilling Program (ODP) Leg 181 legacy cores, collected from the New Zealand region, to examine how changes in ice sheet and sea ice extent in Antarctica and the Southern Ocean over the last five million years are manifested in the ocean/climate system farther afield. This record will document how changes in Antarctica affected the New Zealand region, and should be important for predicting future changes as the climate warms.

Rob will receive $50,000 of the prize money, with the remainder used to support his ongoing research. He intends to use some of the award to

fund a Ph.D. student. In addition, the money will help to support lab work with international collaborators, concentrating on the IODP Exp. 318 Wilkes Land cores and the archive ODP Leg 181 cores from the New Zealand region.

## IODP-MI Announces Upgraded IT Systems

IODP-MI is pleased to announce the development of several IT systems to support mission critical tasks in IODP science. More information can be found at the newly re-designed IODP.org website.

As of 1 April 2012 drilling proposal submission deadline, proponents will be able to submit electronic proposals to the new Proposal Database version 2.0 (PDB 2.0). The PDB 2.0 (proposals.iodp.org) is a user-friendly, webforms system for compiling the many documents required for a drilling proposal. The PDB 2.0 offers significant benefits to proponents and in managing the proposal process from submission through SAS and to operations and scheduling.

Planned developments of Scientific Earth Data Information System (SEDIS: sedis.iodp.org) are completed for the user interface. An exciting functionality now available is parameter queries of the data across IOs and Expeditions. Using the SEDIS search interface to retrieve a subset from the SEDIS catalogue, the user can then click the Data Warehouse link where they can then select specific parameters (and methods) and construct aggregated data sets by retrieving rows matching the specific criteria from all of the data sets preselected by the initial search. The SEDIS catalogue currently comprises approximately 100,000 data sets and 30,000 publications from IODP/ODP/DSDP.

The Sample Materials Request Management (SDRM v.2) system has also been updated by IODP-USIO. The new SDRM v.2 will be launched soon.

The Site Survey Data Bank (SSDB) has been upgraded to use the INT Enterprise Seismic Viewer for web-based visualization of SEGY data.

IODP-MI appreciates feedback on these new systems.

## Joint IODP/ICDP Activities at the 34th IGC in Brisbane, Australia

Several ICDP and IODP activities are scheduled for the forthcoming 34th International Geological Congress in Brisbane, Australia (6–10 August 2012). IODP and ICDP will run a joint booth to provide information about current drilling activities and latest scientific results. An introduction to the International Continental Scientific Drilling Program (Workshop #20) will be offered to inform delegates who are not familiar with the ICDP about the ways of support and funding for continental scientific drilling through ICDP. Symposium #25.1 will report on important results of deep ocean drilling of the Integrated Ocean Drilling Program. Times, dates and locations will be announced in the near future on the congress website (http://www.34igc.org/)

## Peering into the Cradle of Life: Scientific Drilling in the Barberton Greenstone Belt

by N.T. Arndt, A. Wilson, A. Hofmann, P. Mason, D. Lowe, M. Tice, G. Byerly, and G. Chunnett

icdp

The Barberton greenstone belt in South Africa is one of the best-preserved successions of mid-Archean (3.5–3.2 Ga) supracrustal rocks in the world, and, as such, a remarkable natural laboratory where conditions and processes at the surface of the Archean Earth can be studied in detail. A scientific drilling program supported by the International Continental Drilling Program (ICDP) started in August 2011 under the direction of the international team listed on two web sites: http://www.icdp-online.org/front_content.php?idart=2709 and http://www.peeringintobarberton.com/. Updates on drilling progress are also found on these sites.

Planning of this project—the choice of targets, drilling strategies and scientific goals—started in October 2006 at University of the Witwatersrand in Johannesburg, and continued during meetings at San Francisco, Berlin and Vienna, and a one-week field conference in the Barberton belt. A research networking program of the European Science Foundation, "Archean Environment: the Habitat of Early Life", supported this part of the project. ICDP approved the drilling proposal in August 2009, but problems related to the procurement of drilling permission from the South African Department of Mineral Resources held up the start of drilling for more than a year.

Two main drilling targets were identified:

(1) Sedimentary sequences will provide information about erosion and sedimentation on the early Earth, the composition and temperature of Archean seawater, and one possible site where life may have emerged and evolved;

(2) Successions of ultramafic to mafic volcanic rocks will provide new insights into volcanic processes, dynamics of the crust and mantle, interaction between oceanic volcanic crust and the hydrosphere and biosphere.

At the time of writing in mid-January 2012, two holes had been drilled at the Tjakastad komatiite site in the southern part of the belt. The first—BARB1, 425 m in length—intersected an unusual tumulus structure composed of highly magnesian, vesicular komatiite basalts and 300 m of overlying spinifex-textured and massive komatiitic and basaltic flows. The hole BARB2 intersects an adjacent sequence of komatiic flows. Petrological and geochemical study of the core will address questions such as the nature of komatiitic volcanism, the origin of komatiite magmas, alteration of volcanic rocks following their eruption, and the composition and evolution of the Archean mantle.

The second site is in the Buck Reef, a thick sequence of cherts and shales. Hole BARB3 is currently at a depth of 709 m, about 100 m from its target of 800 m. This hole has sampled over 500 m of banded cherts with remarkable sedimentary and diagenetic structures such as those illustrated in Fig. 1a. Study of these samples and interlayered ferruginous and black shale will focus on the other origin of the sedimentary rocks, the evidence of early life that they contain, and the composition and temperature of Archean seawater.

The fourth hole, BARB4 in the middle Fig Tree Group in the eastern part of the belt, is currently at 352 m, 200 m short of its target. It has sampled turbiditic graywackes, mudstones, sideritic banded ferruginous chert and banded iron formations (Fig. 1b). Drilling at the last two sites, one in barite-rich sediments of the Barite Valley and another in silicified komatiites on the Mendon Formation, should be completed before April.

The distribution of samples and post-drilling research will be coordinated by a steering committee comprising representatives from all major participating countries. Compositions of the drilling teams and steering committee are given on the web sites. Once the core has been described and archived, a call will be made for proposals to work on the core. This call will be open to all scientists but priority will be given to those involved in the drilling or working in South African institutions.

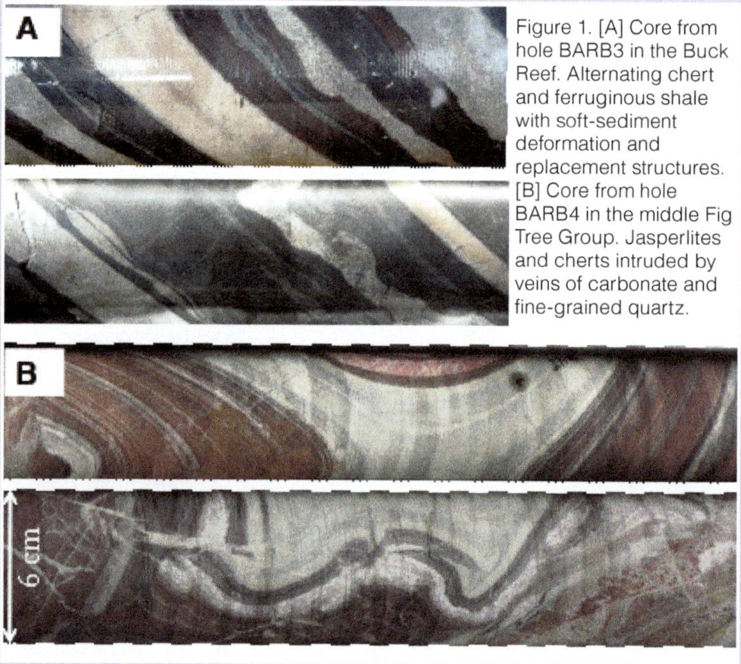


Figure 1. [A] Core from hole BARB3 in the Buck Reef. Alternating chert and ferruginous shale with soft-sediment deformation and replacement structures. [B] Core from hole BARB4 in the middle Fig Tree Group. Jasperlites and cherts intruded by veins of carbonate and fine-grained quartz.

# 36

# IODP Expedition 317: Exploring the Record of Sea-Level Change Off New Zealand

by Craig S. Fulthorpe, Koichi Hoyanagi, Peter Blum,
and IODP Expedition 317 Scientists

## Abstract

Expedition 317 investigated the record of global sea-level change (eustasy) within continental margin sedimentary sequences and how eustasy interacts with local forcing to produce preserved sedimentary architectures. The Canterbury Basin, on the eastern margin of the South Island of New Zealand, was selected to study these complex interactions because of high rates of Neogene sediment supply from the uplifting Southern Alps. This sediment input results in a high-frequency (~0.1–0.5 My periods) record of depositional cyclicity that is modulated by the presence of strong ocean currents. The expedition recovered sediments as old as Eocene but focused on the sequence stratigraphy of the late Miocene to Recent, when global sea-level change was dominated by glacioeustasy. A transect of three sites was drilled on the continental shelf (Sites U1353, U1354, and U1351), plus one on the continental slope (Site U1352). The transect samples the shallow-water environment most directly affected by relative sea-level change. Lithologic boundaries, provisionally correlative with seismic sequence boundaries, have been identified in cores from each site. Continental slope Site U1352 provides a record of ocean circulation and fronts during the last ~35 My. The early Oligocene (~30 Ma) Marshall Paraconformity was the deepest target of Expedition 317 and is hypothesized to represent intensified current erosion or non-deposition associated with the initiation of thermohaline circulation in the region. Expedition 317 involved operational challenges for *JOIDES Resolution*, including shallow-water, continental-shelf drilling and deep penetrations. Despite these challenges, Expedition 317 set a number of records for scientific ocean drilling penetration and water-depth.

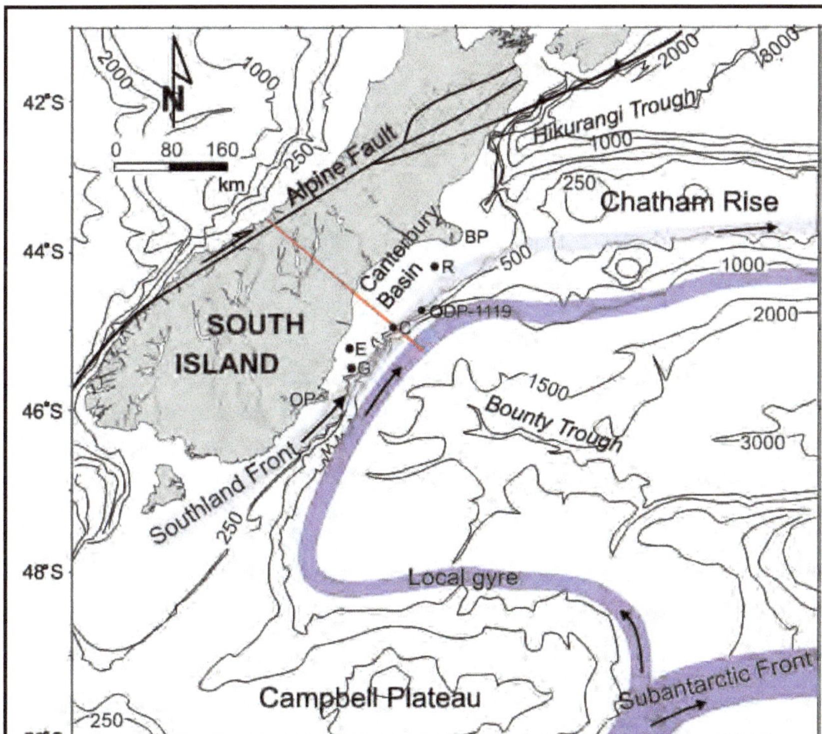


Figure 1. The Canterbury Basin underlies the present-day onshore Canterbury Plains and offshore continental shelf. It is bounded by the Miocene volcanic centers of the Banks Peninsula (BP) to the northeast and the Otago Peninsula (OP) to the southwest, and faces the Bounty Trough to the southeast. The Alpine Fault is the boundary between the Australasian and Pacific plates. Locations of the modern Southland Front and Subantarctic Front, together with local gyre within the Bounty Trough, are shown (Chiswell, 1996; Shipboard Scientific Party, 1999; Morris et al., 2001). Also shown are exploration wells in the offshore basin (Resolution, R; Clipper, C; Endeavour, E; and Galleon, G) as well as ODP Site 1119. The red line marks the approximate location of the schematic cross-section in Fig. 3. Bathymetric contours are in meters.

## Introduction and Goals

In order to evaluate predictions of future changes in global sea level and shoreline location, it is vital to constrain the range of past variability. Throughout the "Icehouse" period of the last ~30–40 million years, the changing volume of continental ice has been the principal driver of high-amplitude and high-frequency (~1 My and less) global sea-level change (eustasy). The last ~1 million years have been dominated by ~100,000-year glacial/interglacial cycles, and the sea-level rise since the last glacial maximum provides particularly valuable constraints on potential future rates of rise. However, earlier periods—notably the early Pliocene (~5.3–3.0 Ma) when $CO_2$ levels were similar to today's, but temperatures were ~3°C warmer and sea level ~25 m higher—provide glimpses of the possible future state of the Earth system. In addition, the geological record shows that climate and sea level do not always respond linearly to forcing and that abrupt events can disrupt gradual trends. If we are to develop reliable predictions of future climate and sea levels, we must look to the geologic record to ensure that the models on which we base forecasts can also reproduce the past.

Continental margin sediments retain long records of global sea-level change. This stratigraphic record comprises stacked sedimentary units (sequences) separated by unconformities. Sequence stratigraphy highlights the cyclic nature of the continental margin record (Mitchum et al., 1977; Vail et al., 1991) and has led to the theory of eustatic control of sequences and the resultant eustatic cycle chart (Haq et al., 1987). However, basin subsidence, changes in the rate of sediment supply, and other local processes can superimpose their signatures on this record. Therefore, the application of sequence stratigraphy to sea-level studies has been contentious, largely because of uncertainty surrounding the complex interaction of global and local processes (Carter, 1985; Karner, 1986; Carter et al., 1991; Christie-Blick, 1991; Miall and Miall, 2001). Understanding how these processes interact to form preserved stratigraphy is a fundamental problem in sedimentary geology. Its solution will yield both a record of eustatic change and a greatly enhanced ability to read the record, covering many tens of millions of years of Earth history, beneath the world's continental shelves.

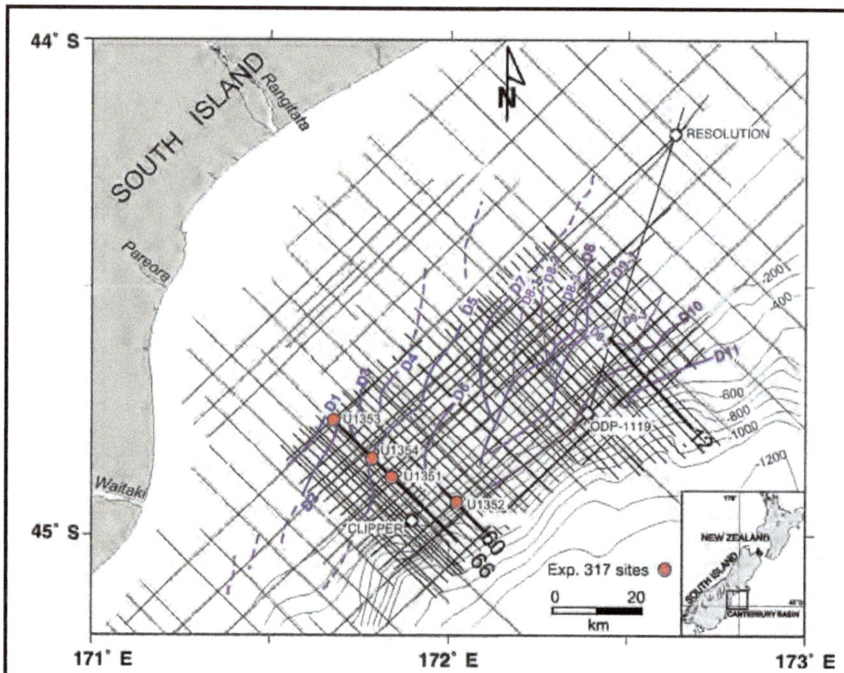


Figure 2. Expedition 317 sites (red), together with EW00-01 high-resolution (frequencies up to 300 Hz) seismic survey grid (dense grid of black straight lines), low-resolution CB-82 commercial seismic profiles (gray straight lines), exploration wells Clipper and Resolution, and ODP Site 1119. The EW00-01 survey was designed to provide improved vertical resolution (~5 m in the upper 1 s) to enhance definition of high-frequency sedimentary sequences. Also shown is the distribution of seismically resolvable sediment drifts (D1 to D11; blue curved lines mark crests of the drift mounds). Dashed blue lines represent drifts identified on CB-82 profiles. Locations of dip profiles shown in Figs. 4, 5, 6 (profiles 12, 66, and 60, respectively) are shown as thicker black lines. Bathymetric contours are in meters.

Scientific ocean drilling of globally coordinated borehole transects across continental margins provides the best way to distinguish local effects and extract the sea-level signal. Such drilling targets the geological environment directly affected by sea-level change as the shoreline migrates back and forth across the continental shelf. Coring the resulting sequence stratigraphic record provides information on ages, depositional environment, and past water depths during sea-level cycles from coastal plain to continental slope settings. Since boreholes provide information at only a few locations, integration with seismic imaging is crucial. Such integration performs three vital functions: 1) placing drilling results within a two- and three-dimensional context, 2) constraining the influence on sequence architecture of along-margin changes in sediment input and basin morphology, and 3) providing paleogeomorphological constraints on sedimentary processes and paleoenvironments.

Expedition 317 applied this approach in the offshore Canterbury Basin, on the eastern margin of the South Island of New Zealand (Figs. 1–3), where high rates of Neogene sediment supply preserved a high-frequency (0.1–0.5 My periods) seismically resolvable record of depositional cyclicity (Fulthorpe et al., 2011). The transect strategy was first applied and has been most thoroughly tested on the New Jersey Mid-Atlantic Transect (MAT) by ODP Legs 150, 150X, 174A, and 174AX. Most recently, IODP Expedition 313 used a mission-specific platform to drill on the New Jersey continental shelf (Mountain et al., 2010). Prior to Expedition 317, only ODP Leg 174A (Austin et al., 1998) had employed the *JOIDES Resolution* to drill shelf sites for sequence stratigraphic and sea-level objectives. Expedition 317, therefore, provided a rare opportunity to investigate the facies, paleoenvironments, and depositional processes associated with the sequence stratigraphic model on a prograding continental margin where large-scale depositional geometries and sequence architectures are well constrained by seismic data. Canterbury Basin complements MAT because the Middle Miocene to Recent Canterbury sequences are mainly younger than those calibrated so far on the New Jersey margin (Fig. 4). Furthermore, in line with the global approach to sea-level change advocated by previous planning groups, the Canterbury Basin allows expanded study of complex processes of sequence formation because of the following two reasons.

1) The stratigraphy records the development of the Antarctic Circumpolar Current and related southern oceanographic fronts (Fig. 1). Currents have strongly

influenced deposition, modifying sequence architecture and locally leading to the deposition of large sediment drifts, which aggraded to near shelf depths, within the prograding Neogene section.

2) Rifting is younger (Cretaceous) than the New Jersey margin (Jurassic), and from the earliest Miocene copious terrigenous sediment was supplied from a rapidly uplifting, nearby mountain range (the Southern Alps). Regional tectonic and geological histories have been intensively studied, allowing evaluation of the influence of sediment supply on sequence formation and of the tectonic evolution of the Alpine Fault plate boundary.

The principal scientific objective of Expedition 317 was to date clinoformal seismic sequence boundaries and sample associated facies to provide information (e.g., paleo-water depths, porosities) necessary for estimation of eustatic amplitudes using backstripping. A proper test of sequence stratigraphy necessitates drilling in shallow water (~100 m or less) on continental shelves. A related objective was to understand the interplay of along-strike and downslope sedimentary processes on this strongly current-influenced margin. Accordingly, our deepest drilling target was the Marshall Paraconformity, a regional unconformity thought to represent the mid-Oligocene initiation of ocean circulation on this margin. The final objective was to expand our knowledge of the erosion history of the Southern Alps using provenance studies and offshore sediment accumulation rates.

The Canterbury Basin is part of the Eastern New Zealand Oceanic Sedimentary System (ENZOSS; Carter et al., 1996). The distal (up to 4460 m water depth) component of ENZOSS was targeted by ODP Leg 181, which focused on drift development in the Southwest Pacific Gateway, principally under the influence of the evolving Antarctic Circumpolar Current and the Deep Western Boundary Current (Shipboard Scientific Party, 1999). Expedition 317 complements Leg 181 drilling by focusing on the landward part of ENZOSS.

## Geological Setting

The eastern margin of the South Island of New Zealand is part of a continental fragment, the New Zealand Plateau, that rifted from Antarctica beginning at ~80 Ma (Anomaly 33). The Canterbury Basin lies at the landward edge of the rifted continental fragment and underlies the present-day onshore Canterbury Plains and offshore continental shelf (Fig. 1; Field and Browne, 1989). The plate tectonic history of the New Zealand Plateau is recorded in the stratigraphy of the South Island. The post-rift, Cretaceous to Recent sedimentary history of the Canterbury Basin comprises a first-order (80 My), tectonically-controlled, transgressive-regressive cycle (Fig. 3; Carter and Norris, 1976; Field and Browne, 1989).

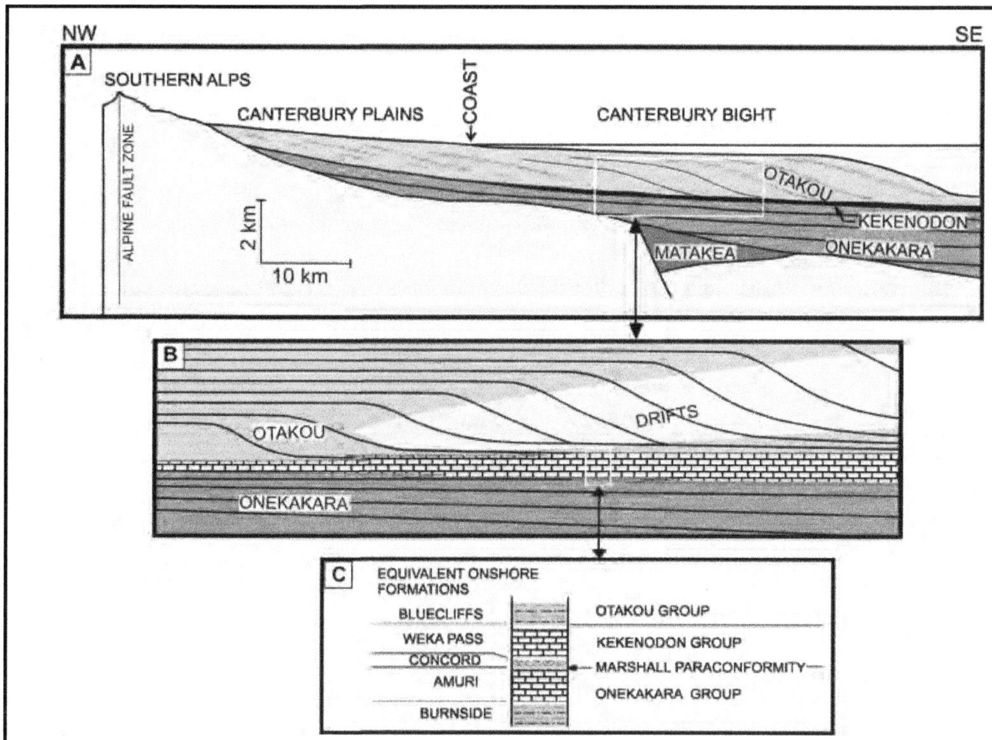


Figure 3. Schematic stratigraphy of the Canterbury Basin at three different scales. [A] Large-scale, post-rift stratigraphy (see profile location in Fig. 1). Onekakara, Kekenodon, and Otakou groups were deposited during regional transgressive, highstand, and regressive phases, respectively. [B] Seismic-scale stratigraphy. Sediment drifts occur within the Otakou Group. [C] Outcrop-scale stratigraphy across the Marshall Paraconformity. Modified from Fulthorpe et al. (1996).

The post-rift transgressive phase terminated during the late Eocene, when flooding of the land mass was at a maximum (Fleming, 1962). Reduced terrigenous influx resulted in deposition of regionally widespread pelagic limestone (Amuri Limestone), which ranges in age up to early Oligocene (~33 Ma). The sequence is then interrupted by the Marshall Paraconformity (Fig. 3; Carter and Landis, 1972). The paraconformity is recognized at drill sites throughout the region east of the Tasmanian gateway and is hypothesized to represent the initiation of thermohaline circulation upon opening of the seaway between Antarctica and Australia (~33.7 Ma; Carter et al., 2004). New Zealand lay directly in the path of the developing current system. Regression commenced in the late Oligocene or early Miocene in response to an increase in sediment supply provided by the initiation of Alpine Fault movement (Carter and Norris, 1976; Kamp, 1987). The Alpine Fault formed as a dextral transpressive zone with 500-km lateral displacement since the earliest Miocene (~23 Ma; Wellman, 1971; Kamp, 1987; King, 2000). The sediment influx following fault initiation is distinct from the later pulse related to uplift that culminated in the present-day Southern Alps. Uplift of the Southern Alps accelerated at ~8–5 Ma (Tippett and Kamp, 1993a; Batt et al., 2000) or ~10–8 Ma (Carter and Norris, 1976; Norris et al., 1978; Adams, 1979; Tippett and Kamp, 1993b), indicating an increased component of convergence along the fault and leading to further increase in the rate of sediment supply to the offshore Canterbury Basin (Lu et al., 2005).

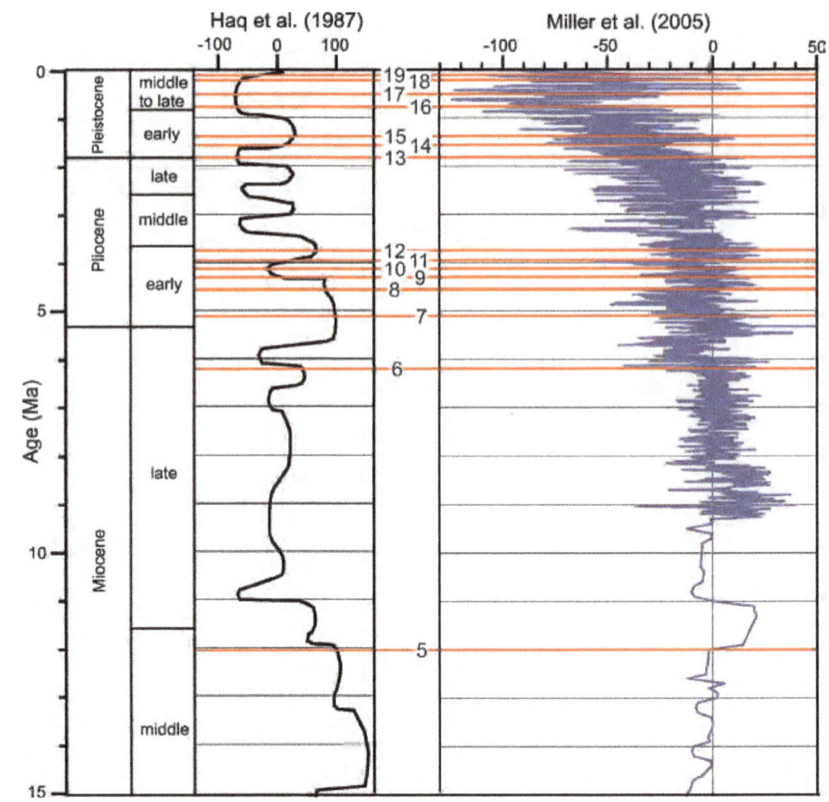


Figure 4. Global sea-level curves of Haq et al. (1987) and Miller et al. (2005) from the middle Miocene to Recent. Unconformities U5 to U19 are labeled and shown as red lines (shipboard ages are approximate). Note that much of the upper part of the Pliocene section is missing at Expedition 317 sites and a 5-My hiatus occurs in the upper Miocene, accounting for the large time gaps between some unconformities. Prior to 9 Ma, the Miller et al. (2005) curve is derived from backstripping of New Jersey MAT sites. However, the record since 9 Ma is a benthic foraminiferal $\delta^{18}O$ record, calibrated to represent eustasy, because the MAT record there is incomplete.

This sediment influx was deposited as prograding clinoforms (Otakou Group; Fig. 3). However, currents continued to influence deposition. At present, the core of the northward-flowing Southland Current, inboard of the Southland Front (part of the Subtropical Front; STF) is over the ~300-m isobath (Chiswell, 1996). In deeper water, to at least 900 m, a local gyre of the Antarctic Circumpolar Current circulates clockwise within the head of Bounty Trough parallel to the Southland Current (Fig. 1; Morris et al., 2001). Large sediment drifts within the prograding section (Fig. 5) show that similar currents, probably strengthened during glacial periods, existed throughout much of the Neogene (Fulthorpe and Carter, 1991; Lu et al., 2003; Carter et al., 2004).

## Drilling and Coring

Expedition 317 cored upper Miocene to Recent sedimentary sequences in a transect of three sites on the continental shelf (landward to basinward: Sites U1353, U1354, and U1351) and one on the continental slope (Site U1352; Fulthorpe et al., 2011). A transect of sites is required because of the need to drill each of the multiple target sequences in at least two locations: 1) landward of clinoform breaks or rollovers, presumed to represent paleo-shelf edges, where paleo-depth indicators are most reliable; and 2) drilling on the slope where increased abundance of pelagic microfossils provides the best age control.

Favorable weather conditions on arrival allowed us to follow the planned drilling strategy by drilling first at shelf Site U1351 at the deep-water end (121 m) of the shelf transect (Figs. 2, 6). This provided experience in shelf sediment drilling before moving to sites in even shallower water. The ship then moved ~15 km to slope site U1352, designed to provide age control for sequences drilled at site U1351 (Figs. 2, 7). An additional target at Site U1352 was penetration and recovery of the Marshall Paraconformity. On completion of slope drilling, the ship moved back to the shelf to drill the two additional shelf sites U1353 and U1354 (Fig. 6) to provide spatial control of facies within sequences and to recover the lowermost unconformities landward of their rollovers.

Because of the requirements to drill in shallow water and to achieve deep penetrations, a number of scientific ocean drilling records were set.

- Deepest hole drilled in a single expedition (Hole U1352C; 1927 m), also the second deepest hole in DSDP/ODP/IODP history
- Deepest hole drilled by *JOIDES Resolution* on the continental shelf (Hole 1351B; 1030 m) and also the second deepest (Hole 1353B; 614 m)

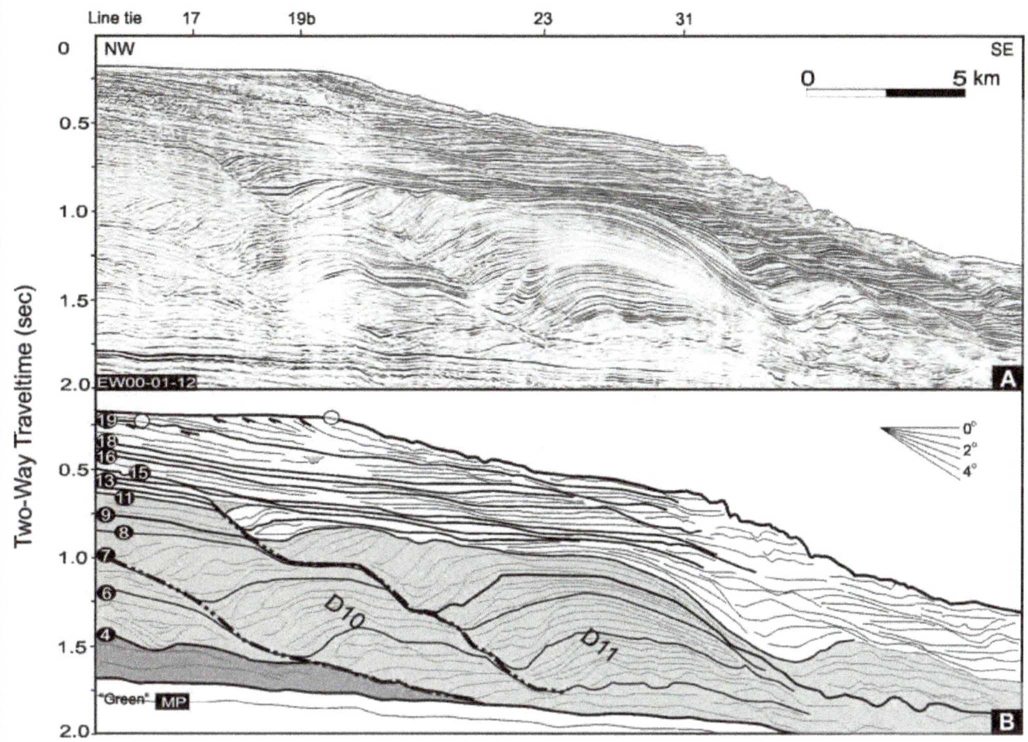


Figure 5. [A] Uninterpreted dip profile EW00-01-12 across the slope in the northern part of the survey grid showing mounded sediment drifts with adjacent (landward) moats (see Fig. 2 for location). [B] Interpretation showing sequence boundaries and selected locations of reflection truncation (arrows). Drifts D10 and D11 are the last of the succession of large elongate drifts of the Canterbury Basin (Fig. 2). Drifts migrate northwestward (landward) on dip profiles. Erosional unconformities at the landward edges of drift moats are diachronous, and they are intersected by multiple sequence boundaries. The underlying Marshall Paraconformity (MP) is also shown.

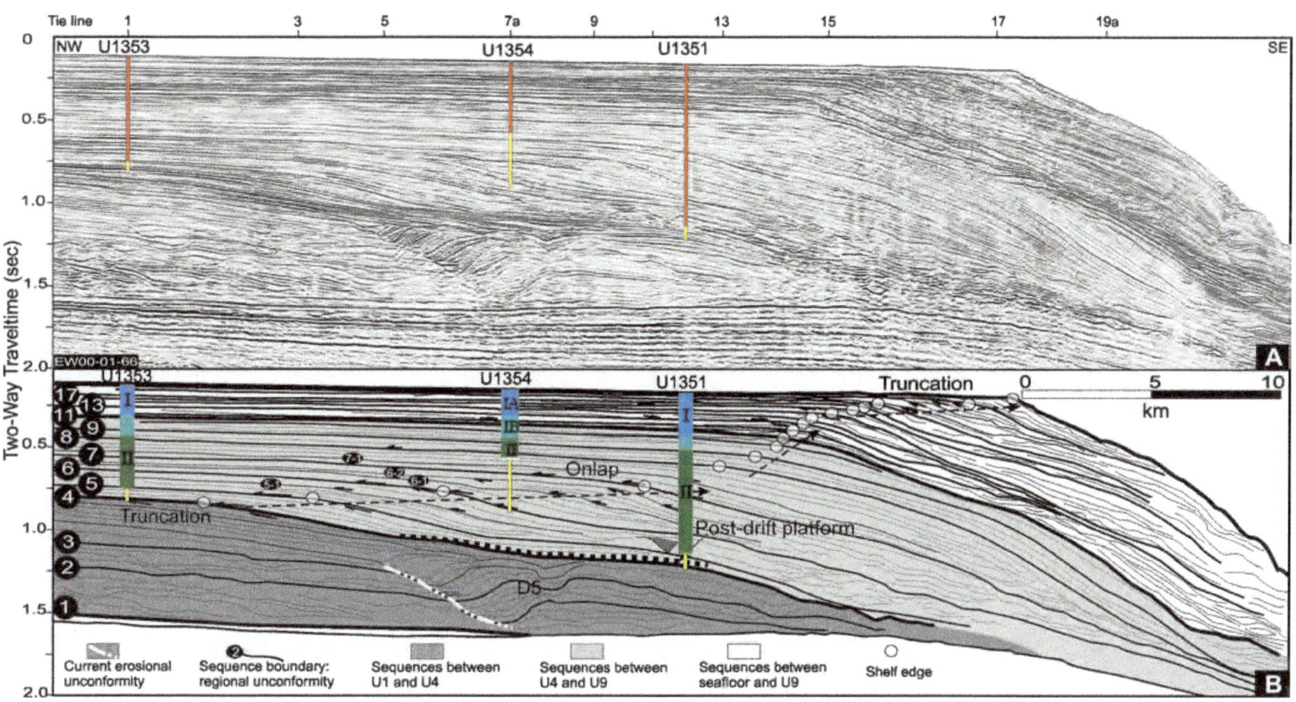


Figure 6. [A] Uninterpreted MCS dip profile EW00-01-66 from the southern part of the survey grid showing locations of sites U1351, U1353, and U1354. Actual penetrations shown in red, proposed penetrations in yellow (see Fig. 2 for location). [B] Interpretation showing sequence boundaries and selected locations of onlap, truncation, and downlap (arrows). Lithostratigraphic units defined at shelf Sites U1353, U1354, and U1351 are also shown. Clinoform rollovers, or breaks, representing paleoshelf edges, are marked by circles. Paleoshelf edges from U4 to U8 prograde steadily. The amount of progradation decreases from U8 to U12, increasing again from U13 to U19. U4–U10 are onlapped and truncate underlying reflections; internal reflection geometries from U4 to U10 are mainly sigmoidal. In contrast, U11–U19 are downlapped on paleoshelves, but also truncate underlying reflections, and internal reflection geometries from U11 to seafloor are oblique. Site U1351 sampled paleoslopes of U6–U7 and sampled U8–U19 landward of their paleoshelf edges. Sites U1353 and U1354 sampled sequence boundaries from U5 to U18 landward of their paleoshelf edges in more proximal settings. Sediment drift development in this area had largely ceased by U4; only drift D5 is present. Prograding clinoforms dominate later sequences.

- Shallowest water depth for a site drilled by *JOIDES Resolution* for scientific purposes (Site U1353, 84.7 m water depth)
- Deepest sample taken by scientific ocean drilling for microbiological studies (1925 m at Site U1352)

## Sequence Stratigraphy and Sea-Level Change

Seismic sequence boundaries provide the large-scale understanding of subsurface architecture necessary for placing drilling results within a broader two- and three-dimensional context. The seismic stratigraphy of the Canterbury Basin was developed using high-resolution (~5-m vertical resolution for two-way traveltimes <1 second) multichannel seismic (MCS) profile collected by R/V *Maurice Ewing* in 2000 (EW00-01 profiles; Figs. 2, 5, 6, 7). These were augmented by commercial MCS data collected in 1982 (Fig. 2; CB-82 profiles).

Nineteen regional seismic sequence boundaries (U1–U19) are identified in the middle Miocene to Recent shelf-slope sediment prism (Fig. 4; Lu and Fulthorpe, 2004). Expedition 317 provides ground truth for sequences above U5. Two larger seismic units are defined, based on seismic architecture and facies, within the cored interval (Fig. 6).

1. U5–U10 feature rounded shelf-slope rollovers. Internal reflection geometries are predominantly sigmoid, and paleoshelves are smooth and defined by onlap and truncation.
2. U11-U19 are downlapped on paleoshelves. They truncate underlying reflections near paleoshelf edges, which tend to be angular. Internal reflection geometries are oblique and U- and V-shaped channels that incise paleoshelves indicate exposure during sea-level lowstands.

The influence of sea-level change is illustrated at several scales. At the largest scale, the boundaries between Lithostratigraphic Units I and II at shelf sites separate overlying heterolithic facies of Unit I from more uniform facies below in Unit II (Figs. 6, 7). Unit I contains a wide variety of facies and many green marl or calcareous beds with sharp (or bioturbated) bases. On the shelf, the more homogeneous Unit II is dominated by mud or muddy sand, with lower percentages of carbonate components and less frequent greenish calcareous beds, whereas on the slope this unit is represented by homogeneous sandy marlstone.

The location of the unit boundary is difficult to identify precisely because the transition between units is gradational and also because of low recovery near some unit boundaries. However, the Unit I to Unit II boundary generally conforms to the interval around the predicted depths of seismic sequence boundaries U10 to U12 (late Pliocene), in a zone where high-amplitude seismic reflections related to clinoforms with angular shelf-slope rollovers transition downward to clinoforms with rounded rollovers (Fig. 6). The lithological contrast between Units I and II and the contrasting characteristics of sequence boundaries and seismic facies above and below ~U10 are inferred to occur because paleoshelves of U10 and underlying sequence boundaries were not subaerially exposed at lowstand, whereas those above U10, were exposed, probably because of increasing eustatic amplitudes during the Pleistocene.

At a finer scale, correlation of specific lithological features with seismic sequence boundaries is also possible. Erosion surfaces with associated overlying sediment facies were

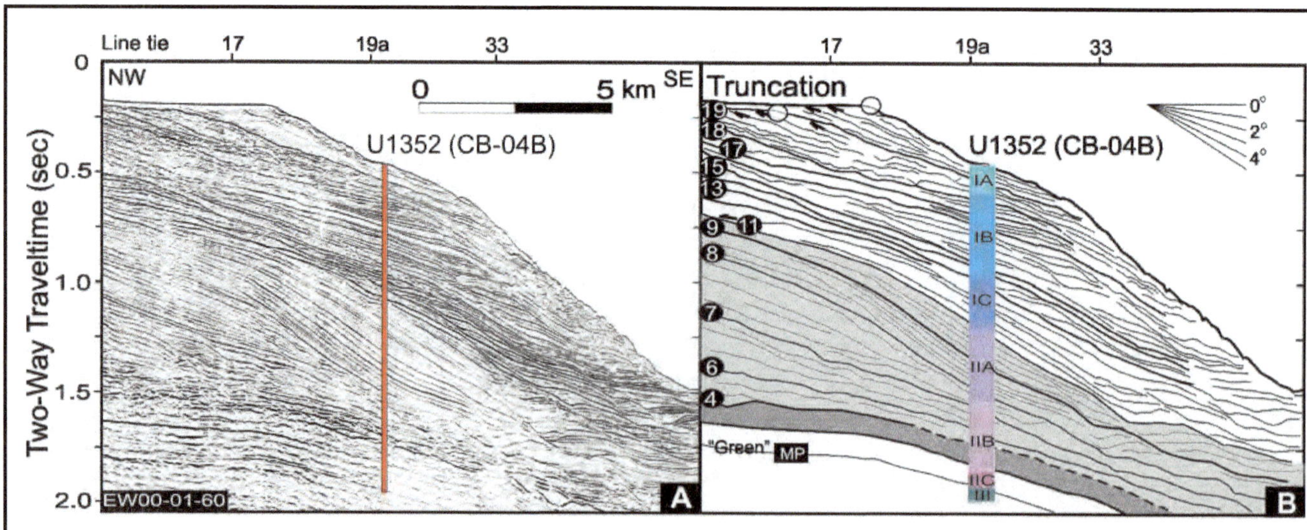


Figure 7. [A] Uninterpreted dip profile EW00-01-60 across the slope in the southern part of the survey grid showing location of Site U1352 (see Fig. 2 for location). [B] Interpretation showing sequence boundaries and selected locations of reflection truncation (arrows). Lithostratigraphic units are also shown. Site U1352 sampled slope facies of sequences, particularly important for age control, as well as the underlying Marshall Paraconformity (MP).

identified near the predicted depths of seismic sequence boundaries at all sites (Fig. 8). Identification is based on changes in lithologic composition, presence of transported sediments, and the nature of the contacts. When basal erosion surfaces were not recovered, surfaces may be inferred from the presence of sand—or the assumption of washed-out sands—and correlation to wireline logs. These lithologic surfaces and associated sediments are believed to be correlative between sites across the transect.

Candidate sequence boundaries are best identified in Lithostratigraphic Unit I at each site, where recovery was highest and where lithologic contrasts are most pronounced. Unit I erosion surfaces have been correlated with seismic sequence boundaries U19–U11 and, at Site U1351, are overlain by a series of lithofacies associations comprising upward-fining shelly sandy mud topped by coarsening-upward sandy mud (Figs. 8, 9). Together, the package is provisionally interpreted as a transgressive, wave-eroded ravinement surface overlain by a lag deposit corresponding to a transgressive systems tract. Subsequent highstand mud units are topped by the regressive, coarsening-upward sediments (Type 2 facies assemblage; Fig. 9). The facies assemblage includes decimeter- to centimeter-thick clay beds similar to those attributed to rapid deposition by flood events on modern shelves near sediment-rich fluvial systems. Truncated versions of this facies association, lacking the regressive highstand phase, occur in the upper 30 m at Site U1352, suggesting more frequent erosive episodes (Type 1 facies assemblage; Fig. 9). There are more such truncated facies associations than there are seismic sequence boundaries, and it is therefore possible that some of the truncated facies associations may represent higher-order cycles or autocyclicity. The facies associations, both complete and truncated, are similar to those known from the Wanganui Basin (Abbott and Carter, 1999; Saul et al., 1999; Abbott et al., 2005; Naish et al., 2005).

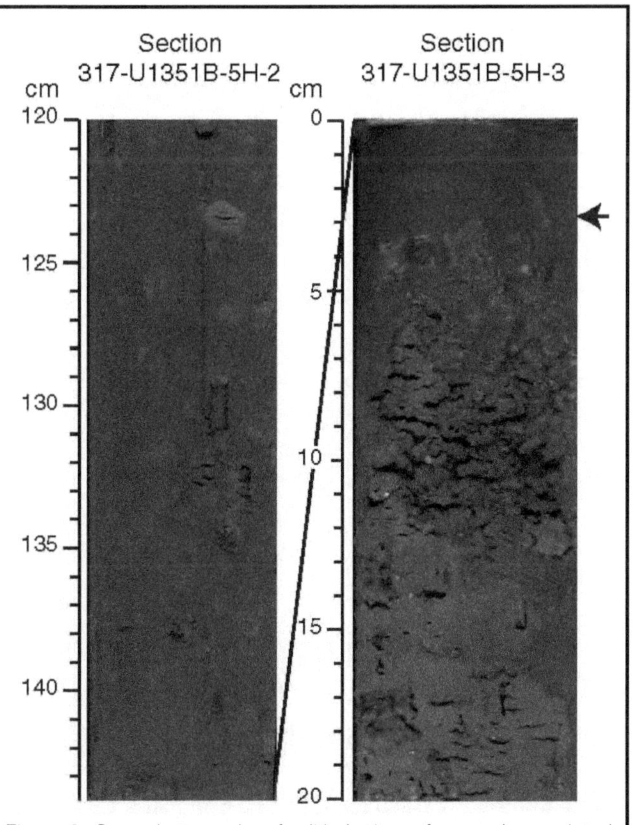


Figure 8. Core photographs of a lithologic surface and associated sediments in Unit I at outermost shelf Site U1351. Gray sand overlies greenish mud at a burrowed contact (arrow). This contact correlates with the predicted depth of seismic sequence boundary U18 (from Fulthorpe et al., 2011).

Erosion surfaces have not been identified in Unit II, which is more homogeneous and where recovery beneath the shelf was poorest. However, indurated intervals of *in situ* carbonate cementation may correlate with seismic sequence boundaries. Sediment recovered from Unit II was predominantly muddy, suggesting deposition on a shelf below fair-weather wave base. The lack of abundant shell debris suggests that the setting was more distal than during deposition of Unit I.

Fitting seismically resolvable sequences within the hierarchy of known global forcing has long been problematic. The relationship between high-frequency Milankovitch cycles and longer, seismic-scale sequences remains an unresolved area of research. Saul et al. (1999) showed that, for the Wanganui Basin, classic fourth- and fifth-order cyclothems are grouped together into third-order cycles that were controlled by basin-wide tectonic rhythms. Alternatively, Miller et al. (2005) have pointed out that some third-order cyclicity may be manifestations of longer wavelength Milankovitch periodicities of 1.2 My and 2.4 My, or of interference between these cycles.

The high accumulation rates of the Canterbury Basin provide the opportunity to link seismic sequences to specific Milankovitch frequencies, as revealed by oxygen isotopic records (Fig. 4). This will allow us to evaluate how predominant forcing frequencies change through time and to constrain the depositional parameters required to generate and preserve seismically resolvable sequences. The seismic sequences span the range between Milankovitch-scale cyclicity and longer period, third-order, seismically resolvable cycles. The two upper, seismically delimited sequences beneath the shelf (corresponding to the sequences overlying seismic sequence boundaries U18 and U19) almost certainly represent 100-k.y. cycles corresponding to marine isotope stages (MIS) 1–5 and 6–7, respectively, based on magnetic susceptibility and natural gamma ray time series. Correlations with older MIS stages are also likely as ages are refined. In contrast, Miocene to early Pliocene sequences represent longer cycles and may correlate with the 400-k.y. eccentricity cycle.

## Influence of Ocean Currents on Deposition

Northeastward flowing currents (Southland Current, inboard of the Subtropical Front, and a gyre of the

Subantarctic Front) sweep the margin today to water depths of ~900 m (Chiswell, 1996; Morris et al., 2001; Fig. 2). The buried sediment drifts in the Canterbury Bight, as well as outcrop evidence of current-related deposition, suggest that similar currents have been active for the last 30 My (Fig. 5). However, the influence of currents on deposition at seismic scale varies markedly along strike, with large drifts developing in the northeast near ODP Site 1119 (Leg 181), while coeval sequences to the southwest at the Expedition 317 transect are clinoformal (compare Figs. 5 and 6). The role of along-strike processes in sequence stratigraphy has been unclear; sediment transport in the sequence stratigraphic model is viewed as essentially two-dimensional. Expedition 317 results provide insight into the interaction of along-strike and downslope processes within a sequence stratigraphic framework in a location where a current is known to have been present, but whose seismic stratigraphic signature varies greatly both spatially and temporally.

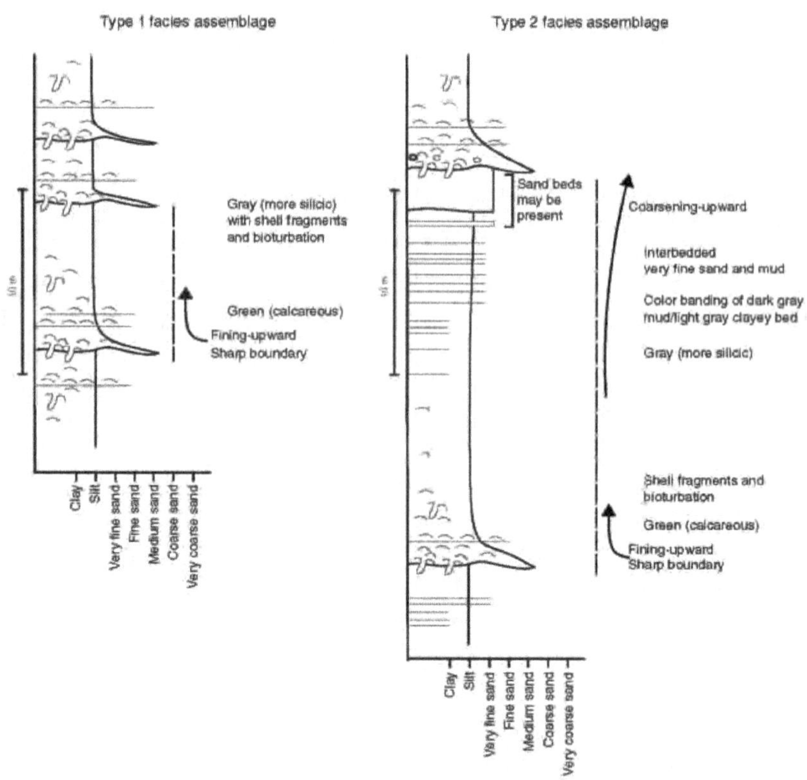


Figure 9. Schematic model of the repetitive facies assemblages that characterize lithologic Unit I. The Type 2 facies assemblage is composed of greenish-gray shelly bioturbated mud with a sharp lower contact (ravinement surface), followed by a transition uphole into dark gray micaceous very fine sandy mud with color banding and alternations of sand and mud showing upward coarsening. The Type 1 facies assemblage is the basal part of the Type 2 assemblage (from Fulthorpe et al., 2011).

Beginning at the seafloor, ODP Site 1119 sediments are interpreted as products of current reworking, corresponding to the clear seismic signatures of sediment drift development. In contrast, Site U1352 features a complex interplay of along-strike and downslope depositional processes. For example, the uppermost interval (Subunit IA; 0–98 m, Pleistocene to Holocene) is interpreted as a series of lowstand shelf-edge deltas fed by sediment gravity flows represented by sharp-based gray very fine to fine sands, whose mineralogy indicates direct offshore and downslope transport from sources in the Torlesse Terrane, located onshore adjacent to the drilling area. This interpretation is supported by the presence of reworked nannofossils consisting of mixed, poorly-preserved Eocene-Oligocene assemblages derived from onshore. However, the rest of Site U1352 (Subunit IB and below) and the slope interval of Site U1351 include sediment from the Otago Schist to the south, indicating increasing downhole importance of along-strike transport. Nevertheless, downslope transport is also interpreted in Subunit IB. Interbedded, sharp-based greenish sands that occur within dark gray hemipelagic mud are interpreted as mass flows; abundant shells in one sand suggest transport from a shallow marine environment. Therefore, Subunit IB is a transitional interval of increasing (downhole) contourite deposition culminating in Subunit IC, which contains a well-sorted, very fine-grained sand fraction consistent with considerable transport distance. Subunit IC also contains reworked nannofossils. However, in contrast to those in Subunit IA, these reworked assemblages are well-preserved and contain cold-water taxa from a monospecific age consistent with an along-strike source. This interpretation is more consistent with that of the coeval interval at Site 1119. Hemipelagic sediments of Units IIA and IIB at Site U1352 may have been deposited as sediment drifts; subtle seismic geometries suggest the influence of along-strike processes (e.g., paleoslopes on Figs. 6 and 7). Diagnostic sedimentary structures are rare, partly because of extensive bioturbation, but intermittent wavy laminations occur in Unit IIB. Alternations of light-colored marlstone with darker mud-stone may reflect switching between hemipelagic carbonate deposition and increased terrigenous influx corresponding to alternating along-strike and downslope sedimentary processes, respectively. Expedition 317 drilling therefore provides insights into the development of ocean circulation in this region, as well as the interaction of along-strike and downslope processes with implications for both sequence stratigraphy and models of clinoform development.

## Marshall Paraconformity

The Marshall Paraconformity is a regional unconformity terminating the post-rift transgressive phase and capping the widespread Amuri Limestone formation. The paraconformity has been dated at its onshore type section using strontium isotopes as representing a minimum hiatus of

32.4–29 Ma (Fulthorpe et al., 1996). There, it is overlain by mid-late Oligocene cross-bedded glauconitic sand (Concord Formation) and calcarenite limestone (Weka Pass Formation). The paraconformity was the deepest target of Expedition 317 drilling and is hypothesized to represent intensified current erosion or non-deposition associated with the initiation of thermohaline circulation upon opening of the seaway between Antarctica and Australia (~33.7 Ma), prior to opening of the Drake passage (Carter, 1985; Fulthorpe et al., 1996; Carter et al., 2004).

The Marshall Paraconformity was represented by a lithologic change at 1853 m in Hole 1352B (Fig. 10). The paraconformity marks the boundary between overlying lower Miocene glauconitic limestone and underlying lower Oligocene recrystallized pelagic nannofossil limestone containing trace fossils and stylolites, equivalent to the onshore Amuri Limestone. Recovery was low across the paraconformity, which was represented by a decimeter-scale rubble zone in recovered core. No equivalent to the Concord Formation glauconitic sand was recovered, although logs from nearby Clipper-1 exploration well (Fig. 2) suggest that a boundary sand layer is present. One possibility is that sediment loading caused the Concord-Formation-equivalent greensand to be injected into the overlying sandy marlstones and limestones to produce distinctive glauconitic sand layers (Fig. 10). The hiatus at the paraconformity is estimated to be 11–12 My at Site U1352, conforming with previous observations that the hiatus is longer offshore than at the type section, perhaps owing to greater current-induced erosion at such deeper-water locations (Shipboard Scientific Party, 1999; Carter et al., 2004). Hole 1352B terminated in upper Eocene (35.2–36.6 Ma) limestone.

## The Future: a Global Array of Sea-Level Transects

The study of global sea-level change requires a global approach. Worldwide correlation is essential to establish the degree to which sequence boundaries are synchronous and to confirm that eustatic amplitude estimates from individual margins are globally consistent. The Canterbury Basin is therefore one element in a worldwide array of scientific ocean drilling transects, some of which have been drilled while others remain in the proposal stage. New Jersey MAT drilling on the coastal plain, shelf, and slope has already yielded a record of global sea-level timing and amplitudes since the Late Cretaceous (Fig. 4; Miller et al., 2005; Kominz et al., 2008). Such records are derived using one- and two-dimensional backstripping, requiring as inputs sequence-boundary ages, paleobathymetry from benthic microfossils, porosity profiles (for decompaction calculations), and tectonic history. Drilling for long-term sea-level objectives has also been carried out in the Bahamas and on the Marion Plateau (ODP Legs 166 and 194, respectively).

Expedition 317 was designed using the strategy developed for MAT. We are following a parallel approach in refining Expedition 317 results with the ultimate goal of producing a late Miocene-Recent eustatic record for correlation both with MAT results and with the oxygen isotopic record of global climatic variation. Additional transects are required to cover both Greenhouse and Icehouse records. This approach will necessitate integration of onshore and offshore drilling and use of Mission Specific Platforms in

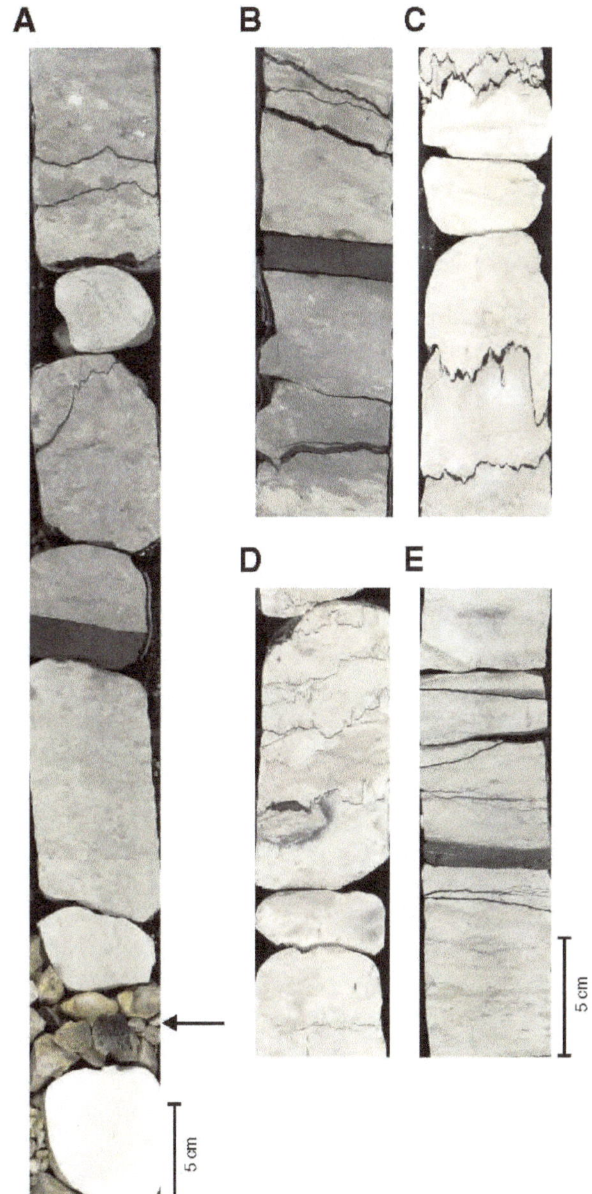


Figure 10. [A] Marshall Paraconformity (arrow) at contact between Units II and III at Site U1352. Only two limestone pieces and some rubble were recovered. However, unconsolidated glauconitic sand may have been present originally, but it was injected into the overlying glauconitic marlstone and limestone to produce the observed layers (see also B), some of which cross cut primary stratigraphy and/or display branching. [B] Glauconitic limestone above the paraconformity. [C] and [D] Stylolitic white limestone (Amuri Limestone equivalent) below the paraconformity. [E] Muddy limestone with layers of marlstone and purple banding below the paraconformity. B–E are at the same vertical scale (from Fulthorpe et al., 2011).

shallow-water settings, as has been done off New Jersey, Tahiti, and the Great Barrier Reef (IODP Expeditions 313, 310, and 325, respectively). Additional drilling of carbonate clinoforms is also required to evaluate how the response of such systems to sea-level change differs from that of siliciclastic systems. Finally, drilling on active margins should be explored to take advantage of very high rates of sediment supply that create expanded sections and allow study of eustatic cycles at orbital forcing frequencies, as well as illuminating the interplay of eustatic and tectonic forcing.

## Acknowledgements

We thank the crew of the D/V *JOIDES Resolution* and all drilling operations personnel, who overcame the considerable difficulties associated with drilling deep boreholes in shallow water to ensure the success of Expedition 317. Many thanks are also due to the U.S. Implementing Organization technicians who are such a vital part of the drilling program. The authors also thank three anonymous reviewers for their helpful and insightful input.

## IODP Expedition 317 Scientists

Craig Fulthorpe (Co-Chief Scientist), Koichi Hoyanagi (Co-Chief Scientist), Peter Blum (Staff Scientist), Stacie Blair, Gregory Browne, Robert Carter, Maria-Cristina Ciobanu, George Claypool, Martin Crundwell, Jaime Dinarès-Turrell, Xuan Ding, Simon George, Gilles Guèrin, Daniel Hepp, John Jaeger, Shungo Kawagata, David Kemp, Young-Gyun Kim, Michelle Kominz, Helen Lever, Julius Lipp, Kathleen Marsaglia, Cecilia McHugh, Naomi Murakoshi, Takeshi Ohi, Laura Pea, Julie Pollard, Mathieu Richaud, Angela Slagle, Itsuki Suto, Susumu Tanabe, Kirsteen Tinto, Goichiro Uramoto, and Toshihiro Yoshimura

## References

Abbott, S.T., and Carter, R.M., 1999. Stratigraphy of the Castlecliffian type section: Ten mid-Pleistocene sequences from the Wanganui coast, *New Zeal. J. Geol. Geop.*, 42:91–111, doi:10.1080/00288306.1999.9514834.

Abbott, S.T., Naish, T.R., Carter, R.M., and Pillans, B.J., 2005. Sequence stratigraphy of the Nukumaru stratotype (Pliocene-Pleistocene, c. 2.08–1.63 Ma) Wanganui Basin, New Zealand. *J. Roy. Soc. New Zeal.*, 35:123–150, doi:10.1080/03014223.2005.9517779.

Adams, C.J.D., 1979. Age and origin of the Southern Alps. *In* Walcott, R.I., and Cresswell, M.M. (Eds.), *The Origin of the Southern Alps, Royal Society of New Zealand Bulletin*, 18:73–78.

Austin, J.A., Jr., Christie-Blick, N., Malone, M.J., Mountain, G.S., and the Leg 174A Shipboard Party, 1998. *Proc. ODP, Init. Repts.*, 174A: College Station, TX (Ocean Drilling Program).

Batt, G.E., Braun, J., Kohn, B.P., and McDougall, I., 2000. Thermochronological analysis of the dynamics of the Southern Alps, New Zealand. *Geol. Soc. Am. Bull.*, 112:250–266, doi:10.1130/0016-7606(2000)112<250:TAOTDO>2.0.CO;2.

Carter, L., Carter, R.M., McCave, N.I., and Gamble, J., 1996. Regional sediment recycling in the abyssal Southwest Pacific Ocean. *Geology*, 24:735–738, doi:10.1130/0091-7613(1996)024<0735:RSRITA>2.3.CO;2.

Carter, R.M., 1985. The mid-Oligocene Marshall Paraconformity, New Zealand: Coincidence with global eustatic fall or rise? *J. Geol.*, 93:359–371, doi:10.1086/628957.

Carter, R.M., and Landis, C.A., 1972. Correlative Oligocene unconformities in southern Australasia. *Nature*, 237:12–13.

Carter, R.M., and Norris, R.J., 1976. Cainozoic history of southern New Zealand: An accord between geological observations and plate-tectonic predictions. *Earth Planet. Sci. Lett.*, 31:85–94, doi:10.1016/0012-821X(76)90099-6.

Carter, R.M., Abbott, S.T., Fulthorpe, C.S., Haywick, D.W., and Henderson, R.A., 1991. Application of global sea-level and sequence stratigraphic models in southern hemisphere Neogene strata from New Zealand. *In* MacDonald, D. (Ed.), *Sea-level and Active Plate Margins. International Association of Sedimentologists Spec. Pub.*, 12:41–65.

Carter, R.M., McCave, I.N., and Carter, L., 2004. Leg 181 synthesis: Fronts, flows, drifts, volcanoes, and the evolution of the southwestern gateway to the Pacific Ocean, Eastern New Zealand. *In* Richter, C. (Ed.), *Proc. ODP Scientific Results*, 181, College Station, TX (Ocean Drilling Program), 1–111.

Chiswell, S.M., 1996. Variability in the Southland Current, New Zealand. *New Zeal. J. Mar. Fresh. Res.*, 30:1–17, doi:10.1080/00288330.1996.9516693.

Christie-Blick, N., 1991. Onlap, offlap, and the origin of unconformity-bounded depositional sequences. *Mar. Geol.*, 97:35–56, doi:10.1016/0025-3227(91)90018-Y.

Field, B.D., and Browne, G.H., 1989. Cretaceous and Cenozoic sedimentary basins and geological evolution of the Canterbury Region, South Island, New Zealand. *New Zeal. Geol. Surv. Bull.*, 2, 94 p.

Fleming, C.A., 1962. New Zealand biogeography: A paleontologist's approach. *Tuatara*, 10:53–108.

Fulthorpe, C.S., and Carter, R.M., 1991. Continental shelf progradation by sediment drift accretion. *Geol. Soc. Am,. Bull.*, 103:300–309, doi:10.1130/0016-7606(1991)103<0300:CSPBSD>2.3.CO;2.

Fulthorpe, C.S., Carter, R.M., Miller, K.G., and Wilson, J., 1996. Marshall Paraconformity: A mid-Oligocene record of inception of the Antarctic Circumpolar Current and coeval glacio-eustatic lowstand? *Mar. Petrol. Geol.*, 13:61–77, doi:10.1016/0264-8172(95)00033-X.

Fulthorpe, C.S., Hoyanagi, K., Blum, P., and the Expedition 317 Scientists, 2011. *Proc. IODP*, 317: Tokyo (Integrated Ocean Drilling Program Management International, Inc.), doi:10.2204/iodp.proc.317.2011.

Haq, B.U., Hardenbol, J., and Vail, P.R., 1987. Chronology of fluctuating sea levels since the Triassic. *Science*, 235:1156–1167, doi:10.1126/science.235.4793.1156.

Kamp, P.J.J., 1987. Age and origin of the New Zealand orocline in relation to Alpine Fault movement. *J. Geol. Soc. London*, 144:641–652, doi:10.1144/gsjgs.144.4.0641.

Karner, G.D., 1986. Effects of lithospheric in-plane stress on sedimentary basin stratigraphy. *Tectonics*, 5:573–588, doi:10.1029/TC005i004p00573.

King, P.R., 2000. Tectonic reconstructions of New Zealand: 40 Ma to the Present. *New Zeal. J. Geol. Geop.*, 43:611–638, doi:10.1080/00288306.2000.9514913.

Kominz, M.A., Browning, J.V., Miller, K.G., Sugarman, P.J., Mizintseva, S., and Scotese, C.R., 2008. Late Cretaceous to Miocene sea-level estimates from the New Jersey and Delaware coastal plain coreholes: An error analysis. *Basin Res.*, 20:211–226, doi:10.1111/j.1365-2117.2008.00354.x.

Lu, H., and Fulthorpe, C.S., 2004. Controls on sequence stratigraphy of a middle-Miocene to Recent, current-swept, passive margin: Offshore Canterbury Basin, New Zealand. *Geol. Soc. Am. Bull.*, 116:1345–1366, doi:10.1130/B2525401.1.

Lu, H., Fulthorpe, C.S., and Mann, P., 2003. Three-dimensional architecture of shelf-building sediment drifts in the offshore Canterbury Basin, New Zealand. *Mar. Geol.*, 193:19–47, doi:10.1016/S0025-3227(02)00612-6.

Lu, H., Fulthorpe, C.S., Mann, P., and Kominz, M., 2005. Miocene-Recent tectonic controls on sediment supply and sequence stratigraphy: Canterbury Basin, New Zealand. *Basin Res.*, 17:311–328, doi:10.1111/j.1365-2117.2005.00266.x.

Miall, A.D, and Miall, C.E., 2001. Sequence stratigraphy as a scientific enterprise: The evolution and persistence of conflicting paradigms. *Earth Sci. Rev.* 54:321–348, doi:10.1016/S0012-8252(00)00041-6.

Miller, K.G., Kominz, M.A., Browning, J.V., Wright, J.D., Mountain, G.S., Katz, M.E., Sugarman, P.J., Cramer, B.S., Christie-Blick, N., and Pekar, S.F., 2005. The Phanerozoic record of global sea-level change. *Science*, 310:1293–1298, doi:10.1126/science.1116412.

Mitchum, R.M., Vail, P.R., and Thompson, S., 1977. Seismic stratigraphy and global changes of sea level, part 2: The depositional sequence as a basic unit for stratigraphic analysis. *In* Payton, C.E. (Ed.), *Seismic Stratigraphy - Applications to Hydrocarbon Exploration, AAPG Memoir*, 26:53–62.

Morris, M., Stanton, B., and Neil, H., 2001. Subantarctic oceanography around New Zealand: Preliminary results from an ongoing survey. *New Zeal. J. Mar. Fresh. Res.*, 35:499–519, doi:10.1080/00288330.2001.9517018.

Mountain, G., Proust, J.-N., and the Expedition 313 Science Party, 2010. The New Jersey margin scientific drilling project (IODP Expedition 313): Untangling the record of global and local sea-level changes. *Sci. Drill.*, 10:26–34, doi:10.2204/iodp.sd.10.03.2010.

Naish, T.M., Wehland, F., Wilson, G.S., Browne, G.H., Cook, R.A., Morgans, H.E., Rosenberg, M., King, P.R., Smale, D., Nelson, C.S., Kamp, P.J.J., and Ricketts, B., 2005. An integrated sequence stratigraphic, paleonenvironmental, and chronostratigraphic analysis of the Tangahoe Formation, southern Taranaki coast, with implications for mid-Pliocene (c. 3.4-3.0 Ma) glacio-eustatic sea-level changes. *J. Roy. Soc. New Zeal.*, 35:151–196, doi:10.1080/03014223.2005.9517780.

Norris, R.J., Carter, R.M., and Turnbull, I.M., 1978. Cainozoic sedimentation in basins adjacent to a major continental transform boundary in southern New Zealand. *J. Geol. Soc. London*, 135:191–205, doi:10.1144/gsjgs.135.2.0191.

Saul, G., Naish, T.R., Abbott, S.T., and Carter, R.M., 1999. Sedimentary cyclicity in the marine Plio-Pleistocene of Wanganui Basin (N.Z.): Sequence stratigraphic motifs characteristic of the last 2.5 Ma. *Geol. Soc. Am. Bull.*, 111:524–537, doi:10.1130/0016-7606(1999)111<0524:SCITMP>2.3.CO;2.

Shipboard Scientific Party, 1999. Leg 181 summary: Southwest Pacific paleoceanography. *In* Carter, R.M., McCave, I.N., Richter, C., Carter, L., et al., *Proc. ODP Init. Repts.*, 181, College Station, TX (Ocean Drilling Program), 1–80.

Tippett, J.M., and Kamp, P.J.J., 1993a. Fission track analysis of the Late Cenozoic vertical kinematics of continental Pacific crust, South Island, New Zealand. *J. Geophys. Res.*, 98:16119–16148, doi:10.1029/92JB02115.

Tippett, J.M., and Kamp, P.J.J., 1993b. The role of faulting in rock uplift in the Southern Alps, New Zealand. *New Zeal. J. Geol. Geop.* 36:497–504, doi:10.1080/00288306.1993.9514595.

Vail, P.R., Audemard, F., Bowman, S.A., Eisner, P.N., and Perez-Cruz, C., 1991. The stratigraphic signatures of tectonics, eustasy and sedimentology—an overview. *In* Einsele, G., Ricken, W., and Seilacher, A. (Eds.), *Cycles and Events in Stratigraphy:* Berlin (Springer-Verlag), 617–659.

Wellman, H.W., 1971. Age of the Alpine Fault, New Zealand. *Proc. 22nd Int. Geol. Congr.*, 4:148–162.

## Authors

**Craig S. Fulthorpe**, Institute for Geophysics, John A. and Katherine G. Jackson School of Geosciences, University of Texas at Austin, 10100 Burnet Road (R2200), Building 196 (ROC), Austin, TX 78758-4445, U.S.A., E-mail: craig@ig.utexas.edu.

**Koichi Hoyanagi**, Department of Geology, Faculty of Science, Shinshu University, 3-1-1 Asahi, Matsumoto 390-8621, Japan, E-mail: hoya101@shinshu-u.ac.jp.

**Peter Blum**, United States Implementing Organization, Integrated Ocean Drilling Program, Texas A&M University, 1000 Discovery Drive, College Station, TX 77845, U.S.A., E-mail: blum@iodp.tamu.edu.

**and the IODP Expedition 317 Scientists**

37

# IODP Expedition 318: From Greenhouse to Icehouse at the Wilkes Land Antarctic Margin

by Carlota Escutia, Henk Brinkhuis, Adam Klaus, and the IODP Expedition 318 Scientists

## Abstract

Integrated Ocean Drilling Program (IODP) Expedition 318, Wilkes Land Glacial History, drilled a transect of sites across the Wilkes Land margin of Antarctica to provide a long-term record of the sedimentary archives of Cenozoic Antarctic glaciation and its intimate relationships with global climatic and oceanographic change. The Wilkes Land drilling program was undertaken to constrain the age, nature, and paleoenvironment of the previously only seismically inferred glacial sequences. The expedition (January–March 2010) recovered ~2000 meters of high-quality middle Eocene–Holocene sediments from water depths between 400 m and 4000 m at four sites on the Wilkes Land rise (U1355, U1356, U1359, and U1361) and three sites on the Wilkes Land shelf (U1357, U1358, and U1360).

These records span ~53 million years of Antarctic history, and the various seismic units (WL-S4–WL-S9) have been successfully dated. The cores reveal the history of the Wilkes Land Antarctic margin from an ice-free "greenhouse" Antarctica, to the first cooling, to the onset and erosional consequences of the first glaciation and the subsequent dynamics of the waxing and waning ice sheets, all the way to thick, unprecedented "tree ring style" records with seasonal resolution of the last deglaciation that began ~10,000 y ago. The cores also reveal details of the tectonic history of the Australo-Antarctic Gulf from 53 Ma, portraying the onset of the second phase of rifting between Australia and Antarctica, to ever-subsiding margins and deepening, to the present continental and ever-widening ocean/continent configuration.

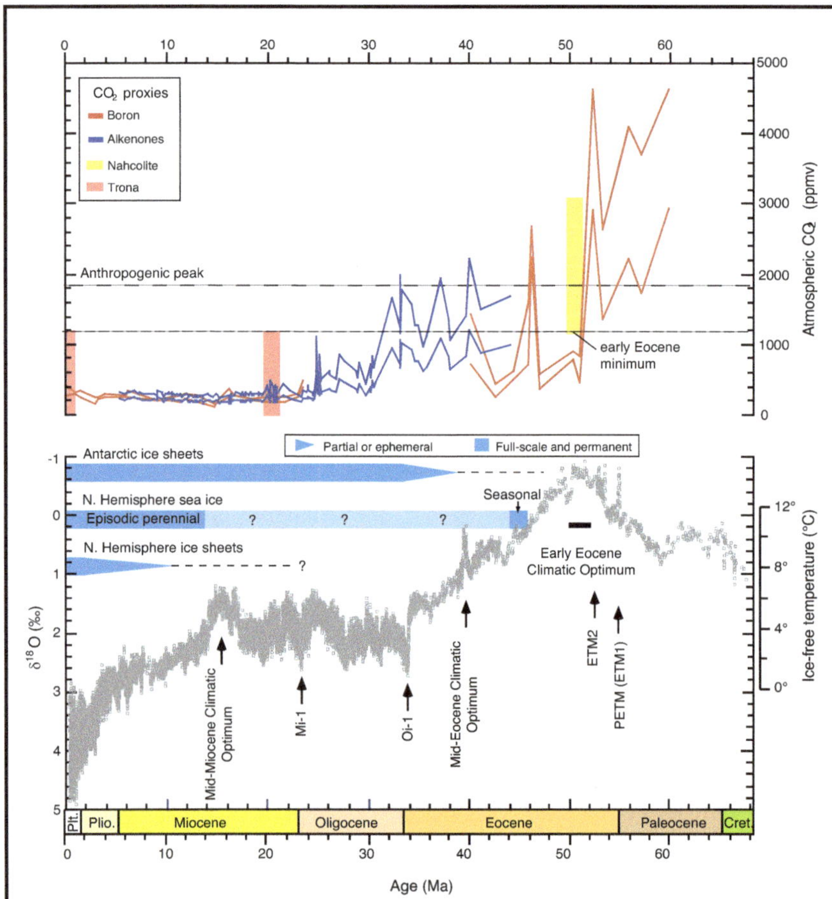


Figure 1. Updated Cenozoic pCO2 and stacked deep-sea benthic foraminifer oxygen isotope curve for 0 Ma to 65 Ma. NB: Updated from Zachos et al. (2008) and converted to the "Gradstein timescale" (Gradstein et al., 2004). Mi-1 = Miocene isotope Event 1, Oi-1 = Oligocene isotope Event 1, ETM2 = Eocene Thermal Maximum 2, PETM = Paleocene/Eocene Thermal Maximum, ETM1 = Eocene Thermal Maximum 1.

## Introduction

Polar ice is an important component of the modern climate system, affecting among other things global sea level, ocean circulation and heat transport, marine productivity, air-sea gas exchange, and planetary albedo. The modern ice caps are, geologically speaking, a relatively young phenomenon. Since mid-Permian times (~270 Ma), parts of Antarctica became reglaciated only ~34 m.y. ago, whereas full-scale, permanent Northern Hemisphere continental ice began only ~3 m.y. ago (Zachos et al., 2008; Fig. 1). The record of Antarctic glaciation, from the time of first ice-sheet inception through the significant periods of climate change during the Cenozoic, is not only of scientific interest but also is of great importance for society. State-of-the-art climate models (DeConto and Pollard, 2003a, 2003b; Huber et al., 2004; DeConto et al., 2007; Pollard and DeConto, 2009) combined with paleoclimatic proxy data (Pagani et al., 2005) suggest that the main triggering mechanism for inception and development of the

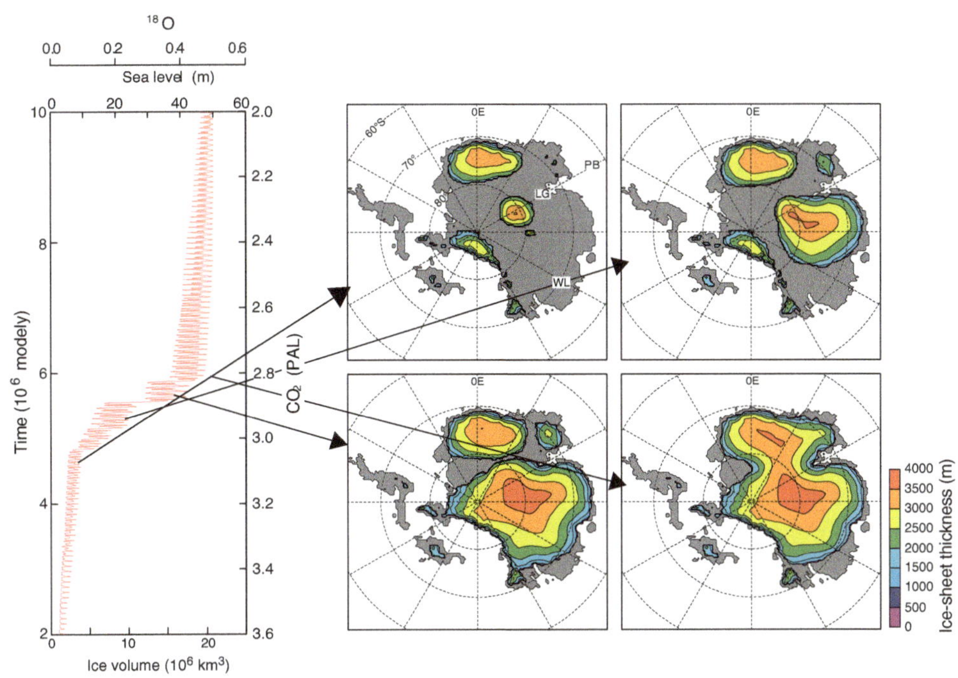


Figure 2. Simulated initiation of East Antarctic glaciation in the earliest Oligocene using a coupled general circulation model (GCM) ice-sheet model (from DeConto and Pollard, 2003a). These results are principally forced by gradual lowering of atmospheric levels of $CO_2$ in the simulated atmosphere. Note that the glaciation takes place in a "two-step" fashion reminiscent of the two-step $\delta^{18}O$ increase recorded in benthic foraminiferal carbonates across the Eocene–Oligocene transition (Coxall et al., 2005). The first step results in glaciation in the Antarctic continental interior, discharging mainly through the Lambert Graben (LG) to Prydz Bay (PB). The second step results in the initial connection and subsequent rapid expansion of the ice sheet, reaching sea level in the Wilkes Land (WL) at a later stage. PAL = Preindustrial Atmospheric Level.

Antarctic ice sheet was the decreasing levels of $CO_2$ (and other greenhouse gases) concentrations in the atmosphere (DeConto and Pollard, 2003a, 2003b; Figs 1, 2). The opening of critical Southern Ocean gateways played only a secondary role (Kennett, 1977; DeConto and Pollard, 2003a; Huber et al., 2004). With current rising atmospheric greenhouse gases resulting in rapidly increasing global temperatures (Intergovernmental Panel on Climate Change [IPCC], 2007; www.ipcc.ch/), studies of polar climates are prominent on the research agenda. Understanding Antarctic ice-sheet dynamics and stability is of special relevance because, based on IPCC (2007) forecasts, atmospheric $CO_2$ doubling and a 1.8°C–4.2°C temperature rise is expected by the end of this century. The lower values of these estimates have not been experienced on our planet since 10–15 Ma, and the higher estimates have not been experienced since before the ice sheets in Antarctica formed.

Since their inception, the Antarctic ice sheets appear to have been very dynamic, waxing and waning in response to global climate change over intermediate and even short (orbital) timescales (Wise et al., 1991; Zachos et al., 1997; Pollard and DeConto, 2009). However, not much is known about the nature, cause, timing, and rate of processes involved. Of the two main ice sheets, the West Antarctic Ice Sheet (WAIS) is mainly marine based and is considered less stable (Florindo and Siegert, 2009). The East Antarctic Ice Sheet (EAIS), which overlies continental terrains that are largely above sea level, is considered stable and is believed to respond only slowly to changes in climate (Florindo and

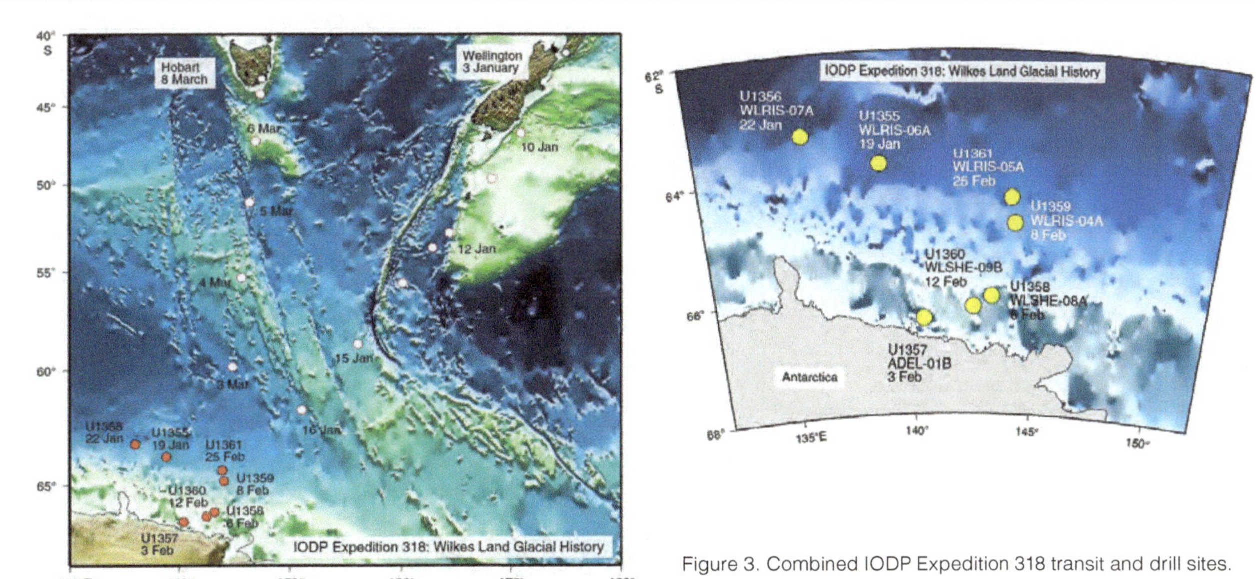


Figure 3. Combined IODP Expedition 318 transit and drill sites.

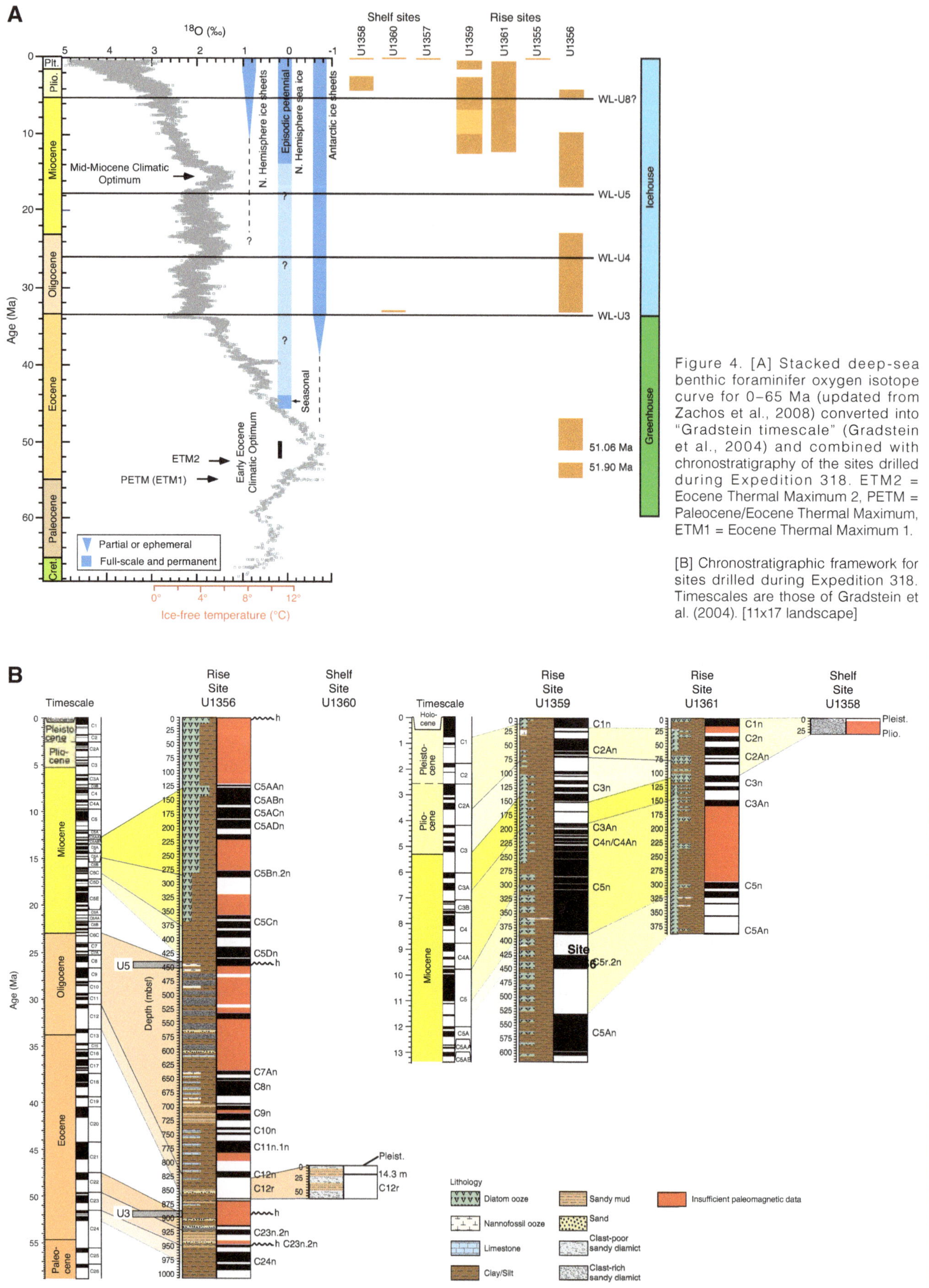


Figure 4. [A] Stacked deep-sea benthic foraminifer oxygen isotope curve for 0–65 Ma (updated from Zachos et al., 2008) converted into "Gradstein timescale" (Gradstein et al., 2004) and combined with chronostratigraphy of the sites drilled during Expedition 318. ETM2 = Eocene Thermal Maximum 2, PETM = Paleocene/Eocene Thermal Maximum, ETM1 = Eocene Thermal Maximum 1.

[B] Chronostratigraphic framework for sites drilled during Expedition 318. Timescales are those of Gradstein et al. (2004). [11x17 landscape]

Siegert, 2009). However, reports of beach gravel deposited 20 m above sea level in Bermuda and the Bahamas from 420 ka to 360 ka indicate the collapse of not only the WAIS (6 m of sea-level equivalent, SLE) and Greenland ice sheet (6 m of SLE), but possibly also 8 m of SLE from East Antarctic ice sources (Hearty et al., 1999). Therefore, during episodes of global warmth, with likely elevated atmospheric $CO_2$ conditions, the EAIS may contribute just as much or more to rising global sea level as the Greenland ice sheet. In the face of rising $CO_2$ levels (Pachauri, R.K., and Reisinger, A., 2007), a better understanding of the EAIS dynamics is therefore urgently needed from both an academic as well as a societal point of view.

A key region for analysis of the long- and short-term behavior of the EAIS is the eastern sector of the Wilkes Land margin, located at the seaward termination of the largest East Antarctic subglacial basin, the Wilkes subglacial basin. The base of the portion of the EAIS draining through the Wilkes subglacial basin is largely below sea level, suggesting that this portion of the EAIS can potentially be less stable than other areas of the EAIS (Escutia et al., 2005). Numerical models of ice-sheet behavior (Huybrechts, 1993; DeConto and Pollard, 2003a, 2003b; DeConto et al., 2007; Pollard and DeConto, 2009) provide a basic understanding of the climatic sensitivity of particular Antarctic regions for early ice-sheet formation, connection and expansion, and eventual development of the entire ice sheet. For example, in these models glaciation is shown to have begun in the East Antarctic interior, discharging mainly through the Lambert Graben to Prydz Bay. These models imply that the EAIS did not reach the Wilkes Land margin until a later stage. These models can only be validated through drilling and obtaining direct evidence from the sedimentary record.

## Scientific Objectives

The overall objectives of Expedition 318 were to date the identified seismic units and to obtain long-term records of Antarctic glaciation to better understand its relationships with global paleoclimate and paleoceanographic changes. Of particular interest is testing the sensitivity of the EAIS to episodes of global warming and detailed analysis of critical periods in Earth's climate history, such as the Eocene–Oligocene and Oligocene–Miocene glaciations, late Miocene, Pliocene, and the last deglaciation. During these times, the Antarctic cryosphere evolved in a step-wise fashion to ultimately assume its present-day configuration, characterized by a relatively stable EAIS. Conceivably even more important than the history of the Antarctic glaciations are past lessons of deglaciations and periods of exceptional warmth. We therefore planned to core several sequences from the Pleistocene and Pliocene that formed during interglacial intervals of exceptional warmth, periods that may provide valuable information about Antarctica's response to warming predicted in the centuries ahead. Furthermore, seismic reflection and shallow coring data indicate that the Wilkes Land margin also includes sites with ultrahigh accumulation rates of sediments that document the Holocene deglaciation and subsequent climate and sedimentological variability extending over the past 10,000 y. In general, our strategy was to core and analyze sedimentary records along the inshore-offshore transect to constrain the age, nature, and environments of deposition, until now only inferred from seismic surveys of the Wilkes Land continental shelf, rise, and abyssal plain (Escutia et al., 1997; De Santis et al., 2003; Escutia et al., 2005).

## The Expedition

IODP Expedition 318 (January–March 2010; Wellington, New Zealand, to Hobart, Australia), occupied seven sites (Fig. 3) across the Wilkes Land Margin at water depths between ~400 mbsl and 4000 mbsl. Together, we retrieved ~2000 meters of high-quality upper Eocene–Quaternary sedimentary cores (Fig. 4a, b). Sites U1355, U1356, U1359, and U1361 are on the Wilkes Land rise, and Sites U1358, U1360, and U1357 are on the Wilkes Land shelf. The cores span ~53 m.y. of Antarctic history, revealing the history of the Wilkes Land Antarctic margin from an ice-free "greenhouse Antarctica," to the first cooling, to the onset and erosional consequences of the first glaciation and the subsequent dynamics of the waxing and waning ice sheets (Fig. 4). Furthermore, we also were able to capture the record of the last deglaciation in terms of thick, unprecedented "tree ring style" records with annual to seasonal resolution taken in the Adélie depression (U1357) (Fig. 5).

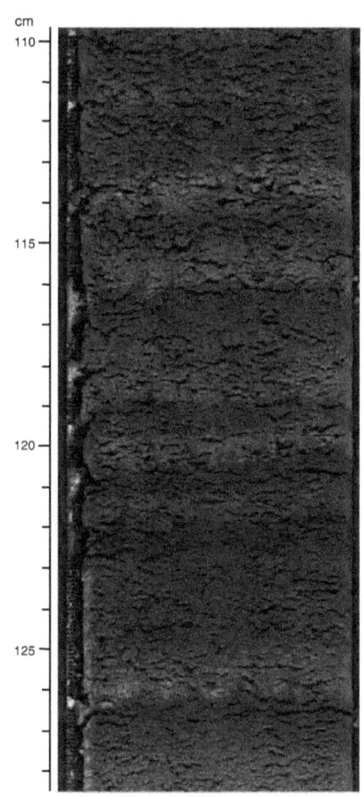


Figure 5. Example of a core section from Holocene diatomaceous ooze, Site U1357. Note the distinct seasonal laminations.

Initial studies now also reveal details of the tectonic history of the Australo-Antarctic Gulf (at 53 Ma), the onset of the second phase of rifting between Australia and Antarctica (Colwell et al., 2006; Close et al., 2009), ever-subsiding margins and deepening, to the present ocean/continent configuration. Tectonic and climatic change turned the initially shallow, broad subtropical Antarctic Wilkes Land offshore shelf into a deeply subsided basin with a narrow ice-infested margin (Fig. 6).

## "Pre-Glacial" Regional Unconformity WL-U3 and the Timing, Nature, and Consequences of the First Major Phase of EAIS Growth

Prior to the expedition, the prominent unconformity WL-U3 had been interpreted to separate pre-glacial (Eocene) strata from (Oligocene) glacial-influenced deposits (Escutia et al., 1997, 2005). Drilling and dating of WL-U3 at continental rise Site U1356 and shelf Site U1360 (Fig. 3) confirmed that this surface represents major erosion related to the onset of glaciation at ~34 Ma (early Oligocene), with immediately overlying deposits dated as 33.6 Ma (Fig. 7). Below unconformity WL-U3 at Site U1356, we recovered a record that is late early to early middle Eocene in age and that includes peak greenhouse conditions and likely some of the early Eocene hyperthermals (Fig. 8). We infer subtropical shallow-water depositional environments for this section based on dinocysts, pollen and spores, and the chemical index of alteration, among other indicators. A hiatus spanning ~2 m.y. separates the lower Eocene from the middle Eocene record at Site U1356 according to dinocyst and magnetostratigraphic evidence. This hiatus may be related to tectonic activity related to the commencement of rapid seafloor spreading in the Australia-Antarctic Basin (AAB), reported to initiate around the same time (~50 Ma; Colwell et al., 2006). Also, combined Site U1356 and ODP Leg 189 dinocyst distribution patterns suggest earliest through-flow of South Pacific Antarctic waters through the Tasmanian Gateway to be coeval with this tectonic phase. Sedimentological and microfossil information from this interval from Hole U1356A also suggest some deepening during the early middle Eocene.

At Site U1356, the upper middle Eocene to the basal Oligocene is conspicuously missing in a ~19-m.y. hiatus at ~890 bsf (~47.9–33.6 Ma) marking unconformity WL-U3 based on dinocyst and paleomagnetic evidence. Despite ongoing tectonic reorganizations, it appears likely that the erosive nature of unconformity WL-U3 is notably related to the early stages of EAIS formation. The impact of ice-sheet growth, including crustal and sea-level response, and major erosion by the ice sheets, is proposed as the principal mechanism that formed unconformity WL-U3. This is supported by the abrupt increase in benthic foraminiferal $\delta^{18}O$ values and coeval sea-level change globally recorded in complete marine successions (Oligocene isotope event Oi-1; Miller et al., 1985; Coxall et al., 2005). Progressive subsidence—the large accommodation space created by erosion in the margin (300–600 m of missing strata; Eittreim et al., 1995)—and partial eustatic recovery allowed sediments of early Oligocene age to accumulate above unconformity WL-U3.

Microfossils, sedimentology, and geochemistry of the Oligocene sediments from Site U1356, at present occupying a distal setting (i.e., lowermost rise-abyssal plain) and immediately above unconformity WL-U3, unequivocally reflect icehouse environments with evidence of iceberg activity (dropstones) and at least seasonal sea-ice cover. The sediments, dominated by hemipelagic sedimentation with bottom current and gravity flow influence, as well as biota, indicate deeper water settings relative to the underlying middle Eocene environments. These findings imply significant crustal stretching, subsidence of the margin, and deepening of the Tasman Rise and the Adélie Rift Block (ARB) between 47.9 Ma and 33.6 Ma (Fig. 6).

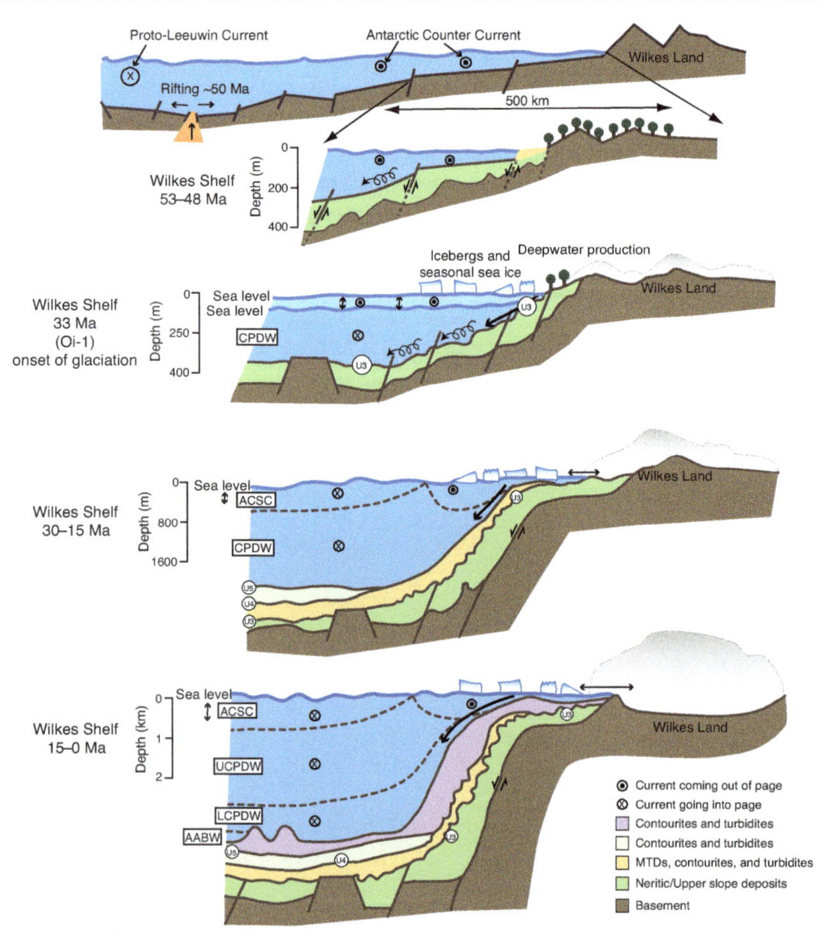


Figure 6. Conceptual illustration of tectonic, geological, sedimentological, and climatic evolution of the Wilkes Land margin since the middle early Eocene (~54 Ma). U3, U4, and U5 refer to seismic unconformities WL-U3, WL-U4, and WL-U5, respectively. Oi-1 = Oligocene isotope Event 1, CPDW = Circumpolar Deep Water, ACSC = Antarctic Circumpolar Surface Water, UCPDW = Upper Circumpolar Deep Water, LCPDW = Lower Circumpolar Deep Water, AABW = Antarctic Bottom Water. MTD = mass transport debris flows.

## Record of EAIS Variability

Drilling at continental rise Site U1356 also recovered a thick section of Oligocene to upper Miocene sediments (Figs. 4, 8) indicative of a relatively deep-water, sea-ice-influenced setting. Oligocene to upper Miocene sediments are indicative of episodically reduced oxygen conditions either at the seafloor or within the upper sediments prior to ~17 Ma. From the late early Miocene (~17 Ma) onward, progressive deepening and possible intensification of deep-water flow and circulation led to a transition from a poorly oxygenated low-silica system (present from the early to early middle Eocene to late early Miocene) to a well-ventilated silica-enriched system akin to the modern Southern Ocean. This change coincides with one of the major regional unconformities in the Wilkes Land margin, unconformity WL-U5, which represents a ~3 m.y. latest Oligocene–early Miocene hiatus (Figs. 4, 8). This unconformity marks a change in the dominant sedimentary processes at this site, which are dominated by mass transport processes below the unconformity and by hemipelagic, turbidity flow, and bottom-current deposition above.

A complete record with good recovery of late Miocene to Pleistocene deposits was achieved at continental rise Sites U1359 and U1361 (Figs. 4, 9), drilled on levee deposits bounding turbidity channels. We successfully dated the seismic units between unconformities WL-U6 and WL-U8, and the sedimentological, wireline logging, and magnetic susceptibility data exhibit relatively high amplitude variations, indicating strong potential for this record to reveal EAIS dynamics down to orbital timescales (100 k.y. and 40 k.y. cyclicity) (Fig. 10). This cyclicity likely documents the successive advances and retreats of the ice-sheet and sea-ice cover, as well as the varying intensity of cold saline density flows related to bottom water production at the Wilkes Land margin (e.g., high-salinity shelf water flowing from the shelf into the deep ocean to form Antarctic Bottom Water [AABW]). In general, typical Southern Ocean open cold-water taxa, with variable abundances of sea-ice-associated diatoms were recovered, indicating a high-nutrient, high-productivity sea-ice-influenced setting throughout the Neogene. Combined sedimentological and microfossil information indicates the ever-increasing influence of typical Antarctic Counter Current surface waters and intensifying AABW flow. Furthermore, the preservation of calcareous microfossils in several intervals indicates times when bottom waters were favorable to the preservation of calcium carbonate. These observations point to a very dynamic ice-sheet/sea-ice regime during the late Miocene through the Pleistocene. Detailed postcruise studies in sediments from the late Neogene will provide a history of glacial-interglacial climate and paleoceanographic variability, including a history of AABW production that can be linked to sea-ice variations in this margin.

## Ultrahigh Resolution Holocene Record of Climate Variability

Coring at Site U1357 yielded a 186-m section of continuously laminated diatom ooze as well as a portion of the underlying Last Glacial Maximum diamict. Based on much shorter piston cores recovered from adjacent basins and banks, the onset of marine sedimentation during the deglacial interval began between 10,400 y and 11,000 y ago. The site was triple cored, providing overlapping sequences that will aid in the construction of a composite stratigraphy spanning at least the last 10,000 y. The Site U1357 sediments are unusual for Antarctic shelf deposits because of their high accumulation rate (2 cm $y^{-1}$), lack of bioturbation, and excellent preservation of organic matter as well as calcareous, opaline, phosphatic, and organic fossils. The sediments are profoundly anoxic, with levels of $H_2S$ as high as 42,000 ppm at 20 mbsf. Larger burrowing organisms are completely

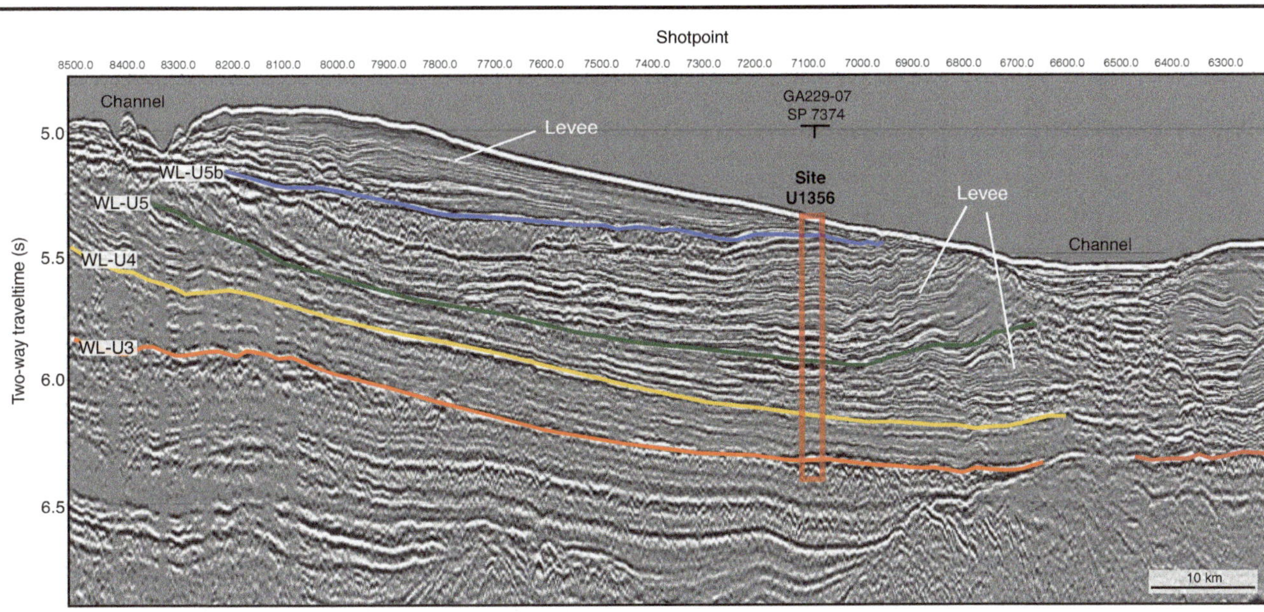


Figure 7. Multichannel seismic reflection profile across Site U1356 showing regional unconformities WL-U3, WL-U4, and WL-U5. Red rectangle = approximate penetration achieved at Site U1356.

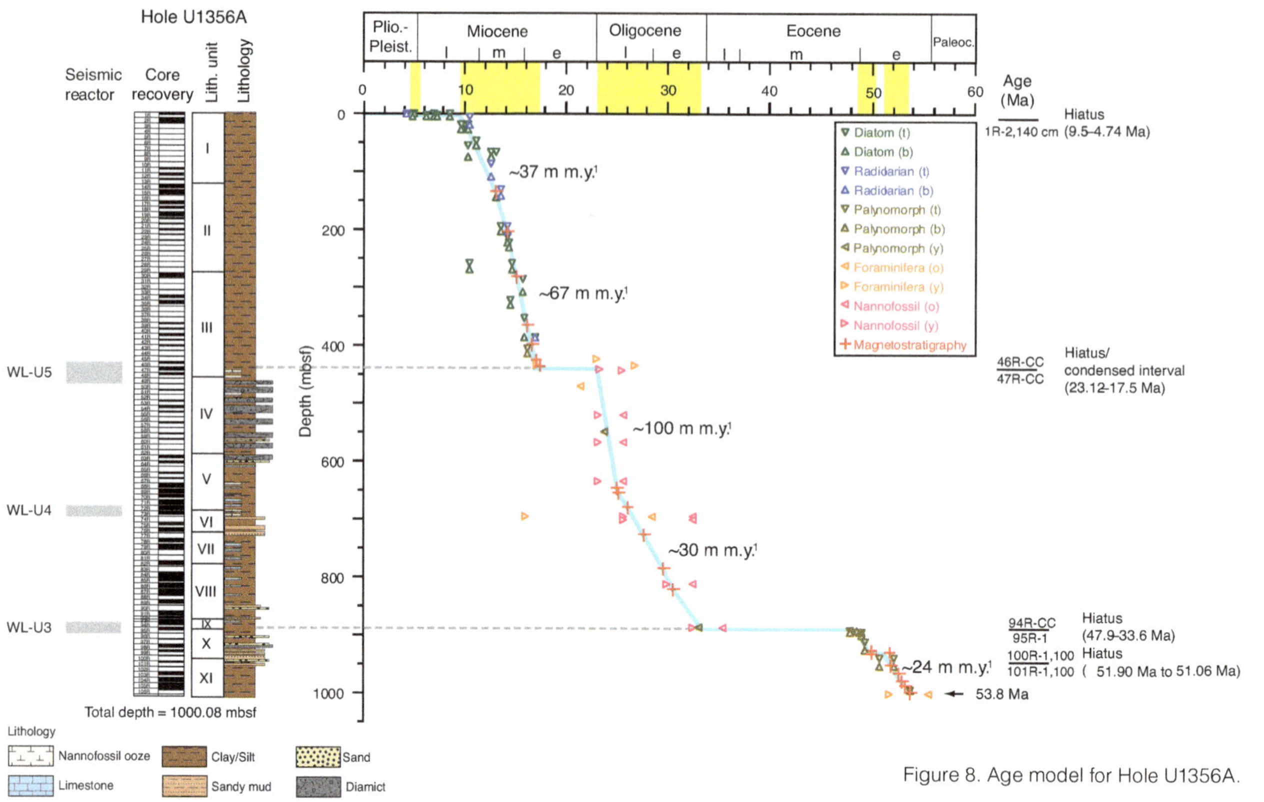


Figure 8. Age model for Hole U1356A.

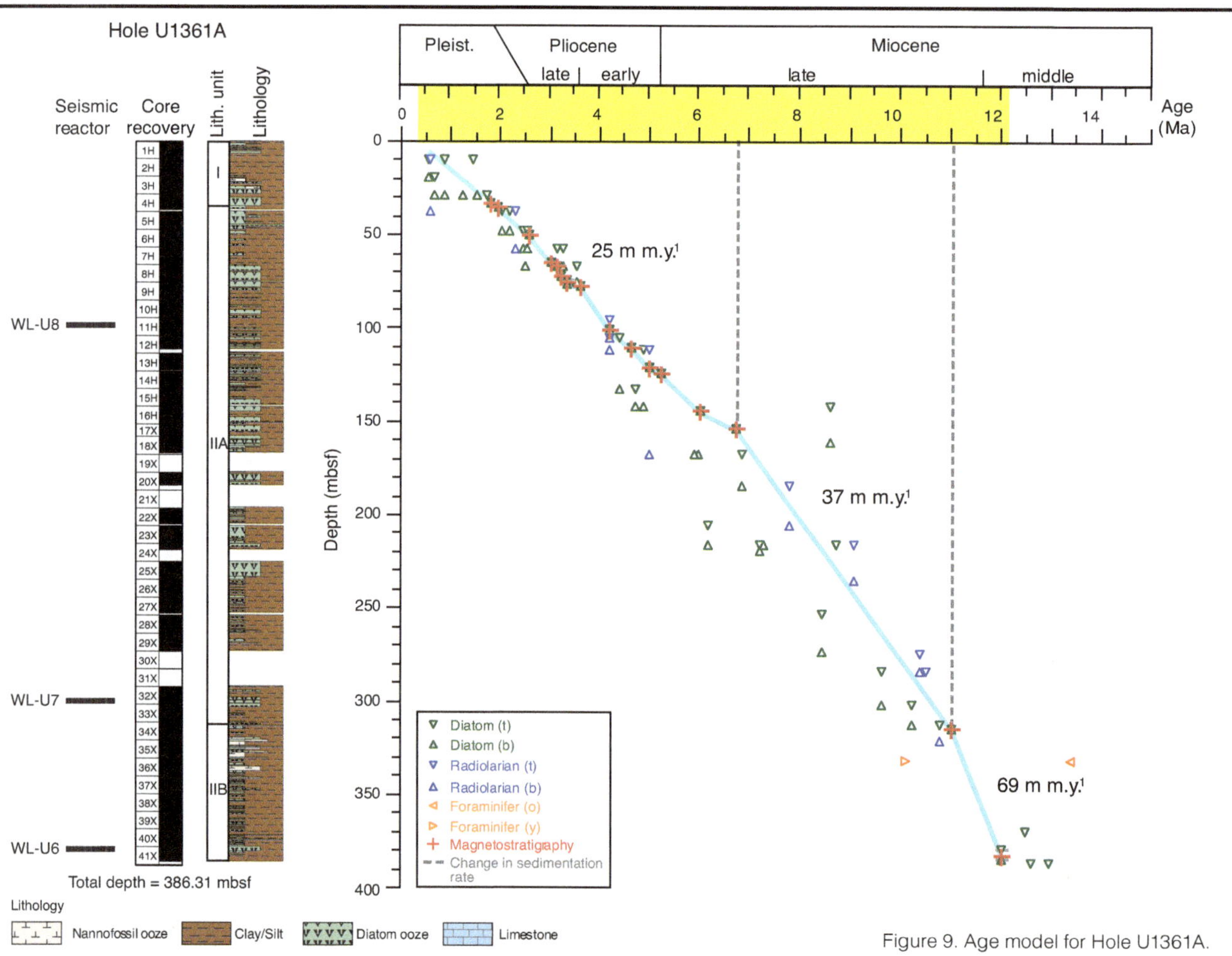


Figure 9. Age model for Hole U1361A.

excluded from this ecosystem, yet the regular occurrence of benthic foraminifers suggests that some oxygen is present at the sediment-water interface. These sediments provide an excellent sample set for geomicrobiology and sedimentary geochemistry studies. In fact, the upper 20 m of one of the three holes was intensively sampled for integrated pore water and microbiological studies.

A paramount achievement from a paleoclimatic standpoint was the retrieval of this continuously laminated deposit (Fig. 5). Spot checks of laminae from top to bottom of the split Hole U1357A sections suggest that paired light-dark laminae sets range in thickness from ~1 cm to 3 cm. Based on radiocarbon dating of a piston core taken earlier from this site (Costa et al., 2007), our own preliminary secular paleomagnetic findings, and the thickness of the deposit combined with the expected age at its base, it is very likely each laminae pair represents one year. If supported by our shore-based research, this will be the first varved sedimentary sequence extending through the Holocene recovered from the Southern Ocean. Analysis at the annual timescale will permit us to examine decadal to subdecadal variability in sea ice, temperature, and wind linked to the Southern Annual Mode (SAM), Pacific Decadal Variability, and possibly ENSO. We will also be able to address questions regarding rates of change during the Hypsithermal Holocene neoglacial events and the time immediately following the first lift-off and pull-back of ice at the end of the last glacial interval. In addition, we now have an excellent opportunity for ultrahigh resolution correlation to the nearby Law Dome Ice Core, one of the most important Holocene ice cores in Antarctica.

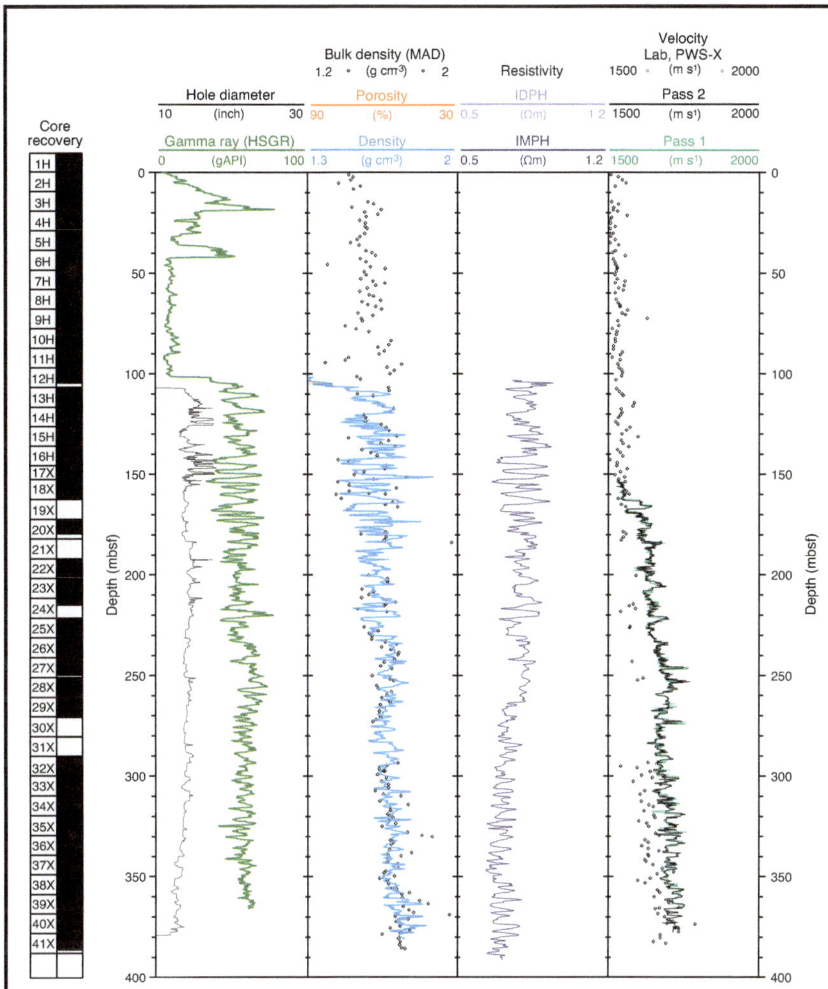


Figure 10. Downhole geophysical logs, Hole U1361A. HSGR = total spectral gamma ray, MAD = moisture and density, IDPH = deep induction phasor-processed resistivity, IMPH = medium induction phasor-processed resistivity, PWS-X = x-direction caliper.

## Acknowledgements

We thank the captain and crew of the *JOIDES Resolution*, the IODP 318 operation superintendent, ice pilot, weatherman, all technicians, and videographer who were instrumental in the success of Expedition 318. They allowed and assisted us in drilling, documenting, sampling, and onboard sample analyses during Expedition 318. Numerous people at IODP-TAMU as well as USIO provided their dedicated effort supporting us, including preparation of the expedition and publication of the proceedings. The USIO curatorial team provided us with their able support during the sampling party at College Station. We also thank the co-PIs of the Proposal 482 and APL 638 and the master of the R/V *Astrolabe* for his support.

## The IODP Expedition 318 Scientists

C. Escutia (Chief Scientist), H. Brinkhuis (Chief Scientist), A. Klaus (Staff Scientist), J.A.P. Bendle, P.K. Bijl, S.M. Bohaty, S.A. Carr, R.B. Dunbar, J.J. Gonzàlez, A. Fehr, T.G. Hayden, M. Iwai, F.J. Jimemez-Espejo, K. Katsuki, G.S. Kong, R.M. McKay, M. Nakai, M.P. Olney, S. Passchier, S.F. Pekar, J. Pross, C. Riesselman, U. Röhl, T. Sakai, P.K. Shrivastava, C.E. Stickley, S. Sugisaki, L. Tauxe, S. Tuo, T. van de Flierdt, K. Welsh, T. Williams, M. Yamane .

## References

Close, D.I., Watts, A.B., and Stagg, H.M.J., 2009. A marine geophysical study of the Wilkes Land rifted continental margin, Antarctica. *Geophys. J. Int.*, 177(2):430–450, doi:10.1111/j.1365-246X.2008.04066.x.

Colwell, J.B., Stagg, H.M.J., Direen, N.G., Bernander, G., and Borisova, I., 2006. The structure of the continental margin off Wilkes Land and Terre Adelie Coast, East Antarctica. *In* Futterer, D.K., Damaske, D., Kleinschmidt, G., Miller, H., and Tessensohn, F. (Eds.), *Antarctica: Contributions to Global Earth Sciences,* Berlin (Springer-Verlag), 327–340.

Costa, E., Dunbar, R.B., Kryc, K.A., Mucciarone, D.A., Brachfeld, S., Roark, E.B., Manley, P.L., Murray, R.W., and Leventer, A., 2007. Solar forcing and El Niño-Southern Oscillation (ENSO) influences on productivity cycles interpreted from a late Holocene high-resolution marine sediment record, Adélie Drift, East Antarctic margin. *In* Cooper, A.K., and Raymond, C.R., and the ISAES Editorial Team. (Eds.), *Antarctica: A Keystone in a Changing World - Proceedings for the Tenth International Symposium on Antarctic Earth Sciences.* USGS Open File Rep., 2007–1047. Washington, DC (The National Academies Press), 1–6, doi:10.3133/of2007-1047.srp036.

Coxall, H.K., Wilson, P.A., Pälike, H., Lear, C.H., and Backman, J., 2005. Rapid stepwise onset of Antarctic glaciation and deeper calcite compensation in the Pacific Ocean. *Nature,* 433(7021):53–57, doi:10.1038/nature03135.

DeConto, R.M., and Pollard, D., 2003a. A coupled climate–ice sheet modeling approach to the early Cenozoic history of the Antarctic ice sheet. *Palaeogeogr., Palaeoclimatol., Palaeoecol.,* 198(1–2):39–52, doi:10.1016/S0031-0182(03)00393-6.

DeConto, R.M., and Pollard, D., 2003b. Rapid Cenozoic glaciation of Antarctica induced by declining atmospheric $CO_2$. *Nature,* 421(6920):245–249, doi:10.1038/nature01290.

DeConto, R., Pollard, D., and Harwood, D., 2007. Sea ice feedback and Cenozoic evolution of Antarctic climate and ice sheets. *Palaeoceanography,* 22(3):PA3214, doi:10.1029/2006PA001350.

De Santis, L., Brancolini, G., and Donda, F., 2003. Seismo-stratigraphic analysis of the Wilkes Land Continental Margin (East Antarctica): influence of glacially driven processes on the Cenozoic deposition. *Deep Sea Res. Part II,* 50(8–9):1563–1594, doi:10.1016/S0967-0645(03)00079-1.

Eittreim, S.L., Cooper, A.K., and Wannesson, J., 1995. Seismic stratigraphic evidence of ice-sheet advances on the Wilkes Land margin of Antarctica. *Sediment. Geol.,* 96(1–2):131–156, doi:10.1016/0037-0738(94)00130-M.

Escutia, C., De Santis, L., Donda, F., Dunbar, R.B., Cooper, A.K., Brancolini, G., and Eittreim, S.L., 2005. Cenozoic ice sheet history from East Antarctic Wilkes Land continental margin sediments. *Global Planet. Change,* 45(1–3):51–81, doi:10.1016/j.gloplacha.2004.09.010.

Escutia, C., Eittreim, S.L., and Cooper, A.K., 1997. Cenozoic sedimentation on the Wilkes Land continental rise, Antarctica. *In* Ricci, C.A. (Ed.), *The Antarctic Region: Geological Evolution and Processes. Proc. Int. Symp. Antarct. Earth Sci.,* 7:791–795.

Florindo, F., and Siegert, M., 2009. *Antarctic Climate Evolution. Developments in Earth and Environmental Sciences, Vol. 8:* Amsterdam, The Netherlands (Elsevier).

Gradstein, F.M., Ogg, J.G., and Smith, A., 2004. *A Geologic Time Scale 2004*: Cambridge (Cambridge University Press).

Hearty, P.J., Kindler, P., Cheng, H., and Edwards, R.L., 1999. A +20 m middle Pleistocene sea-level highstand (Bermuda and the Bahamas) due to partial collapse of Antarctic ice. *Geology,* 27(4):375–378, doi:10.1130/0091-7613(1999)027<0375:MMPSL>2.3.CO;2.

Huber, M., Brinkhuis, H., Stickley, C.E., Döös, K., Sluijs, A., Warnaar, J., Schellenberg, S.A., and Williams, G.L., 2004. Eocene circulation of the Southern Ocean: was Antarctica kept warm by subtropical waters? *Paleoceanography,* 19(4):PA4026, doi:10.1029/2004PA001014.

Huybrechts, P., 1993. Glaciological modelling of the Late Cenozoic East Antarctic Ice Sheet: stability or dynamism? *Geograf. Ann.*, 75(4):221–238m, doi:10.2307/521202.

Intergovernmental Panel on Climate Change (IPCC), The AR4 Synthesis Report 2007: http://www.ipcc.ch/

Kennett, J.P., 1977. Cenozoic evolution of Antarctic glaciation, the circum-Antarctic Ocean, and their impact on global paleoceanography. *J. Geophys. Res.,* 82(27):3843–3860, doi:10.1029/JC082i027p03843.

Miller, K.G., Aubry, M.-P., Kahn, M.J., Melillo, A.J., Kent, D.V., and Berggren, W.A., 1985. Oligocene–Miocene biostratigraphy, magnetostratigraphy, and isotopic stratigraphy of the western North Atlantic. *Geology,* 13(4):257–261, doi:10.1130/00917613(1985)13<257:OBMAIS>2.0.CO;2.

Pagani, M., Zachos, J.C., Freeman, K.H., Tipple, B., and Bohaty, S., 2005. Marked decline in atmospheric carbon dioxide concentrations during the Paleogene. *Science,* 309(5734):600–603, doi:10.1126/science.1110063.

Pollard, D., and DeConto, R.M., 2009. Modelling West Antarctic ice sheet growth and collapse through the past five million years. *Nature,* 458(7236):329–332, doi:10.1038/nature07809.

Wise, S.W., Jr., Breza, J.R., Harwood, D.M., and Wei, W., 1991. Paleogene glacial history of Antarctica. *In* Müller, D.W., McKenzie, J.A., and Weissert, H. (Eds.), *Controversies in Modern Geology: Evolution of Geological Theories in Sedimentology, Earth History and Tectonics:* Cambridge (Cambridge University Press), 133–171.

Zachos, J.C., Dickens, G.R., and Zeebe, R.E., 2008. An early Cenozoic perspective on greenhouse warming and carbon-cycle dynamics. *Nature,* 451(7176):279–283, doi:10.1038/nature06588.

Zachos, J.C., Flower, B.P., and Paul, H., 1997. Orbitally paced climate oscillations across the Oligocene/Miocene boundary. *Nature,* 388(6642):567–570, doi:10.1038/41528.


## Authors


**Carlota Escutia,** Instituto Andaluz de Ciencias de la Tierra, CSIC-Universidad de Granada, Campus de Fuentenueva s/n, 18002 Granada, Spain, e-mail: cescutia@ugr.es.

**Henk Brinkhuis,** Biomarine Sciences, Institute of Environmental Biology, Laboratory of Palaeobotany and Palynology, Utrecht University, Budapestlaan 4, 3584 CD Utrecht, The Netherlands, e-mail: h.brinkhuis@uu.nl.

**Adam Klaus,** Staff Scientist/Expedition Project Manager, United States Implementing Organization, Integrated Ocean Drilling Program, Texas A&M University, 1000 Discovery Drive, College Station, TX 77845-9547, U.S.A., e-mail: aklaus@iodp.tamu.edu.

**and the IODP Expedition 318 Scientists**


## Related Web Link

http://publications.iodp.org/preliminary_report/318/
http://www.stratigraphy.org/

# 38

## IODP Expedition 324: Ocean Drilling at Shatsky Rise Gives Clues about Oceanic Plateau Formation

by William W. Sager, Takashi Sano, Jörg Geldmacher, and the IODP Expedition 324 Scientists

### Abstract

Integrated Ocean Drilling Program (IODP) Expedition 324 cored Shatsky Rise at five sites (U1346–U1350) to study processes of oceanic plateau formation and evolution. Site penetrations ranged from 191.8 m to 324.1 m with coring of 52.6 m to 172.7 m into igneous basement at four of the sites. Average recovery in basement was 38.7%–67.4%. Cored igneous sections consist mainly of variably evolved tholeiitic basalts emplaced as pillows or massive flows. Massive flows are thickest and make up the largest percentage of section on the largest and oldest volcano, late Jurassic age Tamu Massif; thus, it may have formed at high effusion rates. Such massive flows are characteristic of flood basalts, and similar flows were cored at Ontong Java Plateau. Indeed, the similarity of igneous sections at Site U1347 with that cored on Ontong Java Plateau implies similar volcanic styles for these two plateaus. On younger, smaller Shatsky Rise volcanoes, pillow flows are common and massive flows thinner and fewer, which might mean volcanism waned with time. Cored sediments from summit sites contain fossils and structures implying shallow water depths or emergence at the time of eruption and normal subsidence since. Summit sites also show pervasive alteration that could be due to high fluid fluxes. A thick section of volcaniclastics cored on Tamu Massif suggests that shallow, explosive submarine volcanism played a significant role in the geologic development of the plateau summit. Expedition 324 results imply that Shatsky Rise began with massive eruptions forming a huge volcano and that subsequent eruptions waned in intensity, forming volcanoes that are large, but which did not erupt with unusually high effusion rates. Similarities of cored sections on Tamu Massif with those of Ontong Java Plateau indicate that these oceanic plateaus formed in similar fashion.

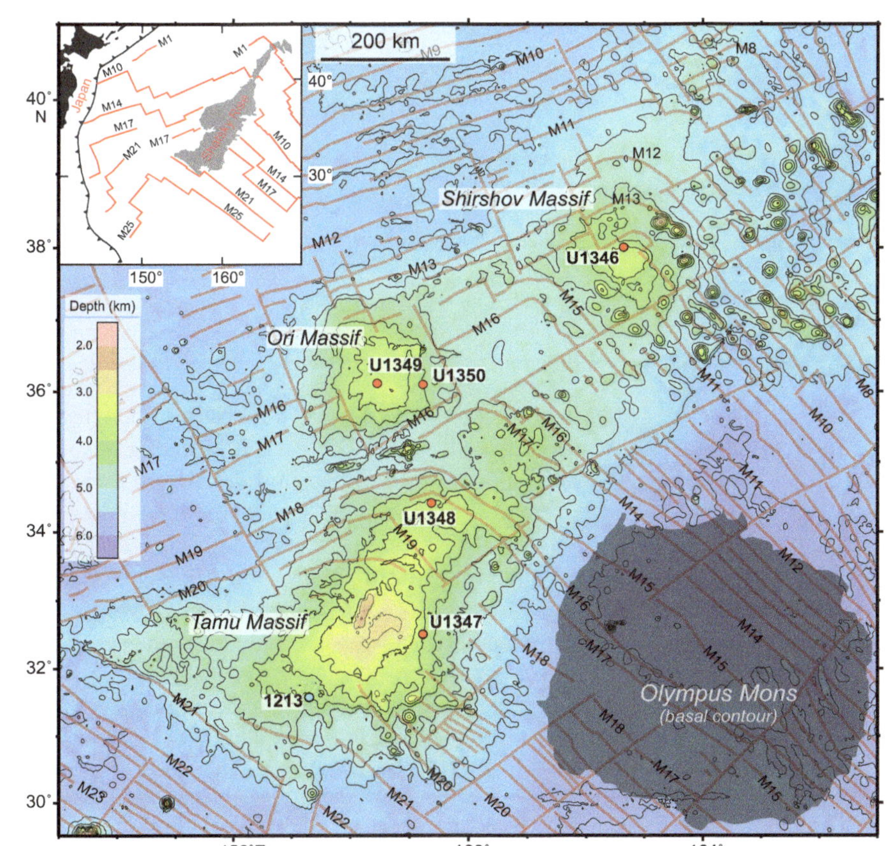


Figure 1. Shatsky Rise, surrounding magnetic lineations and fracture zones (heavy red lines; Nakanishi et al., 1999), Expedition 324 drill sites (red dots) and ODP Site 1213 (blue dot). Dark area at lower right shows the basal contour of Olympus Mons (Mars) at the same scale. Inset shows the location of Shatsky Rise relative to Japan.

### Introduction

Large outpourings of basalt have occurred widely around the globe (Coffin and Eldholm, 1994). The greatest of these "large igneous provinces" (LIPs) produced continental flood basalts on land and oceanic plateaus under the sea (Duncan and Richards, 1991), the largest with volumes in millions of cubic kilometers. Plateau-building magmas must come from the mantle, but source depth and emplacement mechanisms are unclear. A widely held view is that plateaus form when the head of a nascent mantle plume—a huge thermal diapir from the lower mantle—rises to the surface, causing massive volcanic eruptions and sub-crustal intrusion (Duncan and Richards, 1991). An alternative explanation is that plateau eruptions come from decompression melting of unusually fusible upper mantle at plate edges (Foulger, 2007). This explanation is bolstered by the observation that many plateaus formed near ocean ridges, especially at triple junctions, a situation that is unlikely if the volcanism results from deep mantle convection independent of plate tectonics (Sager, 2005).

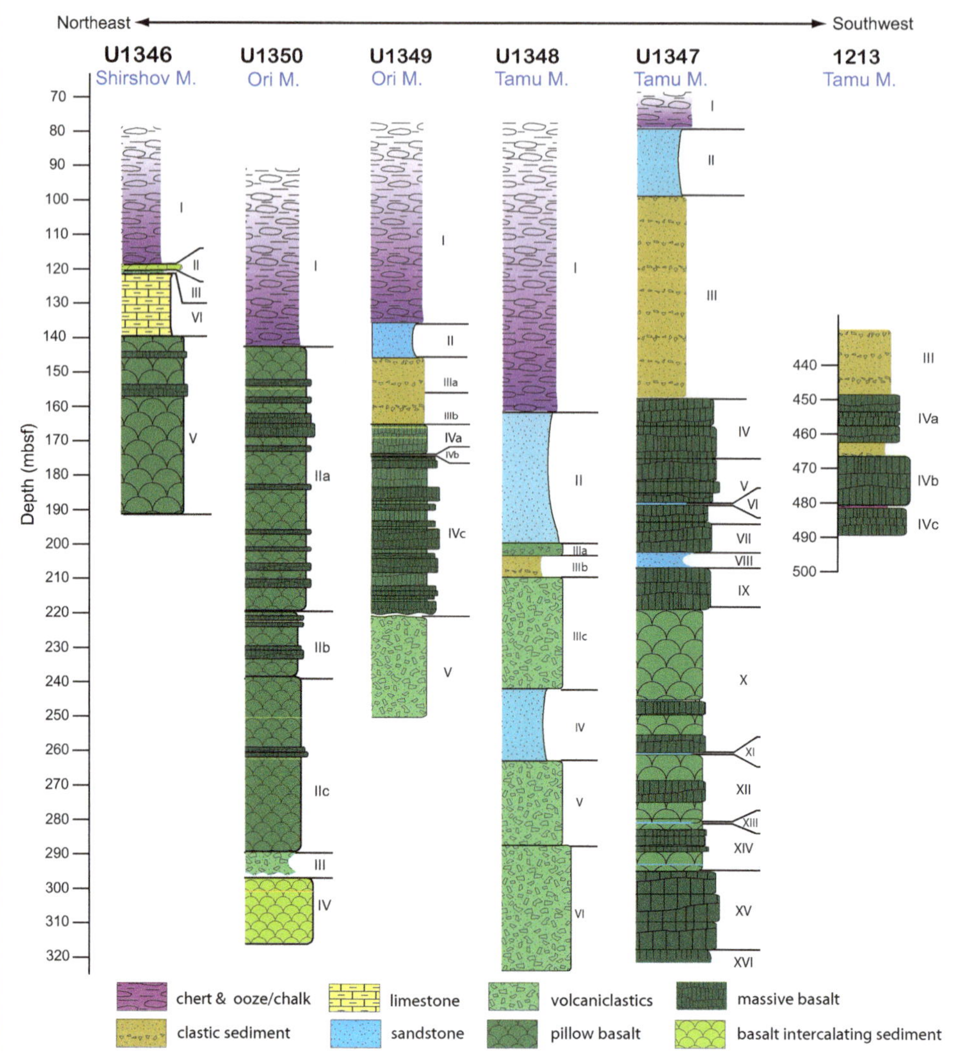


Figure 2. Lithology of Shatsky Rise drill sites, showing both basal sediment and igneous sections. Sites are arranged in transect order, from northeast (left) to southwest (right). Site 1213 from ODP Leg 198 (Shipboard Scientific Party, 2002) is shown at the same scale for completeness. Roman numerals indicate lithologic units, which are described in the Expedition Reports (Sager et al., 2010).

Aside from uncertainty about source mechanism, the processes of ocean plateau volcanism are poorly understood because few plateaus have been sampled extensively. Massive eruptions are deduced mainly from the size of plateau edifices coupled with inferred short eruption periods (Coffin and Eldholm, 1994). In addition, it is suggested that low oceanic plateau flank slopes imply high eruption rates and long flows similar to continental flood basalts (Keszthelyi and Self, 1998; Sager et al., 1999). Ocean Drilling Program (ODP) cruises have sampled the largest oceanic plateaus, Kerguelen (Leg 183; Coffin et al., 2002) and Ontong Java (Leg 192; Mahoney et al., 2001), and they have found that massive lava flows are common. A problem for these studies is that these plateaus formed mostly during the Cretaceous Quiet Period when there were no magnetic reversals to record the positions of mid-ocean ridges, so it is unclear how plate boundaries may have factored into their formation (Sager et al., 1999).

In contrast, Shatsky Rise (Fig. 1) is a large Pacific plateau (area ~4.8 x $10^5$ km$^2$; Sager et al., 1999) with a clear tectonic setting. It erupted at a triple junction during the late Jurassic and Early Cretaceous and stretches over ~1700 km with three principal volcanic massifs: Tamu, Ori, and Shirshov (Fig. 1; Sager et al., 1999). Tamu Massif erupted at ~145 Ma (Mahoney et al., 2005). Other volcanic edifices must be younger than Tamu Massif because the lithosphere on which they repose is younger. Furthermore, isostatic compensation and interaction with spreading ridges imply that these edifices formed at the time the lithosphere was created, which means they are younger toward the northeast (Sager et al., 1999). Although some aspects of Shatsky Rise appear consistent with the plume head hypothesis, others are not. Emplacement apparently began with the construction of a huge volcanic edifice, Tamu Massif (~2.5 x $10^6$ km$^3$), which possibly erupted in <1 Myr, coincident with an ~800-km jump of the triple junction (Nakanishi et al., 1999; Sager et al., 1999; Sager, 2005). Tamu Massif appears to be a large, central "supervolcano" with an area similar to that of Olympus Mons on Mars (Fig. 1), which is widely considered the largest volcano in the solar system. Subsequent volcanoes (Ori and Shirshov massifs) are significantly smaller (each ~25% the size of Tamu Massif), and together with a low volcanic ridge at the north end of the plateau, they may imply a transition from plume head to tail (Sager et al., 1999). Apropos of plate-edge genesis, Shatsky Rise displays a connection with triple junction tectonics that is repeated elsewhere in the Pacific (Sager, 2005). Moreover, basalts from ODP Site 1213 (ODP Leg 198), on Tamu Massif (Fig. 1), have isotopic characteristics similar to mid-ocean ridge basalt (MORB) (Mahoney et al., 2005).

Shatsky Rise was cored on IODP Expedition 324 because it has characteristics that can be attributed to both a mantle plume and ridge tectonics; thus, it is an excellent location to investigate the formation of oceanic plateaus. Here, we present a review of initial findings and insights from cores and logs collected during Expedition 324.

## Drilling Overview

Expedition 324 drilled at five sites on Shatsky Rise (Fig. 1; Sager et al., 2010) to examine changes in volcanism along the presumed age trend. Site U1347 was cored to 317.5 m below the seafloor (mbsf), including 159.9 m of igneous basement consisting of submarine basalt flows (Fig. 2). This site and Site 1213, which penetrated 46.6 m of basalt on ODP Leg 198 (Shipboard Scientific Party, 2002), both sampled the upper flanks of Tamu Massif, the oldest edifice. Drilling at Site U1348, on the north flank of Tamu Massif, targeted a large basement high, recovering a ~120-m-thick section of highly altered volcaniclastic sediments. Farther north, sites U1349 and U1350 were situated to sample the summit and flank of Ori Massif. Drilling at Site U1349 penetrated to 250.4 mbsf and 85.3 m into igneous basement, recovering highly altered subaerial or shallow-water lava flows. At Site U1350, coring proceeded to 315.8 mbsf, with the last 172.7 m being in a section of slightly to moderately altered submarine lava flows. On the Shirshov Massif summit, Site U1346 penetrated to 191.8 mbsf, coring 52.6 m of moderately to highly altered basalt flows.

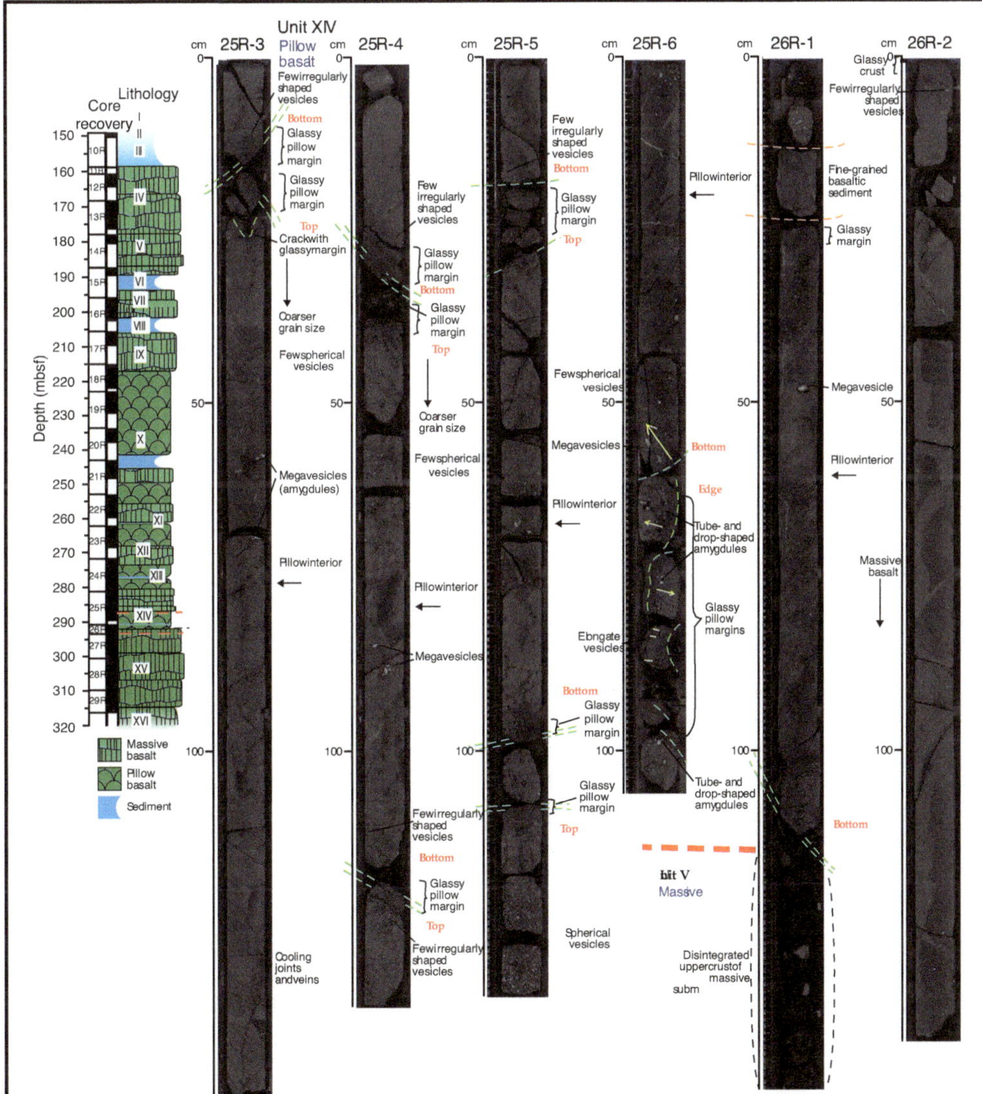


Figure 3. Core images showing pillow lavas and massive flows from Site U1347. The images are annotated with volcanological features and show a section spanning the transition between the lower pillow basalts (Unit XIV) and upper part of a massive basalt flow (Unit XV) from Sections 324-U1347A-25R-3 through 26R-2, Hole U1347A. Red dashed lines = lithologic unit boundaries, red text = top and bottom of pillow lobes, dashed green lines = glassy chilled margins. Dashed lines in lithologic column (left) show location of core sections portrayed.

## Igneous Rocks

The signature structures of igneous rock sections cored on Shatsky Rise are pillow lavas 0.2–1.0 m in diameter and massive inflation flows up to 23 m thick (Figs. 2, 3). Both types of cooling units were distinguished in cores by the presence of chilled margins and internal vesicle patterns. Pillows are typical of submarine eruptions at low effusion rates and are often found on seamounts and mid-ocean ridge flanks (Ballard et al., 1979). Massive flows are characteristic of continental flood basalt provinces, and they indicate a high effusion rate (Jerram and Widdowson, 2005). They are formed by inflation of the massive interior as an elastic skin develops on the upper and lower chilled margins (Self et al., 1997). Similar massive flows were cored at ODP Sites 1185 and 1186 on the Ontong Java Plateau (Mahoney et al., 2001). In addition, we reinterpret three igneous units cored at Site 1213 on Tamu Massif, previously thought to be sills (Shipboard Scientific Party, 2002), as massive flows 8–15 m thick (Fig. 2; Koppers et al., 2010).

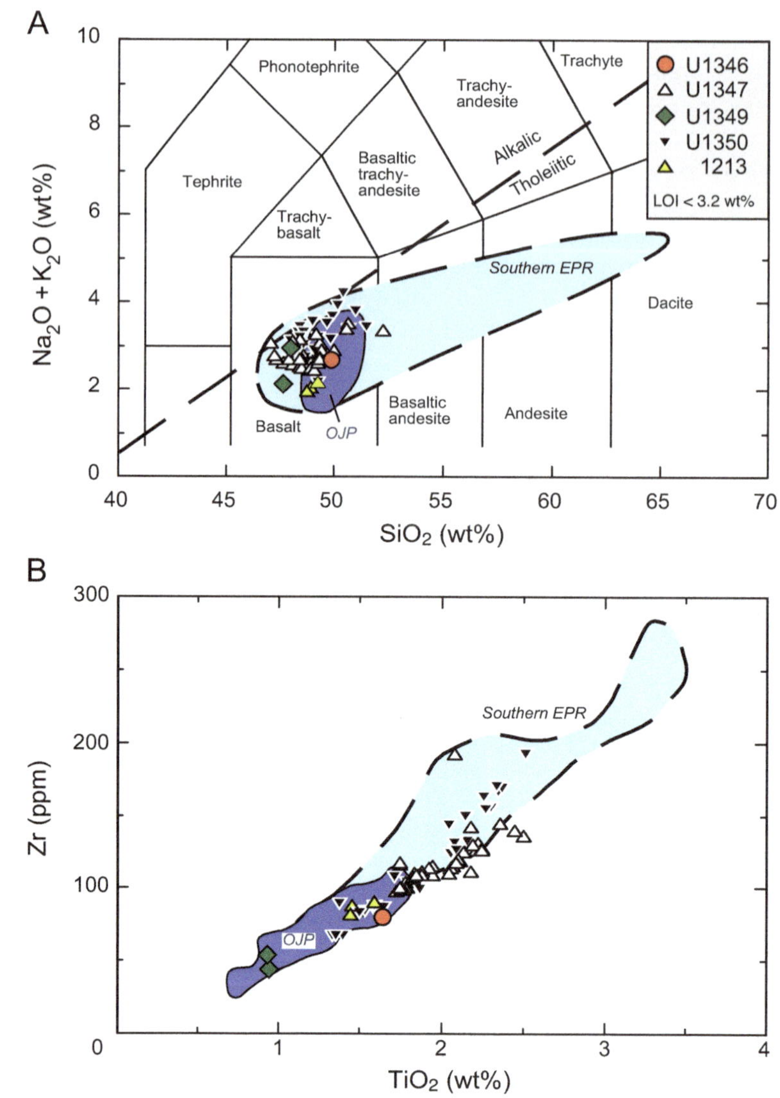


Figure 4. Chemical characteristics of Shatsky Rise basalt samples. [A] Total alkalis vs. silica for least-altered samples with rock classification of Le Maitre et al. (1989). Dashed line divides Hawaiian tholeiitic and alkalic lavas. All data are normalized to 100 wt% totals. [B] $TiO_2$ vs. Zr. Site 1213 data are from Mahoney et al. (2005). OJP = Ontong Java Plateau (data of Tejada et al., 1996, 2002; Fitton and Godard, 2004), EPR = East Pacific Rise (data of Sinton et al., 1991; Bach et al., 1994; Mahoney et al., 1994).

The pattern of igneous activity appears to change across Shatsky Rise. At Tamu Massif, massive flows dominate. The Site U1347 section consists of two pulses of massive flows separated by a ~75 m interval of mostly pillow flows. In total the igneous section contains ~33% pillows. This section is almost indistinguishable from that cored on Ontong Java Plateau at Site 1185. Approximately 220 km away at Site 1213, only massive flows were recovered. On the north side of Shatsky Rise at Shirshov Massif, the Site U1346 section is 90% pillows, with only two larger cooling units (1.9 m and 3.2 m thick) recognized. In between, at Ori Massif, the proportion of massive units appears intermediate. No pillows were recognized in the short Site U1349 section, but the thin massive flows there were emplaced near or above sea level, and none approaches the thickness of those on Tamu Massif. The Site U1350 section contains a mixture of flow types but consists of ~86% pillow units, and massive units are intermediate in thickness (≤6 m). It thus seems that volcanism at Tamu Massif was unusually effusive, with activity typical of the largest ocean plateaus and flood basalts, but the effusion rate apparently waned with time such that flow characteristics on Ori and Shirshov Massifs were similar to those of large seamounts, such as the Emperors (Tarduno et al., 2002).

Cores from Site U1348 imply that explosive volcanism played an important role in building Shatsky Rise. The ~120-m-thick igneous section at that site consists almost entirely of compacted, highly altered volcanic glass fragments, ranging from principally submarine hyaloclastite to redeposited volcaniclastic turbidites (Fig. 2). This finding suggests that other basement highs and cones are composed of similar volcaniclastic materials.

The majority of igneous rocks from Sites U1346, U1347, U1348, and U1350 are plagioclase-clinopyroxene (micro) phyric basalts. The least-altered basalts (recognized by <3.2 wt% volatile loss on ignition, LOI) are from Sites U1346, U1347, and U1350, and are variably evolved tholeiites (Fig. 4). Similar to basalts from Site 1213 (Mahoney et al., 2005), their compositional ranges overlap those of MORB and Ontong Java Plateau basalts. Overall, they are slightly enriched in incompatible elements compared to normal MORB, showing some resemblance to enriched-type MORB. In contrast, lavas from Site U1349 are Cr-spinel-bearing, olivine-phyric basalts that appear to represent significantly less differentiated magmas than those recovered from other sites (Fig. 4).

Alteration of Shatsky Rise igneous rocks follows a pattern tied to basement elevation. The three sections from the rise flanks (Sites 1213, U1347, U1350) all exhibit low to moderate levels of alteration resulting from low-temperature interaction with seawater-derived fluids in anoxic to sub-oxic conditions. In contrast, the three sections cored at high points (Sites U1346, U1348, U1349) are highly altered, and in each case the alteration appears different. Site U1346 alteration is similar to that at the flank sites, but greater in degree. Although most samples are affected by low-temperature alteration, a few samples contain pyrite, which may be indicative of interaction with S-rich hydrothermal fluids. Site U1348 hyaloclastites are all highly altered, mostly converted to palagonite and cemented with calcite and/or zeolites. Basalts recovered at Site U1349 were affected by extensive water-rock interactions at varying temperatures and redox conditions with depth. Some samples from this site appear to record alteration under subaerial, oxidative conditions. The common aspect of all three sites, aside from being at bathymetric highs, is that the alteration implies significant fluid flow.

## Sediments and Sea Level

In previous drilling, Shatsky Rise sediments were found to be dominantly pelagic carbonate sediments (ooze, chalk, limestone) with chert in the Mesozoic section. Expedition 324 drilling recovered a variety of sediment types from four of five sites (Figs. 2, 5) providing clues about depositional environments during and shortly after the formation of the volcanic basement, including the finding that some sites were near sea level during deposition. This is true for both sites on Tamu Massif. Radiolarian-bearing volcaniclastic siltstone and sandstones were cored in the 77 m above the top of basement at Site U1347. These sediments are marked by high content of zeolite and volcaniclastics; they display cross-bedding and intense bioturbation and contain a neritic (<200 m water depth) assemblage of benthic foraminifera. At Site U1348, calcareous sandstones with bioclasts (including material from reef-building fauna), volcanic fragments, and zeolitic clay were recovered. Both sections imply deposition in shallow water near a volcanic source. These findings are consistent with shallow water fossils that were recovered in a dredge haul at a similar depth ~140 km southwest of Site U1347 (Sager et al., 1999).

Site U1349 on the Ori Massif summit yielded volcanic sandstone, siltstone, claystone, and breccia in the ~30 m above igneous basement. At the bottom of this interval, a yellow-red clay layer of intensely-weathered basalt fragments, interpreted as a paleosol, indicates probable subaerial exposure under tropical conditions. The presence of montmorillonite and halloysite in this layer also support the interpretation of subaerial conditions. Only ~10 m deeper in the section, an oolitic limestone between lava flows implies a shallow marine environment. No significant non-pelagic sediment was recovered above basement at Site U1350 on the deep Ori Massif flank. Evidently, this site was bypassed by all but pelagic sediment.

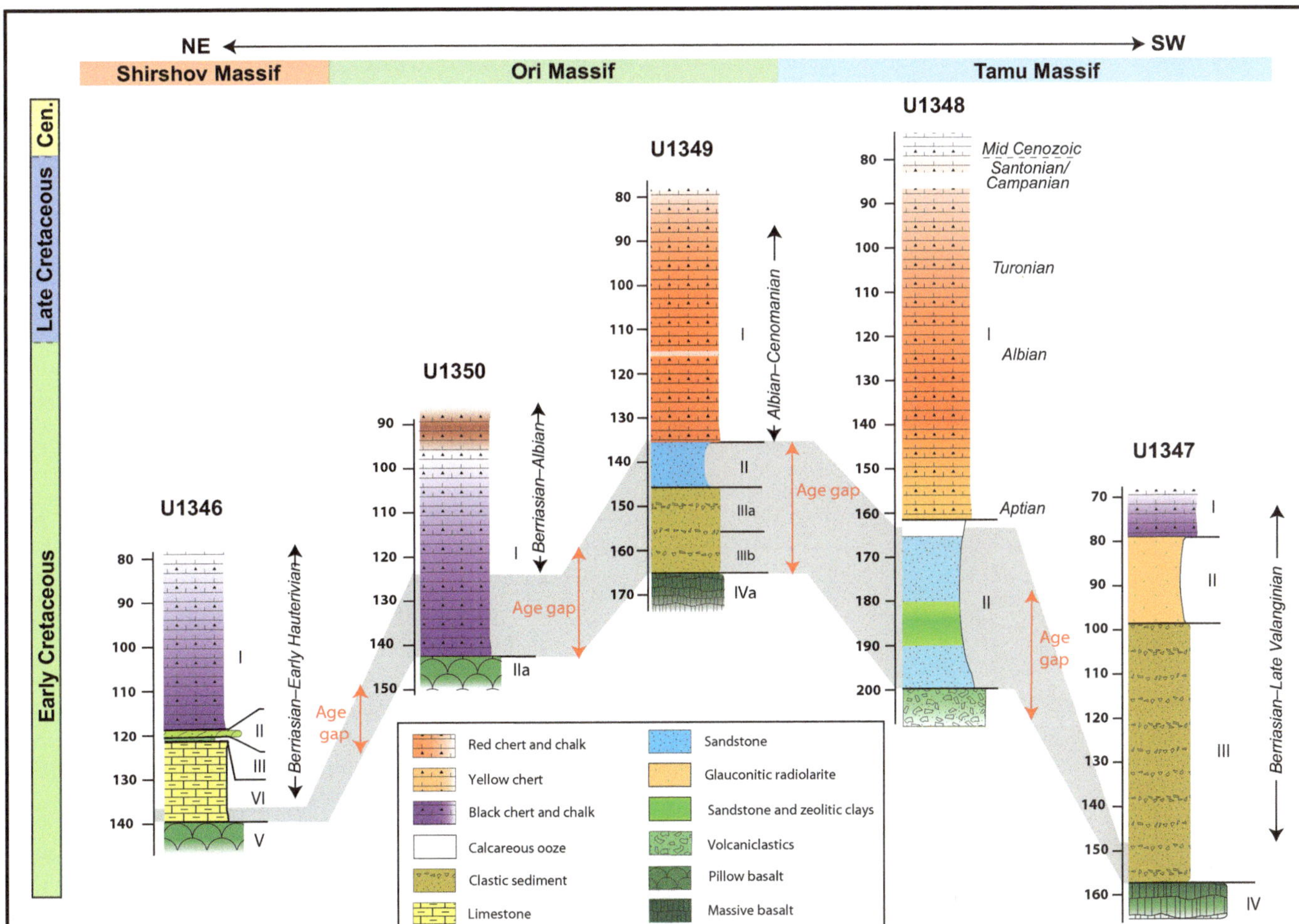


Figure 5. Overview of basal sediment lithology at Expedition 324 sites. Columns are arranged in geographical order from northeast (left) to southwest (right). Ages were deduced from biostratigraphy, as indicated to the right of each column. Vertical position is an indication of biostratigraphic age (left column). Gray band shows the gap between the oldest dated sediment and basement at each site.

At Site U1346 on Shirshov Massif the bottom ~17 m of the sediment column consists of clayey limestones and calcareous mudstone containing abundant shell fragments, a neritic (<500 m water depth) benthic foraminiferal assemblage, and small, probably wood fragments. These sediments are overlain by a short interval of fine-grained volcaniclastic turbidites and a debris flow of intermingled basalt and limestone showing soft-sediment deformation. Altogether, these observations are consistent with deposition in shallow water near forested, volcanic land and subsequent progressive deepening as the volcano subsided.

Sedimentary layers were found intercalated with basalt flows at Sites U1347, U1349, U1350 and at ODP Site 1213 (Shipboard Scientific Party, 2002). At Site U1347, three layers are ~4.5–5.0 m thick, but all other sedimentary interbeds at this and other sites are much thinner. In general, the interbeds consist of carbonate debris or volcaniclastic sandstone or siltstone, often containing fossil radiolarians. Although we have no reliable age progression data to define sedimentation rates of any interbeds, it appears that all were deposited over relatively short periods, implying that hiatuses between lava flows were of short duration, consistent with rapid emplacement of the igneous sections.

## Implications

Although cores from Expedition 324 and prior drilling represent only a small sampling, they provide a clearer picture of Shatsky Rise's geology and give insights for other plateaus. Prior to Expedition 324, we knew that Shatsky Rise must be largely basaltic, but the style of volcanism was unknown. Virtually all sediments previously cored from Shatsky Rise were pelagic, so there was little information about other sedimentary environments or the role of explosive volcanism. Scant evidence existed that the summit of Tamu Massif was originally near sea level. Modestly-altered samples of igneous basement were available from only one site (1213); they exhibited isotopic characteristics similar to MORB (Mahoney et al., 2005), and were described as representing sills. Such characteristics seemed unusual, and their significance was unclear.

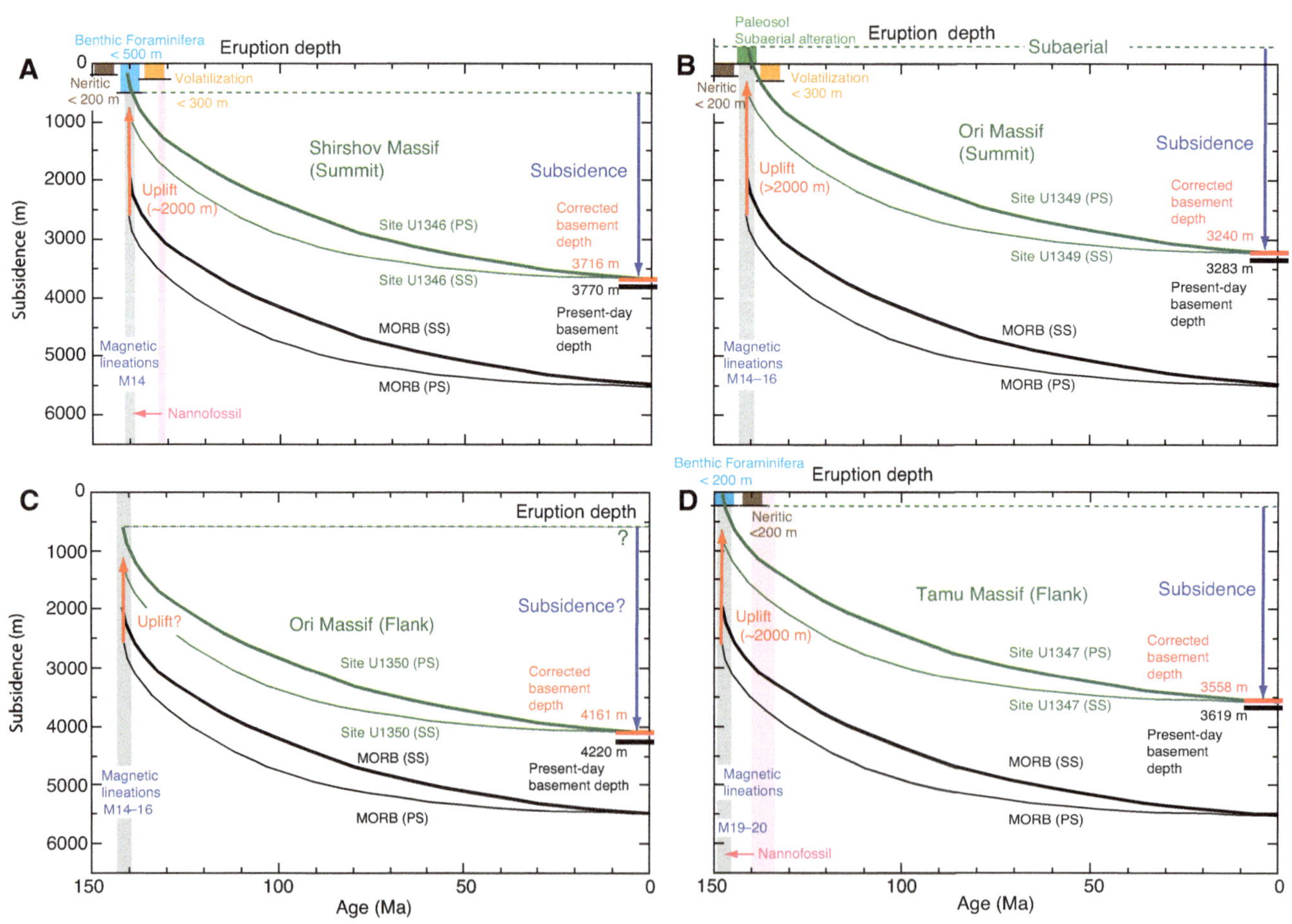


Figure 6. Subsidence curves calculated for Expedition 324 Sites. [A] Site U1346 (Shirshov Massif summit). [B] Site U1349 (Ori Massif summit). [C] Site U1350 (Ori Massif flank). [D] Site U1347 (Tamu Massif flank). Colored curves = extrapolated site depth at the time of eruptions, calculated by backtracking from the present depth (corrected for sediment loading; Crough, 1983) using thermal subsidence models. PS = Parsons and Sclater (1977) model, SS = Stein and Stein (1992) model. Paleodepth estimates from backtracking are compared with paleodepth indicators from Expedition 324 cores. Black curves = subsidence models for normal mid-ocean ridges, red arrow = difference between normal ridge and observed paleodepth, taken as the amount of uplift caused by volcano construction. Magnetic lineation ages are from Nakanishi et al. (1999) using the time scale of Gradstein et al. (2004).

Expedition 324 provided samples of igneous basement at four new sites, and broadly MORB-like characteristics appear general. Massive flows are found at several locations, and the thickest and greatest percentage of total section are found on Tamu Massif, which is characterized by massive flows similar to those found in continental flood basalt provinces (Jerram and Widdowson, 2005) and Ontong Java Plateau (Mahoney et al., 2001). Because of its large size, shape, and style of volcanism, Tamu Massif appears to be a "supervolcano". It suggests that other oceanic plateaus may have formed from similarly colossal volcanoes. In contrast, massive flows are thinner and less abundant at Ori and Shirshov massifs, whose size and eruption style were probably similar to those of the present island of Hawaii—that is , big but not deserving of superlatives.

Expedition 324 cores suggest that the summits of Shatsky Rise volcanoes were near sea level or perhaps emergent. The large interior of the Tamu Massif summit may have been emergent, perhaps with significant elevation because the basal sediments of Site U1347 were deposited in shallow water, but the top of the volcano is ~800 m shallower. Site U1349, with evidence of both shallow water and emergence, suggests that the summit of Ori Massif was at the water's edge. Site U1346, located at the edge of a summit platform on Shirshov Massif, also indicates shallow-water sedimentation and implies that volcanic cones in the interior may have been emergent. It is likely that the plateau formed one or more islands through the late Jurassic and Early Cretaceous, but their size and duration are unclear. Backtracked subsidence curves for the sites (Fig. 6) show that the present depth of the top of igneous basement is consistent with that predicted by normal lithospheric subsidence. If there was dynamic uplift during plateau formation, for example, owing to the arrival of a plume head (Coffin and Eldholm, 1994), there is no evidence of it now.

Site U1348 provided a surprise with a thick pile of highly altered hyaloclastite. Prior to Expedition 324, no significant accumulation of such sediment had been recovered from Shatsky Rise. The finding of this material should not be a surprise, given that it is common on seamounts (Smith and Batiza, 1989). The large basement ridge at Site U1348, thought to be the top of a fault block, consists wholly or in part of the products of explosive submarine volcanism. With this discovery, it seems likely that many of the basement highs imaged on seismic sections (Sager et al., 1999) have a similar origin.

The extent of alteration of igneous basement was also a surprise. We thought it was possible to obtain relatively unaltered samples by drilling through the weathered carapace, but heavy alteration occurred throughout the three higher-elevation igneous sections (U1346, U1348, U1349). This finding implies that fluid circulation at the summits of the volcanic edifices was pervasive. Whether this is a result of proximity to sea level, to heat concentration, or of the volcanic structure is unclear. In contrast, flank flows appear to have been armored by subsequent flows, restricting fluid flow and consequent alteration.

Results now available from Expedition 324 are mainly physical descriptions of the processes of volcanism, sedimentation, and evolution of the Shatsky Rise plateau. Resolution of what caused Shatsky Rise formation is not yet feasible but should be closer after careful studies of the recovered samples, especially those involving high-precision age determination and isotopic and geochemical characterization of the igneous basement. Whatever mechanism is called upon to generate this plateau, it must explain the fact that initial eruptions built an enormous volcanic edifice with massive lava flows and rose to sea level during edifice building and subsided normally thereafter. Expedition 324 results also indicate that different parts of oceanic plateaus can have different styles of eruption, with some having characteristics similar to that of normal seamount volcanism elsewhere.

## IODP Expedition 324 Scientists

W.W. Sager (Co-Chief Scientist), T. Sano (Co-Chief Scientist), J. Geldmacher (Staff Scientist), R. Almeev, A. Ando, C. Carvallo, A. Delacour, H.A. Evans, A.R. Greene, A.C. Harris, S. Herrmann, K. Heydolph, N. Hirano, N. Idrissi, A. Ishikawa, G. Iturrino, M.-H. Kang, A.A.P. Koppers, S. Li, K. Littler, J.J. Mahoney, N. Matsubara, M. Miyoshi, D.T. Murphy, J.H. Natland, M. Ooga, J. Prytulak, K. Shimizu, M. Tominaga, Y. Uchio, M. Widdowson, and S. C. Woodard.

## References


Bach, W., Hegner, E., Erzinger, J., and Satir, M., 1994. Chemical and isotopic variations along the superfast spreading East Pacific Rise from 6°S to 30°S. *Contrib. Mineral. Petrol.*, 116:365–380, doi:10.1007/BF00310905.

Ballard, R.D., Holcomb, R.T., and van Andel, Tj.H., 1979. The Galapagos Rift at 86°W: Sheet flows, collapse pits, and lava lakes of the rift valley. *J. Geophys. Res.*, 84:5407–5422, doi:10.1029/JB084iB10p05407.

Coffin, M.F., and Eldholm, O., 1994. Large igneous provinces: Crustal structure, dimensions, and external consequences. *Rev. Geophys.*, 32:1–36, doi:10.1029/93RG02508.

Coffin, M.F., Pringle, M.S., Duncan, R.A., Gladczenko, T.P., Storey, M., Müller, R.D., and Gahagan, L.A., 2002. Kerguelen hotspot magma output since 130 Ma. *J. Petrol.*, 43:1121–1139. doi:10.1093/petrology/43.7.1121.

Crough, S.T., 1983. The correction for sediment loading on the seafloor. *J. Geophys. Res.*, 88:6449–6454, doi:10.1093/petrology/43.7.1121.

Duncan, R.A., and Richards, M.A., 1991. Hotspots, mantle plumes, flood basalts, and true polar wander. *Rev. Geophys.*, 29:31–50. doi:10.1029/90RG02372.

Fitton, J.G., and Godard, M., 2004. Origin and evolution of magmas on the Ontong Java Plateau. *In* Fitton, J.G., Mahoney, J.J., Wallace, P.J., and Saunders, A.D. (Eds.), *Origin and Evo-*

*lution of the Ontong Java Plateau, Geol. Soc. Spec. Publ.*, 229:151–178.

Foulger, G.R., 2007. The "plate" model for the genesis of melting anomalies. *In* Foulger, G.R., and Jurdy, D.M. (Eds.), *Plates, Plumes, and Planetary Processes. Spec. Pap. – Geol. Soc. Am.*, 430:1–28.

Gradstein, F.M., Ogg, J.G., and Smith, A., 2004. *A Geologic Time Scale 2004*: Cambridge (Cambridge University Press).

Jerram, D.A., and Widdowson, M., 2005. The anatomy of continental flood basalt provinces: Geological constraints on the process and products of flood volcanism. *Lithos*, 79:385–405, doi:10.1016/j.lithos.2004.09.009.

Keszthelyi, L., and Self, S., 1998. Some physical requirements for the emplacement of long basaltic lava flows. *J. Geophys. Res.*, 103:27,447–27,464, doi:10.1029/98JB00606.

Koppers, A.A.P., Sano, T., Natland, J.H., Widdowson, M., Almeev, R., Greene, A.R., Murphy, D.T., Delacour, A., Miyoshi, M., Shimizu, K., Li, S., Hirano, N., Geldmacher, J., and the Expedition 324 Scientists, 2010. Massive basalt flows on the southern flank of Tamu Massif, Shatsky Rise: A reappraisal of ODP Site 1213 basement. *In* Sager, W.W., Sano, T., Geldmacher, J., and the Expedition 324 Scientists, *Proc. IODP*, 324: Tokyo (Integrated Ocean Drilling Progam Management International, Inc.). doi:10.2204/iodp.proc.324.109.2010.

Le Maitre, R.W., Bateman, P., Dudek, A., Keller, J., Lameyre, J., Le Bas, M.J., Sabine, P.A., Schmid, R., Sorensen, H., Streckeisen, A., Wolley A.R., and Zanettin, B., 1989. *A Classification of Igneous Rocks and Glossary of Terms:* Oxford (Blackwell Scientific Publications).

Mahoney, J.J., Duncan, R.A., Tejada, M.L.G., Sager, W.W., and Bralower, T.J., 2005. Jurassic-Cretaceous boundary age and mid-ocean-ridge-type mantle source for Shatsky Rise. *Geology*, 33:185–188, doi:10.1130/G21378.1.

Mahoney, J.J., Fitton, J.G., Wallace, P.J., et al., 2001. *Proc. ODP, Init.* Repts., 192: College Station, TX (Ocean Drilling Program). doi:10.2973/odp.proc.ir.192.2001.

Mahoney, J.J., Sinton, J.M., Kurz, M.D., Macdougall, J.D., Spencer, K.J., and Lugmair, G.W., 1994. Isotope and trace element characteristics of a super-fast spreading ridge: East Pacific Rise 13–23°S. *Earth Planet. Sci. Lett.*, 121:173–193, doi:10.1016/0012-821X(94)90039-6.

Nakanishi, M., Sager, W.W., and Klaus, A., 1999. Magnetic lineations within Shatsky Rise, northwest Pacific Ocean: Implications for hot spot-triple junction interaction and oceanic plateau formation. *J. Geophys. Res.*, 104:7539–7556, doi:10.1029/1999JB900002.

Parsons, B., and Sclater, J.G., 1977. An analysis of the variation of ocean floor bathymetry and heat flow with age. *J. Geophys. Res.*, 82:803–827. doi:10.1029/JB082i005p00803.

Sager, W.W., 2005. What built Shatsky Rise, a mantle plume or ridge tectonics? *In* Foulger, G.R., Natland, J.H., Presnall, D.C., and Anderson, D.L. (Eds.), *Plates, Plumes, and Paradigms. Spec. Pap.-Geol. Soc. Am.*, 388:721–733, doi:10.1029/1998JB900009.

Sager, W.W., Kim, J., Klaus, A., Nakanishi, M., and Khankishieva, L.M., 1999. Bathymetry of Shatsky Rise, northwest Pacific Ocean: Implications for ocean plateau development at a triple junction. *J. Geophys. Res.*, 104:7557–7576, doi:10.1029/1998JB900009.

Sager, W.W., Sano, T., Geldmacher, J., and the Expedition 324 Scientists, 2010. *Proc. IODP*, 324: Tokyo (Integrated Ocean Drilling Program Management International, Inc.), doi:10.2204/iodp.proc.324.2010.

Self, S., Thordarson, T., and Keszthelyi, L., 1997. Emplacement of continental flood basalt lava flows. *In* Mahoney, J.J., and Coffin, M.F. (Eds.), *Large Igneous Provinces. Geophys. Monogr. – Am. Geophys. Union*, 100:381–410.

Shipboard Scientific Party, 2002. Site 1213. *In* Bralower, T.J., Premoli Silva, I., Malone, M.J., et al., *Proc. ODP, Init. Repts.*, 198: College Station, TX (Ocean Drilling Program), 1–110, doi:10.2973/odp.proc.ir.198.109.2002.

Sinton, J.M., Smaglik, S.M., Mahoney, J.J., and Macdonald, K.C., 1991. Magmatic processes at superfast spreading mid-ocean ridges: Glass compositional variations along the East Pacific Rise 13°–23°S. *J. Geophys. Res.*, 96:6133–6155, doi:10.1029/90JB02454.

Smith, T.L., and Batiza, R., 1989. New field and laboratory evidence for the origin of hyaloclastite flows on seamount summits. *Bull. Volcanol.*, 51:96–114, doi:10.1007/BF01081979.

Stein, C.A., and Stein, S., 1992. A model for the global variation in oceanic depth and heat flow with lithospheric age. *Nature*, 359:123–129, doi:10.1038/359123a0.

Tarduno, J.A., Duncan, R.A., Scholl, D.W., et al., 2002. *Proc. ODP, Init. Repts.*, 197: College Station, TX (Ocean Drilling Program).

Tejada, M.L.G., Mahoney, J.J., Duncan, R.A., and Hawkins, M.P., 1996. Age and geochemistry of basement and alkalic rocks of Malaita and Santa Isabel, Solomon Islands, southern margin of Ontong Java Plateau. *J. Petrol.*, 37:361–394, doi:10.1093/petrology/37.2.361.

Tejada, M.L.G., Mahoney, J.J., Neal, C.R., Duncan, R.A., and Petterson, M.G., 2002. Basement geochemistry and geochronology of Central Malaita, Solomon Islands, with implications for the origin and evolution of the Ontong Java Plateau. *J. Petrol.*, 43:449–484, doi:10.1093/petrology/43.3.449.

## Authors

**William W. Sager**, Department of Oceanography, Texas A&M University, College Station, TX 77843-3146, U.S.A., e-mail: wsager@tamu.edu.
**Takashi Sano**, Department of Geology and Paleontology, National Museum of Nature and Science, 4-1-1 Amakubo, Tsukuba 305-0005, Japan, e-mail: sano@kahaku.go.jp.
**Jörg Geldmacher**, Integrated Ocean Drilling Program, Texas A&M University, 1000 Discovery Drive, College Station, TX 77845-9547, U.S.A., e-mail:geldmacher@iodp.tamu.edu.
**and IODP Expedition 324 Scientists**

# IODP Expedition 325: Great Barrier Reefs Reveals Past Sea-Level, Climate and Environmental Changes Since the Last Ice Age

by Yusuke Yokoyama, Jody M. Webster, Carol Cotterill, Juan Carlos Braga, Luigi Jovane, Heath Mills, Sally Morgan, Atsushi Suzuki, and the IODP Expedition 325 Scientists

## Abstract

The timing and courses of deglaciations are key components in understanding the global climate system. Cyclic changes in global climate have occurred, with growth and decay of high latitude ice sheets, for the last two million years. It is believed that these fluctuations are mainly controlled by periodic changes to incoming solar radiation due to the changes in Earth's orbit around the sun. However, not all climate variations can be explained by this process, and there is the growing awareness of the important role of internal climate feedback mechanisms. Understanding the nature of these feedbacks with regard to the timing of abrupt global sea-level and climate changes is of prime importance. The tropical ocean is one of the major components of the feedback system, and hence reconstructions of temporal variations in sea-surface conditions will greatly improve our understanding of the climate system. The Integrated Ocean Drilling Program (IODP) Expedition 325 drilled 34 holes across 17 sites in the Great Barrier Reef, Australia to recover fossil coral reef deposits. The main aim of the expedition was to understand the environmental changes that occurred during the last ice age and subsequent deglaciation, and more specifically (1) establish the course of sea-level change, (2) reconstruct the oceanographic conditions, and (3) determine the response of the reef to these changes. We recovered coral reef deposits from water depths down to 126 m that ranged in age from 9,000 years to older than 30,000 years ago. Given that the interval of the dated materials covers several paleoclimatologically important events, including the Last Glacial Maximum, we expect that ongoing scientific analyses will fulfill the objectives of the expedition.

## Introduction and Goals

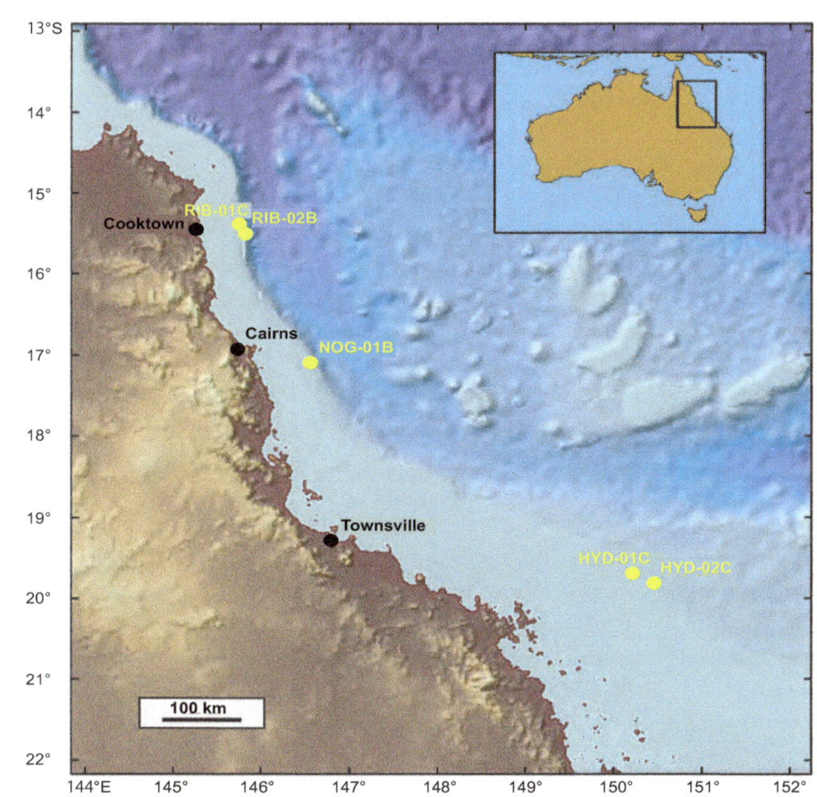


Figure 1. Map showing the locations of drilling sites.

The most prominent feature of the current geological era (Quaternary) is the reoccurrence of glacial and interglacial periods. Waxing and waning of the ice sheets across the North American continent as well as northern Eurasia have been recorded. The Antarctic ice sheet, currently the largest ice sheet on Earth, was larger during the glacial times, with an estimated 10–30 m sea-level equivalent stored in global ice volume during the last glacial maximum at about 20 ka (CLIMAP, 1981; Denton and Hughes, 1981; Nakada and Lambeck, 1987; Yokoyama et al., 2001a, 2001b; Ivins and James, 2005). These ice sheets have been a key component in the global climate system due to their locations and size, as well as their ability to release freshwater into the high latitude ocean during melting. Global climate is regulated by ocean circulation, namely thermohaline circulation (THC), which starts in the high latitude oceans where dense and cold deepwater is formed due to the rapid cooling of the saline and warm Gulf Stream. In turn, the input of freshwater from the melting of ice sheets can strongly influence the strength of thermohaline circulation and

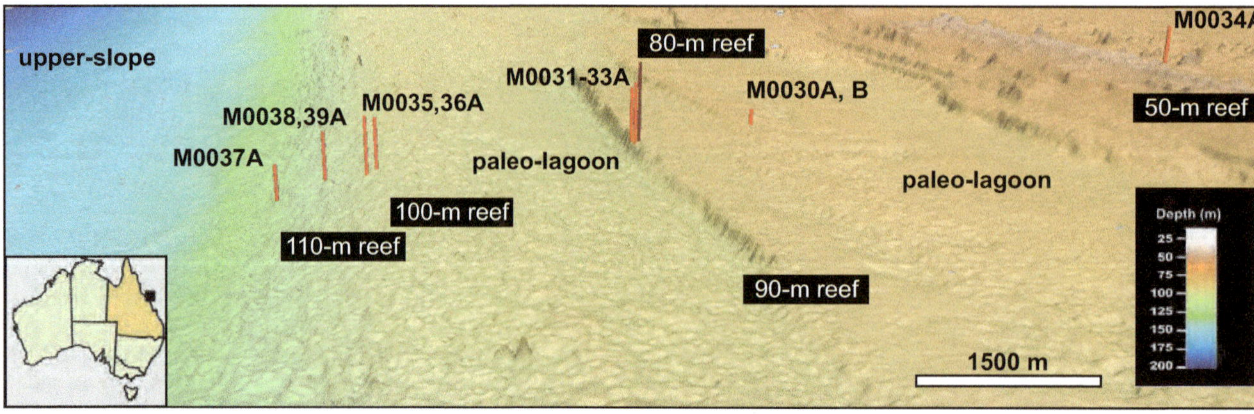


Figure 2. High-resolution multibeam image of Hydrographers Passage transect HYD-01C (multibeam data after Bridge et al., 2011; Abbey et al., submitted). The bathymetry data is gridded at 5 m, and the vertical red lines represent the positions and approximate penetration depths of sites M0030A–39A.

thus drive global climate perturbations (Broecker, 1994; Alley, 1998).

Sophisticated, fully coupled Atmosphere and Ocean General Circulation Models (AOGCM) now predict significant increases in sea surface temperature (SST) by year 2100 due to the release of anthropogenic greenhouse gases and associated atmospheric temperature increase, while ocean acidification, due to higher atmospheric $CO_2$ concentrations and increased solubility in the oceans, may influence large areas of the ocean basins (Solomon et al., 2007). Ice-sheet instability due to increasing temperatures is also anticipated. However, significant uncertainties associated with these model projections remain, and improvement using paleo-data is crucial. Particularly important are the Last Glacial Maximum (LGM: ca. 20 ka; Mix et al., 2001) and subsequent deglaciation (starting from 19 ka; Yokoyama et al., 2000b; Clark et al., 2009), including Bølling-Allerød, Younger Dryas and Heinrich events (Yokoyama, 2011a, 2011b), as the changes observed from LGM to the present day represent the most recent major global climate reorganization under natural climate forcing (Yokoyama and Esat, 2011).

Corals offer the opportunity to reconstruct past environmental information including sea-level and paleoceanographic changes. Reef building corals live in shallow water, and as they secrete their calcium carbonate skeletons, they encapsulate important oceanographic information such as sea-surface temperatures (SST), sea-surface salinity (SSS), and upwelling intensities. Corals are also the only marine calcifying organisms with a closed uranium series system (Stirling and Andersen, 2009), allowing the precise timing of past environmental events to be established (Edwards et al., 1987; Stirling et al., 1995; Gallup et al., 2002). Coupled with precise radiocarbon dating, these U/Th measurements allow the calibration of radiocarbon chronological data to calendar years (Bard et al., 1990; Edwards et al., 1993; Yokoyama et al., 2000a; Reimer et al., 2004; Fairbanks et al., 2005; Esat and Yokoyama, 2008). Massive *Porites* coral grow 1–2 cm annually, and their growth bands allow high temporal resolution (monthly to seasonally) to be made on environment reconstructions (Gagan et al., 1998; Cole et al., 2000; Tudhope et al., 2001; Abram et al., 2008; Felis et al., 2009). Geochemical analyses (stable isotopes, trace elements) of these growth bands allow us to reconstruct seasonal to multi-decadal ENSO signals that can contribute to understanding the natural variability of the climate system.

The Great Barrier Reef (GBR) is the world's largest reef, extending 2000 km laterally northwest-southeast (Davies et al., 1989). Timing of the reef building is proposed to have started around 0.4–0.5 million years ago (Alexander et al., 2001), which coincides with the Mid Pleistocene Climate transition (MPT) (Hays et al., 1976; Clark et al., 2006) or Marine isotope stage (MIS 11; Webster and Davies, 2003). MIS 11 is known as one of the "warmest" interglacials during the Quaternary, and a period when the 100-ka climate cyclicity between glacial to interglacial states was fully established along with higher sea-level amplitudes. Previous deep drilling investigations through the modern reef confirm that the evolution of the GBR has been strongly influenced by these environmental changes (Webster et al., 2008). Thus, investigating the shelf edge, where the GBR has tracked sea-level and climate change since the last ice age, will provide information about how the reef responded to environmental changes over millennial time scales.

The GBR is ideally situated for sea-level studies, as it is located on a passive margin and is far from former ice sheets (Fig. 1). Therefore, GBR sea-level records can be accurately translated to the ambient ice volume variations due to smaller solid earth deformations caused by glacio-hydro-isostatic adjustments (GIA; Nakada and Lambeck, 1987; Yokoyama et al., 2001a, 2001b; 2006; Lambeck et al., 2002; Milne et al., 2009; Yokoyama and Esat, 2011). After correcting the GIA effect, changes in sea level can be directly translated to ice volumes, meaning that the GBR provides the opportunity to remotely monitor the magnitude and timing of ice sheet fluctuations. Located in the southern margin of the Western Pacific Warm Pool (WPWP), the GBR is also well positioned to investigate the role of the tropical Pacific in the climate system. Fossil coral samples from the region can be used to constrain the temporal variations of size of the WPWP and mean temperature with regards to the global climate change.

Figure 3. Representative litholostratigraphic succession from the fossil coral reef terraces (M0033A). The section was obtained from R1 (~90 mbsl) and contains coralgal boundstones typically found at the top of the reef structure [A]. Coralgal-microbialite boundstones were recovered from R5 (98 mbsl) from the same hole [B] and packstone together with unlithfied sediments sequence obtained from deeper part of the section (R21) at approximately 129 mbsl [C].

A diverse suite of fossil coral reef features were observed along the shelf edge of the GBR at 40–130 m depth during the site survey cruise (Fig. 1). The surface distribution and morphology of these reefs are described in detail elsewhere (Webster et al., 2008; Abbey and Webster, 2011; Abbey et al., submitted; Bridge et al., 2011; Webster et al., 2011). However, we present here a brief summary of the representative morphologic features defining transect HYD-01C on the Mackay shelf in Hydrographer's Passage and their drill sites (Fig. 2) to highlight their potential of recording reef growth back to at least the LGM. At this location a double-fronted barrier reef 200 m long and 100 m wide is observed, which is separated by a paleo-lagoon 2 km wide and as deep as 70 m. The barrier reefs occur at 55–51 m and were sampled by Hole M0034A. A steeply sloping 500-m-wide terrace with a sharp break in slope at 80 m defines the seaward expression of this feature. Following this, a complex 1-km-wide paleo-lagoon and reef terrace system is observed with prominent reefs at 80 m and 90 m that were sampled by M0030A, M0030B, and M0031-33A. Another 700-m-wide paleo-lagoon is observed grading into a complex system of reef pinnacles, terraces, and reefs down to the major break in slope at 100 m that defines the shelf edge. Here, M0035A and M0036A sampled the 100-m reef, and seaward of this M0037-39A intersected a series of smaller seaward reef pinnacles and terraces between 110 mbsl (meter below sea level) and 120 mbsl, before the seafloor grades into a gentle upper slope characterized by fore-reef slope sediments.

Following the success of the IODP 310 Tahiti drilling expedition (Camoin et al., 2007), detailed reconstructions have emerged of sea level (Deschamps et al., 2009; Thomas et al., 2009; Fujita et al., 2010), paleoclimate (Asami et al., 2009; DeLong et al., 2010; Inoue et al., 2010) and reef responses (Camoin et al., 2010; Abbey et al., submitted; Seard et al., 2011). However, due to the geometry of the flanks of Tahiti, IODP 310 did not recover a full range of the last deglaciation sequence older than ~16,000 years ago, in particular the LGM fossil reefs (Camoin et al., 2007). Therefore, details of reconstructions of sea level and paleoenvironments since the LGM are yet to be obtained.

The three main scientific objectives of Exp. 325 are as follows:

1) establish the course of sea-level rise during the last deglaciation (~20–10 ka) with a particular focus on the interval (>16 ka) not sampled by Exp. 310

Table 1. Operational summary of IODP 325

| Hole | Latitutde | Longtitude | Water Depth (LAT from corrected EM 300 echo sounder data (m)) | Water Depth (Drill string tagging seabed including predicted tidal variations (m)) | Number of core runs | Interval cored (m) | Core recovered (m) | Core Recovery (%) | Penetration Depth SDF-A (m) | Hole recovery (%) | Time on Site (days) |
|---|---|---|---|---|---|---|---|---|---|---|---|
| **M0030A** | 19.6819 | 150.2379 | 83.47 | 85.00 | 2 | 6.00 | 0.24 | 4.00 | 6.00 | 4.00 | 0.85 |
| **M0030B** | 19.6819 | 150.2379 | 83.47 | 85.00 | 3 | 9.00 | 0.55 | 6.11 | 9.00 | 6.11 | 0.40 |
| **M0031A** | 19.6790 | 150.2396 | 90.05 | 92.00 | 17 | 43.00 | 5.68 | 13.21 | 43.00 | 13.21 | 2.09 |
| **M0032A** | 19.6788 | 150.2397 | 91.58 | 93.00 | 20 | 36.70 | 5.99 | 16.32 | 36.70 | 16.32 | 1.15 |
| **M0033A** | 19.6789 | 150.2399 | 91.32 | 91.50 | 23 | 32.80 | 13.41 | 40.88 | 32.80 | 40.88 | 1.15 |
| **M0034A** | 19.6923 | 150.2303 | 51.00 | 55.00 | 16 | 23.10 | 6.71 | 29.05 | 23.10 | 29.05 | 3.19 |
| **M0035A** | 19.6726 | 150.2438 | 100.05 | 103.00 | 23 | 29.90 | 12.23 | 40.90 | 29.90 | 40.90 | 2.98 |
| **M0036A** | 19.6724 | 150.2440 | 103.21 | 103.00 | 22 | 34.00 | 8.91 | 26.21 | 34.00 | 26.21 | 1.77 |
| **M0037A** | 19.6707 | 150.2463 | 122.29 | 129.17 | 14 | 21.00 | 7.52 | 35.81 | 21.00 | 35.81 | 0.83 |
| **M0038A** | 19.6716 | 150.2449 | 107.04 | 108.58 | 1 | 1.50 | 0.18 | 12.00 | 1.50 | 12.00 | 0.53 |
| **M0039A** | 19.6716 | 150.2449 | 107.04 | 108.58 | 21 | 28.40 | 10.39 | 36.58 | 28.40 | 36.58 | 1.02 |
| **M0040A** | 19.7963 | 150.4814 | 126.07 | 132.67 | 12 | 21.50 | 11.73 | 54.56 | 21.50 | 54.56 | 0.61 |
| **M0041A** | 19.7963 | 150.4815 | 126.58 | 132.67 | 12 | 22.10 | 10.06 | 45.52 | 22.10 | 45.52 | 0.55 |
| **M0042A** | 19.8440 | 150.4480 | 50.78 | 56.32 | 29 | 46.40 | 10.94 | 23.58 | 46.40 | 23.58 | 2.19 |
| **M0043A** | 19.7989 | 150.4794 | 102.93 | 107.88 | 23 | 35.00 | 6.04 | 17.26 | 35.00 | 17.26 | 0.85 |
| **M0044A** | 19.7985 | 150.4796 | 105.25 | 104.31 | 9 | 11.00 | 1.67 | 15.18 | 11.00 | 15.18 | 0.72 |
| **M0045A** | 19.7984 | 150.4976 | 105.25 | 105.01 | 4 | 14.60 | 0.00 | 0.00 | 14.60 | 0.00 | 0.24 |
| **M0046A** | 19.7985 | 150.4796 | 117.49 | 120.41 | 11 | 20.40 | 2.52 | 12.35 | 20.40 | 12.35 | 0.45 |
| **M0047A** | 19.7998 | 150.4789 | 99.12 | 100.51 | 14 | 33.20 | 3.79 | 11.42 | 33.20 | 11.42 | 0.56 |
| **M0048A** | 19.8012 | 150.4777 | 97.47 | 102.31 | 4 | 7.10 | 0.69 | 9.72 | 7.10 | 9.72 | 0.28 |
| **M0049A** | 15.4724 | 145.8237 | 97.63 | 98.61 | 2 | 3.50 | 0.77 | 22.00 | 3.50 | 22.00 | 0.24 |
| **M0049B** | 15.4724 | 145.8237 | 97.63 | 100.00 | 13 | 15.60 | 2.79 | 17.88 | 15.60 | 17.88 | 0.52 |
| **M0050A** | 15.4723 | 145.8237 | 97.63 | 98.20 | 6 | 10.50 | 1.87 | 17.81 | 10.50 | 17.81 | 0.27 |
| **M0051A** | 15.4721 | 145.8230 | 78.13 | 80.90 | 2 | 2.50 | 0.15 | 6.00 | 2.50 | 6.00 | 0.35 |
| **M0052A** | 17.1011 | 146.5763 | 97.63 | 103.70 | 1 | 1.40 | 1.30 | 92.86 | 1.40 | 92.86 | 0.25 |
| **M0052B** | 17.1011 | 146.5763 | 97.63 | 103.70 | 4 | 6.90 | 0.46 | 6.67 | 6.90 | 6.67 | 0.30 |
| **M0052C** | 17.1011 | 146.5763 | 97.63 | 106.80 | 2 | 1.90 | 0.10 | 5.26 | 8.80 | 5.26 | 0.22 |
| **M0053A** | 17.1012 | 146.5763 | 97.87 | 104.60 | 33 | 37.30 | 12.18 | 32.65 | 37.30 | 32.65 | 1.59 |
| **M0054A** | 17.1007 | 146.5767 | 107.23 | 110.33 | 6 | 9.30 | 2.23 | 23.98 | 18.72 | 11.91 | 1.06 |
| **M0054B** | 17.1007 | 146.5767 | 107.23 | 110.33 | 12 | 27.84 | 8.25 | 29.63 | 33.20 | 24.85 | 2.26 |
| **M0055A** | 17.1019 | 146.5747 | 87.33 | 93.06 | 10 | 28.50 | 10.00 | 35.09 | 31.29 | 31.96 | 0.89 |
| **M0056A** | 17.1022 | 146.5741 | 81.22 | 85.56 | 16 | 40.20 | 12.73 | 31.67 | 41.29 | 30.83 | 1.05 |
| **M0057A** | 17.1050 | 146.5640 | 42.27 | 47.67 | 16 | 40.60 | 19.00 | 46.80 | 41.78 | 45.48 | 1.30 |
| **M0058A** | 17.0973 | 146.5893 | 167.14 | 172.41 | 15 | 41.40 | 33.94 | 81.98 | 41.40 | 91.98 | 1.06 |

2) reconstruct the nature and magnitude of seasonal to millennial scale climate variability (e.g., SST, SSS)

3) determine the biologic and geologic response of the GBR to abrupt sea-level and climate changes.

In this report we present a first summary of the operations, preliminary sedimentologic, chronologic, petrophysical, paleomagnetic, and geochemistry results and their implications and discuss the future plans for the post-cruise science.

## Operations

The drilling platform chosen for Exp. 325 was the *Greatship Maya*, an IMO Class II dynamically positioned vessel with geotechnical coring capability. Operations were conducted between 12 February and 6 April 2010 (Table 1). Cyclone Ului, combined with several serious technical difficulties (Webster et al., 2011), meant that total percent recovery of the GBR cores was lower than IODP 310 Tahiti Sea Level (57.5% for Tahiti, whereas 27.2% for GBR; Camoin et al., 2007; Webster et al., 2011).

The vessel was equipped with a large moon pool and Bluestone TT150 derrick, capable of handling 9-m string lengths, with a Foremost Hydraulic top drive and relative motion-compensating heave (2.5-m stroke cylinders). Wireline operation of the core barrel was conducted through the top drive. Deployment of the wireline logging tools was conducted through the mud valve at the top of the top drive, from the rooster box, providing a heave compensated platform from which to zero in the tools.

Two types of QDTech coring tools were carried: an extended nose corer (EXN; equivalent to the extended core barrel [XCB]), and a standard rotary corer (ALN; equivalent to the rotary core barrel [RCB]), along with four bottom-hole

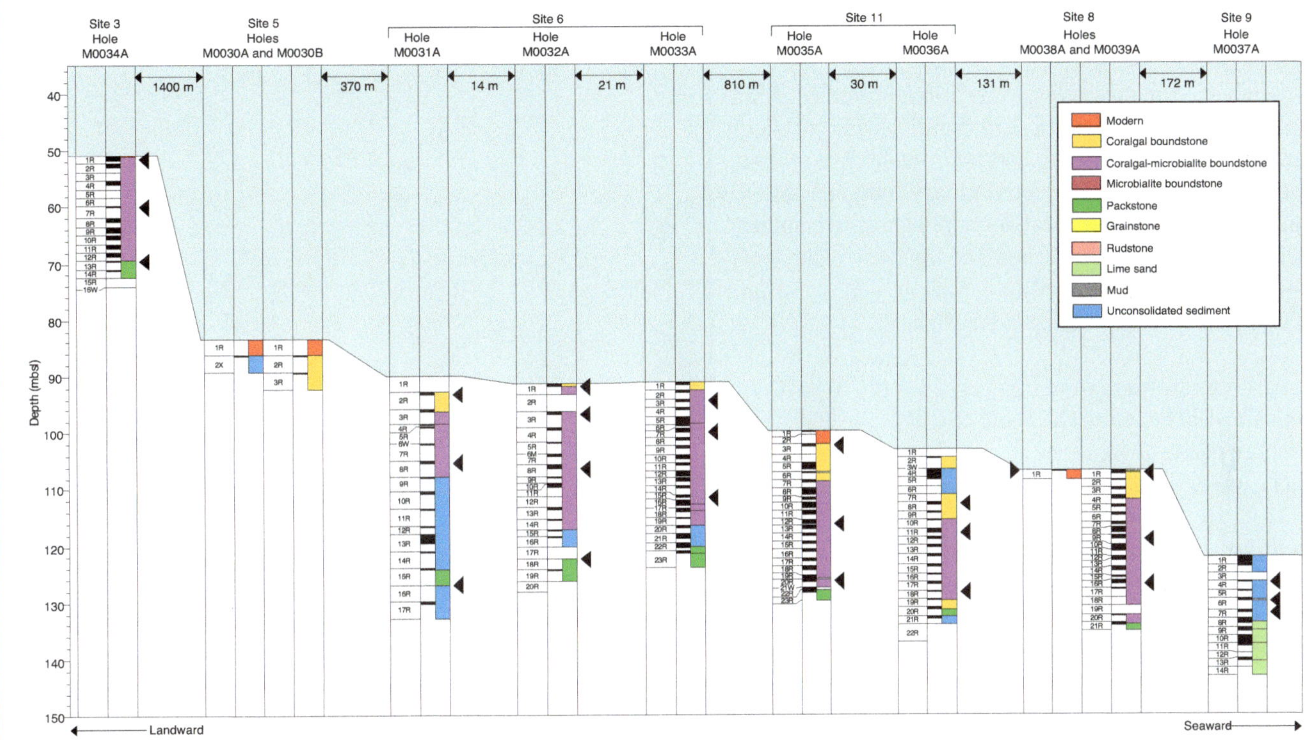


Figure 4. Lithostratigraphic summary of holes on the Transect HYD-02 (Sites from landward to seaward; M0034A, M0030A, M0030B, M0031A, M0032A, M0033A, M0035A, M0036A, M0038A, M0039A, and M0037A).

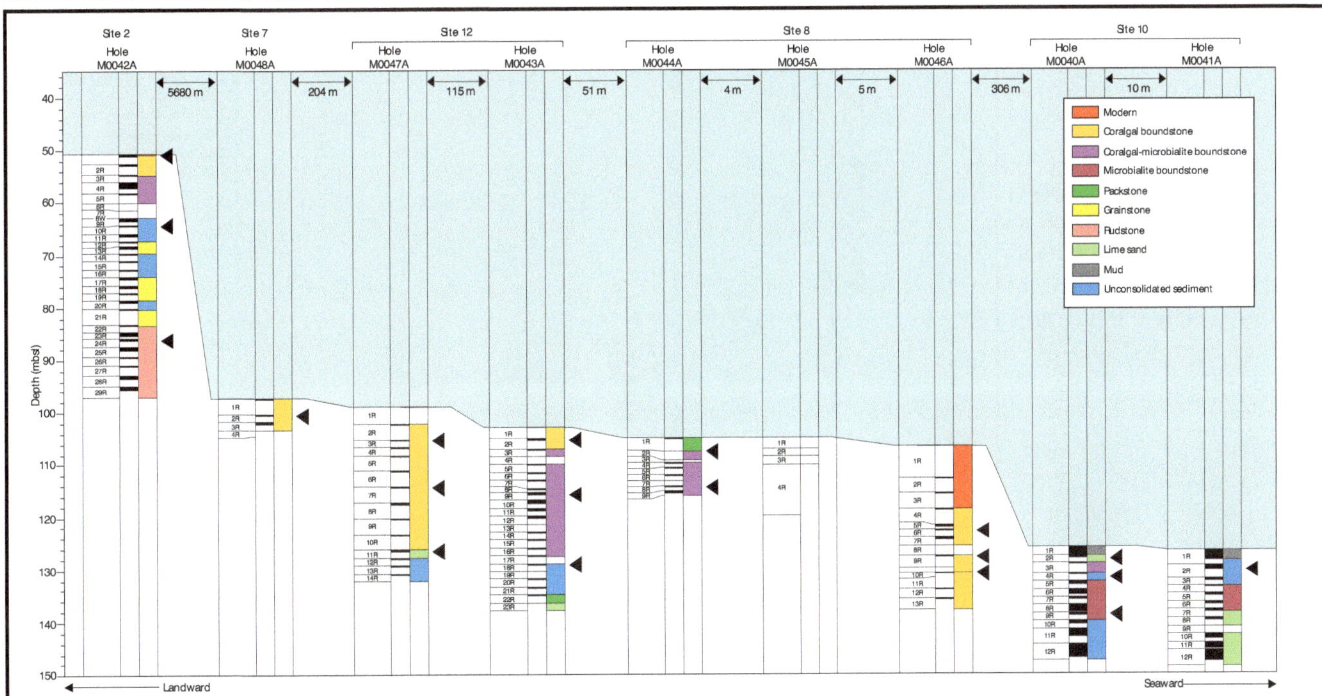


Figure 5. Lithostratigraphic summary of holes on the Transect HYD-01 (Sites from landward to seaward; M0042A, M0048A, M0047A, M0043A, M0044A, M0045A, M0046A, M0040A, and M0041A).

assemblies (BHAs), two each to fit the American Petroleum Institute (API) 4-inch-bore inner diameter string, and the Longyear HQ mining drill string.

At each site a pre-coring downpipe camera survey was conducted as part of the Environmental Management Plan agreed with the Great Barrier Reef Marine Park Authority (GBRMPA). Very strong bottom currents required the use of a seabed template to stabilize the drill string.

Offshore, the cores were carefully curated by ESO staff before ephemeral physical (multi-sensor core logger, MSCL), geochemical, and microbiological properties were measured and preliminary samples taken. Initial lithological and coral descriptions were conducted by visual inspection through the liner and using core catcher materials. No further sampling, core splitting, or analysis work was undertaken offshore.

All cores and data were transferred to the IODP Core Repository at Bremen, Germany (BCR) at the end of the offshore phase. Prior to the start of the Onshore Science Party (OSP), additional thermal conductivity measurements and computed tomography (CT) scans were undertaken on selected whole core sections. All the cores were assessed for the presence of massive corals, so as to instruct the core splitting procedure used. The complete Science Party plus ESO and BCR personnel and student helpers met at the BCR on 2–16 July to split, analyze, and sample the cores according to standard IODP procedures.

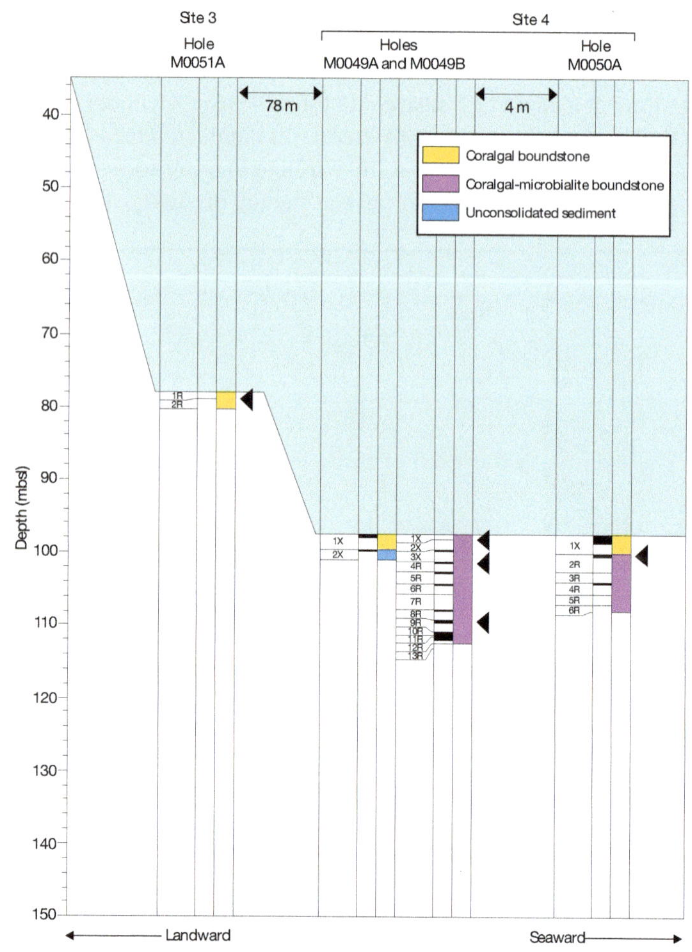


Figure 6. Lithostratigraphic summary of holes on the Transect RIB-02 (Sites from landward to seaward; M0051A, M0049A, M0049B, and M0050A).

## Sedimentology

Sedimentological investigation of the cores was conducted during both the offshore and onshore phases of operations. Nine lithological types were recognized, and their main characteristics and distributions as distinct lithostratigraphic units are shown in Figs. 4–7 and summarized below.

Modern or sub-recent deposits consisting of lime sand to pebbles, locally in a muddy matrix, cover the boundstone lithologies at the top of a number of holes. Pebbles are made up of coralgal boundstone, coralline algae (some of them with reddish-pinkish color, indi-

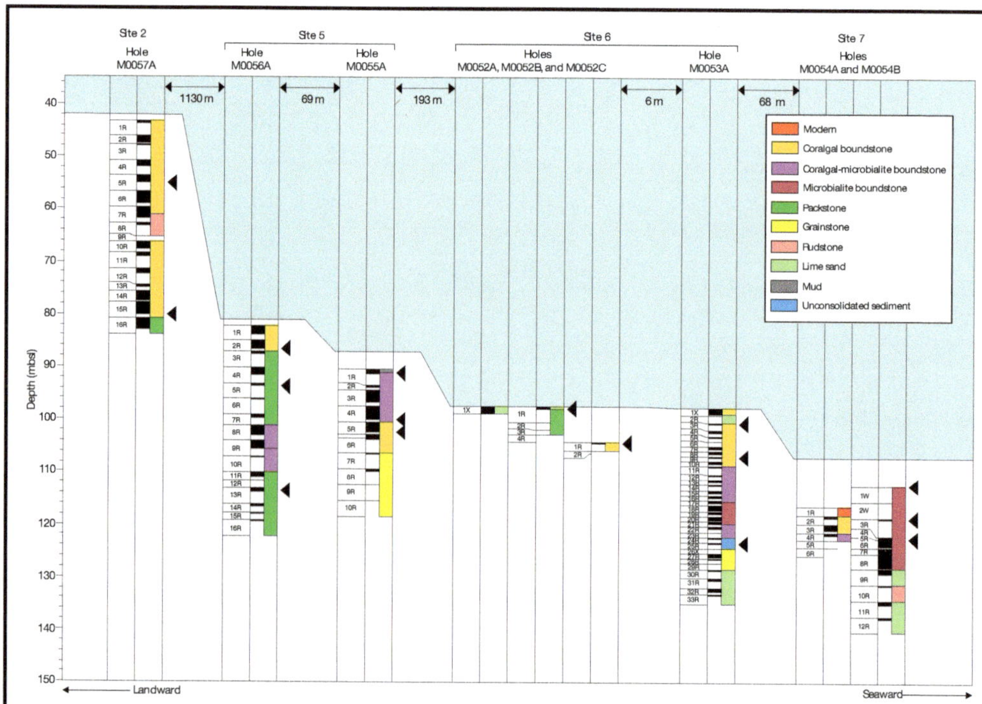


Figure 7. Lithostratigraphic summary of holes on the Transect NOG-01 (Sites from landward to seaward; M0057A, M0056A, M0055A, M0052A, M0052B, M0052C, M0053A, M0054A, and M0054B).

cating they were alive when recovered), serpulid worm tubes, mollusk shells and bryozoans. Some pebbles have brown staining. *Halimeda* have been identified, as well as larger benthic foraminifera (abundant and well-preserved or stained specimens of *Alveolinella, Amphistegina, Cycloclypeus,* Elphidiidae, *Heterostegina, Operculina,* and *Sphaerogypsina*).

Coralgal-microbialite and coralgal boundstones are the dominant lithologies in the recovered deposits (Fig. 3). The coralgal-microbialite boundstones, 10–30 m thick, are composed of corals partially coated by coralline algae and vermetids encrusted by microbialites, which volumetrically are the major component (Figs. 4–7). *Halimeda*, mollusks, benthic foraminifera, red algae, bryozoans, and echinoderms occur as pockets of internal sediment in the coralgal-microbialite frameworks. Coral assemblages are dominated by massive *Isopora,* branching *Acropora,* and *Seriatopora,* but massive *Porites* and *Faviidae* are locally abundant. *Hydrolithon onkodes* is the most abundant coralline alga, together with *Lithophyllum prototypum* and *Neogoniolithon fosliei. Amphistegina, Operculina, Heterostegina,* and *Alveolinella* are the most common larger benthic foraminifera. The relative proportions of coral, coralline algae, and microbialite vary within this lithology. The coralgal lithologies, ranging from 1 m to 24 m in thickness, only differ from the coralgal-microbialite boundstone in containing little or no microbialite.

Stratigraphically, the coralgal boundstones consistently overlie coralgal-microbialite units. However, in two holes (M0034A, M0055A) a thin unit of coralgal boundstone underlies a thicker coralgal-microbialite interval, and in five other holes (Holes M0046A, M0047A, M0048A, M0052A, M0057A) a coralgal framestone is the only recovered lithology bound by organisms (Figs. 5, 7).

Unconsolidated sediment <1–19 m thick underlies the coralgal-microbialite and coralgal boundstone units and is composed of bioclastic lime sand to pebbles containing mollusks, larger benthic foraminifera, *Halimeda,* fragments of corals and red algae, bryozoans, and echinoderms. In Hole M0036A, an unconsolidated unit 6 m in thickness is bracketed by coralgal boundstone, whereas in Hole M0037A—the most distal and deepest site (122 m) on the HYD-01A transect (Fig. 4)—only unconsolidated sediment was recovered. These unconsolidated sediments were likely to have been partly disturbed by coring operations.

Skeletal packstone to grainstone, up to 13 m thick, underlies the unconsolidated sediment or the coralgal/coralgal-microbialite boundstones in most holes. These lithologies are variably cemented and composed of fragments of shells, coral, coralline algae, *Halimeda*, and abundant larger benthic foraminifera. The top of the cemented lithologies is commonly bored by worms and sponges. Features indicating subaerial exposure, such as calcrete deposits, brownish staining, and rhizoliths, appear at the top of grainstones in the shallowest holes on transect NOG-01B. In Hole M0036A, a dark-colored boundstone, about 1.5 m thick, overlies the packstone. This blackened boundstone is made up of encrusting coral and thin coralline algae in its upper part, and a boundstone of coral, thin foliose coralline algae, and worm tubes in the lowest 10 cm. The dominant coral is massive *Goniopora*, with fragments of massive Faviidae and fine-branching *Seriatopora*.

Below the uppermost packstone-grainstone, a variety of lithologies were encountered. In the shallowest three holes on NOG-01B (Fig. 7) transect coralgal and coralgal-microbialite boundstones alternate with packstone-grainstone units, reaching up to 25 m in total thickness. All lithologies show dissolution surfaces, brownish stainings, and dissolution of aragonitic components. These features suggest several phases of emersion and weathering. In other holes, unconsolidated sediment underlies or alternates with cemented inter-vals of packstones and grainstones. In the deep holes of transects HYD-01C

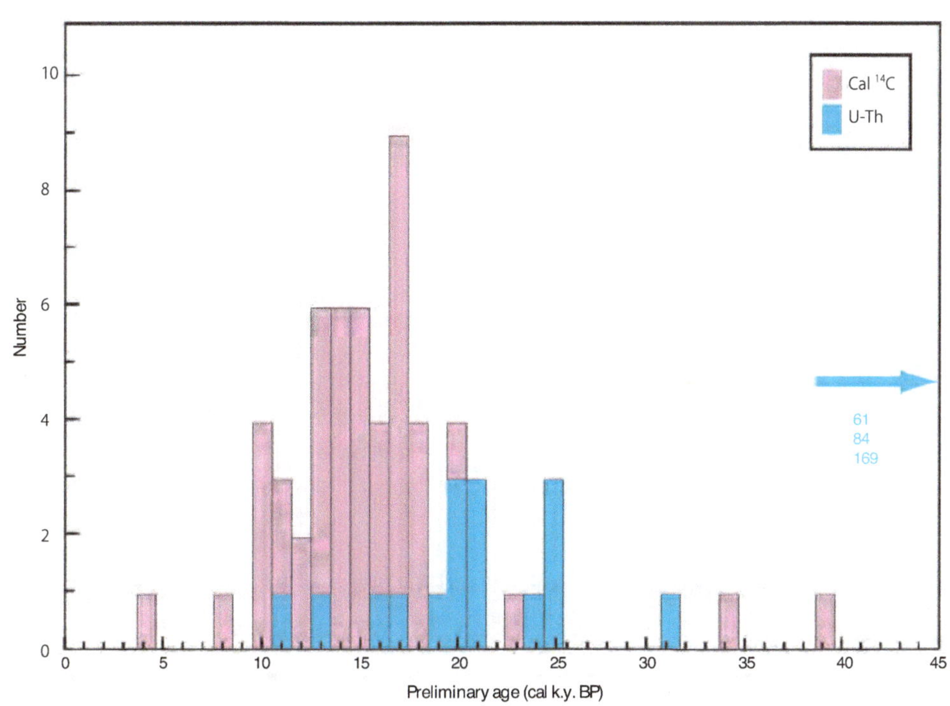


Figure 8. Age histograms of preliminary dating samples (1-ky rounded). All ages are in years before AD 1950. Numbers in blue listed below the arrow indicate 3 U/Th ages obtained for older samples.

(M0037A) and NOG-01B (M0053A and M0054A-B) lime sand rich in larger foraminifera and mollusks underlies the consolidated lithologies. Although there is clear evidence of downhole contamination in their upper part, these deposits appear to be undisturbed and are probably *in situ*, with minimal disturbance from downhole contamination. No consistent pattern has yet been extracted in the succession of these units.

Finally, the deepest hole of Exp. 325 was M0058A at a depth 167 m in the fore-reef slope; it recovered 41 meters of mainly unconsolidated green mud with two intercalated units of fine to medium sand and a few grainstone intervals. The three mud units in M0058A are characterized by a lack of bedding and scattered small fragments of mollusk shells and benthic foraminifera tests. The sand/grainstone units are up to 7 m thick and consist of fine to medium sand with fragments of well-cemented grainstone, mollusks, bryozoa, coralline algae, echinoids, larger foraminifera, and serpulids.

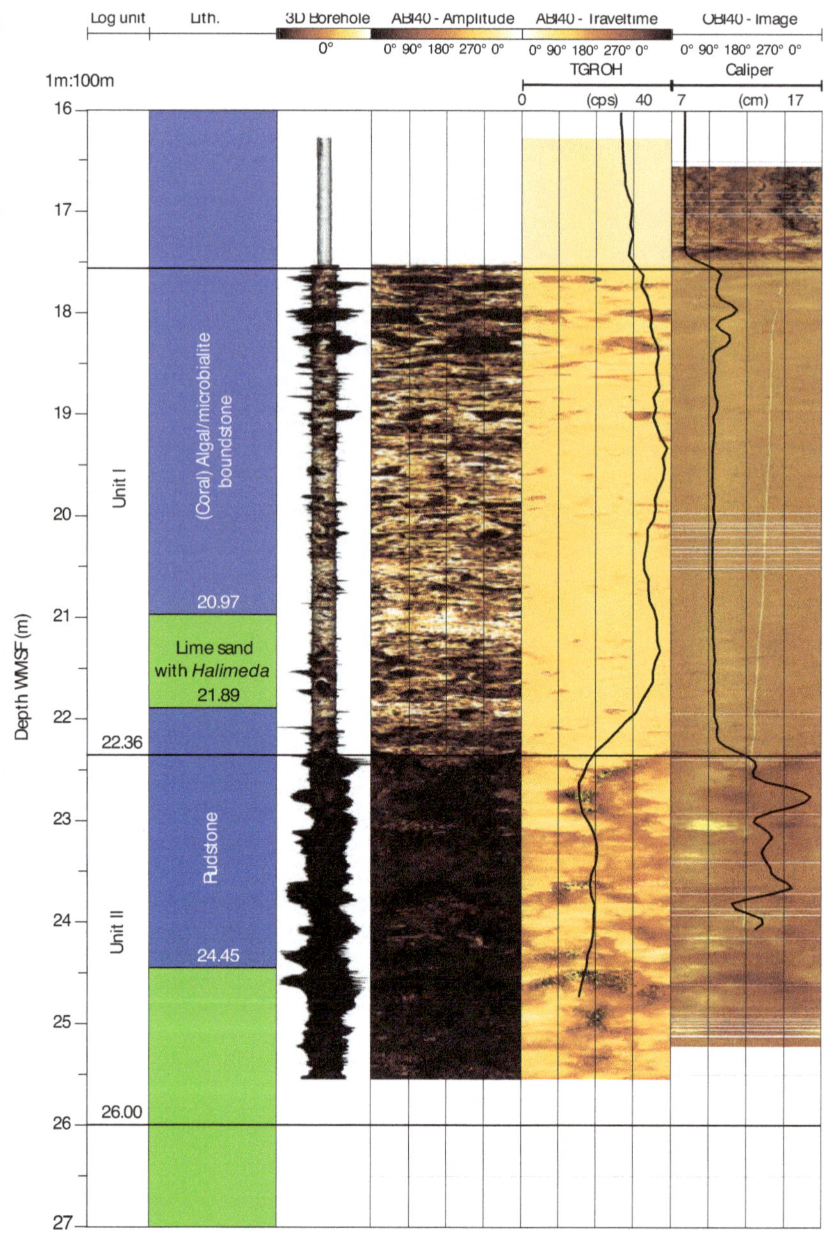


Figure 9. Examples of borehole logging data. The figure illustrates the detail obtained from the ABI40 tool and 3D virtual borehole, transect NOG-01B. (3D Borehole = 3D borehole visualization, ABI 40 traveltime image is overlain by total gamma ray (TGR), OH = open hole). Depth given as wireline log matched depth below sea floor (WMSF).

## Chronology

Preliminary dating provided an important overview of the age of the material recovered during the offshore phase of Exp. 325. Subsamples of core catcher material (coral and mollusk) were taken during the offshore phase for U-Th or radiocarbon dating. The dated samples were free of visible diagenetic features of detrital contamination. Additional diagenetic screening of these samples using XRD and SEM is ongoing. To constrain the basic chronology of each hole, we collected samples from representative core sections from the top, middle, and bottom of the hole. A total of sixty-eight samples were sent to the University of Tokyo (Japan) and the University of Oxford (United Kingdom) for dating. Radiocarbon samples, after being graphitized at University of Tokyo, were transferred to the Australian National University for analysis by AMS (Accelerator Mass Spectrometry). To ensure that the chronological control on the lower portions of holes was not limited by the range of the radiocarbon chronometer (~50 cal k.y. BP [~50 thousand calendar years BP]), the deeper samples from each hole were selected for U-Th analysis, and the shallower samples were analyzed for radiocarbon (see Webster et al., 2011 for detailed information).

Except for Hole M0037A, all sample ages in all of the holes are in stratigraphic order. This provides confidence in the accuracy of the dates, and it ensures that the cores have sampled *in situ* reef framework, consistent with the sedimentologic observations. The majority of the deeper holes have U-Th dates near their bases that indicate the Last Glacial Maximum (LGM) was sampled, ranging between 20 cal k.y. BP and 25 cal k.y. BP (Fig. 8). Shallower samples from these cores (dated by radiocarbon) indicated dates as recent as 13–14 cal k.y. BP, which suggests that the early portion of the deglacial has also been captured by Exp. 325 drill cores. Holes drilled in shallower water have ages as young as 10 cal k.y. BP, indicating that early Holocene coral reef samples have been collected. This data opens up the possibility of comparison between these sites in the GBR with other localities that were drilled from onshore rigs. There appears to be a sharp decline in the number of samples postdating 10 cal k.y. BP (Fig. 8). This may reflect a reef drowning event at this time or may simply be an artifact of the relatively small dataset and/or a sampling bias.

U-Th ages prior to the LGM show promise for recovering high-quality material for dating of earlier periods in the Pleistocene. Holes M0032A, M0056A, and potentially M0033A have material from Marine Isotope Stages 4 and 3, whereas Hole M0042A may provide material from the transition between marine isotope Stages 7 and 6. Hole M0057A has U-Th isotope ratios that suggest that the age of the samples is older than the LGM and, although not yielding a closed system U-Th age due to its intermediate depth, this hole may enable insights into glacial-interglacial transitions older than the last interglacial.

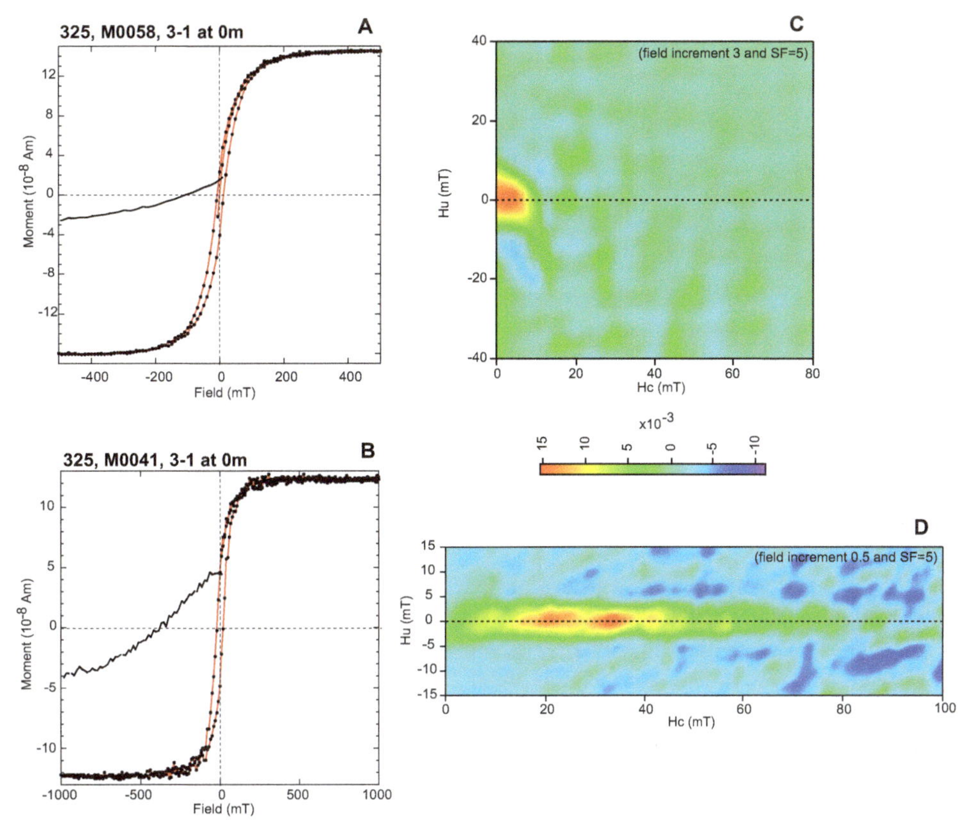


Figure 10. Hysteresis loops, back-field curve ([A], [B]) and first-order reversal curve (FORC) ([C], [D]) of samples from Hole M0058A and M0041A performed with software package FORCinel (Harrison and Feinberg, 2008). The sample from Hole M0058A (field increment 3 mT and SF=5) shows rock magnetic properties mainly of a low coercivity material terrigenous magnetite. The sample from M0041A shows the presence of non-interactive SD magnetite that may represent the presence of magnetosomes.

## Petrophysics

A set of slimline borehole logging probes was chosen on the basis of the scientific objectives and geological setting of the expedition. The tool suite used during Exp. 325 comprised probes with the capability of yielding a variety of data including high resolution borehole images (optical [OBI40] and acoustic [ABI40] borehole televiewers; Fig. 9), borehole fluid characterization (IDRONAUT), borehole diameter (CAL3), and a variety of petrophysical measurements such as electrical conductivity (DIL45), acoustic velocity (SONIC), spectral natural gamma radiation (ASGR), and magnetic susceptibility (EM51). Downhole logging was performed in a total of four boreholes, with the majority of measurements being taken in open hole. The only exception to this was a preliminary spectral gamma log conducted through-pipe in each hole.

Downhole logging units were identified on the basis of combined data signatures. These logging divisions were found to be largely coincident with the lithostratigraphic units defined by core description (Fig. 9). A combination of whole core, split core, and discrete sampling petrophysical measurements were taken on the expedition cores. These measurements include density and porosity, resistivity, P-wave velocity, magnetic susceptibility, thermal conductivity, and color reflectance spectrophotometry. All cores recovered during the expedition were measured where appropriate and possible to do so. Multivariate analysis relating the physical properties data with different coralgal assemblage compositions (Lado-Insua et al., 2010) indicate that it is possible to infer important composition information from the petrophysical dataset. This has the potential for use as a proxy for the identification of sample types from the nondestructive, offshore physical properties measurements in future research. Future work planned by the Exp. 325 Science Party will move towards improving and furthering the integration of the different petrophysical and lithological datasets.

## Paleomagnetics

A total of thirty cores were provided for paleomagnetic measurements. The AF demagnetization of the U-channels and of discrete samples from the fore-reef slope Hole M0058A suggests the presence of magnetic impurities and potential deformation during drilling operations. However, the magnetic susceptibility and artificial remnant magneti-

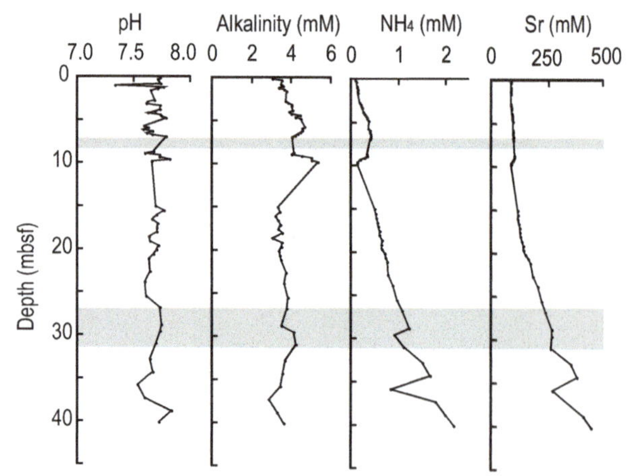


Figure 11. Porewater geochemistry data from Hole M0058A. Gray bars represent the two coarse sand/grainstone units

zations such as anhysteretic (ARM), isothermal (IRM), and indirect parameters (ARM/IRM, S-ratio, and HIRM) show clear separations, with two horizons of high concentrations of a low coercivity magnetic mineral (Webster et al., 2011). Rock magnetic properties confirm that the magnetic carrier is magnetite (Fig. 10a, 10c) with the two horizons characterized by a mixture of low and high-coercivity minerals. The magnetic horizons may correspond to two periods of higher terrigenous sediment input that likely correlate with the two sand/grainstone lithologic units observed in the core. While in the shelf edge fossil reef cores (e.g., M0041), a non-interactive single domain (SD) low coercivity mag-netic mineral (probably magnetite) was measured in discrete samples from fossil coral materials. We attribute that to the presence of magnetosomes of magnetotactic bacteria (Fig. 10b, 10d; Egli et al., 2010). Further studies of rock magnetic properties will provide more detailed information and help establish the nature of relationships between magnetic properties, coral formation, and climate change.

## Geochemistry

A total of 115 interstitial water (IW) samples were acquired during Exp. 325, from transects of HYD-01C (16), HYD-02A (20), RIB-02A (2), and NOG-01B (77). (Numbers in parentheses indicate the number of samples obtained from each transect.) The majority of the IW samples was collected from the holes drilled into the carbonate reef complex of the GBR. The pH, alkalinity, and ammonium concentrations of IW collected from the holes drilled into the carbonate reef complex did not indicate any apparent depth-related or transect-specific variation, probably due to the scarcity of IW samples at each transect.

However, Hole M0058A (NOG-01B transect) consisted of fine to coarse sediments, unlike other holes, and therefore continuous IW sampling was possible. While there was no systematic vertical variation in the pH, alkalinity, and

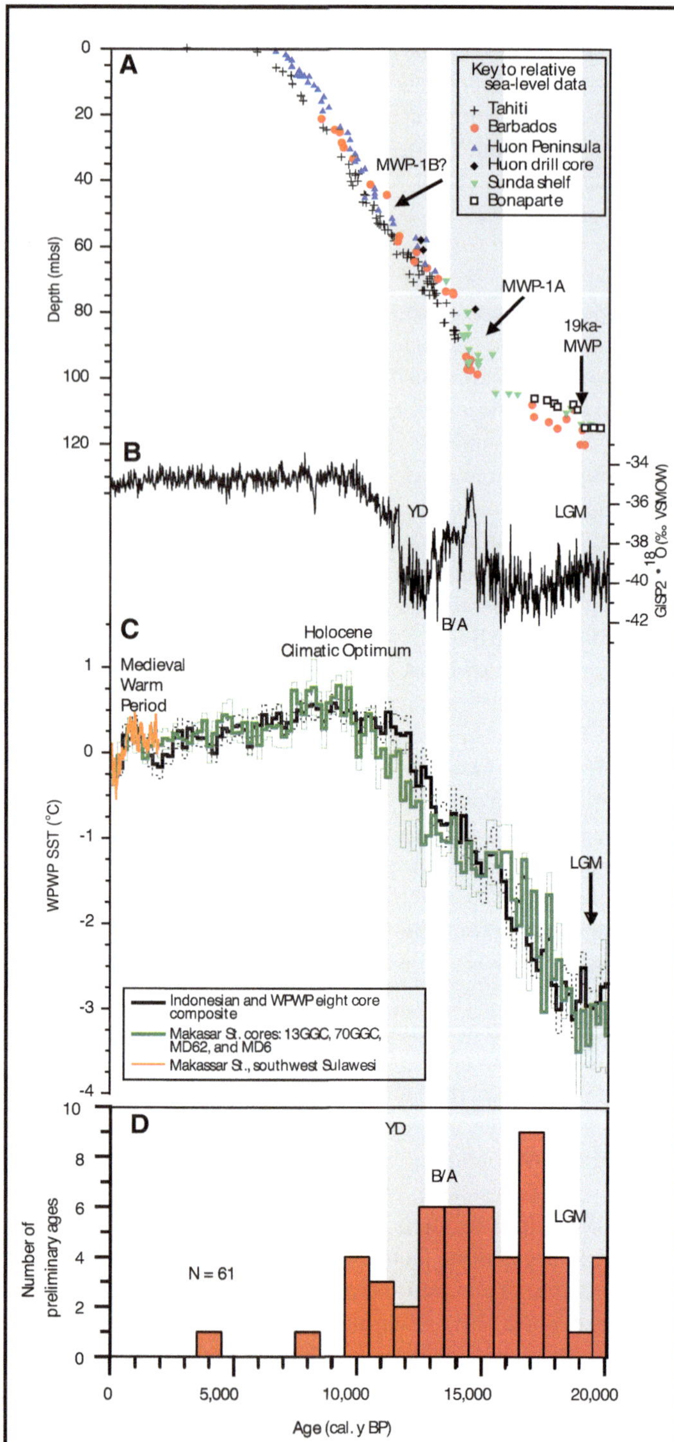


Figure 12. Comparisons of preliminary dating results and previously published sea-level and climatological data. [A] Previously published data on relative sea level from 20 cal k. y. BP through present. MWP= meltwater pulse. [B] GISP2 $\delta^{18}O$ proxy for temperature over Greenland. [C] Sea-surface temperature variation in the Western Pacific Warm Pool (WPWP) (Linsley et al., 2010). [D] Histogram showing preliminary dating results on core catcher samples. Age distribution clearly indicates that the recovered fossil coral reef cores cover key intervals of interest for sea-level changes and environmental reconstruction, including the last glacial maximum (LGM), Bölling-Alleröd (B/A), and Younger Dryas (YD). Sources of data: Tahiti = Bard et al., 1996, 2010; Huon Peninsula = Chappell and Polach, 1991; Edwards et al., 1993; Huon drill core = Cutler et al., 2003; Sunda shelf = Hanebuth et al., 2000; Barbados = Fairbanks, 1989; Bard et al., 1990; Bonaparte = Yokoyama et al., 2000b; 2001a; DeDeckker and Yokoyama, 2009. GISP2 = Stuiver and Grootes, 2000.

chloride, the ammonia and strontium concentrations increased with depth (Fig. 11). The notable characteristic of IW from Hole M0058A is that two large anomalies occur coincident with the two sand/grainstone lithologic units, in the profiles for total iron and manganese concentrations. The percentage quartz profile of the sediments displayed the opposite trend to that of percentage carbonate. These litho-logic units also contain low total organic carbon content, with an average value of 0.25%, compared to the rest of the core. Consistent with the paleomagnetic data, these findings suggest that the two sand/grainstone units may correspond to the increased input of terrestrial material during their deposition. Further investigations are needed to fully understand the cause of lithologic changes found at Hole M0058A.

## Microbiology

The subsurface microbial ecology will be characterized using multiple molecular based techniques. Fine grain sediment cores collected from transect NOG-01B, Hole M0058A represent the bulk of microbiological sampling. Sediment was collected for cell enumeration and phylogenetic analysis during shipboard and onshore operations. Shipboard samples were taken at multiple depths downcore, immediately frozen at -80°C for the duration of the cruise and then shipped on dry ice to labs in the United States (Texas A&M University) and China (China University of Geosciences). Onshore, sediment was collected from locations adjacent to the samples collected during the offshore phase. These samples are designed to test shifts in microbial community activity and structure during 4°C transit to, and storage in, Bremen, Germany against those samples collected immediately after coring that were frozen offshore. Samples remained at 4°C for approximately three months prior to shore-based sampling. Microbial communities in both shipboard and onshore sediment samples will be described for total structure and function using DNA- and RNA-based molecular targets, respectively. For cell enumeration, onshore sediment samples were preserved in 4% formaldehyde and stored at 4°C. By combining all microbial data, the total, potential, and active communities can be determined. The analysis will provide a unique advance in understanding subsurface microbial ecology, as well as a description of potential sediment composition and chemistry altering biological processes that may occur during standard IODP 4°C storage.

## Concluding Remarks and Future Plans

During the course of IODP Expedition 325 offshore phase, thirty-four holes were drilled, obtaining 225 meters of core materials from water depths ranging from 42 mbsl to 167 mbsl (Fig. 12). The preliminary chronology of core catcher materials dated by both radiocarbon and U-series methods prior to the OSP provided a firm chronological framework for conducting further scientific analyses. Fulfillment of the Exp. 325 scientific objectives can be seen as follows:

1. Reconstruct the course of postglacial sea-level change in GBR from 20 ka to 10 ka. Coral-algal-microbialite lithologies of ***in situ*** and robust *Isopora* and *Acropora* assemblages are indicative of very shallow reef environments. They will contribute to the reconstruction of a robust sea-level curve from the LGM to 10 ka.

2. Establish sea-surface variations in the GBR from 20 ka to 10 ka. preliminary chronologic results clearly show that important coral reef deposits were recovered consisting of key paleoenvironmental intervals including LGM, Bölling-Alleröd, Younger Dryas, and Heinrich events. Massive coral colonies suitable for paleoclimate studies will help define SST variations and aid paleoceanographic reconstruction for the region.

3. Investigate the response of the GBR to environmental changes caused by sea-level and climate changes. Cores were recovered from various locations and water depths along the GBR. Their large geographical spread and shelf edge position will allow the results to be interpreted in both a broad temporal and spatial context, allowing a better understanding of the development of the GBR in response to environmental changes.

In addition to the main scientific outcomes summarized above, we expect three additional scientific outcomes will be achieved.

1. It is believed that sea-level and paleoclimate information can be extended back to the LGM and pre-LGM periods based on the preliminary age determinations. This will enable a reconstruction of the evolution of the GBR during these earlier periods.

2. High-resolution paleoenvironmental information obtained from a nearly continuous 41-m sediment core in the fore-reef slope of Noggin Pass (M0058A) will complement the sea-level and paleoenvironmental records obtained from reef cores recovered from shallower shelf sites.

3. The microbial community structure and function will also be assessed within the Hole M0058 sediments using direct count microscopy and RNA/DNA gene targets. Correlations will be made between offshore and onshore geochemical characterizations in order to better describe the subsurface biosphere ecology.

Future work will now focus on more detailed sedimentological, petrophysical, geochemical, and geophysical investigations in the various laboratories of the IODP Exp. 325 Science Party members. The results obtained from these studies will improve our understanding of sea-level and climate change as well as coral reef response since the LGM.

## Acknowledgements

We are grateful to Great Barrier Reef Marine Park Authority (GBRMPA), to the captains and crew of the RV *Great Ship Maya*, to Dan Evans, David Smith, Colin Graham, David McInroy, Dave Wallis, Graham Tulloch, Lee Baines, Mary Mowat, Martin Kölling, Richard le Provost, Jan Hoffman, Simon Barry and the rest of the ESO support staff of the mission-specific operations for ECORD, and to Ursula Röhl, Holger Kuhlmann, Alex Wülbers and the BCR staff in Bremen. Without these individuals and the many others left unnamed, the Great Barrier Reef Coral Reef operations would not have been possible. The work presented here was partly supported by JSPS (NEXT program GR031), GCOE, and the Australian Research Council (DP109400).

## IODP Expedition 325 Scientists

Y. Yokoyama (Co-Chief Scientist), J. M. Webster (Co-Chief Scientist), C. Cotterill (Staff Scientist), L. Anderson, S. Green, R. Bourillot, J.C. Braga, A. Droxler, T.M. Esat, T. Felis, K. Fujita, M. Gagan, E. Gischler, E. Herrero-Bervera, J. Hongchen, M. Humblet, M. Inoue, T. Lado Insua, Y. Iryu, L. Jovane, H. Kan, B. Linsley, D. Loggia, H. Mills, D. Potts, C. Seard, A. Suzuki, A. Thomas, W. Thompson, M. Tiwari, and A. Tudhope.

## References

Abbey, E., Webster, J.M., and Beaman, R., submitted. Morphology and origin of submerged features along the shelf edge of the Great Barrier Reef, Australia. *Mar. Geol.*

Abbey, E.A., and Webster, J.M., 2011. Submerged reefs. *In* Hopley, D. (Ed.), *Encyclopaedia of Modern Coral Reefs:* New York (Springer-Verlag), 1236.

Abram, N., Gagan, M.K., Cole, J.E., Hantoro, W., and Mudelsee, M., 2008. Coral evidence for a recent intensification of the Indian Ocean Dipole. *Nature Geosci.*, 1:849–853, doi:10.1038/ngeo357.

Alexander, I., Wilson, P.A., Cooper, M.J., and International Consortium Great Barrier Reef Drilling, 2001. New constraints on the origin of the Australian Great Barrier Reef: Results from an international project of deep coring. *Geology*, 29(6):483–486, doi:10.1130/0091-7613(2001)029<0483:NCOTOO>2.0.CO;2.

Alley, R.B., 1998. Palaeoclimatology-Icing the north Atlantic. *Nature*, 392(6674):335, doi:10.1038/32781.

Asami, R., Felis, T., Deschamps, P., Hanawa, K., Iryu, Y., Bard, E., Durand, N., and Murayama, M., 2009. Evidence for tropical South Pacific climate change during the Younger Dryas and the Bølling-Allerød from geochemical records of fossil Tahiti corals. *Earth Planet. Sci. Lett.*, 288:96–107, doi:10.1016/j.epsl.2009.09.011.

Bard, E., Hamelin, B., and Delanghe-Sabatier, D., 2010. Deglacial Meltwater Pulse 1B and younger Dryas sea levels revisited with boreholes at Tathiti. *Nature*, 327:1235–1237.

Bard, E., Hamelin, B., and Fairbanks, R.G., 1990. U-Th ages obtained by mass spectrometry in corals from Barbados: Sea level during the past 130,000 years. *Nature*, 346:456–458, doi:10.1038/346456a0.

Bard, E., Hamelin, B., Arnold, M., Montaggioni, L.F., Cabioch, G., Faure, G., and Rougerie, F., 1996. Deglacial sea-level record from Tahiti corals and the timing of global meltwater discharge. *Nature*, 382:241–244, doi:10.1038/382241a0.

Bridge, T.C.L., Done, T.J., Beaman, R.J., Friedman, A., Williams, S.B., Pizarro, O., and Webster, J.M., 2011. Topography, substratum and benthic macrofaunal relationships on a tropical mesophotic shelf margin, central Great Barrier Reef, Australia. *Coral Reefs*, 30:143–153, doi:10.1007/s00338-010-0677-3.

Broecker, W.S., 1994. Massive iceberg discharges as triggers for global climate change. *Nature*, 372:421–424, doi:10.1038/372421a0.

Camoin, G., Seard, C., Deschamps, P., Webster, J., Abbey, E., Braga, J.-C., Iryu, Y., Durand, N., Bard, E., Hamelin, B., Yokoyama, Y., and Thomas, A., 2010. Reef response to sea-level and environmental changes during the last deglaciation. IODP Expedition 310 "Tahiti Sea Level". Geophys. Res. Abstr., EGU2010-14988.

Camoin, G.F., Iryu, Y., McInroy, D.B., and the IODP Expedition 310 Scientists, 2007. IODP Expedition 310 reconstructs sea level, climatic, and environmental changes in the South Pacific during the last degraciation. *Sci. Drill.*, 5:4–12, doi:10.2204 /iodp.sd.5.01.2007.

Chappell, J., and Polach, H., 1991. Post-glacial sea-level rise from a coral record at Huon Peninsula, Papua New Guinea. *Nature*, 349:147–149, doi:10.1038/349147a0.

Clark, P.U., Archer, D., Pollard, D., Blum, J.D., Rial, J.A., Brovkin, V., Mix, A.C., Pisias, N.G., and Roy, M., 2006. The middle Pleistocene transition: Characteristics, mechanisms, and implications for long-term changes in atmospheric pCO2. *Quat. Sci. Rev.*, 25:3150–3184, doi:10.1016/j.quascirev.2006.07.008.

Clark, P.U., Dyke, A.S., Shakun, J.D., Carlson, A.E., Clark, J., Wohlfarth, B., Mitrovica, J.X., Hostetler, S.W., and McCabe, A.M., 2009. The last glacial maximum. *Science*, 325:710–714, doi:10.1126/science.1172873.

CLIMAP, 1981. Seasonal reconstruction of the Earth's surface at the last glacial maximum. Geological Society of America, Map and Chart Series, C36.

Cole, J.E., Dunbar, R.B., McClanahan, T.R., and Muthiga, N., 2000. Tropical Pacific forcing of decadal SST variability in the western Indian Ocean over the past two centuries. *Science*, 287:617–619, doi:10.1126/science.287.5453.617.

Cutler, K.B., Edwards, R.L., Taylor, F.W., Cheng, H., Adkins, J., Gallup, C.D., Cutler, P.M, Burr, G.S., and Bloom, A.L., 2003. Rapid sea-level fall and deep ocean temperature change since the last interglacial period. *Earth Planet. Sci. Lett.*, 206(3–4):253–271, doi:10.1016/S0012-821X(02)01107-X.

Davies, P.J., Symonds, P.A., Feary, D.A., and Pigram, C.J., 1989. The evolution of the carbonate platforms of northeast Australia, controls on carbonate platform and basin development. *Society of Economic Paleontologists and Mineralogists Special Publication*, 44:233–258.

DeDeckker, P., and Yokoyama, Y., 2009. Micropalaeontological evidence for Late Quaternary sea-level changes in Bonaparte Gulf, Australia. *Global Planet. Change*, 66:85–92, doi:10.1016/j.gloplacha.2008.03.012.

DeLong, K., Quinn, T.M., Shen, C.C., and Lin, K., 2010. A snapshot of climate variability at Tahiti 9.5 ka using a fossil coral from IODP Expedition 310. *Geochem. Geophys. Geosyst.*, 11:Q06005, doi:10.1029/2009GC002758.

Denton, G.H., and Hughes, T.J., 1981. *The Last Great Ice Sheets:* New York (John Wiley and Sons).

Deschamps, P., Durand, N., Bard, E., Hamelin, B., Camoin, G., Thomas, A.L., Henderson, G.M., and Yokoyama, Y., 2009. Synchroneity of Meltwater Pulse 1A and the Bolling onset: New evidence from the IODP "Tahiti Sea-Level" Expedition. *Geophys. Res. Abstr.*, 11(EGU2009-10233).

Edwards, R.L., Beck, J.W., Burr, G.S., Donahue, D.J., Chappell, J.M.A., Bloom, A.L., Druffel, E.R.M., and Taylor, F.W., 1993. A large drop in atmospheric 14C/12C and reduced melting in the Younger Dryas, documented with 230Th ages of corals. *Science*, 260:962–968, doi:10.1126/science.260.5110.962.

Edwards, R.L., Chen, J.H., and Wasserburg, G.J., 1987. $^{238}U$-$^{234}U$-$^{230}Th$-$^{232}Th$ systematics and the precise measurement of time over the past 500,000 years. *Earth Planet. Sci. Lett.*, 81:175–192, doi:10.1016/0012-821X(87)90154-3.

Egli, R., Chen, A.P., Winklhofer, M., Kodama, K.P., and Horng, C.-S., 2010. Detection of noninteracting single domain particles using first-order reversal curve diagrams. *Geochem. Geophys. Geosyst.*, 11, Q01Z11, doi:10.1029/2009GC002916.

Esat, T.M., and Yokoyama, Y., 2008. Issues in radiocarbon and U-series dating of corals from the last glacial period. *Quat. Geochronol.*, 3:244–252, doi:10.1016/j.quageo.2007.11.004.

Fairbanks, R.G., 1989. A 17,000-year glacio-eustatic sea level record: Influence of glacial melting dates on Younger Dryas event and deep ocean circulation. *Nature*, 342:637–642, doi:10.1038/342637a0.

Fairbanks, R.G., Mortlock, R.A., Chiu, T.-C., Cao, L., Kaplan, A., Guilderson, T.P., Fairbanks, T.W., Bloom, A.L., Grootes, P.M., and Nadeau, M.-J., 2005. Radiocarbon calibration curve spanning 0 to 50,000 years BP based on paired $^{230}Th$/$^{234}U$/$^{238}U$ and $^{14}C$ dates on pristine corals. *Quat. Sci. Rev.*, 24:1781–1796, doi:10.1016/j.quascirev.2005.04.007.

Felis, T., Suzuki, A., Kuhnert, H., Dima, M., Lohmann, G., and Kawahata, H., 2009. Subtropical coral reveals abrupt early-twentieth-century freshening in the western North Pacific Ocean. *Geology*, 37:527–530, doi:10.1130/G25581A.1.

Fujita, K., Omori, A., Yokoyama, Y., Sakai, S., and Iryu, Y., 2010. Sea-level rise during Termination II inferred from large benthic foraminifers: IODP Expedition 310, Tahiti Sea Level. *Mar. Geol.*, 271:149–155, doi:10.1016/j.margeo.2010.01.019.

Gagan, M.K., Ayliffe, L.K., Hopley, D., Cali, J.A., Mortimer, G.E., Chappell, J., McCulloch, M.T., and Head, M.J., 1998. Temperature and surface-ocean water balance of the mid-Holocene tropiacl western Paficic. *Science*, 279:1014–1018, doi:10.1126/science.279.5353.1014.

Gallup, C.D., Cheng, H., Taylor, F.W., and Edwards, R.L., 2002. Direct determination of the timing of sea level change during Termination II. *Science*, 295:310–313, doi:10.1126/science.1065494.

Hanebuth, T., Stattegger, K., and Grootes, P.M., 2000. Rapid flooding of the Sunda Shelf: A late-glacial sea-level record. *Science*, 288:1033–1035, doi:10.1126/science.288.5468.1033.

Harrison, R.J., and Feinberg, J.M., 2008. FORCinel: An improved algorithm for calculating first-order reversal curve distributions using locally weighted regression smoothing, *Geochem. Geophys. Geosyst.*, 9, Q05016, doi:10.1029/2008GC001987.

Hays, J.D., Imbrie, J., and Shackelton, N.J., 1976. Variations in the Earth's orbit: Pacemaker of the Ice Ages. *Science*, 194:1121–1132, doi:10.1126/science.194.4270.1121.

Inoue, M., Yokoyama, Y., Harada, M., Suzuki, A., Kawahata, H., Matsuzaki, H., and Iryu, Y., 2010. Trace element variations in fossil corals from Tahiti collected by IODP Expedition 310: Reconstruction of marine environments during the last deglaciation (15 to 9 ka). *Mar. Geol.*, 271:303–306, doi:10.1016/j.margeo.2010.02.016.

Ivins, E.R., and James, T.S., 2005. Antarctic glacial isostatic adjustment: A new assessment. *Antarct. Sci.*, 17:537-549, doi:10.1017/S0954102005002968.

Lado-Insua, T., Moran, K., Anderson, L., Webster, J.M., Morgan, S., Fehr, A., Lofi, J., Lukies, V., Loggia, D., and IODP Expedition 325 Scientists, 2010. Are physical properties able to differentiate glacial and interglacial coral identity? [American Geophysical Union, Fall Meeting, San Francisco, 13–17 December 2010] Abstract PP11E-1476.

Lambeck, K., Yokoyama, Y., and Purcel, A., 2002. Into and out of the Last Glacial Maximum: Sea-level change during oxygen isotope stages 3 and 2. *Quat. Sci. Rev.*, 21:343–360, doi:10.1016/S0277-3791(01)00071-3.

Linsley, B.K., Rosenthal, Y., and Oppo, D.W., 2010. Holocene evolution of the Indonesian throughflow and the western Pacific warm pool. *Nature Geosci.*, 3(8):578–583, doi:10.1038/ngeo920.

Milne, G.A., Gehrels, W.R., Hughes, C.W., and Tamisiea, M.E., 2009. Identifying the causes of sea-level change. *Nature Geosci.*, 2:471–478, doi:10.1038/ngeo544.

Mix, A.C., Bard, E., and Schneider, R., 2001. Environmental processes of the ice age: Land, oceans, glaciers (EPILOG). *Quat. Sci. Rev.*, 20:627–657, doi:10.1016/S0277-3791(00)00145-1.

Nakada, M., and Lambeck, K., 1987. Glacial rebound and relative sea-level variations: A new appraisal. *Geophys. J. Roy. Astr. Soc.*, 90:171–224.

Reimer, P.J., Baillie, M.G.L., Bard, E., Bayliss, A., Beck, J.W., Bertrand, C.J.H., Blackwell, P.G., Buck, C.E., Burr, G.S., Cutler, K.B., Damon, P.E., Edwards, R.L., Fairbanks, R.G., Friedrich, M., Guilderson, T.P., Hogg, A.G., Hughen, K.A., Kromer, B., McCormac, G., Manning, S., Ramsey, C.B., Reimer, R.W., Remmele, S., Southon, J.R., Steiver, M., Talamo, S., Taylor, F.W., van der Plicht, J., and Weyhenmeyer,

C.E., 2004. Intcal04 terrestrial radiocarbon age calibration, 0-26 cal kyr BP. *Radiocarbon*, 46:1029–1058.

Seard, C., Camoin, G., Yokoyama, Y., Matsuzaki, H., Durand, N., Bard, E., Sepulcre, S., and Deschamps, P., 2011. Microbialite development patterns in the last deglacial reefs from Tahiti (French Polynesia; IODP Expedition #310): Implications on reef framework architecture. *Mar. Geol.*, 279:63–86, doi:10.1016/j.margeo.2010.10.013.

Solomon, S., Qin, D., Manning, M., Chen, Z., Marquis, M., Averyt, K.B., Tignor, M., and Miller, H.L., 2007. *Climate Change 2007 - The Physical Science Basis. Contribution of Working Group I to the Fourth Assessment Report of the Intergovernmental Panel on Climate Change:* U.K. and New York (Cambridge University Press).

Stirling, C.H., and Andersen, M.B., 2009. Uranium-series dating of fossil coral reefs: Extending the sea-level record beyond the last glacial cycle. *Earth Planet. Sci. Lett.*, 284:269–283, doi:10.1016/j.epsl.2009.04.045.

Stirling, C.H., Esat, T.M., McCulloch, M.T., and Lambeck, K., 1995. High-precision U-series dating of corals from Western Australia and implications for the timing and duration of the Last Interglacial. *Earth Planet. Sci. Lett.*, 135:115–130.

Stuiver, M., and Grootes, P.M., 2000. GISP2 oxygen isotope ratios. *Quat. Res.*, 53:277–283, doi:10.1006/qres.2000.2127.

Thomas, A.L., Henderson, G., Deschamps, P., Yokoyama, Y., Mason, A.J., Bard, E., Hamelin, B., Durand, N., and Camoin, G., 2009. Penultimate deglacial sea level timing from uranium/thorium dating of Tahitian corals. *Science*, 324:1186–1189, doi:10.1126/science.1168754.

Tudhope, A.W., Chilcott, C.P., McCulloch, M.T., Cook, E.R., Chappell, J., Ellan, R.M., Lea, D.W., Lough, J.M., and Shimmield, G.B., 2001. Variability in the El Niño-Southern Oscillation through a glacial-interglacial cycle. *Science*, 291(5508):1511–1517, doi:10.1126/science.1057969.

Webster, J.M., and Davies, P.J., 2003. Coral variation in two deep drill cores: Significance for the Pleistocene development of the Great Barrier Reef. *Sediment. Geol.*, 159:61–80, doi:10.1016/S0037-0738(03)00095-2.

Webster, J.M., Davies, P.J., Beaman, R.J., Williams, S., and Byrne, M., 2008. Evolution of drowned shelf edge reefs in the GBR; implications for understanding abrupt climate change, coral reef response and modern deep water benthic habitats. CSIRO Marine National Facility, RV *Southern Surveyor* voyage summary SS07/2007. http://www.marine.csiro.au/nationalfacility/voyagedocs/2007/summarySS07-2007.pdf.

Webster, J.M., Yokoyama, Y., Cotterill, C., and the Expedition 325 Scientists, 2011. IODP Expedition 325. *Proc. IODP*, 325: Washington, DC (Integrated Ocean Drilling Program Management International, Inc.). doi:10.2204/iodp.proc.325.106.2011.

Yokoyama, Y., 2011a. Last glacial interstadials. *In* Hopley, D. (Ed.), *Encyclopaedia of Modern Coral Reefs*. New York (Springer-Verlag), 1236.

Yokoyama, Y., 2011b. Last glacial lowstand and shelf exposure. *In* Hopley, D. (Ed.) *Encyclopaedia of Modern Coral Reefs*. New York (Springer-Verlag), 1236.

Yokoyama, Y., and Esat, T.M., 2011. Global climate and sea level: Enduring variability and rapid fluctuations over the past 150,000 years. *Oceanography*, 24:54–67, doi:10.5670/oceanog.2011.27.

Yokoyama, Y., DeDeckker, P., Lambeck, K., Johnston, P., and Fifield, L.K., 2001a. Sea-level at the last glacial maximum: Evidence from northwestern Australia to constrain ice volumes for oxygen isotope stage 2. *Palaeogeogr. Palaeoclim. Palaeoecol.*, 165:281–297, doi:10.1016/S0031-0182(00)00164-4.

Yokoyama, Y., Esat, T.M., Lambeck, K., and Fifield, L.K., 2000a. Last ice age millennial scale climate changes recorded in Huon Peninsula corals. *Radiocarbon*, 42:383–401.

Yokoyama, Y., Lambeck, K., DeDeckker, P., Johnston, P., and Fifield, L.K., 2000b. Timing of the Last Glacial Maximum from observed sea-level minima. *Nature*, 406:713–716, doi:10.1038/35021035.

Yokoyama, Y., Purcell, A., Lambeck, K., and Johnston, P., 2001b. Shore-line reconstruction around Australia during the Last Glacial Maximum and Late Glacial Stage. *Quat. Int.*, 83–85:9–18, doi:10.1016/S1040-6182(01)00028-3.

Yokoyama, Y., Purcell, A., Marshall, J.F., and Lambeck, K., 2006. Sea-level during the early deglaciation period in the Great Barrier Reef, Australia. *Global Planet. Change*, 53:147–153, doi:10.1016/j.gloplacha.2006.01.014.

## Authors

**Yusuke Yokoyama**, Atmosphere and Ocean Research Institute, University of Tokyo, 5-1-5 Kashiwanoha, Kashiwa, Chiba 277-8564 Japan, e-mail: yokoyama@aori.u-tokyo.ac.jp.

**Jody M. Webster**, Geocoastal Reseach Group, School of Geosciences, The University of Sydney, Sydney, NSW 2006, Australia.

**Carol Cotterill**, British Geological Survey, Murchison House, Edinburgh, Scotland, EH9 3LA, U.K.

**Juan Carlos Braga**, Departamento de Estratigrafia y Paleontologia, Facultad de Ciencias, Universidad de Granada, Campus Fuentenueva, 18002 Granada, Spain.

**Luigi Jovane**, Geology Department, Western Washington University, 516 High Street, MS 908, Bellingham, WA 98225, U.S.A.

**Heath Mills**, Department of Oceanography, Texas A&M University, 716A Eller, O&M Building, College Station, TX 77843, U.S.A.

**Sally Morgan**, University of Leicester, Department of Geology, University Road, Leicester, LE1 7RH, U.K.

**Atsushi Suzuki**, Geological Survey of Japan, National Institute of Advanced Industrial Science and Technology (AIST), AIST Tsukuba Central 7, 1-1-1 Higashi, Tsukuba, Ibaraki 305-8567, Japan.

**and the IODP Expedition 325 Scientists**

# Initial Feasibility Study to Drill and Core the Ocean Mantle

by Nicolas Pilisi and Bill Whitney

## Introduction

An initial feasibility study (Pilisi and Whitney, 2011) of drilling through the Mohorovičić discontinuity (Moho) into the oceanic mantle specifically focused on future requirements for planning, drilling and coring a hole 500 m into the oceanic mantle from three candidate locations in the Pacific Ocean (Cocos Plate, Baja California, and offshore Hawaii). The study points out some of the critical issues that need to be resolved before embarking upon such a challenging project. It was conducted on the basis of data provided by the Integrated Ocean Drilling Program–Management International (IODP-MI), the Center for Deep Earth Exploration (CDEX) operating the drilling vessel *Chikyu* within IODP, public domain information, and past experience that Blade Energy Partners (hereafter mentioned as "Blade"; http://www.blade-energy.com/) has had with frontier projects in the offshore deepwater oil and gas and geothermal industries.

## Project Challenges

The challenges associated with drilling/coring a hole through the Moho and into the mantle have been understood for some time and have been comprehensively discussed in previous IODP and legacy program reports and documents. Blade's focus was therefore not to re-examine the issues, but rather to evaluate them in terms of current and trending technologies in oilfield and geothermal industries to determine how difficult it will be to resolve the issues. The obvious constraints for this project versus "normal" offshore drilling operations are the extreme water depths where operations need to be conducted, the extreme high temperatures present in very hard igneous rocks that push the limit of all the drilling, and coring tools that are routinely used in less demanding environments. The main challenges that need to be overcome include the following:

1. Riser drilling in ultra-deepwater environments with water depths around 4000 m (which will set a new world record)
2. Drilling and coring in very high temperature igneous rocks with bottom-hole temperatures that are estimated to be as high as 250°C
3. Drilling and coring a very deep hole having a total drilled/cored interval of up to 6700 m in the oceanic crust below the Pacific Ocean seafloor

## Feasibility Study Results

The study (Pilisi and Whitney, 2011) included a detailed analysis of the drilling riser design requirements for the environmental conditions expected at the three candidate locations (Fig. 1). The analysis looked at the operational envelope of the current riser configuration on riser equipped drillship *Chikyu* (Fig. 2), as well as alternative designs for five other possible riser configuration options (current riser with lighter buoyancy modules, titanium riser, slim riser, hybrid riser, current riser with two additional riser tensioners). The analysis concluded that there are existing available technologies, equipment, and materials in the ultra-deepwater industry that should enable the *Chikyu* to conduct operations in the expected water depth range (3650–4300 m) at the candidate locations.

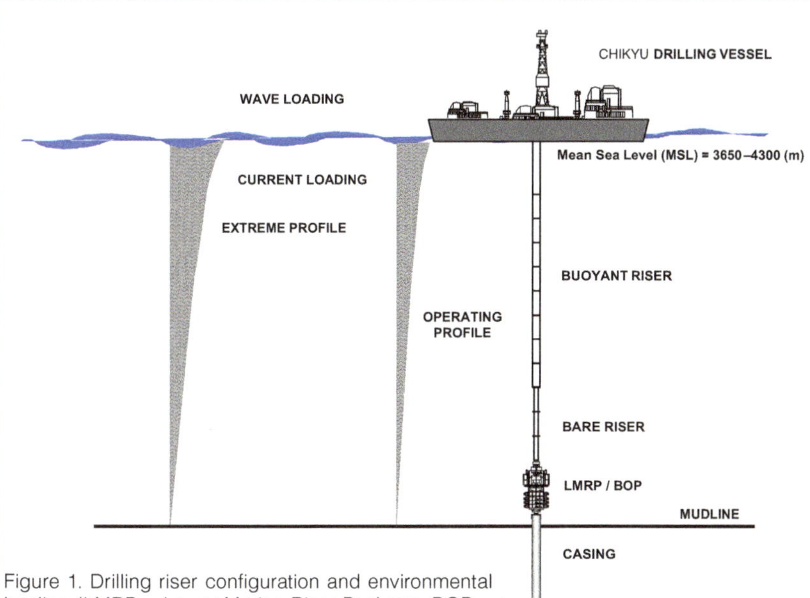

Figure 1. Drilling riser configuration and environmental loading (LMRP = Lower Marine Riser Package; BOP = Blowout Preventer)

The study also looked at different well design scenarios in order to further define the key operational and design issues that will need to be resolved prior to drilling into the mantle well. Most of the information about the downhole conditions is presently unknown. Therefore, assumptions had to be made about the fundamental downhole conditions that impact well design. For example, the assumed downhole temperature profiles for the candidate locations were based on previous models of formation burial depth and age provided by IODP-MI and from operations conducted at the Cocos location in the IODP Hole 1256D (Alt et al., 2007). In addition, a downhole pressure estimate was developed in order to develop a casing program. In most deepwater wells the presence of abnormal pressure is a fundamental criterion for determining casing points and the drilling mud density required to reach total depth. Since the Moho wellbore is expected to be entirely normally pressured (1.03 specific gravity [SG]), abnormal pressure is not an issue. Therefore, the selection of casing points and mud weights will be based on wellbore stability considerations. For example, if the mud weight is too low, the hole will essentially collapse due to a compressive shear failure in the rock. However, a trade-off must be made between the allowable mud weights and the number of casing strings that are used. There are only so many casing strings that can fit in a well; running multiple strings is time-consuming and costly, and it complicates the geometry of the well. It is certainly advantageous to minimize the number of casing strings used for the Moho well, if only to minimize the sizes of the rotary core barrel bits that would need to be developed.

This effort resulted in the development of a base case wellbore configuration that assumes that significant parts of the wellbore will need to be cased off in order to reach total depth (Fig. 3). With the wellbore thus defined, the expected downhole circulating temperature profiles could be modeled, and the issues around the design of a drill string could be examined. It was determined that the circulating temperatures exceed the temperature ratings of most downhole tools that are commercially available today. Increasing the downhole tools' temperature rating is a significant issue that will need to be addressed. On the other hand, it was determined that the Moho wells could be drilled today with a

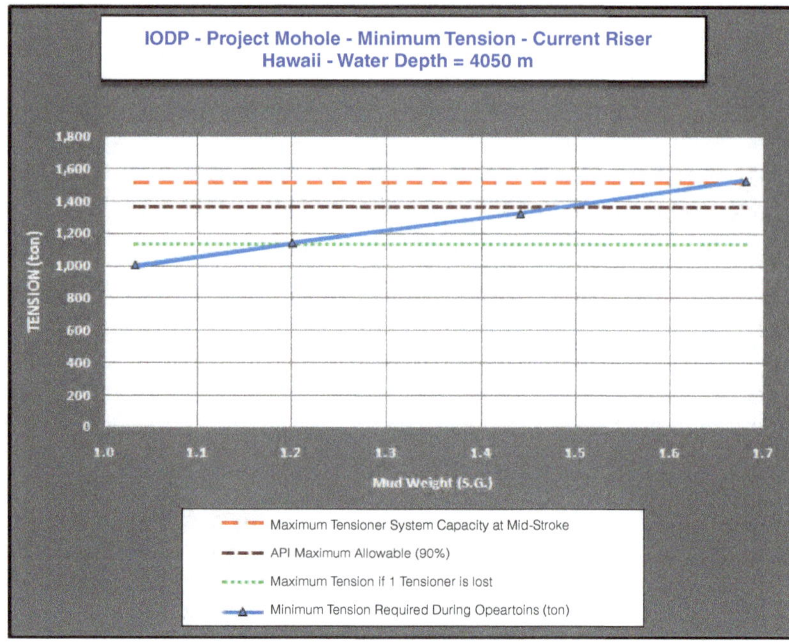


Figure 2. Riser tension requirements vs. mud weight for the current *Chikyu* drilling riser at the Hawaii location (*Chikyu* Spec. 2008.pdf).

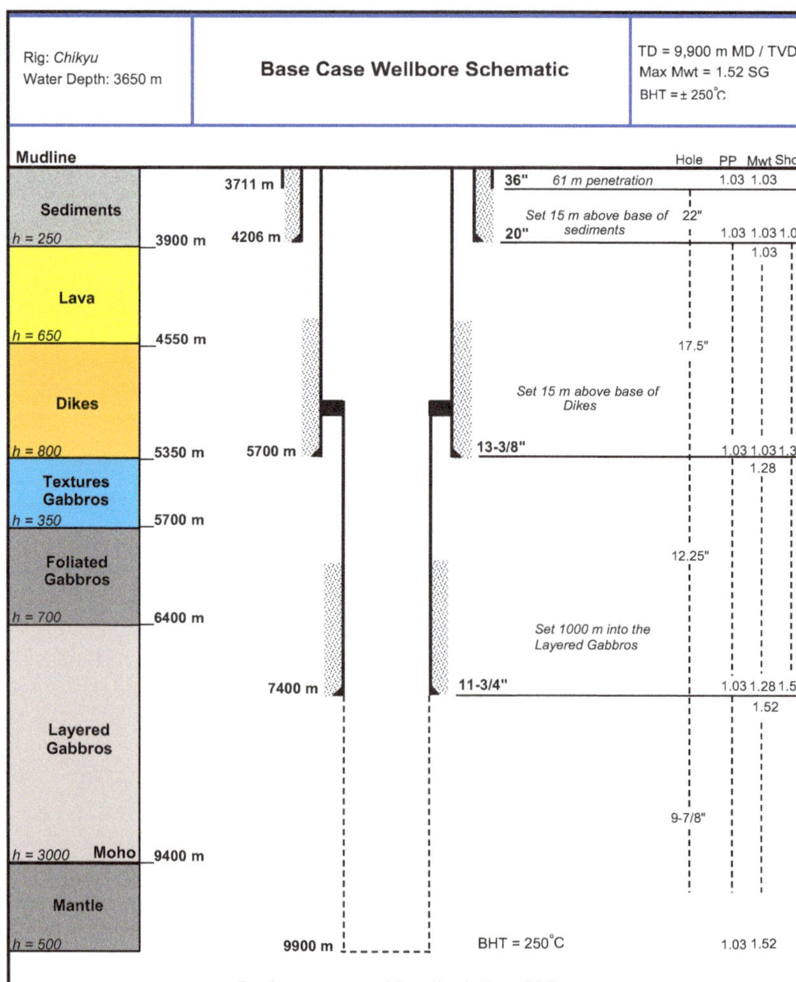


Figure 3. Base case wellbore configuration.

combination of readily available oilfield drill pipe and the *Chikyu*'s existing drill pipe. Therefore a non-standard, special high strength drill string is not needed.

Operational time estimates for four different scientific drilling cases were developed for each of the candidate locations (Table 1). Case 1 assumed that that the hole is continuously cored to total depth. Case 2 assumed long sections of continuous core are taken across the major lithologic and geophysical transition intervals of key sections. Case 3 assumed that only spot coring is done during the last 10 m of hole before each bit trip. And , Case 4 assumed that the hole is drilled to the Moho and that only the mantle is cored. Operations time curves were developed for each different case and location (Fig. 4). It was determined from this work that the time spent tripping for a new drill bit was as much as 40% of the overall operational time. The development of drill bits with enhanced operation hours will therefore have a tremendous impact on the time and cost it will take to drill/core the Moho well.

Table 1. Operational time estimates (based on current technologies).

| Candidate Location | Water Depth (m) | Total Depth Below (m): | | Drill/ Core Time (days) | Total Project Time (days) |
|---|---|---|---|---|---|
| | | Rig Floor | Sea Floor | | |
| Cocos Location | | | | | |
| Case 1 | 3650 | 9900 | 6250 | 696 | 756 |
| Case 2 | 3650 | 9900 | 6250 | 564 | 617 |
| Case 3 | 3650 | 9900 | 6250 | 433 | 480 |
| Case 4 | 3650 | 9900 | 6250 | 374 | 418 |
| Baja Location | | | | | |
| Case 1 | 4300 | 10,400 | 6100 | 807 | 866 |
| Case 2 | 4300 | 10,400 | 6100 | 642 | 693 |
| Case 3 | 4300 | 10,400 | 6100 | 405 | 445 |
| Case 4 | 4300 | 10,400 | 6100 | 386 | 425 |
| Hawaii Location | | | | | |
| Case 1 | 4050 | 10,750 | 6700 | 876 | 934 |
| Case 2 | 4050 | 10,750 | 6700 | 688 | 737 |
| Case 3 | 4050 | 10,750 | 6700 | 448 | 485 |
| Case 4 | 4050 | 10,750 | 6700 | 422 | 443 |

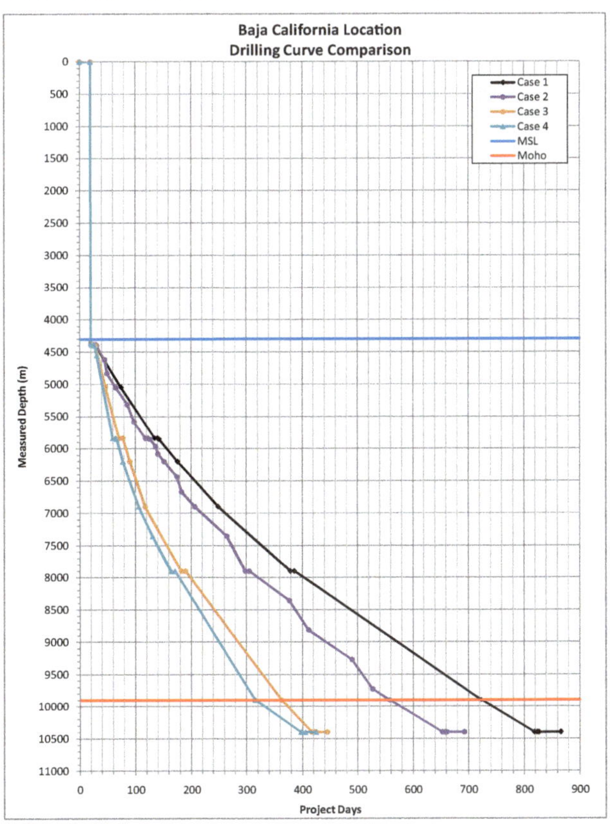


Figure 4. Depth vs. duration of operations. Example for Baja location comparing the four scientific drilling cases

## Conclusions

The results of this study show that drilling/coring a scientific hole into the upper mantle is certainly feasible, and that existing solutions are currently available to many of the technological challenges based on work being done in the oilfield and geothermal industries. In addition, technologies and techniques are continuously advancing and can be expected to continue to close the gap between what is required for the "Moho Project" and what is currently possible.

## References

Alt, J.C., Teagle, D.A.H., Umino, S., Miyashita, S., Banerjee, N.R., Wilson, D.S., the IODP Expeditions 309 and 312 Scientists, and the ODP Leg 206 Scientific Party, 2007. IODP Expeditions 309 and 312 drill an intact section of upper oceanic basement into gabbros. *Sci. Drill.*, 4:4–10. doi:10.2204/iodp.sd.4.01.2007

Pilisi, N., and Whitney, B., 2011. *IODP - Project Mohole - Initial Blade Feasibility Study.* Report to IODP by Blade Energy Partners, Frisco, Texas, 7 June 2011. Available online at http://www.iodp.org/weblinks/Featured-Publications-HOME-PAGE/IODP-Project-MoHole-Initial-Feasibility-Study-PDF-

For full references, see the full report at http://www.iodp.org/weblinks/Featured-Publications-HOME-PAGE/IODP-Project-MoHole-Initial-Feasibility-Study-PDF-

## Authors

**Nicolas Pilisi and Bill Whitney**, Blade Energy Partners, 2600 Network Boulevard, #550, Frisco, TX 75034-6036, U.S.A., e-mail: bwhitney@blade-energy.com, npilisi@blade-energy.com.

# International Drilling to Recover Aquifer Sands (IDRAs) and Arsenic Contaminated Groundwater in Asia

by Alexander van Geen

A group of forty-four scientists and students from fourteen countries met on 25–27 April 2011 in Hanoi, Vietnam to set the stage for a new continental drilling program focused on the groundwater arsenic (As) problem in Asia. Half of the participants were from low and lower-middle income countries directly facing the major health threat of elevated groundwater As. This issue affects over 100 million rural inhabitants across Bangladesh, Cambodia, China, India, Myanmar, Nepal, Pakistan, and Vietnam who rely on shallow (typically <100-m-deep) wells as their main source of drinking water (Ravenscroft et al., 2009). Groundwater from these areas can contain As at levels that have been shown to cause deadly cancers and cardiovascular disease and to inhibit the mental development of children (Smith, A.H. et al., 2000; Wasserman et al., 2004; Kapaj et al., 2006; Argos et al., 2010).

There is broad agreement within the international scientific community that reductive dissolution of iron (Fe) oxyhydroxides in river basins across Asia is a key factor leading to widespread occurrence of As in shallow groundwater (Kinniburgh and Smedley, 2001; Fendorf et al., 2010). The underlying reason is the strong affinity of As for Fe oxyhydroxides that results in As accumulation on the coatings of suspended particles. Upon burial, As is released to groundwater by microbially mediated dissolution of these Fe oxyhydroxides. There is a lack of necessary data, however, to discern the relative importance of other mechanisms, including the role of local vs. advected sources of reactive organic carbon that drive aquifers towards reduction. There is also a lack of critical information required to identify the cause of the extraordinary variability of groundwater As concentrations across a wide range of spatial scales (Fig. 1.) It is clear that this spatial variability is not directly controlled by the bulk As content of the sediment which, although not particularly elevated, is sufficient to increase groundwater As concentrations to over 1000 times the World Health Organization's guideline of 10 μg $L^{-1}$ for As in drinking water if released.

Recent work reviewed at the International Drilling to Recover Aquifer Sands (IDRAs) workshop highlighted the importance of understanding the interactions between geology, hydrology, geochemistry, and microbiology that must be understood to understand and predict the release of As that occurs mostly in aquifers <100 m deep (Fig. 1a). Because of the difficulty of distinguishing some of these processes, participants agreed to focus their efforts on the issue of greatest societal relevance, which is the vulnerability of those aquifers that are currently low in As (von Brömssen et al., 2007; Michael and Voss, 2008; Winkel et al., 2011). Tapping those aquifers, shallow or deep, that are low in As has had a much bigger impact in terms of lowering exposure than any other form of mitigation, including water treatment or rainwater harvesting, and it will likely continue to do so for the foreseeable future (Ahmed et al., 2006).

Workshop participants agreed that an effective approach to gaining a better understanding of the vulnerability of low-As aquifers is to study pronounced lateral and vertical concentration gradients and determine whether such transitions could shift in response to massive groundwater pumping in some regions. In this context, one key question is whether a low-As aquifer is more likely to become contaminated because of inflow from a high-As zone or because of *in situ* As release triggered by a change in groundwater composition induced by massive pumping for irrigation or municipal and industrial use (Fig. 2). This is a critical distinction. A number of plausible mechanisms have been proposed by which changes in the flow or composition of groundwater in response to pumping could either increase or decrease the As levels in groundwater (Klump et al., 2006; Neumann et al., 2010; McArthur et al., 2010), but none have been demonstrated across a broad range of settings.

Workshop participants agreed that a decade-long integrated research program is required in order to identify the mechanisms that are most relevant to predicting the fate of low-As aquifers. A guiding hypothesis for such a program was formulated at the workshop as follows:

"The distribution of As in reducing groundwater is anchored to the local geology across a range of spatial scales and therefore naturally evolves on 100–1000 year time scales. The scale of human perturbations of the natural groundwater flow that threaten aquifers that are currently low in As on 10-yr timescales can be predicted by considering a limited set of parameters."

Initial studies would be focused on a subset of previously studied sites from which background information is already available, as well as a on series of new transects that are scientifically relevant but also cross the borders between affected countries, some of which are politically sensitive (e.g., Pakistan-India, India-Bangladesh, and Cambodia-Vietnam). The objective of in-depth study of selected sites would be to identify which minimum set of geological, hydrological, geochemical, and microbial parameters must be considered to make reasonably accurate predictions of the vulnerability of low-As aquifers across the affected region in the absence of a full-scale study.

A key factor that has impeded progress in understanding the processes regulating groundwater As levels is that matched samples of uncompromised groundwater and aquifer sediment from precisely the same interval are typically not available for detailed analysis and incubation. Participants agreed that a transportable drilling facility with new technology for collecting and processing paired groundwater and sediment samples under sterile and anoxic conditions would be an important catalyst for a long-term collaborative interdisciplinary research program across the affected region. The new technology is based on isolating a coring tube until the desired depth has been reached, followed by sealing the end of the coring tube before retrieval by *in situ* freezing.

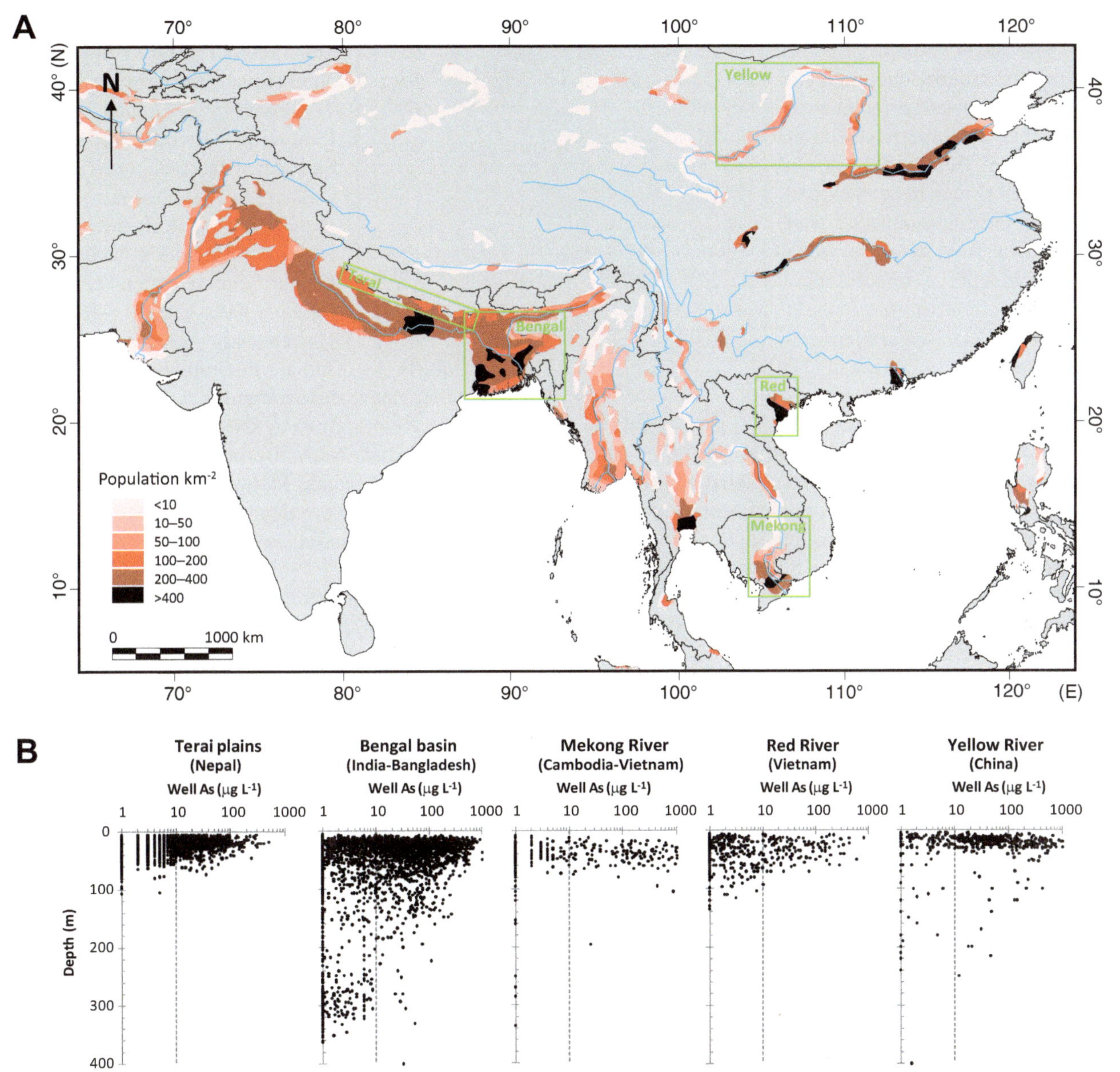


Figure 1. [A] Map of population density within regions where groundwater As concentrations are potentially elevated. [B] Geographic information system (GIS) files with population density from Ravenscroft (2007). Distribution of As in groundwater as a function of depth within five areas indicated by green rectangles in [A]. The vertical dashed line marks the WHO guideline of 10 µg $L^{-1}$. Sources include Kinniburg and Smedley (2001), Guo et al. (2003), Shrestha et al. (2004), Nickson et al. (2007), Buschmann et al. (2008), Guo et al. (2008), Deng et al. (2009), Gao et al. (2010), Han et al. (2010), and Winkel et al. (2011). These regions were outlined by considering a limited set of hydrogeological and climatic factors. Whereas groundwater contamination with As has been confirmed by actual measurements in many of the predicted regions, groundwater As concentrations have also been measured and found not to be elevated in some, including the lower reaches of the Yellow River in China and Chao Praya Delta in Thailand. In other areas such as the lower Mekong River, groundwater As levels may be elevated but the local population does not drink the water because it is affected by salt intrusion (Ravenscroft et al., 2009).

Such a drill rig, sampling tools, and a machine shop for repair and maintenance included with the mobile facility would fit in four 6-m shipping containers.

The proposed drilling facility would be complemented with a fifth containerized laboratory equipped to sterilize sampling tools, extrude aquifer samples anaerobically, analyze labile properties, and prepare samples for shipping to laboratories for further analyses. The facility would have the ability to test for sample contamination with drilling fluids using techniques developed under the Integrated Ocean Drilling Program (Smith, D.C. et al., 2000). The facility has been designed for sampling both groundwater and aquifer sediment from depths of up to 300 m, which will be important to determine the fate of the particularly deep low-As aquifers that are increasingly relied on in the Bengal Basin (Fig. 1).

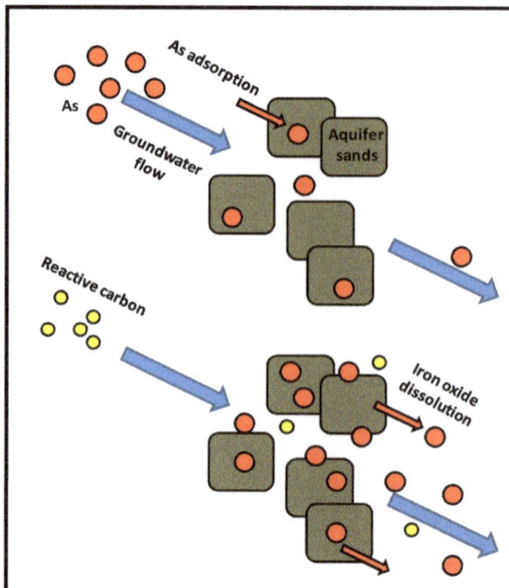


Figure 2. Conceptual representation of processes regulating the As content of groundwater. A distinction is drawn between a low-As aquifer that is contaminated by As released elsewhere (top) and locally (bottom). Groundwater flow could play a role by supplying As directly or indirectly by supplying reactive organic carbon that triggers reductive dissolution of iron oxides.

Over the next several months, workshop participants as well as interested scientists who could not join this particular meeting will prepare a full International Continental Scientific Drilling Program (ICDP) proposal for the January 2012 deadline. Key features were outlined at the Hanoi workshop, including identifying a viable management structure for prioritizing study sites, ensuring broad participation, standardizing methods, and securing supplemental funding. The group's likely strategy will be to focus during a first five-year phase on sites with sufficient geophysical and hydrological background information that are either unperturbed by groundwater pumping (e.g., portions of Vietnam or Cambodia) or so strongly perturbed by municipal pumping (e.g., around Kolkata, Dhaka, or Hanoi) that groundwater flow patterns are fairly predictable. Based on the new knowledge generated during the first phase, the group expects to be ready to study sites that are affected by more complex patterns of irrigation pumping in other regions in a second five-year phase. This multi-year and multi-country project is unique in scientific drilling, and it will require strong management, communication, and coordination. Participants agreed that annual workshops and support for scientists and students from countries affected by As to work in countries other than their own will be key to maintaining the momentum generated in Hanoi by providing a forum for planning and reviewing the latest findings.

## IDRAs Workshop Participants

Australia: Bibhash Nath (University of Sydney); Bangladesh: Kazi Matin Ahmed (University of Dhaka), Ashraf Ali Seddique (Jessore Science & Technology University), Mohammad Tauhid-Ur-Rahman (Shahjalal University of Science & Technology), Anwar Zahid (Bangladesh Water Development Board); Canada: Roger Beckie (University of British Columbia), Mario Bianchin (Lorax Environmental); China: Huaming Guo (China University of Geosciences); France: Laurent Charlet (University of Grenoble); Germany: Sebastian Behrens, Andreas Kappler (Eberhard-Karls-University Tuebingen), Harald Neidhardt, Thomas Neumann (Karlsruhe Institute of Technology), Thomas Wiersberg (Scientific Drilling ICDP), India: Sudipta Chakraborty (Kanchrapara College), Debashis Chatterjee (University of Kalyani), Syed Hillal Farooq (St.Xavier's College), Manoj Kumar, Sonnenchien Kushagra, Al Ramanathan, Chander Kumar Singh (Jawaharlal Nehru University), Abhijit Mukherjee (Indian Institute of Technology (IIT)-Kharagpur); Japan: Harue Masuda (Osaka City University); Nepal: Jaya Kumar Gurung (Nepal Development Research Institute), Bishal Nath Upreti (Tribhuvan University); Pakistan: Abida Farooqi, Mehwish Ramzan (Fatima Jinnah Women University), Sweden: Prosun Bhattacharya (Royal Institute of Technology); Switzerland: Michael Berg, Rolf Kipfer, Lenny Winkel (Eawag); U.S.A.: Benjamin Bostick (Lamont-Doherty Earth Observatory), Karen Johannesson (Tulane University), Natalie Mladenov, Diana Nemergut (University of Colorado), Dennis Nielson (DOSECC), Peggy O'Day (University of California, Merced), David Smith, Athur Spivack (University of Rhode Island), Alexander van Geen (Lamont-Doherty Earth Observatory); Vietnam: Ngoc Son Nguyen (Vietnam National Assembly Committee), Thi Ha Nguyen (Centre of Water Resources Monitoring and Forecast), Pham Thi Kim Trang, Pham Hung Viet (Hanoi University of Science).

## Acknowledgments

The ICDP and the U.S. National Science Foundation's Division of Earth Sciences provided partial support for this workshop.

## References

Ahmed, M.F., Ahuja, S., Alauddin, M., Hug, S.J., Lloyd, J.R., Pfaff, A., Pichler, T., Saltikov, C., Stute, M., and van Geen, A., 2006. Ensuring safe drinking water in Bangladesh. *Science*, 314:1687–1688, doi:10.1126/science.1133146.

Argos, M., Kalra, T., Rahouz, P.J., Chen, Y., Pierce, B., Parvez, F., Islam, T., Ahmed, A., Rakibuz-Zaman, M., Hasan, R.,

Sarwar, G., Slavkovich, V., van Geen, A., Granziano, J., and Ahsan, H., 2010. Arsenic exposure from drinking water, and all-cause and chronic-disease mortalities in Bangladesh (HEALS): A prospective cohort study. *The Lancet*, 376 (9737):252–258, doi: 10.1016/S0140-3736(10)60481-3.

Buschmann, J., Berg, M., Stengel, C., Winkel, L., Sampson, M.L., Trang, P.T.K., and Viet, P.H., 2008. Contamination of drinking water resources in the Mekong delta floodplains: Arsenic and other trace metals pose serious health risks to population. *Environ. Int.*, 34:756–764, doi:10.1016/j.envint.2007.12.025.

Deng, Y., Wang, Y., and Ma, T., 2009. Isotope and minor element geochemistry of high arsenic groundwater from Hangjinhouqi, the Hetao plain, Inner Mongolia. *Appl. Geochem.*, 24:587–599.

Fendorf, S., Michael, H., and van Geen, A., 2010. Spatial and temporal variations of groundwater arsenic in South and Southeast Asia. *Science*, 328:1123–1127, doi:10.1126/science.1172974.

Gao, C.R., Liu, W.B., Li, J.F., and Li, F., 2010. Modes of occurrence of arsenic in Quaternary sediments of the Hetao Plain. *Geol. China*, 37(3):760–770.

Guo, H., Wang, Y., Shpeizer, G.M., and Yan, S., 2003. Natural occurrence of arsenic in shallow groundwater, Shanyin, Datong Basin, China. *J. Environ. Sci. Health, Part A*, 38(11):2565–2580.

Guo, H., Yang, S., Tang, X., Li, Y., and Shen, Z., 2008. Groundwater geochemistry and its implications for arsenic mobilization in shallow aquifers of the Hetao Basin, Inner Mongolia. *Sci. Total Environ.*, 393:131–144, doi:10.1016/j.scitotenv.2007.12.025.

Han, S.B., Zhang, F.C., Zhang, H., Jia, X.F., He, J., and Li, X.F., 2010. An analysis of the distribution and formation of high arsenic groundwater in northern China. *Geol. China*, 37(3):747–753.

Kinniburgh, D.G., and Smedley, P.L., 2001. Arsenic contamination of ground water in Bangladesh, Final Report, vol. 2, *BGS Technical Report* WC/00/19, British Geological Survey, Keyworth, U.K., http://www.bgs.ac.uk/arsenic/bphase2/reports.htm.

Klump, S., Kipfer, R., Cirpka, O.A., Harvey, C.F., Brennwald, M.S., Ashfaque, K.N., Badruzzaman, A.B.M., Hug, S.J., and Imboden, D.M., 2006. Groundwater dynamics and arsenic mobilization in Bangladesh assessed using noble gases and tritium. *Environ. Sci. Technol.*, 40:243–250, doi:10.1021/es051284w.

Kapaj, S., Peterson, H., Liber, K., and Bhattacharya, P., 2006. Human health effects from chronic arsenic poisoning – A review. *J. Environ. Sci. Health, Part A*, 41(10):2399–2428.

McArthur, J.M., Banerjee, D.M., Sengupta, S., Ravenscroft, P., Klump, S., Sarkar, A., Disch, B., and Kipfer, R., 2010. Migration of As, and $^3H/^3He$ ages, in groundwater from West Bengal: Implications for monitoring. *Water Res.*, 44:4171–4185, doi:10.1016/j.watres.2010.05.010.

Michael, H.A., and Voss, C.I., 2008. Evaluation of the sustainability of deep groundwater as an arsenic-safe resource in the Bengal Basin. *Proc. Natl. Acad. Sci. USA*, 105:8531–8536, doi:10.1073/pnas.0710477105.

Neumann, R.B., Ashfaque, K.N., Badruzamman, A.B.M., Ali, M.A., Shoemaker, J.K., and Harvey, C.F., 2010. Anthropogenic influences on groundwater arsenic concentrations in Bangladesh. *Nature Geosci.*, 3:46–52. doi: 10.1038/NGEO685.

Nickson, R., Sengupta, C., Mitra, P., Dave, S.N., Banerjee, A.K., Bhattacharya, A., Basu, S., Kakoti, N., Moorthy, N.S., Wasuja, M., Kumar, M., Mishra, D.S., Ghosh, A., Vaish, D.P., Srivastava, A.K., Tripathi, R.M., Singh, S.N., Prasad, R., Bhattacharya, S., and Deverill, P., 2007. Current knowledge on the distribution of arsenic in groundwater in five states of India. *J. Environ. Sci. Health A Tox. Hazard. Subst. Environ. Eng.*, 42:1707–1718, doi:10.1080/10934520701564194.

Ravenscroft, P., 2007. Predicting the global extent of arsenic pollution of groundwater and its potential impact on human health. Unpublished report prepared for UNICEF, December 2007. Accessible online at: http://phys4.harvard.edu/~wilson/arsenic/references/Ravenscroft_Prediction.pdf.

Ravenscroft, P., Brammer, H., and Richards, K., 2009. *Arsenic Pollution: A Global Synthesis:* Chichester, U.K. (Wiley-Blackwell). doi:10.1002/9781444308785.

Shrestha, C.R., Whitney, J.W., and Shrestha, K.B., 2004. *The State of Arsenic in Nepal-2003*: Kathmandu, Nepal (National Arsenic Steering Committee, Environment and Public Health Organization), 126 pages and CD-ROM.

Smith, A.H., Lingas, E.O., Rahman, M., 2000. Contamination of drinking-water by arsenic in Bangladesh: a public health emergency. *Bulletin* WHO 2000, 78(9):1093–1103.

Smith, D.C., Spivack, A.J., Fisk, M.R., Haveman, S.A., Staudigel, H., and Ocean Drilling Program Leg 185 Shipboard Scientific Party, 2000. Tracer-based estimates of drilling-induced microbial contamination of deep sea crust. *Geomicrobiol. J.*, 17:207–219, doi:10.1080/01490450050121170.

von Brömssen, M., Jakariya, M., Bhattacharya, P., Ahmed, K.M., Hasan, M.A., Sracek, O., Jonsson, L., Lundell, L., and Jacks G., 2007. Targeting low-arsenic aquifers in groundwater of Matlab Upazila, Southeastern Bangladesh. *Sci. Total Environ.*, 379:121–132, doi:10.1016/j.scitotenv.2006.06.028.

Wasserman, G.A., Liu, X., Parvez, F., Ahsan, H., Factor-Litvak, P., van Geen, A., Cheng, Z., Slavkovich, V., Hussain, I., Momotaj, H., and Graziano, J.H., 2004. Water arsenic exposure and children's intellectual function in Araihazar, Bangladesh. *Environ. Health Perspect.*, 112:1329–1333, doi:10.1289/ehp.6964.

Winkel, L.H.E., Trang, P.T.K., Lan, V.M, Stengel, C., Amini, M., Ha, N.T., Viet, P.H., and Berg, M., 2011. Arsenic pollution of groundwater in Vietnam exacerbated by deep aquifer exploitation for more than a century. *Proc. Natl. Acad. Sci. USA*, 108:1246–1251, doi:10.1073/pnas.1011915108.

## Author

**Alexander van Geen,** Lamont-Doherty Earth Observatory (LDEO), Columbia University, 131 Corner, 61 Route 9W – P.O. Box 1000, Palisades, N.Y. 10964-8000, U.S.A., e-mail: avangeen@ldeo.columbia.edu.

## Related Web Link

http://hanoi.icdp-online.org

# 42

# Deep Scientific Drilling to Study Reservoir-Triggered Earthquakes in Koyna, Western India

by Harsh Gupta, Shailesh Nayak and the Koyna Workshop Committee

## Introduction

During 21–25 March 2011, the International Workshop on Scientific Deep Drilling in the Koyna region in western India was held. It was organized by the National Geophysical Research Institute (NGRI), Hyderabad. The main objective of the workshop was to discuss a proposal for deep scientific drilling down to earthquake focal depths of ~7 km at Koyna, a classical site of Reservoir Triggered Seismicity (RTS) in an intra-plate setting, and to design a comprehensive experiment through discussions with national and international experts. Participants of the workshop included seismologists, geologists, and drilling experts, many of them associated with active fault zone drilling projects worldwide, such as the San Andreas Fault Observatory at Depth (SAFOD) in California, the Chelungpu Fault Drilling Project in Taiwan, the Nojima Fault Drilling in Japan, the Gulf of Corinth in Greece, and the Latur Fault of India. There were twenty-six international participants from Canada, France, Germany, Italy, Japan, New Zealand, Poland, Taiwan, and the U.S.A., and fifty Indian participants represented major earth science organizations, institutes, and universities within the country. The workshop was supported by the Ministry of Earth Sciences (MoES), Government of India, and the International Continental Scientific Drilling Program (ICDP).

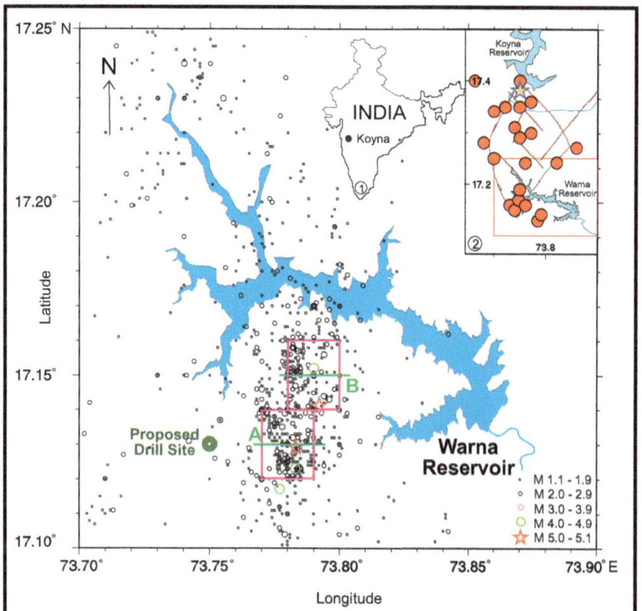


Figure 1. Seismicity in the Warna region south of Koyna, western India, during 2009–10 (red rectangle in inset 2). Inset 1: Map of India indicating the Koyna region; Inset 2: Earthquakes of magnitude 5 and above in the Koyna-Warna region since 1967.

The Koyna region near the west coast of India is globally the premier site of RTS, where induced earthquakes have been occurring in a restricted area of 20 x 30 km since the impoundment of Shivajisagar Lake in 1962. These include the largest triggered earthquake of M~6.3 on 10 December 1967 and twenty-one earthquakes of M>5 since 1962 (Fig. 1). The RTS was further enhanced by impoundment of the nearby located Warna reservoir in 1993. The seismic zone is quite isolated with no other source of activity within 50 km of the Koyna Dam. The continued seismicity in an isolated zone that can be easily monitored provides a unique opportunity to directly measure the physical and mechanical properties of rocks, pore fluid pressure, hydrology, temperature, and other parameters of an intra-plate active fault zone in the "near-field" of earthquakes before, during, and after their occurrence. The focal depths, especially in the Warna region to the south, are mostly within 7 km (Fig. 1), and they can be accessed by drilling with available expertise. This active seismic zone, therefore, forms an ideal site to set up a borehole observatory which will permit direct and continuous monitoring of an intra-plate seismic zone at depth, leading to a better understanding of the mechanics of faulting and the physics of reservoir triggered earthquakes, and it will contribute appreciably to earthquake hazard assessment and forecasting.

## Structure of the Workshop

The initial part of the workshop had four thematic sessions including (i) Global Review of Reservoir Triggered Seismicity, Models and Hypotheses, (ii) Geology and Geophysics of the Koyna Region, (iii) Global Status of Drilling into Fault Zones, and (iv) Designing the Koyna Scientific Drilling Experiment. The presentations provided an up-to-date status of expertise on fault drilling projects worldwide and brought out the core issues with respect to deep drilling investigations in the Koyna region. The next phase of the workshop included a field trip to the Koyna region and breakout group discussions for detailed planning of the required investigations. The participants were divided into four breakout groups based on their expertise: (i) Seismology/Borehole Location(s)/RTS, (ii) Drilling/Coring/Geological Logging/Fault Zone Studies, (iii) Geophysical Logging/Petrophysical Properties/Long-term

Monitoring, and (iv) Temperature/Hydrogeology/Fluid and Gas Sampling. The breakout group discussions started at Hyderabad and continued through the field visit to Koyna and for another full day at Karad. The hallmark of the event was the gathering of experts from all major fault zone drilling programs worldwide, all working together to address the major challenges of undertaking the deep drilling investigations at Koyna.

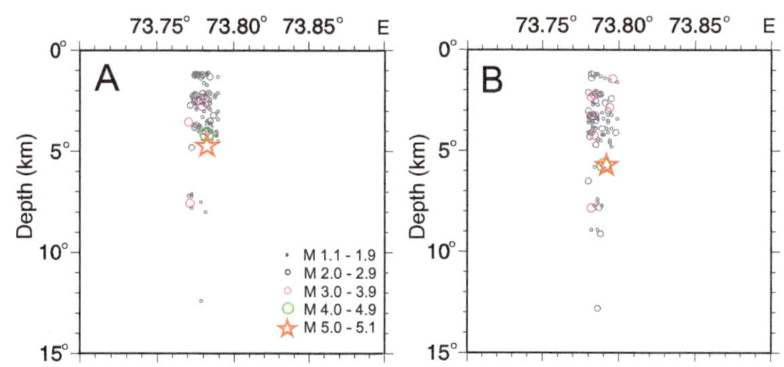


Figure 2. Depth sections of the two blocks shown in Fig. 1 in pink. Number of earthquakes during 2009–2010 in [A]: M1.1–1.9 (81 events), M2–2.9 (26 events), M3–3.9 (5 events), and one earthquake each of M4.3 and M5.0. In [B], M1.1–1.9 (54 events), M2–2.9 (24 events), M3–3.9 (7 events), and one earthquake each of M4.2 and M5.1.

## Suggested Site for Deep Borehole

A borehole site for deep scientific drilling (green dot in Fig. 1, about 2 km west of block) was suggested on the basis of observed seismicity in the past two years (2009 and 2010), accessibility, and other logistic considerations. In Fig. 1, hypocenters are plotted of earthquakes of magnitude M1.1 and larger that occurred during 2009 and 2010 in block A. The seismic activity has been quite intense, as there were 81, 26, and 5 events of M1.1 to M1.9, M2.0 to M2.9, and M3.0 to M3.9, respectively. During 2009 and 2010 there was also one earthquake each of M4.3 and 5.0 in block A. A majority of these events are confined within a focal depth of 1–5 km in block A.

## Outcome of the Workshop

All the participants at the workshop agreed that Koyna is an outstanding world-class geological site and a natural seismic laboratory to conduct a deep borehole experiment for earthquake studies. The MoES has declared its support to the program, and the ICDP has offered to make available its technical expertise in deep drilling and logging, training of manpower, and support towards drilling after receiving a successful drilling proposal. Experts from around the world have offered to bring in new tools and techniques for measurements and modeling.

On the basis of intensive discussions among the participants and important suggestions received from experts from India and abroad, a few key areas were identified for detailed preliminary studies. Foremost among these is the need to understand the hydrology of the region and connectivity between the reservoirs and host country rock. We also need to constrain the fine structure of the seismic zone in the area including detailed mapping of the causative faults, both of which would be critical in locating the deep borehole observatory in the region. The following three-tiered action plan was agreed upon.

1. Revisiting old data and acquiring new data
   - Compile all available earthquake data for Koyna area and apply the most appropriate techniques to improve hypocenter locations.
   - Deploy a larger number of seismic stations, especially in the region close to the active Warna seismic zone.
   - Take up geophysical surveys like seismic reflection, magnetotelluric, deep electrical sounding, gravity and magnetics to constrain the fine structure of the seismic zone.
   - Acquire very high-resolution Light Detection And Ranging (LIDAR) data.
2. Studying hydraulic connectivity
   - Initially drill about four test boreholes around the seismic zone, each penetrating about 200 m into the pre-Trappean basement.
   - Perform suites of measurements on geochemical and geophysical parameters in and across these boreholes.
   - Model regional hydraulic connectivity.
3. Plan the main borehole based on 1 and 2.

The workshop provided an excellent opportunity for the global community of concerned scientists to discuss the ideas and data collected and results obtained from the Koyna site of RTS for over the last four decades. The suggestions and plan of action as summarized above is now being implemented.

## Authors

**Harsh Gupta**, National Geophysical Research Institute (CSIR-NGRI), Uppal Road, Hyderabad, 500 007, India; e-mail: harshg123@gmail.com.
**Shailesh Nayak**, Ministry of Earth Sciences (MoES), C.G.O. Complex, Lodhi Road, New Delhi 110 003, India.

**and the Koyna Workshop Committee:** Y.J. Bhaskar Rao, R.K. Chadha, B.K. Bansal, D. Srinagesh, N. Purnachandra Rao, Sukanta Roy, H.V.S. Satyanarayana, D. Shashidhar, and K. Mallika.

## New Science Plan on Ocean Drilling Published

The science plan for the new ocean drilling program, the International Ocean Discovery Program 2013–2023, was published in June 2011. The title of the science plan is "Illuminating Earth's Past, Present and Future". The science plan (available on www.iodp.org) has been in preparation for more than two years with input from a broad scientific community representing Earth, ocean, climate and life sciences. Numerous workshops were held, and white papers invited from across the community. The planning process culminated with the INVEST Conference in September 2009 at the University of Bremen, an event that was attended by approximately 600 scientists from 21 countries.

Based on the extensive report from the INVEST meeting, a select group of researchers prepared a draft science plan that underwent extensive community review in 2010 before being completed in spring of 2011. It was presented to the international press on 16 June (http://iodp.streamco.tv/IODP_-_Livestream_16th_of_June_2011.html).

The science plan includes four major themes (see below) and within these, 14 specific challenges.

- Climate and Ocean Change: Reading the Past, Informing the Future
- Biosphere Frontiers: Deep Life and Environmental Forcing of Evolution
- Earth Connections: Deep Processes and Their Impact on Earth's Surface Environment
- Earth in Motion: Processes and Hazards on Human Time Scales

## The New Web-Based Journal *Frontiers in Extreme Microbiology* Invites Manuscripts

*Abstract deadline: *15 July 2011*
*Manuscript deadline: *15 September 2011*

The new web-based journal "*Frontiers in Extreme Microbiology*" will mark its launch with a special topic, deep subsurface microbiology, and invites abstracts and manuscripts.

Deep subsurface microbiology focuses on the molecular detection and quantification, cultivation, biogeographic examination, and distribution of bacteria, archaea and eukaryotes that permeate the subsurface biosphere of deep marine sediments and the basaltic ocean crust. The deep subsurface biosphere abounds with uncultured, only recently discovered and (at best) incompletely understood microbial populations. In spatial extent and volume, the subsurface biosphere is only rivaled by the deep sea water column. So far, no deep subsurface sediment has been found that is entirely devoid of microbial life; microbial cells and DNA remain detectable at sediment depths of more than 1 km; microbial life permeates deeply buried hydrocarbon reservoirs, and it is also found several kilometers down in continental crust aquifers. Severe energy limitation, either as electron acceptor or donor shortage, and scarcity of microbially degradable organic carbon sources are among the evolutionary pressures that have shaped the genomic and physiological repertoire of the deep subsurface biosphere. Its biogeochemical importance as a long-term organic carbon repository, inorganic electron and energy source, and subduction recycling engine is a major focus of current research at the interface of microbiology, geochemistry and biosphere/geosphere evolution. The *Frontiers in Extreme Microbiology* special topic will address some of the central research issues in deep subsurface microbiology and biogeochemistry: phylogenetic and physiological microbial diversity in the deep subsurface; microbial activity and survival strategies in severely energy-limited subsurface habitats; cell-specific microbial activity as reflected in process rates and gene expression patterns; biogeographic isolation and connectivity in deep subsurface microbial communities; and the ecological standing of subsurface biospheres in comparison to the surface biosphere. Is this region an independently flourishing biosphere, or merely a site of survivors that tolerate burial (along with organic carbon compounds), or a combimnation of both? Studying these issues on Earth's deep subsurface biosphere has far-ranging implications, not the least being in the field of astrobiology and the search for subsurface life beyond Earth.

*Abstracts and manuscripts will be considered with some fliexibility (one or two months later than the posted deadliens).

Web Link: www.frontiersin.org/extrememicrobiology/specialtopics/deep_subsurface_microbiology.

Edited by Andreas Teske (UNC Chapel Hill), Axel Schippers (BGR), Peter Dunfield (Univ. of Calgary, Virginia Edgcomb (WHOI), and Jennifer Biddle (University of Delaware)

## Call for IODP-Canada Graduate Student Research Awards

*Application deadline: 18 November 2011*

To enhance the experience of graduate students investigating marine geology questions related to the research themes of the Integrated Ocean Drilling Program, IODP-Canada offers merit-based awards of up to $3000 to students enrolled in either an MSc or PhD program or employed as a post-doc at a Canadian institution. The research awards are intended to support projects directed toward the objectives of upcoming or past DSDP/ODP/IODP expeditions, utilizing core material and/or shipboard data.

Please contact the IODP-Canada Coordinator (coordinator@mail.iodpcanada.ca) for more information.

## The Netherlands Joins ICDP

Starting 2011, the Netherlands is a member country of ICDP. Dutch community, ICDP-NL, is an open platform that intends to bundle the expertise of (sometimes rather small) research groups and specialists and to support the ambitions of the Dutch scientists. The community consists of scientists of a broad variety of Earth science disciplines, ranging from archaeology to tectonics. It not only involves universities but also the Netherlands Organisation for Applied Scientific Research (TNO). The Board of ICDP-NL is chaired by Stefan Luthi (TU Delft) and advised by a representative of the drilling industry. The membership fee is financed by the Earth and Life Science division of the Netherlands Organization for Scientific Research (NWO).

A first symposium was held on 14 February 2011, during which also the agreement between NWO and ICDP was signed. ICDP-NL will cooperate closely with IODP-NL, for instance, by organizing joints meetings.

Dutch scientists already have a long track record in continental drilling. ICDP-NL will build on the expertise gained during these collaborative projects and is looking forward to now becoming fully involved in initiating and executing ICDP projects.

Contact details, the press release and the first newsletter can be found on the ICDP website under National Programs.

## Australia and New Zealand in Scientific Drilling

ANZIC IODP activities are proceeding smoothly. The 2010 ANZIC Annual Report is now available on www.iodp.org.au. The report not only covers Australian and New Zealand activities in IODP in 2010, but looks at the future and makes the point that we are fully committed to IODP, now and in the future. Australia now has funding until the end of the present phase of IODP, and New Zealand is actively pursuing such funding. Planning is underway for an IODP session at the International Geological Congress in Brisbane in mid-2012 (www.34igc.org), and a large number of international participants are expected to attend.

In the first six months of 2011, four ANZIC scientists participated in IODP expeditions: two aboard the Louisville Seamount Trail expedition (330), one aboard the CRISP expedition (334) and one aboard the Superfast Spreading expedition (335).

A New Zealand Scientific Drilling Workshop was held in Dunedin in March 2011, and a white paper is currently under development to sketch the concept and importance of the DrillNZ initiative to various stakeholders. A New Zealand-led IODP workshop "Conceptual framework for ocean drilling to unlock the secrets of slow slip events" was held at Gisborne, New Zealand in August 2011. Since beginning of August 2011, NIWA (the Natonal Insitute of Water and Atmosphere) has joined the consortium of New Zealand Universities and Crown reseach Institutes contributing to ANZIC, and has committed funds towards the New Zealand IODP membership fee up to 2013–2014.

An Indian Ocean IODP workshop will be held in Goa, India over the period 16-18 October 2011. The organizers are India and ANZIC. The aim of the workshop is to further develop existing proposals and design new ones, in the hope that *JOIDES Resolution* will be in the Indian Ocean in 2014.

Australia and New Zealand are making important scientific contributions to IODP, and a number of major coring expeditions in our region have improved and will improve the understanding of global scientific questions. Although IODP is now working outside the region, these two countries remain keen supporters of global IODP science.

## 2nd Swiss ICDP Meeting (March 2011, Zürich)

About 40 scientists gathered in March 2011 in Zurich to exchange news about past, present and future ICDP projects with active Swiss involvement. Since Switzerland became an ICDP member through support from the Swiss National Sccience Foundation only three years ago, a series of projects with Swiss participation have been already drilled or are underway, and more are planned, reflecting an active contribution to the overall ICDP program. Scientists from Swiss institutions are acting as as PIs and Co-PIs in several drilling projects targeting lake sediment records, the subsurface biosphere, volcanic domains, sedimentary basins and groundwater contamination.

One of the main goals of the meeting was to address new themes to take further advantage from the Swiss ICDP membership. For example, several ideas were raised in the context of research on geothermal energy, where large commercial projects are currently underway, and it was suggested that a geothermal component could complement any continental drilling project. In the same context, $CO_2$ sequestration gained wide attention and might be further explored in the near future. New ideas will address drilling targets also within Switzerland. For example, a drilling initiative involving scientists from perialpine countries aims to explore the timing and extent of alpine glaciations and related erosion and sedimentation through drilling the sedimentary infill of overdeepened troughs all around the Alpine region. Kick-off meetings with international partners are planned to explore possibilities for an ICDP workshop proposal. In a similar fashion, the Alps themselves provide a series of drilling objectives that may serve as unique records of tectonic and petrologic processes. One example is the highly metamorphic Ivrea Zone (Southern Alps), where lower-crust rocks are outcropping at the surface and where a drill hole could reach upper mantle rocks (i.e., the Moho) in much shallower depth as in its original location.

Altogether, the Swiss scientific drilling community is very active, and the combined IODP and ICDP memberships will continue to provide the ideal platform to conduct and realize these fascinating drilling initiatives. Contact: flavio.anselmetti@eawag.ch.

## D/V *Chikyu* Damaged by Tsunami, Repairs Planned

At 2:47 p.m. JST on 11 March, the Tohoku region in the northern Pacific coast of Japan suffered an M9 earthquake followed by a 7+ m tsunami strike and numerous aftershocks. At that time, D/V *Chikyu* was in port at Hachinohe preparing for IODP Exp. 337. The *Chikyu* immediately evacuated port; however, during the emergency evacuation the vessel suffered damage. All personnel and visitors aboard, including a visiting delegation of local elementary school students, were safely accounted for.

On 20 March, *Chikyu* arrived at the port of Muroran in southern Hokkaido where divers made a provisional assessment of the ship's damage. From 20 April until 18 June, the *Chikyu* was dry-docked for repair at the Yokohama Dockyard and Machinery, and returned to sea for testing on 18 June 2011.

One of DV/ *Chikyu*'s azimuth thrusters suffered damage as tsunami waves buffeted the vessel in the Hachinohe port, and its replacement is currently being manufactured. The new thruster is expected to be installed by the end of March 2012.

## IODP-Canada Summer School Scholarships

IODP-Canada is pleased to announce that five students received scholarships to attend one of the 2011 ECORD summer schools held in Bremen, Germany and Urbino, Italy.

Congratulations to Thi Hao Bui (PhD student, McGill University), Jon Furlong (MSc student, University of Victoria), Fritz Griffith (MSc student, University of Ottawa), Lucie Hubert-Théou (PhD student, McGill University), and Stefan Markovic (PhD student, University of Toronto).

## IODP/ICDP-Canada Booth at GAC-MAC 2011

IODP-Canada and ICDP-Canada shared an exhibition booth at the joint annual meeting of the Geological Association of Canada and the Mineralogical Association of Canada, held at the University of Ottawa on 25–27 May 2011. Over 1050 Earth scientists (including 285 students) from Asia, Canada, Europe, and the U.S. attended the conference.

Anne de Vernal, Chair of IODP-Canada, and the Coordinators of IODP-Canada and ICDP-Canada staffed the booth and met with students, faculty and funding agency representatives. Many ECORD, IODP and ICDP outreach materials were distributed, and subscribers to the Canadian mailing list increased by over 10%. The new IODP-Canada Website (www.iodpcanada.ca) was launched just prior to the start of the conference.

## ICDP Dead Sea Core Opening Party, Part I

Within the framework of the ICDP Dead Sea Deep Drilling Project (DSDDP) a first core opening party took place at the GFZ German Research Centre for Geosciences, Potsdam, in June 2011. A group of 18 involved scientists and student helpers from Israel, Japan, Switzerland, and the U.S., together with the German team around Prof. Dr. Achim Brauer, opened and described around 470 meters of a total of 720 meters of sediment cores that had been obtained during the drilling campaign at the Dead Sea deep basin from November 2010 to March 2011. Additional assistance was given by the ICDP Operational Support Group. Furthermore, modern non-destructive scanning techniques, such as micro-XRF analyses, magnetic susceptibility measurements and photographic line-scanning, have been applied at the sediments, encouraging the scientists to work in a two-shift system seven days a week in order to cope with the enormous amount of Dead Sea sediments from one 450-m deep borehole and several short cores. First estimations based on the succession of salt, laminated sequences and detrital sediment flux point out that the sediments comprise at least two climatic cycles, hence covering the last ~200 thousand years. All scientists are looking forward to meeting again at the GFZ laboratories in October this year to perform the same procedure with the remaining 250 meters from the second deep borehole and to get new insights into this fascinating archive of natural climate dynamics and seismic activity.

## ANDRILL Coulman High Project – Site Survey Outcomes and Future Plans

Field surveys were conducted from November 2010 through January 2011 on the Ross Ice Shelf (RIS), northeast of McMurdo Station and Scott Base, in Antarctica, in preparation for the proposed ANDRILL Coulman High (CH) Project. The survey work is chronicled in a well-illustrated blog provided by Dr. Frank Rack, which is online at <http://www.andrill.org/science/ch/news>.

The CH Project surveys, which involved integrated planning and execution by a combined international team from New Zealand and the United States, achieved all primary and secondary objectives. A transverse route from Ross Island to CH across the RIS was established using a tracked-vehicle-mounted ground-penetrating radar system supplemented by airborne radar. Four GPS stations and a weather station were established to measure lateral and vertical motions of the RIS and to monitor environmental conditions. A series of combined U.S.-NZ field camps on the RIS were occupied, and the ANDRILL hot water drill (HWD) system was used to melt numerous holes through 260–275 m of ice. Oceanographic moorings comprised of inductive oceanographic instruments were deployed through the RIS at two sites, and were recovered to the ice surface after two months; one mooring was then redeployed to accomplish multi-year observations at CH. Video camera observations of the interior and basal surface of the ice shelf were made at several sites and these observations were integrated with conductivity-temperature-depth (CTD) measurements through the ice shelf to the seafloor at each site. The Submersible Capable of under-Ice Navigation and Imaging (SCINI) underwater remotely operated vehicle (ROV) was deployed through the ice to explore the underside of the ice shelf while conducting operational tests. SCINI discovered an unusual biological community living in the ice and recovered biological samples using an improvised suction pump (Rack et al., 2011); these samples and extensive imagery are being further investigated.

ANDRILL is a multinational scientific program to investigate Antarctica's glacial and tectonic history. Of specific interest is Antarctica's past and future role in global environmental change as revealed through stratigraphic drilling along the Antarctic margin and related numerical modeling of climate and ice sheet behavior. ANDRILL is developing the CH Project to investigate Early Miocene and Paleogene paleoenvironmental and tectonic history by drilling at two sites on the Ross Ice Shelf, approximately 150 km east of McMurdo Station, over the subsea geologic structure known as the Coulman High. Further information about the project is available at <http://www.andrill.org/science/ch>.

The ANDRILL Science Committee (ASC) has issued an international call for submission of statements of interest in the CH Project from international scientists. This survey is available online at <http://www.andrill.org/science/ch-survey>. The scientific team for the project will only be selected from nations that belong to the ANDRILL consortium. If you would like to know more about how your nation might join the ANDRILL consortium as a participating member, please contact Dr. Tim Naish, Chair of the ASC, at <tim.naish@vuw.ac.nz> or contact Dr. Frank Rack, Secretary of the ASC and Executive Director, ANDRILL Science Management Office, University of Nebraska-Lincoln, at <frack2@unl.edu>.

### Reference

Rack, F., Levy, R., Falconer, T., Zook, R., Mahacek, P., Carroll, D., Williams, M., Stewart, C., Limeburner, R., and the ANDRILL Coulman High Site Survey Team, 2011. PS14.3: Interdisciplinary outcomes of the ANDRILL Coulman High Site surveys. [paper presented at 11th International Symposium on Antarctic Earth Sciences, Edinburge, Scotland, 10–15 July 2011]

## Bighorn Basin Coring Project (BBCP)

The Bighorn Basin Coring Project (BBCP) initiated drilling operations in the Bighorn Basin of Wyoming (U.S.) on 13 July 2011 and completed drilling on 4 August 2011. The main goal of the project is to recover complete, unweathered core samples from Paleogene continental strata that record rapid global warming events known as hyperthermals (e.g., Paleocene-Eocene Thermal Maximum [PETM] and Elmo). By applying various geochemical, geophysical, and paleontological methods on these high-quality (HQ) cores, the project aims to better understand the causes of these climate events and their effects on continental systems. The project was highly successful in achieving all of its drilling goals with >95% core recovery in each hole. Three sites in the basin were targeted (Basin Substation, Polecat Bench, Gilmore Hill) with two holes drilled at each site to get overlapping cores. A total of 921 meters of HQ core was recovered in liners with all but one of the sites being drilled with municipal water only, thus reducing the potential for core contamination. In mid-August, the cores started their journey from Powell, Wyoming to the IODP Bremen Core Repository at the University of Bremen, Germany, where they will be split, described, and sampled by the BBCP Science Team in early 2012. BBCP web page: http://earth.unh.edu/clyde/BBCP.shtml

Facebook page with real time updates: http://www.facebook.com/pages/Bighorn-Basin-Coring -Project /120607821348013.

## ISAES-11 Antarctic Shallow Scientific Drilling Workshop

Following the 11th International Symposium on Antarctic Earth

## The PASADO Core Processing Strategy: A Flexible Protocol for Sediment Subsampling in Multidisciplinary Drilling Projects

Subsampling of sediment cores is the important first step of almost every study. Sampling procedures were developed within IODP/ODP/DSDP to ensure the reproducibility of core treatment as well as its documentation. In contrast, procedures for lacustrine sediment cores are as manifold as the corresponding scientific questions and communities. A sampling assembly and a flexible sampling protocol are described in the version adapted to the PASADO project. This protocol introduces a core splitter and a sampling assembly consisting of a divider and a D-scoop for fast and efficient subsampling. Detailed information can be found in our recently published paper (Ohlendorf et al., 2011).

The core splitter cuts plastic core liners without contamination of the sediment by liner material. The core sampling assembly is used to sample one core half completely in contiguous steps. The core half is split into slices of similar thickness by inserting thin disks through a comb-like divider, and the D-scoop is used to sample these sediment disks (Fig. 1). The utilization of the sediment is optimized because all sediment is sampled in one step. Subsamples with known volume can then be taken from the exact same stratigraphic level in the core. Quality flags are defined to document the condition of every (sub)sample and to allow a first classification of the sediment character.

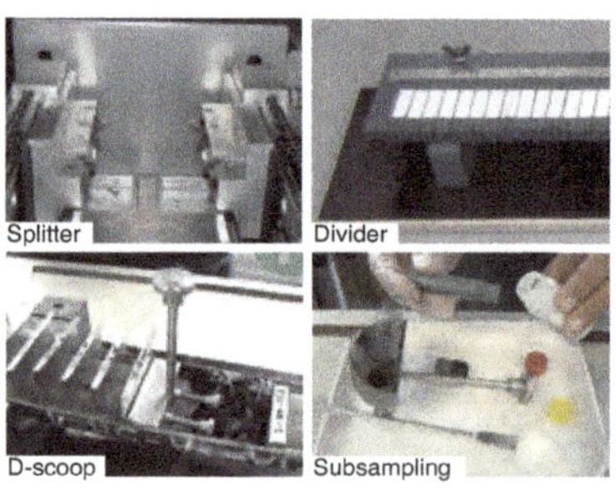


Figure 1. Photographs of core splitter and the sledges with the four blades, divider, and the D-scoop.

### Reference

Ohlendorf, C., Gebhardt, C., Hahn, A., Kliem, P., Zolitschka, B., and the PASADO science team, 2011. The PASADO core processing strategy - A proposed new protocol for sediment core treatment in multidisciplinary lake drilling projects. *Sediment. Geol.*, 239:104–115, doi:10.1016/j.sedgeo.2011.06.007.

### Authors

**Christian Ohlendorf**, Geomorphology and Polar Research (GEOPOLAR), Institute of Geography, University of Bremen, Celsiusstr. FVG-M, D-28359 Bremen, Germany, e-mail: ohlen@uni-bremen.de.
**Catalina Gebhardt**, Alfred Wegener Institute for Polar and Marine Research, Am Alten Hafen 26, D-27568 Bremerhaven, Germany.
**Annette Hahn**, **Pierre Kliem**, and **Bernd Zolitschka**, Geomorphology and Polar Research (GEOPOLAR), Institute of Geography, University of Bremen, Celsiusstr. FVG-M, D-28359 Bremen, Germany.
**and the PASADO science team**

### Related Web Link

http://www.icdp-online.org/front_content.php?idcatart=2794

Sciences (ISAES-11) in Edinburgh, Scotland, a one-day workshop on Antarctic Shallow Scientific Drilling was held in the John McIntyre Conference Centre of the University of Edinburgh (Scotland) on Sunday, 17 July 2011. The workshop, which attracted over thirty-six international attendees from six countries, was co-convened by Frank Rack (University of Nebraska-Lincoln) and Bruce Luyendyk (University of California, Santa Barbara). The presentations explored the use of numerical models and data-model integration strategies to answer important scientific questions, identified potential shallow drilling targets from specific sites around the Antarctic margin, including land-ocean transects, and reviewed the status of existing and emerging technology options that could be used to implement shallow (up to 200 mbsf) scientific drilling/coring of glacial and glacial-marine sediment and rocks in ice-covered seas (e.g., mission specific platforms, including options for either ship-mounted rigs or seafloor drills). Further information about the outcomes of the workshop will be available online in the near future at <http://www.shaldril.org>.

## International Workshop on Scientific Drilling in the Indian Ocean

*17–18 October 2011, Goa, India*

The Indian Ocean has experienced an extremely complicated tectonic history. From the disintegration of Gondwanaland to the collision of the Indian plate with Asia and the subduction of the Indian plate in one of the most prominent subduction zones, the region has a diversity of key tectonic features. Much geo-scientific knowledge regarding the Indian Ocean was gained from the International Indian Ocean Expedition program (IIOE-1960–65). Since then, integrated geophysical and geological studies have considerably improved our understanding of geological evolution of this region. Results from previous studies in association with the ODP/DSDP boreholes in the Indian Ocean have further strengthened our knowledge.

Despite these remarkable efforts, several key questions related to the Indian Ocean with significant global relevance are still unanswered. Decades of geo-scientific work focused on this sector has produced compelling hypotheses on various fronts. Our present understanding vis-a-vis these propositions requires corroboration from direct measurements such as deep ocean cores. The need for scientific drilling in the Indian Ocean sector is highly evident by several active IODP drilling proposals at various stages (e.g. proposal nos. 549, 552, 595, 667, 701, 702, 704, 717, 724, 727, 760, 776, 778, 780 and 783).

IODP-India in association with ANZIC is organizing an international workshop to discuss various scientific proposals for drilling in the Indian Ocean sector as well as to encourage discussions of new proposals in this sector for the new phase of the IODP. The Indian Ocean IODP Workshop will be held in Goa on 17–18 October with about 100 participants from all over the world. The workshop is planned back to back with the seventh Asian Marine Geology Meeting, also taking place in Goa during the previous week.

Because there has been no scientific ocean drilling in the Indian Ocean for nearly a decade, this workshop is vital to developing strong new drilling proposals. Also, India itself has only joined IODP recently, and such a workshop will provide wide exposure to the Indian science community of IODP science and IODP capability. The following four themes have been arranged for the workshop and with planning subcommittees largely formed:

1. Cenozoic oceanography, climate change, gateways and reef development: broad questions related to the Indian Ocean, and narrow ones such as the Indonesian Throughflow, sea-level changes, and the origin of late Pleistocene reefs.
2. The history of the monsoons: tectonics, uplift, weathering and erosion, sediment deposition, climate and oceanography, and discussions/nurturing of proposals that have emerged since the 2008 Detailed Planning Group (DPG) workshop on monsoons.
3. Tectonics and volcanism: tectonism of the Indian Ocean, including plate tectonics, the evolution of the oceanic crust and mid-ocean ridge formation, the formation of large igneous provinces, continental rifting and related deposition, and subduction, arc volcanism and earthquakes.
4. The deep biosphere: studies of the "extremophiles" of the deep biosphere in sediments and basalts focusing on the nature of the oceanography and inputs of organic matter into the Indian Ocean.

More information on this workshop can be found at www.iodp.org/workshops/8/ or www.ncaor.gov.in/iodp/index.html.

Workshop conveners:

Dr. Dhananjai K. Pandey, Program Officer (IODP-India), e-mail: iodp.india@ncaor.org, and Professor Neville Exon: ANZIC Program Scientist, e-mail: neville.exon@anu.edu.au.

## International Review of the ICDP with Outstanding Results

On behalf of the Assembly of Governors, (AOG) and at the request of the main funding partners of the program, the International Continental Scientific Drilling Program was evaluated in May 2011 by an international review committee of independent experts with key expertise in major programs in the geosciences and funding organizations.

The committee concluded that ICDP is a highly successful program, achieving with very modest investments world-class science of global impact. The program has been highly effective in community-building and is driving integration in modern Earth system science. It has demonstrated strong scientific leadership and effective allocation of its limited financial resources. The program holds great promise to further attract new member states, organizations and industrial partners.

One important outcome of the assessment is to start developing a new ICDP Science Plan. The AOG has therefore initiated the planning for an International Symposium on Continental Scientific Drilling in 2013.

## Drilling into the Fault Zone of the Giant Tohoku Earthquake?

Following the 11 March Tohoku earthquake, IODP mobilized a Detailed Planning Group (DPG) to investigate the feasibility of drilling into the seismogenic fault. Box 1 on page 61 is an excerpt from the report produced by the DPG.

A DPG meeting in May led to a full proposal for drilling within an area of 6–7 km water depth near the trench where the fault can be reached within 1 km subseafloor. Later plans are to measure ephemeral properties related to the rupturing of the fault as a baseline; repeat observations are a key target. Properties include an expected friction induced temperature anomaly providing a proxy for fault strength (friction) during rupture. Drilling is required within about 18 months after rupture before a temperature anomaly has dissipated too much. This is a unique opportunity to sample a fault as close in time as possible to a major and large displacement earthquake event, but drilling and casing a hole in these water depths is a technological challenge. The study, if deemed feasible and implemented, will complement ongoing studies in the Nankai Trough, which targets a strong mega-fault expected to rupture within a few decades.

References for figures in Box 1 available in the full DPG Report, which is at http://www.iodp.org/weblinks/Featured-Publications-HOME-PAGE/Tohoku-Rapid-Response-Drilling-DPG-Report/.

Used by permission.

## Box 1: The 2011 Off the Pacific Coast of Tohoku Earthquake ($M_w$9.0)

The March 11, 2011 Tohoku earthquake came as a disheartening surprise to the geophysical community. With a historical record of nearly 500 years including 13 $M_w$7 and 5 $M_w$8 earthquakes, this region was thought to be relatively well understood in terms of the locations and sizes of expected subduction zone earthquakes. An $M_w$9 event breaking through the entire region of many fault segments, with the associated huge tsunami, was not at all anticipated for this thoroughly studied area. This failure highlights the need for a more physically-based understanding of initiation and rupture.

The sequence started with an $M_w$7.2 foreshock, which occurred 2 days before and about 40 km NE of the mainshock hypocenter. In the hour after the mainshock, there were large $M_w$7.9 and $M_w$7.7 aftershocks. In addition to the countless aftershocks in the immediate region, the seismic activity of small earthquakes increased across most of Japan with several $M_w$5 and a few $M_w$6 earthquakes over the following month. Small earthquakes were also triggered at 13 volcanoes according to the Coordinating Committee for Prediction of Volcanic Eruption.

Apart from the foreshocks, there were no clear precursory signs or large pre-slip. The foreshocks themselves were only identified as precursory with hindsight. Current models of earthquake clustering suggest that the probability of having an $M_w$ 9.1 earthquake following an $M_w$ 7.2 earthquake within 2 days is <0.001% and thus a societally useful prediction could not be provided based on the foreshocks alone.

Modeling of seismic, crustal deformation, and tsunami data shows very large slip on the fault plane, with values up to 30–50 meters (e.g. Simons et al., 2011, Sato, et al., 2011, Ammon et al., 2011, Ohta et al., 2011, Lay et al., 2011). The area of maximum slip is on the shallow portions of the fault trenchward of the hypocenter. The large and shallow slip near the trenchward limit of the megathrust caused large deformation of the seafloor, which generated the devastating tsunami. Because of the dense network of observations in the region prior to the earthquake, the extremely large slip has been verified by direct observations of displacement on the seafloor through repeat bathymetry and GPS.

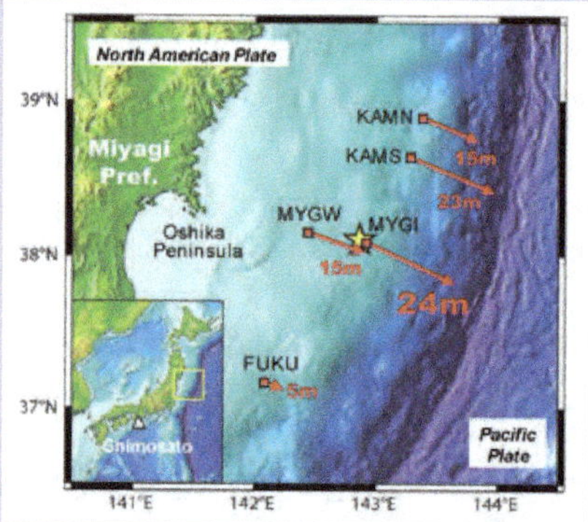


Figure 2. Horizontal coseismic displacements at seafloor benchmarks, associated with the Tohoku earthquake. Red squres are locations of GPS benchmarks and yellow star is the epicenter of the mainshock (Sato et al., 2011b)

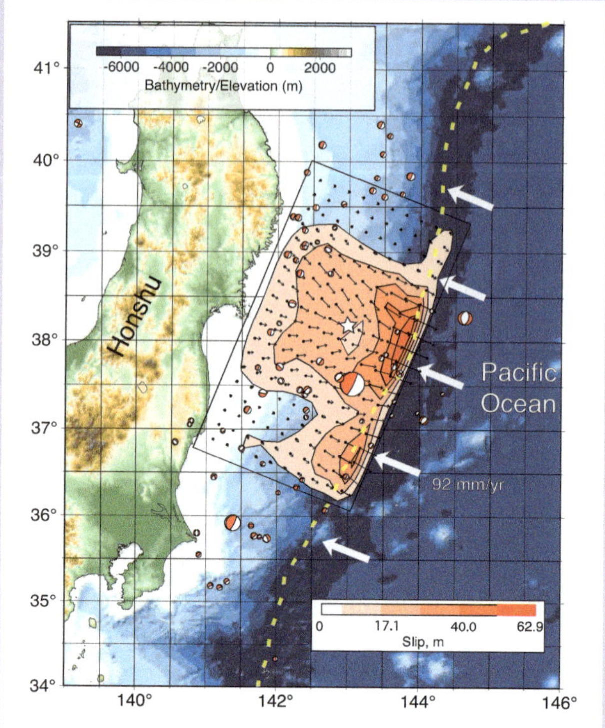


Figure 1. Teleseismic P-wave inversion for slip (Lay et al., 2011). Maximum slip ranges from 40–80 for a suite of inversion models with the majority of models placing large slip near the trench. Most models do not directly incorporate the ocean bottom observations in Figure 2 and 11 which require large slip towards the updip end of the rupture.

The earthquake produced severe strong ground motions with accelerations over 1 g and long durations of about 100 s. However, the large losses of lives and property were mainly due to the tsunami, and shaking damage was relatively limited, considering the size and intensity of the earthquake. Even the well-publicized problems at the Fukushima No. 1 nuclear power plant were caused by loss of power due to the tsunami inundation, and not by the shaking itself. The resilience of most structures throughout the region can be attributed to the high seismic standards of Japanese construction.

During the Tohoku earthquake, the regions of the fault that produced the dominant high-frequency energy are different from the areas of large slip. The large slip is on the shallow updip portion of the fault, while the high-frequency radiation originated from the deeper downdip portions of the fault (e.g. Ide et al., 2011, Wang and Mori, 2011, Koper et al., 2011). The difference in frequency of the radiated energy for different portions of the fault reflects variations in rupture dynamics. The deeper portion appears to be have undergone a more brittle rupture with a higher proportion of radiated energy, while the large shallow slip probably absorbed more energy through dissipated processes. The greater dissipation may be characteristic of tsunami earthquakes and can be studied by sampling this portion of the fault with a borehole.

# Permissions

All chapters in this book were first published in Scientific Drilling, by Copernicus Publications; hereby published with permission under the Creative Commons Attribution License or equivalent. Every chapter published in this book has been scrutinized by our experts. Their significance has been extensively debated. The topics covered herein carry significant findings which will fuel the growth of the discipline. They may even be implemented as practical applications or may be referred to as a beginning point for another development.

The contributors of this book come from diverse backgrounds, making this book a truly international effort. This book will bring forth new frontiers with its revolutionizing research information and detailed analysis of the nascent developments around the world.

We would like to thank all the contributing authors for lending their expertise to make the book truly unique. They have played a crucial role in the development of this book. Without their invaluable contributions this book wouldn't have been possible. They have made vital efforts to compile up to date information on the varied aspects of this subject to make this book a valuable addition to the collection of many professionals and students.

This book was conceptualized with the vision of imparting up-to-date information and advanced data in this field. To ensure the same, a matchless editorial board was set up. Every individual on the board went through rigorous rounds of assessment to prove their worth. After which they invested a large part of their time researching and compiling the most relevant data for our readers.

The editorial board has been involved in producing this book since its inception. They have spent rigorous hours researching and exploring the diverse topics which have resulted in the successful publishing of this book. They have passed on their knowledge of decades through this book. To expedite this challenging task, the publisher supported the team at every step. A small team of assistant editors was also appointed to further simplify the editing procedure and attain best results for the readers.

Apart from the editorial board, the designing team has also invested a significant amount of their time in understanding the subject and creating the most relevant covers. They scrutinized every image to scout for the most suitable representation of the subject and create an appropriate cover for the book.

The publishing team has been an ardent support to the editorial, designing and production team. Their endless efforts to recruit the best for this project, has resulted in the accomplishment of this book. They are a veteran in the field of academics and their pool of knowledge is as vast as their experience in printing. Their expertise and guidance has proved useful at every step. Their uncompromising quality standards have made this book an exceptional effort. Their encouragement from time to time has been an inspiration for everyone.

The publisher and the editorial board hope that this book will prove to be a valuable piece of knowledge for researchers, students, practitioners and scholars across the globe.

# List of Contributors

**Mitchell Lyle**
Department of Oceanography, Texas A&M University, TAMU 3146, College Station, TX 77840-3146, U.S.A

**Heiko Pälike**
School of Ocean and Earth Science, National Oceanography Centre, European Way, Southampton, SO14 3ZH, U.K

**Hiroshi Nishi**
Department of Natural History Sciences, Hokkaido University, Kita 10, Nishi 8, Kita-ku, Sapporo, 060- 0810, Japan

**Isabella Raffi**
Dipartimento di Geotecnologie per l'Ambiente e il Territorio–DiGAT, Universitá "G. D'Annunzio", Campus Universitario, via dei Vestini 31, 66013 Chieti Scalo, Italy

**Kusali Gamage**
Integrated Ocean Drilling Program, Texas A&M University, 1000 Discovery Drive, College Station, TX 77845-9547, U.S.A

**Adam Klaus**
Integrated Ocean Drilling Program, Texas A&M University, 1000 Discovery Drive, College Station, TX 77845-9547, U.S.A

**Andrew T. Fisher**
University of California Santa Cruz, Earth and Planetary Sciences Department and Instsitute for Geophysics and Planetary Physics, 1156 High Street, E&MS A232, Santa Cruz, CA 95064, U.S.A

**Sascha Flögel**
IFM-GEOMAR, Leibniz-Institut für Meereswissenschaften (an der Universität Kiel), Dienstgebäude Ostufer; 8C-110, Wischhofstr. 1-3, 24148 Kiel, Germany

**Wolfgang Kuhnt**
Christian-Albrechts-Universität zu Kiel (CAU), Institute of Geosciences, Marine Micropaleontology, Ludewig-Meyn-Str. 14, D-24118 Kiel, Germany

**Victor A. Melezhik**
Geological Survey of Norway, N-4791, Trondheim, Norway; Centre for Geobiology, University of Bergen, N-5020, Bergen, Norway

**Aivo Lepland**
Melanie Mesli, Tor Erik Finne, Geological Survey of Norway, N-4791, Trondheim, Norway

**Alexander E. Romashkin**
Institute of Geology of Karelian Research Centre, 185610, Petrozavodsk, Russia

**Dmitry V. Rychanchik**
Institute of Geology of Karelian Research Centre, 185610, Petrozavodsk, Russia

**Ronald Conze**
Operational Support Group ICDP, GFZ German Research Center for Geosciences (GeoForschungsZentrum Potsdam), D-14473 Potsdam, Germany

**FAR-DEEP Scientists**
**Ronald Conze**
Operational Support Group ICDP, GFZ German Research Centre for Geosciences, elegrafenberg, D-14473 Potsdam, Germany

**Frank Krysiak**
smartcube GmbH Berlin, Puschkinallee 48, D-12435 Berlin, Germany

**Josh Reed**
Andrill Science Management Office, 126 Bessey Hall, Lincoln, University of Nebraska-Lincoln, P.O. Box 880851, Loncoln, NE 68588-0341, U.S.A

**Yu-Chung Chen**
University of Illinois at Chicago, Electronic Visualization Laboratory, U.S.A

**Hans-Joachim Wallrabe-Adams**
MARUM, University of Bremen, Leobener Str., D-28359 Bremen, Germany

**Colin Graham**
British Geological Survey Edinburgh, ESO - ECORD Science Operator, Murchison House, West Mains Road, Edinburgh EH9 3LA, Scotland, U.K

**Volker Wennrich**
University of Cologne, Institute of Geology and Mineralogy, Zuelpicher Str. 49a, D-50674 Koeln, Germany

**Yuki Morono**
Geomicrobiology Group, Kochi Institute for Core Sample Research (KOCHI), Japan Agency for Marine-Earth Science and Technology (JAMSTEC), Monobe B200, Nankoku, Kochi 783-8502, Japan

**Fumio Inagaki**
Geomicrobiology Group, Kochi Institute for Core Sample Research (KOCHI), Japan Agency for Marine-Earth Science and Technology (JAMSTEC), Monobe B200, Nankoku, Kochi 783-8502, Japan

**Christopher Virgil**
Institut für Geophysik und extraterrestrische Physik, Technische Universität Braunschweig, Mendelssohnstrass 3, D-38106 Braunschweig, Germany

**Andreas Hördt**
Institut für Geophysik und extraterrestrische Physik, Technische Universität Braunschweig, Mendelssohnstrass 3, D-38106 Braunschweig, Germany

**Torsten Klein**
Institut für Geophysik und extraterrestrische Physik, Technische Universität Braunschweig, Mendelssohnstrass 3, D-38106 Braunschweig, Germany

**Jochem Kück**
GFZ - German Research Centre for Geosciences, Telegrafenberg, D-14473, Potsdam, Germany

**Martin Leven**
Institut für Geophysik, Georg-August-Universität Göttingen, Friedrich-Hund-Platz 1, D-37077, Göttingen, Germany

**Erich Steveling**
Institut für Geophysik, Georg-August-Universität Göttingen, Friedrich-Hund-Platz 1, D-37077, Göttingen, Germany

**Katrin Jaksch**
Helmholtz Centre Potsdam - GFZ German Research Centre for Geosciences, Scientific Drilling, Telegrafenberg, F 358, D-14473, Potsdam, Germany

**Rüdiger Giese**
Helmholtz Centre Potsdam GFZ German Research Centre for Geosciences, Scientific Drilling, Telegrafenberg, D-14473, Potsdam, Germany

**Matthias Kopf**
Helmholtz Centre Potsdam GFZ German Research Centre for Geosciences, Scientific Drilling, Telegrafenberg, D-14473, Potsdam, Germany

**Andreas Jurczyk**
Helmholtz Centre Potsdam GFZ German Research Centre for Geosciences, Scientific Drilling, Telegrafenberg, D-14473, Potsdam, Germany

**Stefan Mikulla**
Helmholtz Centre Potsdam GFZ German Research Centre for Geosciences, Scientific Drilling, Telegrafenberg, D-14473, Potsdam, Germany

**Stefan Weisheit**
Helmholtz Centre Potsdam GFZ German Research Centre for Geosciences, Scientific Drilling, Telegrafenberg, D-14473, Potsdam, Germany

**Marco Groh**
Helmholtz Centre Potsdam GFZ German Research Centre for Geosciences, Scientific Drilling, Telegrafenberg, D-14473, Potsdam, Germany

**Kay Krüger**
Helmholtz Centre Potsdam GFZ German Research Centre for Geosciences, Scientific Drilling, Telegrafenberg, D-14473, Potsdam, Germany

**Bernard Coakley**
Geophysical Institute, University of Alaska, Fairbanks, AL, U.S.A

**Ruediger Stein**
Alfred Wegener Institute for Polar and Marine Research, Bremerhaven, Germany

**Fatima Abrantes**
Laboratório Nacional de Energia e Geologia (LNEG), I.P. Estrada da Portela, Bairro do Zambujal - Alfragide Apartado 7586 - 2720-866 Amadora, Portugal

**Gabriella Carrara**
Laboratório Nacional de Energia e Geologia (LNEG), I.P. Estrada da Portela, Bairro do Zambujal - Alfragide Apartado 7586 - 2720-866 Amadora, Portugal

**Luis Batista**
Laboratório Nacional de Energia e Geologia (LNEG), I.P. Estrada da Portela, Bairro do Zambujal - Alfragide Apartado 7586 - 2720-866 Amadora, Portugal

**Henrique Duarte**
Laboratório Nacional de Energia e Geologia (LNEG), I.P. Estrada da Portela, Bairro do Zambujal - Alfragide Apartado 7586 - 2720-866 Amadora, Portugal

**David Hodell**
University of Cambridge - Department of Earth Sciences, Cambridge, U.K

**Johan Etourneau**
Christian-Albrechts-Universität zu Kiel/ Université Bordeaux 1, 24118 Kiel/ 33405 Talence, Germany/ France

**Nabil Khélifi**
Christian-Albrechts-Universität zu Kiel, 24118 Kiel, Germany

**Wolfgang Bach**
Department of Geosciences, University of Bremen, Klagenfurter Strasse, D-28359 Bremen, Germany

**Christina Ravelo**
Ocean Sciences Department, University of California Santa Cruz, CA 95064, U.S.A

**Jan Behrmann**
Leibniz-Institut für Meereswissenschaften, IFM-GEOMAR, Gebäude Ostufer, Wischhofstrasse 1-3, 24148 Kiel, Germany

**Gilbert Camoin**
National Center for Scientific Research (CNRS), CEREGE UMR 6635, CNRS, Europôle Méditerranéen de l'Arbois, BP 80, F-13545 Aix-en-Provence cedex 4, Aix-en-Provence, France

**Robert Duncan**
College of Oceanic and Atmospheric Sciences, Oregon State University, 104 COAS Administration Building, Corvallis, OR 97331-5503, U.S.A

**Katrina Edwards**
Department of Biological Sciences, Hancock Auditorium and Museum (AHF) 107F, University of Southern California, Los Angeles, CA 90089-0371, U.S.A

**Sean Gulick**
The University of Texas at Austin, Department of Geological Sciences, Geology Building 2. 106, 23rd Street at San Jacinto Boulevard., Austin, TX 78705, U.S.A

**Fumio Inagaki**
Kochi Core Center, JAMSTEC, 200 Monobe Otsu, Nankoku City, Kochi Prefecture, 783-8502, Japan

**Heiko Pälike**
National Oceanography Centre Southampton, University of Southampton and Natural Environment Research Council, Southampton, University of Southampton Waterfront Campus, European Way, Southampton, SO14 3ZH, U.K

**Ryuji Tada**
Department of Earth and Planetary Science, University of Tokyo, Sci. Bldg. No.1-734, Hongou 7-3-1, Bunkyo-ku, Tokyo, 113-0033, Japan

**Pierre Henry**
Centre Européen de Recherche et d'Enseignement en Géosciences de l'Environement, Aix- Marseille Université et CNRS, Europôle de l'Arbois, 13545 Aix en Provence Cedex 04, France

**Toshiya Kanamatsu**
Institute for Research on Earth Evolution, Japan Agency for Marine-Earth Science and Technology, 2-15 Natsushima-cho, Yokosuka, Kanagawa 237-0061, Japan

**Kyaw Thu Moe**
Center for Deep Earth Exploration, Japan Agency for Marine-Earth Science and Technology, 3173-25 Showa-machi, Kanazawa-ku, Yokohama, Kanagawa, 236- 0001, Japan

**Michael Strasser**
ETH Zürich, Geologisches Institut, Sonneggstrasse 5, 8092 Zürich, Switzerland

**Georg Heumann**
Steinmann Institute of Geology, Mineralogy and Paleontology, Bonn University, Bonn, Germany

**Thomas Litt**
Steinmann Institute of Geology, Mineralogy and Paleontology, Bonn University, Bonn, Germany

**Nadine Pickarski**
Steinmann Institute of Geology, Mineralogy and Paleontology, Bonn University, Bonn, Germany

**Deniz Cukur**
GEOMAR, Helmholtz Centre for Ocean Research Kiel, Wischhofstr. 1-3, 24148 Kiel, Germany

**Sebastian Krastel**
GEOMAR, Helmholtz Centre for Ocean Research Kiel, Wischhofstr. 1-3, 24148 Kiel, Germany

**Hans-Ulrich**
GEOMAR, Helmholtz Centre for Ocean Research Kiel, Wischhofstr. 1-3, 24148 Kiel, Germany

**Schmincke**
GEOMAR, Helmholtz Centre for Ocean Research Kiel, Wischhofstr. 1-3, 24148 Kiel, Germany

**Mari Sumita**
GEOMAR, Helmholtz Centre for Ocean Research Kiel, Wischhofstr. 1-3, 24148 Kiel, Germany

**Flavio S. Anselmetti**
Eawag, Swiss Federal Institute of Aquatic Science and Technology, Dübendorf and Kastanienbaum, Switzerland

**Jürg Beer**
Eawag, Swiss Federal Institute of Aquatic Science and Technology, Dübendorf and Kastanienbaum, Switzerland

**Gerald Haug**
Eawag, Swiss Federal Institute of Aquatic Science and Technology, Dübendorf and Kastanienbaum, Switzerland

**Rolf Kipfer**
Eawag, Swiss Federal Institute of Aquatic Science and Technology, Dübendorf and Kastanienbaum, Switzerland

**Ola Kwiecien**
Eawag, Swiss Federal Institute of Aquatic Science and Technology, Dübendorf and Kastanienbaum, Switzerland

**Marie-Eve Randlett**
Eawag, Swiss Federal Institute of Aquatic Science and Technology, Dübendorf and Kastanienbaum, Switzerland

**Carsten J. Schubert**
Eawag, Swiss Federal Institute of Aquatic Science and Technology, Dübendorf and Kastanienbaum, Switzerland

**Mona Stockhecke**
Eawag, Swiss Federal Institute of Aquatic Science and Technology, Dübendorf and Kastanienbaum, Switzerland

**Mike Sturm**
Eawag, Swiss Federal Institute of Aquatic Science and Technology, Dübendorf and Kastanienbaum, Switzerland

**Yama Tomonaga**
Eawag, Swiss Federal Institute of Aquatic Science and Technology, Dübendorf and Kastanienbaum, Switzerland

**A. Feray Meydan**
Yüzüncü Yıl Üniversitesi, Mühendislik - Mimarlık Fakültesi Jeoloji Mühendisliği Bölümü, Van, Turkey

**Sefer Orcen**
Yüzüncü Yıl Üniversitesi, Mühendislik - Mimarlık Fakültesi Jeoloji Mühendisliği Bölümü, Van, Turkey

**Namik Cagatay**
Department of Geological Engineering and Eastern Mediterranean Centre for Oceanography and Limnology, Istanbul Technical University, Istanbul, Turkey

**Emre Damci**
Department of Geological Engineering and Eastern Mediterranean Centre for Oceanography and Limnology, Istanbul Technical University, Istanbul, Turkey

**Henrike Baumgarten**
Leibniz- Institute for Applied Geophysics (LIAG), Hannover, Germany

**Thomas Wonik**
Leibniz- Institute for Applied Geophysics (LIAG), Hannover, Germany

**Clemens Glombitza**
Potsdam University, Institute for Earth and Environmental Sciences, Geomicrobiology Group, Potsdam, Germany

**Jens Kallmeyer**
Potsdam University, Institute for Earth and Environmental Sciences, Geomicrobiology Group, Potsdam, Germany

**Luigi Vigliotti**
Istituto di Scienze Marine, ISMAR-CNR, Bologna, Italy

**PALEOVAN Scientific Team**

**Masa (Masataka) Kinoshita**
Japan Agency for Marine- Earth Science and Technology–Kochi Core Center (JAMSTEC-KCC), Monobe-B200, Nankoku-City, Kochi, 783-8502 Japan

**Harold Tobin**
Department of Geology and Geophysics, University of Wisconsin-Madison, 1215 West Dayton Street, Madison, WI 53706, U.S.A

**Nobuhisa Eguchi**
Center for Deep Earth Exploration–Japan Agency for Marine-Earth Science and Technology (CDEX-JAMSTEC), 3173-25 Showa-machi, Kanazawa-ku, Yokohama, Kanagawa 236-0001, Japan

**Simon Nielsen**
Center for Deep Earth Exploration–Japan Agency for Marine-Earth Science and Technology (CDEX-JAMSTEC), 3173-25 Showa-machi, Kanazawa-ku, Yokohama, Kanagawa 236-0001, Japan

**Sean T. Toczko**
Center for Deep Earth Exploration (CDEX), Japan Agency for Marine-Earth Science and Technology (JAMSTEC), Yokohama Institute for Earth Sciences, 3173- 25 Showa-machi, Kanazawa-ku, Yokohama, Kanagawa 236- 0001 Japan

**Achim J. Kopf**
MARUM and Department of Geosciences, University of Bremen, 28359 Bremen, Germany

**Eiichiro Araki**
Institute for Research on Earth Evolution, Japan Agency for Marine-Earth Science and Technology (JAMSTEC), 2-15 Natsushima-cho, Yokosuka, Kanagawa 237-0061 Japan

**C. Geoffrey Wheat**
Global Undersea Research Unit, University of Alaska Fairbanks, P.O. Box 475, Moss Landing, CA 95039, U.S.A

**Katrina J. Edwards**
Department of Biological Sciences, Marine Environmental Biology Section, University of Southern California, Los Angeles, CA 90089, U.S.A

**Tom Pettigrew**
Pettigrew Engineering, 479 Nine Mile Road, Milam, TX 75959, U.S.A

**Hans W. Jannasch**
Monterey Bay Aquarium Research Institute, 7700 Sandholdt Road, Moss Landing, CA 95039, U.S.A

**Keir Becker**
University of Miami, 4600 Rickenbacker Causeway, Miami, FL 33149, U.S.A

**Earl E. Davis**
Pacific Geoscience Centre, Geological Survey of Canada, 9860 West Saanich Road, Sidney, BC V8L 4B2, Canada

**Heiner Villinger**
Department of Geosciences, University of Bremen, Klagenfurter Strasse, 28359 Bremen, Germany

**Wolfgang Bach**
Department of Geosciences, University of Bremen, Klagenfurter Strasse, 28359 Bremen, Germany

**J. Carlos Santamarina**
School of Civil and Environmental Engineering, Georgia Institute of Technology, Mason Building, 790 Atlantic Drive, Atlanta, GA 30332-0355, U.S.A

**Sheng Dai**
School of Civil and Environmental Engineering, Georgia Institute of Technology, Mason Building, 790 Atlantic Drive, Atlanta, GA 30332-0355, U.S.A

**Junbong Jang**
School of Civil and Environmental Engineering, Georgia Institute of Technology, Mason Building, 790 Atlantic Drive, Atlanta, GA 30332-0355, U.S.A

**Marco Terzariol**
School of Civil and Environmental Engineering, Georgia Institute of Technology, Mason Building, 790 Atlantic Drive, Atlanta, GA 30332-0355, U.S.A

**James M. Russell**
Department of Geological Sciences, Brown University, 324 Brook Street, Box 1846, Providence, RI 02912-9019, U.S.A

**Andrew S. Cohen**
Department of Geosciences, University of Arizona, Gould-Simpson Building #77, 1040 East 4th Street, Room 326, Tucson, AZ 85721-0001, U.S.A

**Thomas C. Johnson**
Large Lakes Observatory, University of Minnesota Duluth, 2205 East 5th Street, Research Laboratory Building 204, Duluth, MN 55812-3024, U.S.A

**Christopher A. Scholz**
The College of Arts and Sciences, Department of Earth Sciences, Syracuse University, 204 Heroy Geology Laboratory, Syracuse, NY 13244-1070, U.S.A

**Jian Lin**
Woods Hole Oceanographic Institution, Department of Geology & Geophysics, Woods Hole, MA 02543, U.S.A

**Chun-Feng Li**
Tongji University, State Key Laboratory of Marine Geology, 1239 Siping Road, Shanghai 200092, China

**Pinxian Wang**
Tongji University, State Key Laboratory of Marine Geology, 1239 Siping Road, Shanghai 200092, China

**Dieter Franke**
Federal Institute for Geosciences and Natural Resources (BGR), Stilleweg 2, 30655 Hannover, Germany

**Jun Tian**
Tongji University, State Key Laboratory of Marine Geology, 1239 Siping Road, Shanghai 200092, China

**Stephen J. Gallagher**
The School of Earth Sciences, The University of Melbourne, Victoria 3010, Australia

**Neville Exon**
Research School of Earth Sciences, Australian National University, Canberra 0200, Australia

**Dhananjai Pandey**
National Centre for Antarctic and Ocean Research (NCAOR), Vasco da Gama, Goa, India

**S. Rajan**
National Centre for Antarctic and Ocean Research (NCAOR), Vasco da Gama, Goa, India

**Mike Coffin**
Institute for Marine and Antarctic Studies, University of Tasmania, Private Bag 129,HOBART, TAS 7001, Australia

**Ken Takai**
Subsurface Geobiology Advanced Research (SUGAR) Project, Japan Agency for Marine-Earth Science & Technology (JAMSTEC), 2-15 Natsushima-cho, Yokosuka 237-0061, Japan

**Laura M. Wallace**
GNS Science, P.O. Box 30-368, Lower Hutt, New Zealand
University of Texas, Institute for Geophysics, J.J. Pickle Research Campus, 10100 Burnet Road (R2200) Austin, TX 78758- 4445, U.S.A

**Eli A. Silver**
Earth and Planetary Sciences, University of California Santa Cruz, 1156 High Street, Santa Cruz, CA 95064-1077, U.S.A

**Nathan Bangs**
University of Texas, Institute for Geophysics, J.J. Pickle Research Campus, 10100 Burnet Road (R2200) Austin, TX 78758-4445, U.S.A

**Rebecca Bell**
Department of Earth Science and Engineering, Imperial College London, South Kensington Campus London SW7 2AZ, London, U.K

**Joshu Mountjoy**
National Institute of Water and Atmospheric Research (NIWA), 301 Evans Bay Parade, Hataitai, Wellington 6021, Private Bag 14901, Wellington, New Zealand

**Stuart Henrys**
GNS Science, 1 Fairway Drive, Avalon 5010, P.O. Box 30-368, Lower Hutt 5040, Lower Hutt, New Zealand

**Ingo Pecher**
GNS Science, 1 Fairway Drive, Avalon 5010, P.O. Box 30-368, Lower Hutt 5040, Lower Hutt, New Zealand
University of Auckland, School of Environment, Private Bag 92019, Auckland, New Zealand

**James Russell**
Department of Geological Sciences, Brown University, Box 1846, Providence, RI 02912, U.S.A

**Satria Bijaksana**
Global Geophysics Research Group, Faculty of Mining and Petroleum Engineering, Institut Teknologi Bandung, Jalan Ganesa 10, Bandung, 40132, Indonesia

**Erik T. Brown**
Large Lakes Observatory and Department of Geological Sciences, University of Minnesota Duluth, 2205 East 5th Street, Duluth, MN 55812-3024, U.S.A

**Josef P. Werne**
Large Lakes Observatory and Department of Geological Sciences, University of Minnesota Duluth, 2205 East 5th Street, Duluth, MN 55812-3024, U.S.A

**Socorro Lozano-García**
Instituto de Geología, Universidad Nacional Autónoma de México, Cd. Universitaria, 04510 México D.F., México

**Margarita Caballero**
Instituto de Geofísica, Universidad Nacional Autónoma de México, Cd. Universitaria, 04510 México D.F., México

**Beatriz Ortega-Guerrero**
Instituto de Geofísica, Universidad Nacional Autónoma de México, Cd. Universitaria, 04510 México D.F., México

**Enrique Cabral-Cano**
Instituto de Geofísica, Universidad Nacional Autónoma de México, Cd. Universitaria, 04510 México D.F., México

**Alejandra Arcniega-Ceballos**
Instituto de Geofísica, Universidad Nacional Autónoma de México, Cd. Universitaria, 04510 México D.F., México

**Blas L. Valero-Garces**
Instituto Pirenaico de Ecologia Consejo Superior de Investigaciones Cientificas, E-50080 Zaragoza, Spain

**Antje Schwalb**
Institut für Geosysteme und Bioindikation Technische, Universität Braunschweig, D-38106 Braunschweig, Germany

**Andrew T. Fisher**
Earth and Planetary Sciences Department and Institute for Geophysics and Planetary Physics, University of California at Santa Cruz, 1156 High Street, Santa Cruz, CA 95064, U.S.A

**Takeshi Tsuji**
Graduate School of Engineering, Kyoto University, C1-1-110 Kyotodaigaku-Katsura, Nishikyo-ku, Kyoto 615-8540, Japan

**Katerina Petronotis**
Integrated Ocean Drilling Program, Texas A&M University, 1000 Discovery Drive, College Station, TX 77845-9547, U.S.A

**C. Geoff Wheat**
Global Undersea Research Unit, P.O. Box 475, Moss Landing, CA 95039, U.S.A

**Keir Becker**
Rosenstiel School of Marine and Atmospheric Science, University of Miami, Division of Marine Geology and Geophysics, 4600 Rickenbacker Causeway, Miami, FL 33149-1098, U.S.A

**Jordan F. Clark**
Department of Earth Science and Program of Environmental Studies, University of California Santa Barbara, 1006 Webb Hall, Santa Barbara, CA 93106-9630, U.S.A

**James Cowen**
Department of Oceanography, University of Hawaii at Manoa, 1000 Pope Road, Marine Sciences Building, Honolulu, HI 96822, U.S.A

**Katrina Edwards**
Center for Earth Sciences, University of Southern California, 3651 Trousdale Parkway, Los Angeles, CA 90089-0740, U.S.A

**Hans Jannasch**
Monterey Bay Aquarium Research Institute, 7700 Sandholt Road, Moss Landing, CA, 95039, U.S.A

**Earl Davis**
Pacific Geoscience Centre, Geological Survey of Canada, 9860 West Saanich Road, Sidney, BC V8L 4B2, Canada

**Martin Heesemann**
NEPTUNE-Canada, University of Victoria, P.O. Box 1700 Stn CSC, Victoria, BC V8W 2Y2, Canada

**Ken Takai**
Subsurface Geobiology Advanced Research (SUGAR) Project, Japan Agency for Marine-Earth Science & Technology (JAMSTEC), 2-15 Natsushima-cho, Yokosuka 237-0061, Japan

**Michael J. Mottl**
Department of Oceanography, School of Ocean and Earth Science and Technology (SOEST), University of Hawaii at Manoa, 1000 Pope Road, Honolulu, HI 96822

**Simon H. H. Nielsen**
Center for Deep Earth Exploration (CDEX), Japan Agency for Marine-Earth Science & Technology (JAMSTEC), 3173-25 Showa-machi Kanazawa-ku, Yokohama 236-0001, Japan

**Damon A. H. Teagle**
National Oceanography Centre, Southampton, University of Southampton Waterfront Campus, European Way, Southampton SO14 3ZH, U.K

**Benoit Ildefonse**
Géosciences Montpellier, CNRS, Université Montpellier 2, CC 60, 34095 montpellier cédex 05, France

**Peter Blum**
Integrated Ocean Drilling Program, Texas A&M University, 1000 Discovery Drive, College Station, TX 77845-9547, U.S.A

**Moe Kyaw Thu**
Nobu Eguchi, and Ikuo Sawada, Center for Deep Earth Exploration (CDEX)—Japan Agency for Marine-Earth Science and Technology (JAMSTEC), 3173-25 Showa-machi, Kanazawa-ku, Yokohama, Kanagawa 236-0001, Japan

**Yoshihisa Kawamura**
Ippan Shadan Hojin IODP-MI, Tokyo University of Marine Science and Technology, Office of Liaison and Cooperative Research, 3rd Floor, 2-1-6 Etchujima, Koto-ku, Tokyo, 135-8533 Japan

**Chee Kin Khong**
ECA Services Techniques Schlumberger, Le Palatin 1, 1 cour du Triangle, 92936 La Defense France

**Takatoshi Ito**
Institute of Fluid Science, Tohoku University, 2-1-1 Katahira, Aoba-ku, Sendai, Miyagi 980-8577, Japan

**Weiren Lin**
Kochi Institute for Core Sample Research, Japan Agency for Marine-Earth Science and Technology, 200 Monobe-Otsu, Nankoku, Kochi 783-8502, Japan

**Mai-Linh Doan**
ISTerre Université de Grenoble 1, CNRS, F-38041 Grenoble, France

**David Boutt**
Department of Geosciences, University of Massachusetts Amherst, 611 North Pleasant Street, 233 Morrill Center, Amherst, MA 01002, U.S.A

**Lisa McNeill**
School of Ocean and Earth Science, National Oceanography Centre, Southampton, University of Southampton, Southampton SO14 3ZH, U.K

**Demian Saffer**
The Pennsylvania State University, 0310 Deike Building, University Park, PA 16802, U.S.A

**Eiichiro Araki**
Earthquake and Tsunami Research Project for Disaster Prevention, Japan Agency for Marine-Earth and Technology, 2-15 Natsushima-cho, Yokosuka, Kanagawa 237-0061, Japan

**Timothy Byrne**
Center for Integrative Geosciences, University of Connecticut, U-2045, 354 Mansfield Road, Storrs, CT 06269, U.S.A

**Yasuyuki Kano**
Disaster Prevention Research Institute, Kyoto University, Gokasho, Uji 611-0011, Japan

**Peter Flemings**
Institute for Geophysics, Jackson School of Geosciences, University of Texas at Austin, 10100 Burnet Road (R2200), Austin, TX 78758-4445, U.S.A

**Casey Moore**
University of California Santa Cruz, 1156 High Street, Santa Cruz, CA 95064, U.S.A

**Masataka Kinoshita**
Institute for Research on Earth Evolution, Japan Agency for Marine-Earth Science and Technology, 2-15 Natsushima-cho, Yokosuka, Kanagawa 237-0061, Japan

**Harold Tobin**
Department of Geology and Geophysics, University of Wisconsin-Madison, 1215 West Dayton Street, Madison, WI 53706, U.S.A

**Lallan Gupta**
Kochi Core Center, 200 Monobe Otsu, Nankoku, Kochi 783-8502, Japan

**Toshio Hisamitsu**
Kochi Core Center, 200 Monobe Otsu, Nankoku, Kochi 783-8502, Japan

**Noriaki Masui**
Kochi Core Center, 200 Monobe Otsu, Nankoku, Kochi 783-8502, Japan

**Rodolfo Coccioni**
Dipartimento di Scienze della Terra, della Vita e dell'Ambiente dell'Università degli Studi "Carlo Bo," Campus Scientifico "E. Mattei", Località Crocicchia, 61029 Urbino, Italy

**Luigi Jovane**
Departamento de Oceanografia Física, Instituto Oceanográfico, Universidade de São Paulo, 05508- 900 São Paulo, Brazil

**Giuseppe Bancalà**
Dipartimento di Scienze della Terra, della Vita e dell'Ambiente dell'Università degli Studi "Carlo Bo," Campus Scientifico "E. Mattei", Località Crocicchia, 61029 Urbino, Italy

**Carla Bucci**
Dipartimento di Scienze della Terra, della Vita e dell'Ambiente dell'Università degli Studi "Carlo Bo," Campus Scientifico "E. Mattei", Località Crocicchia, 61029 Urbino, Italy

**Gerson Fauth**
Centro de Ciências Exatas e Tecnológicas, Área de Conhecimento e Aplicação de Geociências, Universidade do Vale do Rio dos Sinos, São Leopoldo, Brazil

**Fabrizio Frontalini**
Dipartimento di Scienze della Terra, della Vita e dell'Ambiente dell'Università degli Studi "Carlo Bo," Campus Scientifico "E. Mattei", Località Crocicchia, 61029 Urbino, Italy

**Liliane Janikian**
Departamento de Geofísica, Instituto de Astronomia, Geofísica e Ciências Atmosféricas, Universidade de São Paulo, 05508-090 São Paulo, Brazil

**Jairo Savian**
Departamento de Geofísica, Instituto de Astronomia, Geofísica e Ciências Atmosféricas, Universidade de São Paulo, 05508-090 São Paulo, Brazil

**Renato Paes de Almeida**
Departamento de Geologia Sedimentar e Ambiental, Instituto de Geociências, Universidade de São Paulo, 05508-090 São Paulo, Brazil

**Grasiane Luz Mathias**
Departamento de Geofísica, Instituto de Astronomia, Geofísica e Ciências Atmosféricas, Universidade de São Paulo, 05508-090 São Paulo, Brazil

**Ricardo Ivan Ferreira da Trindade**
Departamento de Geofísica, Instituto de Astronomia, Geofísica e Ciências Atmosféricas, Universidade de São Paulo, 05508-090 São Paulo, Brazil

**Heather N. Schrum**
Massachusetts Maritime Academy, 101 Academy Drive, Buzzards Bay, MA 02532, U.S.A

**Richard W. Murray**
Department of Earth Sciences, Boston University, 1 Silber Way, Boston, MA 02215, U.S.A

**Britta Gribsholt**
Department of Biological Sciences, Center for Geomicrobiology, University of Aarhus, Ny Munkegade 114, DK-8000 Aarhus C, Denmark

**Hedi Oberhänsli**
Museum of Natural History Berlin, Invalidenstrasse 43, 10115 Berlin, Germany

**Peter Molnar**
Department of Geological Sciences and Cooperative Institute for Research in Environmental Sciences (CIRES), University of Colorado at Boulder, Boulder, CO 80309-0216, U.S.A

**Donald T. Rodbell**
Union College, Schenectady, NY 12308, U.S.A

**Mark B. Abbott**
Department of Geology and Planetary Science, University of Pittsburgh, 4107 O'Hara Street, Room 200, SRCC Building, Pittsburgh, PA 15260-3332, U.S.A

**Cecilia M. G. McHugh**
School of Earth and Environmental Sciences, Queens College, City University of New York, 65-30 Kissena Boulevard, Flushing, NY 11367, U.S.A

**Leonardo Seeber**
Lamont-Doherty Earth Observatory of Columbia University, Route 9W, Palisades, NY 10964, U.S.A

**M. Namik Çağatay**
Geology Department, Istanbul Technical University, Ayazaga 80626, Istanbul, Turkey

**Pierre Henry**
Centre National de la Recherche Scientifique (CNRS), College de France, Bat Trocadero, Europole de l'Arbois, BP80, 13545 Aix en Provence Cedex 04, France

**Christopher Sorlien**
Earth Research Institute, 6832 Ellison Hall, University of California, Santa Barbara, CA 93106-3060, U.S.A

**Michael Steckler**
Lamont-Doherty Earth Observatory of Columbia University, Route 9W, Palisades, NY 10964, U.S.A

**Gülsen Uçarkuş**
Geology Department, Istanbul Technical University, Ayazaga 80626, Istanbul, Turkey

**Craig S. Fulthorpe**
Institute for Geophysics, John A. and Katherine G. Jackson School of Geosciences, University of

Texas at Austin, 10100 Burnet Road (R2200), Building 196 (ROC), Austin, TX 78758-4445, U.S.A

**Koichi Hoyanagi**
Department of Geology, Faculty of Science, Shinshu University, 3-1-1 Asahi, Matsumoto 390-8621, Japan

**Peter Blum**
United States Implementing Organization, Integrated Ocean Drilling Program, Texas A&M University, 1000 Discovery Drive, College Station, TX 77845, U.S.A

**Carlota Escutia**
Instituto Andaluz de Ciencias de la Tierra, CSIC-Universidad de Granada, Campus de Fuentenueva s/n, 18002 Granada, Spain

**Henk Brinkhuis**
Biomarine Sciences, Institute of Environmental Biology, Laboratory of Palaeobotany and Palynology, Utrecht University, Budapestlaan 4, 3584 CD Utrecht, The Netherlands

**Adam Klaus**
Staff Scientist/Expedition Project Manager, United States Implementing Organization, Integrated Ocean Drilling Program, Texas A&M University, 1000 Discovery Drive, College Station, TX 77845-9547, U.S.A

**William W. Sager**
Department of Oceanography, Texas A&M University, College Station, TX 77843-3146, U.S.A

**Takashi Sano**
Department of Geology and Paleontology, National Museum of Nature and Science, 4-1-1 Amakubo, Tsukuba 305-0005, Japan

**Jörg Geldmacher**
Integrated Ocean Drilling Program, Texas A&M University, 1000 Discovery Drive, College Station, TX 77845-9547, U.S.A

**Yusuke Yokoyama**
Atmosphere and Ocean Research Institute, University of Tokyo, 5-1-5 Kashiwanoha, Kashiwa, Chiba 277-8564 Japan

**Jody M. Webster**
Geocoastal Reseach Group, School of Geosciences, The University of Sydney, Sydney, NSW 2006, Australia

**Carol Cotterill**
British Geological Survey, Murchison House, Edinburgh, Scotland, EH9 3LA, U.K

**Juan Carlos Braga**
Departamento de Estratigrafia y Paleontologia, Facultad de Ciencias, Universidad de Granada, Campus Fuentenueva, 18002 Granada, Spain

**Luigi Jovane**
Geology Department, Western Washington University, 516 High Street, MS 908, Bellingham, WA 98225, U.S.A

**Heath Mills**
Department of Oceanography, Texas A&M University, 716A Eller, O&M Building, College Station, TX 77843, U.S.A

**Sally Morgan**
University of Leicester, Department of Geology, University Road, Leicester, LE1 7RH, U.K

**Atsushi Suzuki**
Geological Survey of Japan, National Institute of Advanced Industrial Science and Technology (AIST), AIST Tsukuba Central 7, 1-1-1 Higashi, Tsukuba, Ibaraki 305-8567, Japan

**Alexander van Geen**
Lamont-Doherty Earth Observatory (LDEO), Columbia University, 131 Corner, 61 Route 9W - P.O. Box 1000, Palisades, N.Y. 10964-8000, U.S.A

**Harsh Gupta**
National Geophysical Research Institute (CSIR-NGRI), Uppal Road, Hyderabad, 500 007, India

**Shailesh Nayak**
Ministry of Earth Sciences (MoES), C.G.O. Complex, Lodhi Road, New Delhi 110 003, India